Boolean Valued Analysis

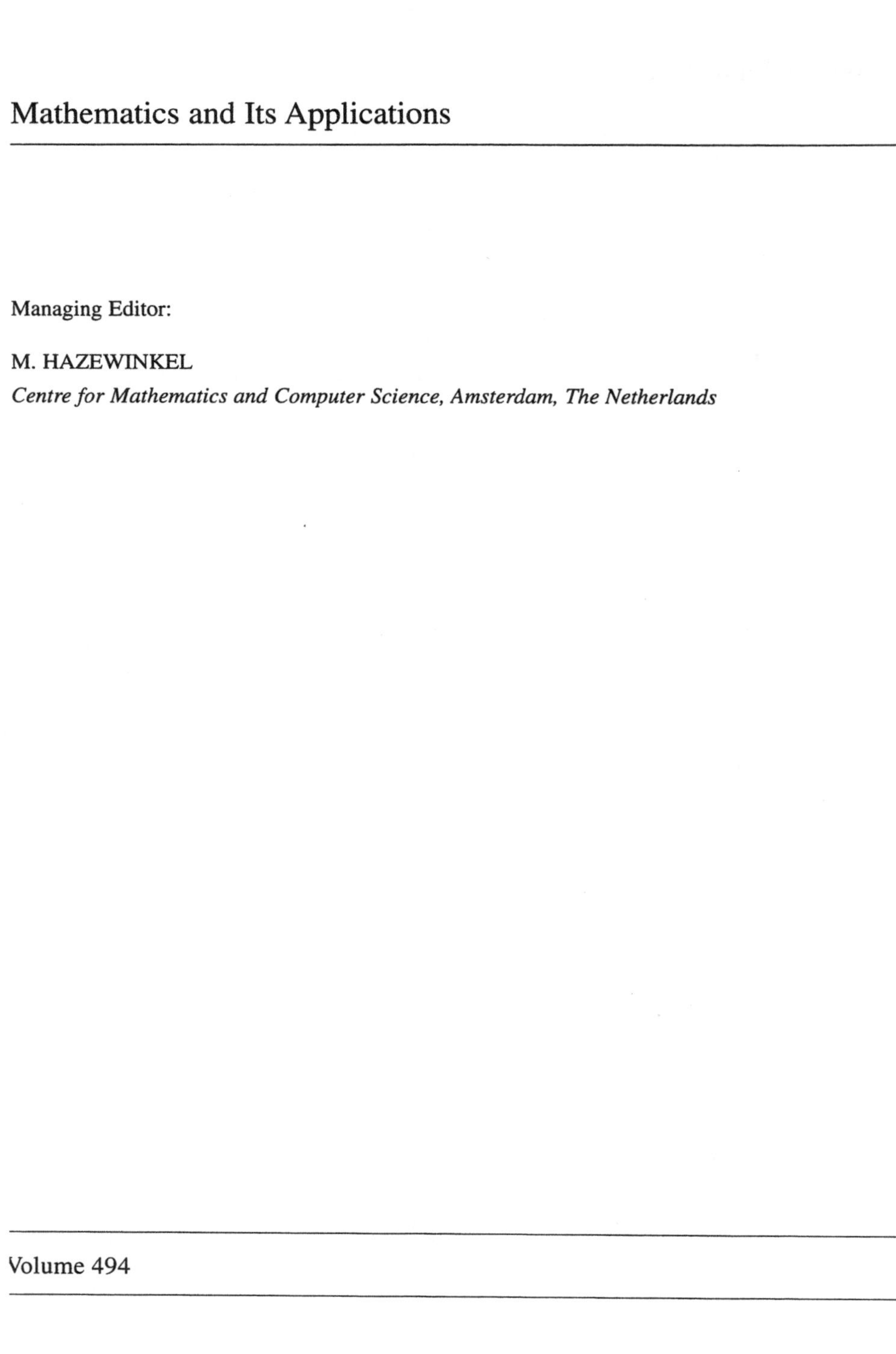

Mathematics and Its Applications

Volume 494

Boolean Valued Analysis

by

A.G. Kusraev

Institute for Mathematics and Informatics,
Vladikavkaz, Russia

and

S.S. Kutateladze

Sobolev Institute of Mathematics,
Siberian Division of the Russian Academy of Sciences,
Novosibirsk, Russia

SPRINGER SCIENCE+BUSINESS MEDIA, B.V.

Library of Congress. Cataloging-in-Publication Data

ISBN 978-0-7923-5921-0 ISBN 978-94-011-4443-8 (eBook)
DOI 10.1007/978-94-011-4443-8

Printed on acid-free paper

This is revised translation of the original Russian work Ъудевозначный анадиз
Sobolev Institute Press, © 1999 Novosibirsk, Russia.
Translated by S.S. Kutateladze.

Contents

Foreword to the English Translation

This is a translation of the book that opens the series "Nonstandard Methods of Analysis" in print by the Sobolev Institute Press at Novosibirsk.

Nonstandard methods of analysis consist generally in comparative study of two interpretations of a mathematical claim or construction given as a formal symbolic expression by means of two different set-theoretic models: one, a "standard" model and the other, a "nonstandard" model. The second half of the twentieth century is a period of significant progress in these methods and their rapid development in a few directions.

The first of the latter appears often under the name coined by its inventor, A. Robinson. This memorable but slightly presumptuous and defiant term, *nonstandard analysis*, often swaps places with the term *Robinson's* or *classical nonstandard analysis*. The characteristic feature of Robinson's nonstandard analysis is a frequent usage of many controversial concepts appealing to the actual infinitely small and infinitely large quantities that have happily resided in natural sciences from ancient times but were strictly forbidden in modern mathematics for many decades. The present-day achievements revive the forgotten term *infinitesimal analysis* which expressively reminds us of the heroic bygones of Calculus.

Infinitesimal analysis expands rapidly, bringing about radical reconsideration of the general conceptual system of mathematics. The principal reasons for this progress are twofold. Firstly, infinitesimal analysis provides us with a novel understanding for the method of indivisibles rooted deeply in the mathematical classics. Secondly, it synthesizes both classical approaches to differential and integral calculuses which belong to the noble inventors of the latter. Infinitesimal analysis finds newer and newest applications and merges into every section of contemporary mathematics. Sweeping changes are on the march in nonsmooth analysis, measure theory, probability, the qualitative theory of differential equations, and mathematical economics.

The second direction, *Boolean valued analysis* distinguishes itself by ample usage of such terms as the technique of ascending and descending, cyclic envelopes and mixings, B-sets and representation of objects in $\mathbf{V}^{(B)}$. Boolean valued analysis originated with the famous works by P. J. Cohen on the continuum hypothesis. Progress in this direction has evoked radically new ideas and results in many sections of functional analysis. Among them we list Kantorovich space theory, the theory of von Neumann algebras, convex analysis, and the theory of vector measures.

The book [135], printed by the Siberian Division of the «Наука» Publishers in 1990 and translated into English by Kluwer Academic Publishers in 1994, gave a first unified treatment of the two disciplines forming the core of the present-day nonstandard methods of analysis.

The reader's interest as well as successful research into the field assigns a task of updating the book and surveying the state of the art. Implementation of the task has shown soon that it is impossible to compile new topics and results in a single book. Therefore, the Sobolev Institute Press decided to launch the series "Nonstandard Methods of Analysis" which will consist of the monographs devoted to various aspects of this direction of mathematical research.

The present book opens the series and treats Boolean valued analysis. The formal technique of the discipline is expounded in detail. The book also pays much attention to studying the classical objects of functional analysis, namely, Banach spaces and algebras by means of Boolean valued models.

This edition was typeset using $\mathcal{A}_\mathcal{M}\mathcal{S}$-TEX, the American Mathematical Society's TEX macro package.

As the editor of the series, I am deeply grateful to Kluwer Academic Publishers for cooperation and support of the new project.

S. Kutateladze

Preface

As the title implies, the present book treats *Boolean valued analysis.* This term signifies a technique for studying the properties of an arbitrary mathematical object by means of comparison between its representations in two different set-theoretic models whose construction utilizes principally distinct Boolean algebras. We usually take as these models the classical Cantorian paradise in the shape of the von Neumann universe and a specially-trimmed Boolean valued universe in which the conventional set-theoretic concepts and propositions acquire bizarre interpretations. Usage of two models for studying a single object is a family feature of the so-called *nonstandard methods of analysis.* For this reason, Boolean valued analysis means an instance of nonstandard analysis in common parlance.

Proliferation of Boolean valued analysis stems from the celebrated achievement of P. J. Cohen who proved in the beginning of the sixties that the negation of the continuum hypothesis, CH, is consistent with the axioms of Zermelo–Fraenkel set theory, ZFC. This result by P. J. Cohen, alongside the consistency of CH with ZFC established earlier by K. Gödel, proves that CH is independent of the conventional axioms of ZFC.

The genuine value of the great step forward by P. J. Cohen could be understood better in connection with the serious difficulty explicated by J. Shepherdson and absent from the case settled by K. Gödel. The crux of J. Shepherdson's observation lies in the impossibility of proving the consistency of (ZFC) + ($\neg$ CH) by means of any standard models of set theory. Strictly speaking, we can never find a subclass of the von Neumann universe which models (ZFC) + ($\neg$ CH) provided that we use the available interpretation of membership. P. J. Cohen succeeded in inventing a new powerful method for constructing noninner, *nonstandard*, models of ZFC. He coined the term *forcing.* The technique by P. J. Cohen invokes the axiom of existence of a standard transitive model of ZFC in company with the forcible and forceful transformation of the latter into an immanently nonstandard model by the method of forcing. His tricks fall in an outright contradiction with the routine mathematical intuition stemming "from our belief into a natural nearly physical model of the mathematical world" as P. J. Cohen phrased this himself [30].

Miraculously, the difficulties in comprehension of P. J. Cohen's results gained a perfect formulation long before they sprang into life. This was done in the famous talk "Real Function Theory: State of the Art" by N. N. Luzin at the All-Russian Congress of Mathematicians in 1927. Then N. N. Luzin said: "The first idea that might leap to mind is that the determination of the cardinality of the continuum is a matter of a free axiom like the parallel postulate of geometry. However, when we vary the parallel postulate, keeping intact the rest of the axioms of Euclidean geometry, we in fact change the precise meanings of the words we write or utter, that is, 'point,' 'straight line,' etc. What words are to change their meanings if we attempt at making the cardinality of the continuum movable along the scale of alephs, while constantly proving consistency of this movement? The cardinality of the continuum, if only we imagine the latter as a set of points, is some unique entity that must reside in the scale of alephs at the place which the cardinality of the continuum belongs to; no matter whether the determination of this place is difficult or even 'impossible for us, the human beings' as J. Hadamard might comment" [159, pp. 11–12].

P. S. Novikov expressed a very typical attitude to the problem: "...it might be (and it is actually so in my opinion) that the result by Cohen conveys a purely negative message and reveals the termination of the development of 'naive' set theory in the spirit of Cantor" [192, p. 209].

Intention to obviate obstacles to mastering the technique and results by P. J. Cohen led D. Scott and R. Solovay to constructing the so-called Boolean valued models of ZFC which are not only visually attractive from the standpoint of classical mathematicians but also are fully capable of establishing consistency and independence theorems. P. Vopěnka constructed analogous models in the same period of the early sixties.

The above implies that the Boolean valued models, achieving the same ends as P. J. Cohen's forcing, must be nonstandard in some sense and possess some new features that distinguish them from the standard models.

Qualitatively speaking, the *notion of Boolean valued model involves a new conception of modeling* which might be referred to as *modeling by correspondence* or *long-distance modeling*. We explain the particularities of this conception as compared with the routine approach. Encountering two classical models of a single theory, we usually seek for a bijection between the universes of the models. If this bijection exists then we translate predicates and operations from one model to the other and speak about isomorphism between the models. Consequently, this conception of isomorphism implies a direct contact of the models which consists in witnessing to bijection of the universes of discourse.

Imagine that we are physically unable to compare the models pointwise. Happily, we take an opportunity to exchange information with the owner of the other model by using some means of communication, e.g., by having long-distance calls. While communicating, we easily learn that our interlocutor uses his model to operate on some objects that are the namesakes of ours, i.e., sets, membership, etc. Since we are interested in ZFC, we ask the interlocutor whether or not the axioms of ZFC are satisfied in his model. Manipulating the model, he returns a positive answer. After checking that he uses the same inference rules as we do, we cannot help but acknowledge his model to be a model of the theory we are all investigating. It is worth noting that this conclusion still leaves unknown for us the objects that make up his universe and the procedures he uses to distinguish between true and false propositions about these objects.[†]

All in all, the *new conception of modeling implies not only refusal from identification of the universes of discourse but also admission of various procedures for verification of propositions.*

To construct a Boolean valued model, we start with a complete Boolean algebra B, a cornerstone of a special Boolean valued universe $\mathbf{V}^{(B)}$ consisting of "B-valued sets" that are defined recursively as B-valued functions over available B-valued sets. This $\mathbf{V}^{(B)}$ will serve as a universe of discourse for ZFC. Also, we appoint B as the target of the truth value sending each formula of ZFC to a member of B. More explicitly, to each formula φ of ZFC whose every variable ranges now over $\mathbf{V}^{(B)}$, we put in correspondence some element $[\![\varphi]\!]$ of the parental Boolean algebra B. The quantity $[\![\varphi]\!]$ is the *truth value* of φ. We use truth values for validating formulas of ZFC. In particular, every theorem φ of ZFC acquires the greatest truth value $\mathbf{1}_B$, and we declare φ holding inside the model $\mathbf{V}^{(B)}$.

This construction is elaborated in Chapters 1–3. Application of Boolean valued models to problems of analysis rests ultimately on the procedures of *ascending and descending*, the two natural functors acting between $\mathbf{V}^{(B)}$ and the von Neumann universe $\mathbf{V}$. Preliminaries to the axiomatics of Zermelo–Fraenkel set theory are gathered in the Appendix in order to alleviate the burden of the reader. This Appendix also contains preliminaries to category theory.

In the concluding chapters we demonstrate the main advantages of Boolean valued analysis: tools for transforming function spaces to subsets of the reals; operators, to functionals; vector functions, to numerical mappings, etc. Surely, selection of analytical topics and objects and the respective applications to functional analysis is mainly determined from the personal utility functions of the authors.

We start with thorough examination of the Boolean valued representations of algebraic systems in Chapter 4. The theory of algebraic systems, propounded in the works by A. I. Maltsev and A. Tarski, ranks among the most vital mathematical

---

[†] The "*E*, *Eir*, and *Em*" of the celebrated Personal Pronoun Pronouncement seems by far a better choice of pronouns for this paragraph (cf. [228]).

achievements of general import. A profusion of algebraic systems makes information on their Boolean valued representation a must for meaningful application to every section of the present-day mathematics.

Of the same high relevance are the constructions of Chapter 5. Mathematics, in any case mathematics as the Science of Infinity, is inconceivable without the reals. Boolean valued analysis has revealed the particular role of a universally complete Kantorovich space. It turns out that each of these spaces serves as a lawful and impeccable model of the reals. Recall that L. V. Kantorovich was the first who introduced *Dedekind complete* (that is, boundedly order complete) vector lattices as fruitful abstraction of the reals in the thirties. These spaces are also called *K-spaces* or *Kantorovich spaces* in memory of Leonid Vital′evich Kantorovich, a great mathematician and a Nobel Laureate in economics. Considering the new objects, L. V. Kantorovich propounded the ***heuristic transfer principle.*** Kantorovich's principle claims that the members of a K-space are analogs of real numbers and to each theorem about functionals there corresponds a similar theorem about operators taking values in a K-space. Time enables us to ascribe a clear and rigorous meaning to this heuristic transfer principle. The relevant tools, including the fundamental theorem by E. I. Gordon, comprise the bulk of Chapter 5. Here we also expatiate upon the problem of Boolean valued representation for Banach space, the central object of classical functional analysis. It turns out miraculously that the so-called *lattice normed spaces*, discovered at the cradle of K-space theory, depict the conventional normed spaces.

Chapter 6 deals with the theory of operator algebras. Boolean valued analysis of these algebras is the direction of research originated with the pioneer works by G. Takeuti. Study in this direction is very intensive in the recent decades. Our exposition leans upon the results of Chapter 4 about Boolean valued representation of lattice normed spaces. This approach enables us to treat in a unified fashion various analytical objects such as involutive Banach algebras, Banach modules, Jordan–Banach algebras, algebras of unbounded operators, etc.

Our book is intended to the reader interested in the modern set-theoretic models as applied to functional analysis. We tried to make the book independent to the utmost limits. However, we are fully aware that our attempts at independence were mostly foiled. Clearly, the topic of exposition needs the mathematical ideas and objects plenty above our ability to devour them. We nevertheless hope that the reader will understand our problems and forgive unintentional gaps and inaccuracies.

A. Kusraev
S. Kutateladze

Chapter 1

Universes of Sets

The credo of naive set theory cherishes a dream about the "Cantorian paradise" which is the universe that contains "any many which can be thought of as one, that is, every totality of definite elements which can be united to a whole through a law" or "every collection into a whole M of definite and separate objects m of our perception or our thought" [26].

The contemporary set theory studies realistic approximations to the ethereal ideal. These are suitable formal systems enabling us to deal with a wide spectrum of particular sets not leaving the comfortable room of soothing logical rigor. The essence of such a formalism lies in constructing a universe that "approximates from below" the world of naive sets so as to achieve the aim of current research. The corresponding axiomatic set theories open up ample opportunities to comprehend and corroborate in full detail the qualitative phenomenological principles that lie behind the standard and nonstandard mathematical models of today. ZFC, Zermelo–Fraenkel set theory, is most popular and elaborate. So, it is no wonder that our exposition proceeds mostly in the realm of ZFC. The reader, who desires to recall the subtleties of the language and axioms of ZFC, will look at the Appendix.

In the present chapter we consider a formal technique for constructing universes of sets by some transfinite processes that lead to the so-called cumulative hierarchies. This technique is vital for Boolean valued analysis. Of profound importance is the detailed description of how the von Neumann universe grows from the empty set. So, we thoroughly analyze the status of classes of sets within the formal system stemming from J. von Neumann, K. Gödel, and P. Bernays and serving as a conservative extension of Zermelo–Fraenkel set theory.

Since the main topic of the book is conspicuously tied with Boolean algebra, we start this chapter with the relevant preliminaries including the celebrated Stone Theorem. For the sake of diversity, we demonstrate it by using the Gelfand transform.

1.1. Boolean Algebras

In this section we sketch the minimum about Boolean algebras which we need in the sequel. A more explicit exposition may be found elsewhere; for instance, cf. [74, 89, 171, 220, 250].

1.1.1. To fix terminology, we recall some well known notions.

An *ordered set* is a pair $(M, \leq)$, where $\leq$ is an order on M (see A.1.10). An ordered set is also called a *partially ordered set* or, briefly, a *poset*. It is in common parlance to apply all names of $(M, \leq)$ to the underlying set M of $(M, \leq)$. We indulge in doing the same elsewhere without further circumlocution.

An *upper bound* of a subset X of a poset M is an element $a \in M$ such that $x \leq a$ for all $x \in X$. A least element of the set of upper bounds of X is called a *least upper bound* or *supremum* of X and denoted by $\sup(X)$ or $\sup X$. In other words, $a = \sup(X)$ if and only if a is an upper bound of X and $a \leq b$ for every upper bound b of X. By reversal, i.e., by passing from the original order $\leq$ on a poset M to the *reverse* or *opposite order* $\leq^{-1}$, define a *lower bound* of a subset X of M and a *greatest lower bound*, $\inf(X)$ of X, also called an *infimum* of X and denoted by $\inf X$. If a least upper or greatest lower bound of a set in M exists then it is unique and so deserves the definite article.

A *lattice* is an ordered set L in which each pair $\{x, y\}$ has the *join* $x \vee y := \sup\{x, y\}$ and *meet* $x \wedge y := \inf\{x, y\}$. Given a subset X of a lattice L, we use the notation:

$$\bigvee X := \sup(X), \quad \bigwedge X := \inf(X),$$

$$\bigvee_{\alpha \in \mathrm{A}} x_\alpha := \bigvee \{x_\alpha : \alpha \in \mathrm{A}\}, \quad \bigwedge_{\alpha \in \mathrm{A}} x_\alpha := \bigwedge \{x_\alpha : \alpha \in \mathrm{A}\},$$

$$\bigvee_{k=1}^{n} x_k := x_1 \vee \ldots \vee x_n := \sup\{x_1, \ldots, x_n\},$$

$$\bigwedge_{k=1}^{n} x_k := x_1 \wedge \ldots \wedge x_n := \inf\{x_1, \ldots, x_n\}.$$

Here $(x_\alpha)_{\alpha \in \mathrm{A}}$ is a family in L, and $x_1, \ldots, x_n$ stand for some members of L.

The binary operations *join* $(x, y) \mapsto x \vee y$ and *meet* $(x, y) \mapsto x \wedge y$ act in every lattice L and possess the following properties:

(1) *commutativity*:

$$x \vee y = y \vee x, \quad x \wedge y = y \wedge x;$$

(2) *associativity*:

$$x \vee (y \vee z) = (x \vee y) \vee z, \quad x \wedge (y \wedge z) = (x \wedge y) \wedge z.$$

By induction, from (2) we deduce that *every nonempty finite set in a lattice has the join and meet.* If every subset of a lattice L has the supremum and infimum then L is a *complete* lattice.

A lattice L is *distributive* provided that the following *distributive laws* hold:

(3) $x \wedge (y \vee z) = (x \wedge y) \vee (x \wedge z)$;

(4) $x \vee (y \wedge z) = (x \vee y) \wedge (x \vee z)$.

If a lattice L has the least or greatest element then the former is called the *zero* of L and the latter, the *unity* of L. The zero and unity of L are solemnly denoted by $\mathbf{0}_L$ and $\mathbf{1}_L$. It is customary to use the simpler symbols $\mathbf{0}$ and $\mathbf{1}$ and nicknames *zero* and *unity* provided that the context prompts the due details. Note also that $\mathbf{0}$ and $\mathbf{1}$ are neutral elements:

(5) $\mathbf{0} \vee x = x, \quad \mathbf{1} \wedge x = x.$

Specifying the general definitions, note also that $\bigvee \varnothing = \sup \varnothing := \mathbf{0}$ and $\bigwedge \varnothing = \inf \varnothing := \mathbf{1}$. A *complement* x^* of a member x of a lattice L with zero and unity is an element x^* of L such that

(6) $x \wedge x^* = \mathbf{0}, \quad x \vee x^* = \mathbf{1}.$

Elements x and y in L are *disjoint* if $x \wedge y = \mathbf{0}$. So, every element x is disjoint from any complement x^*. Recall by the way that a set U is *disjoint* whenever every two distinct members of U are disjoint. Note finally that if each element in L has at least one complement then we call L a *complemented lattice.* It is rather evident that an arbitrary lattice L may fail to have a complement to each element of L.

1.1.2. A *Boolean algebra* is a distributive complemented lattice with zero and unity.

The above definition looks somewhat strange at first sight. Indeed, it does not reveal the reasons for whatever distributive lattice to be called an algebra since the term "algebra" refers to conventional objects (cf. Lie algebra, Banach algebra, C^*-algebra, etc.). The arising ambiguity is easily eliminated because a Boolean algebra is in fact an algebra over the two-element field. The principal importance of this peculiarity is partially reflected in the subsection to follow. At the same time, it is perfectly natural to view Boolean algebras in different contexts at different angles. Below we will however treat a Boolean algebra primarily as a distributive complemented lattice with zero and unity. It is worth emphasizing that the particular Boolean algebras we deal with in functional analysis appear mostly as distributive complemented lattices.

Note also that as a formal example of a Boolean algebra we may take the *one-element lattice*; i.e., the singleton $\{x\}$ with the only order relation $x \leq x$. This algebra is called *degenerate.* A degenerate Boolean algebra is a noble instance of an algebraic system but an unassuming simpleton in the context of Boolean valued analysis we are interested in. The slimmest nondegenerate Boolean algebra $\mathbf{2}$, alternatively denoted by $\mathbb{Z}_2$, is the *two-element lattice* with the underlying set $\{\mathbf{0}, \mathbf{1}\}, \mathbf{0} \neq \mathbf{1}$, and the order $\mathbf{0} \leq \mathbf{1}$, $\mathbf{0} \leq \mathbf{0}$, and $\mathbf{1} \leq \mathbf{1}$. Austerity notwithstanding, the *two-element Boolean algebra* **2** plays an important role in the subsequent chapters.

Therefore, speaking about a Boolean algebra B, we agree always to assume that $\mathbf{0}_B \neq \mathbf{1}_B$, i.e., we eliminate the degenerate algebras from the further consideration.

Each element x of a Boolean algebra B has a unique complement denoted by x^*. This gives rise to the mapping $x \mapsto x^*$ $(x \in B)$ which is idempotent (i.e., $(\forall x \in B)$ $(x^{**} := (x^*)^* = x)$) and presents a *dual isomorphism* or an *anti-isomorphism* of B onto itself (i.e., it is an order isomorphism between $(B, \leq)$ and $(B, \leq^{-1})$). In particular, the *De Morgan laws* hold:

$$\Big(\bigvee_{\alpha \in \mathrm{A}} x_\alpha\Big)^* = \bigwedge_{\alpha \in \mathrm{A}} x_\alpha^*, \quad \Big(\bigwedge_{\alpha \in \mathrm{A}} x_\alpha\Big)^* = \bigvee_{\alpha \in \mathrm{A}} x_\alpha^*,$$

with $x_\alpha \in B$ for all $\alpha \in \mathrm{A}$.

1.1.3. The three entities $\vee$, $\wedge$, and $*$, living in every Boolean algebra B, [illegible] jointly referred to as *Boolean operations*.

Recall that a *universal algebra* is an algebraic system without predicates. This concept makes available another definition of Boolean algebra. Namely, a Boolean algebra B is a universal algebra $(B, \vee, \wedge, *, \mathbf{0}, \mathbf{1})$ with two binary operations $\vee$ and $\wedge$, one unary operation $*$, and two distinguished elements $\mathbf{0}$ and $\mathbf{1}$ obeying the conditions:

(1) $\vee$ and $\wedge$ are commutative and associative;
(2) $\vee$ and $\wedge$ are both distributive relative to one another;
(3) x and x^* complement one another;
(4) $\mathbf{0}$ and $\mathbf{1}$ are neutral for $\vee$ and $\wedge$, respectively.

Conversely, given a universal algebra B of the above type, make B into a poset by letting $x \leq y$ whenever $x \wedge y = x$ for $x, y \in B$. In this event, note that $(B, \leq)$ is a distributive complemented lattice with join $\vee$, meet $\wedge$, complementation $*$, zero $\mathbf{0}$, and unity $\mathbf{1}$.

1.1.4. Using the basic Boolean operations $\vee$, $\wedge$, and $*$, we may define a few other operations:

$$x - y := x \wedge y^*, \quad x \Rightarrow y := x^* \vee y,$$
$$x \vartriangle y := (x - y) \wedge (y - x) = (x \wedge y^*) \vee (y \wedge x^*),$$
$$x \Leftrightarrow y := (x \Rightarrow y) \wedge (y \Rightarrow x) = (x^* \vee y) \wedge (y^* \vee x).$$

We list several easy formulas of constant use in what follows:

(1) $x \Rightarrow y = (x - y)^*, \quad x \Leftrightarrow y = (x \vartriangle y)^*$;

(2) $x \Rightarrow (y \Rightarrow z) = (x \wedge y) \Rightarrow z = (x \wedge y) \Rightarrow (x \wedge z)$;

(3) $x \leq y \Rightarrow z \leftrightarrow x \wedge y \leq z \leftrightarrow y - z \leq x^*$;

(4) $x \leq y \leftrightarrow x \Rightarrow y = \mathbf{1} \leftrightarrow x - y = \mathbf{0}$;

(5) $x = y \leftrightarrow x \Leftrightarrow y = \mathbf{1} \leftrightarrow x \vartriangle y = \mathbf{0}$.

It is worth observing that $\vartriangle$, the so-called *symmetric difference*, has the properties resembling a metric:

(6) $x \vartriangle y = \mathbf{0} \leftrightarrow x = y$;

(7) $x \vartriangle y = y \vartriangle x$;

(8) $x \vartriangle y \leq (x \vartriangle z) \vee (z \vartriangle y)$.

Moreover, the lattice operations become contractive with respect to this "metric," while complementation becomes an isometry:

$$(x \vee y) \vartriangle (u \vee v) \leq (x \vartriangle u) \vee (y \vartriangle v),$$
$$(x \wedge y) \vartriangle (u \wedge v) \leq (x \vartriangle u) \vee (y \vartriangle v),$$
$$x^* \vartriangle y^* = x \vartriangle y.$$

1.1.5. A Boolean algebra B is *complete* (σ*-complete*) if each subset (countable subset) of B has a supremum and an infimum. By tradition, we speak of σ*-algebras* instead of σ-complete algebras.

Associated with a Boolean algebra B, the mappings $\bigvee, \bigwedge : \mathscr{P}(B) \to B$ are available that ascribe to a set in B its supremum and infimum, respectively. These mappings are sometimes referred to as *infinite operations*. The infinite operations obey many important rules among which we mention the *infinite distributive laws*:

(1) $x \vee \bigwedge\limits_{\alpha \in \mathrm{A}} x_\alpha = \bigwedge\limits_{\alpha \in \mathrm{A}} x \vee x_\alpha$;

(2) $x \wedge \bigvee\limits_{\alpha \in \mathrm{A}} x_\alpha = \bigvee\limits_{\alpha \in \mathrm{A}} x \wedge x_\alpha$.

From (1) and (2) the following useful equalities ensue:

(3) $\big(\bigvee\limits_{\alpha \in \mathrm{A}} x_\alpha\big) \Rightarrow x = \bigwedge\limits_{\alpha \in \mathrm{A}} (x_\alpha \Rightarrow x)$;

(4) $\big(\bigwedge\limits_{\alpha \in \mathrm{A}} x_\alpha\big) \Rightarrow x = \bigvee\limits_{\alpha \in \mathrm{A}} (x_\alpha \Rightarrow x)$;

(5) $x \Rightarrow \big(\bigvee\limits_{\alpha \in \mathrm{A}} x_\alpha\big) = \bigvee\limits_{\alpha \in \mathrm{A}} (x \Rightarrow x_\alpha)$;

(6) $x \Rightarrow \big(\bigwedge\limits_{\alpha \in \mathrm{A}} x_\alpha\big) = \bigwedge\limits_{\alpha \in \mathrm{A}} (x \Rightarrow x_\alpha)$.

Ensured are also the commutativity and associativity of suprema and infima, we mentioned earlier in some particular cases, cf. 1.1.1 (1, 2):

(7) $\bigvee_{\alpha\in\mathrm{A}} \bigvee_{\beta\in\mathrm{B}} x_{\alpha,\beta} = \bigvee_{\beta\in\mathrm{B}} \bigvee_{\alpha\in\mathrm{A}} x_{\alpha,\beta}$;

(8) $\bigwedge_{\alpha\in\mathrm{A}} \bigwedge_{\beta\in\mathrm{B}} x_{\alpha,\beta} = \bigwedge_{\beta\in\mathrm{B}} \bigwedge_{\alpha\in\mathrm{A}} x_{\alpha,\beta}$;

(9) $\bigvee(\bigcup_{\alpha\in\mathrm{A}} X_\alpha) = \bigvee_{\alpha\in\mathrm{A}} \bigvee X_\alpha$;

(10) $\bigwedge(\bigcup_{\alpha\in\mathrm{A}} X_\alpha) = \bigwedge_{\alpha\in\mathrm{A}} \bigwedge X_\alpha$,

where $X_\alpha \subset B$ for all $\alpha \in \mathrm{A}$.

Note that (1)–(6) hold in an arbitrary Boolean algebra, whereas (7)–(10) hold in every poset on duly stipulating existence of the suprema and infima in question.

1.1.6. Consider some methods of arranging new Boolean algebras.

(1) A nonempty subset B_0 of a Boolean algebra B is a *subalgebra* of B if B_0 is closed under the Boolean operations $\vee$, $\wedge$, and $*$; i.e., $\{x \vee y, x \wedge y, x^*\} \subset B_0$ for all $x, y \in B_0$.

Under the order induced from B, every subalgebra B_0 is a Boolean algebra with the same zero and unity as those of B. In particular, $B_0 := \{\mathbf{0}_B, \mathbf{1}_B\}$ is a subalgebra of B.

A subalgebra $B_0 \subset B$ is *regular* (*σ-regular*) provided that for every set (countable set) A in B_0 the elements $\bigvee A$ and $\bigwedge A$, if exist in B, belong to B_0.

The intersection of every family of subalgebras is a subalgebra too. The same holds for regular (σ-regular) subalgebras, which makes the definition to follow sound.

The least subalgebra of B containing a nonempty subset M of B is the *subalgebra generated by* M. The *regular* (*σ-regular*) *subalgebra generated by* M is introduced in much the same manner.

(2) An *ideal* of a Boolean algebra B is any nonempty set J in B obeying the conditions:

$$x \in J \wedge y \in J \to x \vee y \in J,$$
$$x \in J \wedge y \leq x \to y \in J.$$

The set $B_a := \{x \in B : x \leq a\}$, with $a \in B$, provides an example of an ideal of B. An ideal of this shape is called *principal.* If $\mathbf{0} \neq e \in B$ then the principal ideal B_e with the order induced from B is a Boolean algebra in its own right. The element e plays the role of unity in B_e. The lattice operations of B_e are inherited from B, and the complementation of B_e has the form $x \mapsto e - x$ for all $x \in B$.

An ideal J is *proper* provided that $J \neq B$. A regular ideal of B is often called a *band* or *component* of B.

(3) Take Boolean algebras B and B' and a mapping $h : B \to B'$. Say that H *isotonic*, or *isotone*, or *monotone* if $(x \leq y \to h(x) \leq h(y))$. (Note by the way that an isotonic mapping from B to B' with the opposite order is *antitonic*.) Say that h is a (*Boolean*) *homomorphism* if for all $x, y \in B$ the following equalities are fulfilled:

$$h(x \vee y) = h(x) \vee h(y),$$
$$h(x \wedge y) = h(x) \wedge h(y),$$
$$h(x^*) = h(x)^*.$$

Every homomorphism h is monotone and the image $h(B)$ of B is a subalgebra of B'. If h is bijective then we call h an *isomorphism* referring to B and B' as *isomorphic* Boolean algebras. An injective homomorphism is a *monomorphism*. A homomorphism h (of B to a complete B') is *complete* if h *preserves suprema and infima*; i.e., $h(\sup(U)) = \sup(h(U))$ and $h(\inf(V)) = \inf(h(V))$ for all $U \subset B$ and $V \subset B$ for which there are $\sup(U)$ and $\inf(V)$.

Assume given a set C and a bijection $h : B \to C$. We may then equip C with an order by putting $h(x) \leq h(y)$ whenever $x \leq y$. In this event C turns into a Boolean algebra and h becomes an isomorphism between B and C.

(4) Let J be a proper ideal of a Boolean algebra B. Define the equivalence $\sim$ on B by the rule

$$x \sim y \leftrightarrow x \bigtriangleup y \in J \quad (x, y \in B).$$

Denote by φ the factor mapping of B onto the factor set $B/J := B/\sim$. Recall that φ is also called *canonical*. Given *cosets* (equivalence classes) u and v, i.e., members of B/J; agree to write $u \leq v$ if and only if there are $x \in u$ and $y \in v$ satisfying $x \leq y$. We have thus defined an order on B/J. In this event B/J becomes a Boolean algebra which is called *factor algebra* of B by J. The Boolean operations in B/J make φ a homomorphism. So, φ is referred to as the *factor homomorphism* of B onto B/J.

If $h : B \to B'$ is a homomorphism then $\ker(h) := \{x \in B : h(x) = \mathbf{0}\}$ is an ideal of B and there is a unique monomorphism $g : B/\ker(h) \to B'$ satisfying $g \circ \varphi = h$, where $\varphi : B \to B/\ker(h)$ is the factor homomorphism. Therefore, each homomorphic image of a Boolean algebra B is isomorphic to the factor algebra of B by a suitable ideal.

(5) Take a family of Boolean algebras $(B_\alpha)_{\alpha \in \mathrm{A}}$. Furnish the product $B := \prod_{\alpha \in \mathrm{A}} B_\alpha$ with the *coordinatewise order* or *product order* by putting $x \leq y$ for $x, y \in B$ whenever $x(\alpha) \leq y(\alpha)$ for all $\alpha \in \mathrm{A}$. In this event B becomes a Boolean algebra.

Each Boolean operation in B consists in implementing the respective operation in every coordinate Boolean algebra B_α, i.e., it is carried out *coordinatewise*. The zero $\mathbf{0}_B$ and unity $\mathbf{1}_B$ of B are as follows: $\mathbf{0}_B(\alpha) := \mathbf{0}_\alpha$ and $\mathbf{1}_B(\alpha) := \mathbf{1}_\alpha$ $(\alpha \in \mathrm{A})$, where $\mathbf{0}_\alpha$ and $\mathbf{1}_\alpha$ are the zero and unity in B_α. The Boolean algebra B is the *Cartesian product* or, simply, *product* of $(B_\alpha)_{\alpha\in\mathrm{A}}$.

(6) We again assume given a family of Boolean algebras $(B_\alpha)_{\alpha\in\mathrm{A}}$. Then, there are a Boolean algebra B and a family of monomorphisms $\imath_\alpha : B_\alpha \to B$ $(\alpha \in \mathrm{A})$ obeying the following conditions:

(a) The family of subalgebras $(\imath_\alpha(B_\alpha))_{\alpha\in\mathrm{A}}$ of B is *independent*; i.e., every collection of finitely many nonzero elements $x_k \in \imath_{\alpha_k}(B_{\alpha_k})$, with $\alpha_k \neq \alpha_l$ for $k \neq l$ and $\alpha_1, \dots, \alpha_n \in \mathrm{A}$, satisfies the condition $x_1 \wedge \dots \wedge x_n \neq \mathbf{0}$;

(b) The subalgebra of B, generated by the union of $\imath_\alpha(B_\alpha)$, coincides with B.

If a Boolean algebra B' and a family of monomorphisms $\imath'_\alpha : B_\alpha \to B'$ $(\alpha \in \mathrm{A})$ obey the same conditions (a) and (b) then there is an isomorphism h of the algebra B onto the algebra B' such that $\imath_\alpha \circ h = \imath'_\alpha$ $(\alpha \in \mathrm{A})$.

We call the pair $(B, (\imath_\alpha)_{\alpha\in\mathrm{A}})$ the *Boolean product* or *tensor product* of $(B_\alpha)_{\alpha\in\mathrm{A}}$ and denoted it by the symbol $\bigotimes_{\alpha\in\mathrm{A}} B_\alpha$.

(7) A *completion* of a Boolean algebra B is a pair $(\imath, A)$ satisfying the following conditions: (a) A is a complete Boolean algebra; (b) $\imath$ is a complete monomorphism from B to A; and (c) the regular subalgebra of A, generated by $\imath(B)$, coincides with A.

Naturally, A itself is also called a "completion" of B. Say that pairs $(\imath, A)$ and $(\imath', A')$ are *isomorphic* if there is an isomorphism $h : A \to A'$ such that $h \circ \imath = \imath'$. All completions of B are isomorphic to one another and so each of them is sometimes referred to as *the* completion of B. Avoiding circumlocution, we exploit this advantage of the English usage to full extend in similar circumstances. The completion of a Boolean algebra may be obtained for instance by using the classical method of Dedekind cuts.

1.1.7. Examples.

(1) Given a nonempty set X, note that the inclusion ordered powerset $\mathscr{P}(X)$ of X is a complete Boolean algebra. This algebra is often the *boolean* of X. The Boolean operations on every boolean are the set-theoretic operations of union, intersection, and complementation.

(2) Let X be a topological space. Recall that a closed and open subset of X is called *clopen*. The collection of all clopen sets in X, ordered by inclusion, is a subalgebra of the boolean $\mathscr{P}(X)$. Denote this subalgebra by $\mathrm{Clop}(X)$. The Boolean operations in $\mathrm{Clop}(X)$ are inherited from $\mathscr{P}(X)$. Hence, they are set-theoretic. However, Clop(X) is not a regular subalgebra $\mathscr{P}(X)$; i.e., the infinite operations in $\mathscr{P}(X)$ and Clop(X) may differ essentially.

(3) A closed subset F of a topological space X is called *regular* if $F = \mathrm{cl}(\mathrm{int}(F))$; i.e., if F coincides with the closure of the interior of F. By analogy, a *regular open set* G is defined by the formula $G = \mathrm{int}(\mathrm{cl}(G))$. Let RC ($X$) and RO ($X$) stand for the collections of all regular closed subsets and all regular open subsets of X.

Equipped with the order by inclusion, RC (X) and RO (X) become complete Boolean algebras. The mapping $F \mapsto \mathrm{int}(F)$ is an isomorphism between RC (X) and RO (X). Despite RC (X) and RO (X) are included in the boolean $\mathscr{P}(X)$, they are not subalgebras of the latter. For instance, the Boolean operations on RC (X) have the form

$$E \vee F = E \cup F, \quad E \wedge F = \mathrm{cl}(\mathrm{int}(E \cap F)), \quad F^* = \mathrm{cl}(X - F).$$

(4) Denote by $\mathscr{B}or(X)$ the Borel σ-algebra of a topological space X (i.e., the σ-regular subalgebra of the boolean $\mathscr{P}(X)$ generated by the open sets of X). Consider the ideal $\mathscr{N}$ of $\mathscr{B}or(Q(X))$ comprising the *meager* subsets of X (also called the first category sets in X). The factor algebra $\mathscr{B}or(Q(X)/\mathscr{N})$ is a complete Boolean algebra called the *algebra of Borel sets modulo meager sets* or briefly *Borel-by-meager algebra.*

We arrive at an isomorphic algebra if instead of $\mathscr{B}or(Q(X))$ we take the σ-algebra of sets with the Baire property. (A subset M of X has the *Baire property* if there is an open set G in X such that the symmetric difference $M \vartriangle G$ is a meager set.) If X is a *Baire* space; i.e., if X lacks nonempty open meager subsets; then the algebra in question is isomorphic to the algebra RC (X) of regular closed sets.

(5) Assume given a σ-complete Boolean algebra, $\mathscr{B}$ and a positive countably additive function $\mu : \mathscr{B} \to \mathbb{R}$. *Countable additivity*, as usual, means that

$$\mu\Bigl(\bigvee_{n=1}^{\infty} x_n\Bigr) = \sum_{n=1}^{\infty} \mu(x_n)$$

for every disjoint sequence (x_n) of $\mathscr{B}$. A function μ with the above properties is called a (finite) *measure.*

Let $\mathscr{N} := \{x \in \mathscr{B} : \mu(x) = 0\}$. Then $\mathscr{N}$ is a σ-complete ideal. There is a unique countably additive function $\bar{\mu}$ on the factor algebra $B := \mathscr{B}/\mathscr{N}$ for which $\mu = \bar{\mu} \circ \varphi$, where $\varphi : \mathscr{B} \to B$ is the factor mapping. The algebra B is complete, and the function $\bar{\mu}$ is strictly positive; i.e., $\bar{\mu}(x) = 0 \to x = \mathbf{0}$. If $\rho(x, y) := \bar{\mu}(x \vartriangle y)$ then ρ is a metric, and the metric space (B, ρ) is complete. Assume that $(X, \mathscr{B}, \mu)$ is a *finite measure space*; i.e., X is a nonempty set, $\mathscr{B}$ is a σ-complete subalgebra in $\mathscr{P}(X)$, and μ is the same as above. The algebra B is called the *algebra of measurable sets by measure zero sets.*

(6) Assume that $(X, \mathscr{B}, \mu)$ is the same as in (5), and denote by $M(\mu)$ or $M(X, \mathscr{B}, \mu)$ the set of the equivalence classes of μ-measurable almost everywhere finite functions on X. Recall that measurable functions are *equivalent* provided that they may differ only on a measure zero set. Furnish $M(\mu)$ with an order by putting $\bar{f} \leq \bar{g}$ if and only if $f(x) \leq g(x)$ for almost all $x \in X$. Here $\bar{f}$ is the coset of f. In this event $M(\mu)$ becomes a lattice. Denote by $\mathbf{1}$ the coset of the identically one function on X. Put $B := \{e \in M(\mu) : e \wedge (\mathbf{1} - e) = \mathbf{0}\}$. Under the order induced from $M(\mu)$, the set B is a complete Boolean algebra with the following Boolean operations:

$$c \vee e = c + e - c \cdot e, \quad c \wedge e = c \cdot e, \quad e^* = \mathbf{1} - e \quad (c, e \in B),$$

where $+$, $\cdot$, and $-$ stand for the addition, multiplication, and complementation of $M(\mu)$.

(7) Assume that H is a complex Hilbert space, and $\mathscr{L}(H)$ stands for the *bounded endomorphism algebra* of H; i.e., the algebra of all bounded linear operators from H to H.

Given a set A in $\mathscr{L}(H)$, define the *commutant* A' of A by the formula $A' := \{T \in \mathscr{L}(H) : (\forall S \in A)\,(TS = ST)\}$. Define the *double commutant* or *bicommutant* of A to be the set $A'' := (A')'$. A subalgebra A of $\mathscr{L}(H)$ is *selfadjoint* whenever $T \in A \to T^* \in A$. A *von Neumann algebra* is a selfadjoint subalgebra A of $\mathscr{L}(H)$ that coincides with the double commutant of A.

Consider a commutative von Neumann algebra A. Denote by $\mathfrak{P}(A)$ the set of all orthoprojections belonging to A. Furnish $\mathfrak{P}(A)$ with the following order:

$$\pi \leq \rho \leftrightarrow \pi(H) \subset \rho(H) \quad (\pi, \rho \in \mathfrak{P}(A)).$$

In this event $\mathfrak{P}(A)$ becomes complete Boolean algebra whose Boolean operations have the form:

$$\pi \vee \rho = \pi + \rho - \pi \circ \rho, \quad \pi \wedge \rho = \pi \circ \rho, \quad \pi^* = I_H - \pi.$$

1.1.8. Comments.

(1) The theory of Boolean algebras originated from the classical work by G. Boole "An Investigation of the Laws of Thought on Which Are Founded the Mathematical Theories of Logic and Probabilities" [16, 17]. The author himself formulated his intentions as follows: "The design of the following treatise is to investigate the fundamental laws of those operations of the mind by which reasoning is performed; to give expression to them in the language of a Calculus, and upon this foundation to establish the science of Logic and construct its method...."

Pursuing this end, G. Boole carried out, in fact, algebraization of the logical system lying behind the classical mathematical reasoning. In a result, he become the author of the algebraic system omnipresent under the name of Boolean algebra.

(2) The principal objects of G. Boole's book are propositions. In the modern language, the ***proposition algebra*** or ***sentence algebra*** is the Boolean algebra resulting from identification of equivalent formulas in the set of all sentences of propositional calculus. We may put this formally as follows:

Let $\mathscr{T}$ be a first-order theory based on classical (two-valued) logic. Furnish the set of all sentences Φ in the theory $\mathscr{T}$ with a preorder by putting $\varphi \leq \psi$ whenever the formula $\varphi \to \psi$ is a theorem of $\mathscr{T}$. Consider the associate equivalence $\sim$ in Φ:

$$\varphi{\sim}\psi \leftrightarrow \varphi \leq \psi \wedge \psi \leq \varphi \quad (\varphi, \psi \in \Phi).$$

Arrange the factor set $\mathfrak{A}(\mathscr{T}) := \Phi/{\sim}$ with the induced order. In more detail, if $|\varphi|$ is the coset of $\varphi \in \Phi$ then $|\varphi| \leq |\psi|$ implies $\varphi \leq \psi$. The resultant poset $\mathfrak{A}(\mathscr{T})$ is a Boolean algebra called sometimes the ***Lindenbaum–Tarski algebra*** of $\mathscr{T}$. The Boolean operations in the algebra $\mathfrak{A}(\mathscr{T})$ have the form

$$|\varphi| \vee |\psi| = |\varphi \vee \psi|,$$
$$|\varphi| \wedge |\psi| = |\varphi \wedge \psi|,$$
$$|\varphi|^* = |\neg\varphi|.$$

Translation of the logical problems of formal theories into the language of the corresponding Boolean algebras, the Lindenbaum–Tarski algebras, is called the *Boolean method.*

(3) The classical ways of deduction (syllogisms, the excluded middle, modus ponens, generalization, etc.) are constructs that originated from abstracting the actual operations of mind in the process of reasoning.

Inevitably displaying the reality in rough, the two-valued logic provides, strictly speaking, only an approximate and incomplete description for the laws of thought, which explains interest in nonclassical logical systems. One of these systems is elaborated within intuitionism. Avoiding details, we briefly describe the corresponding sentence algebra.

A ***pseudo-Boolean algebra*** is a lattice L with zero and unity in which to x, $y \in L$ there corresponds the pseudocomplement $x \Rightarrow y$ of x relative to y.

By definition, the ***pseudocomplement*** $x \Rightarrow y$ is the greatest of the elements $z \in L$ obeying the inequality $z \wedge x \leq y$. Hence, the following equivalence holds (cf. 1.1.4 (3))

$$z \leq x \Rightarrow y \leftrightarrow x \wedge z \leq y \quad (x, y, z \in L)$$

which may also be considered as the definition of $x \Rightarrow y$. A pseudo-Boolean algebra is a distributive lattice. A complete lattice is a pseudo-Boolean algebra if and only if the following distributive laws hold in it:

$$x \wedge \bigvee_{\alpha \in \mathrm{A}} x_\alpha = \bigvee_{\alpha \in \mathrm{A}} x \wedge x_\alpha \quad (x, x_\alpha \in L).$$

The set of all open subsets of a topological space, ordered by inclusion, exhibits an example of a complete pseudo-Boolean algebra.

A *Brouwer lattice* is another name for a pseudo-Boolean algebra. However, the commonest title of a pseudo-Boolean algebra is a *Heyting algebra.* It may be demonstrated that the Lindenbaum–Tarski algebra of intuitionistic logic is a Heyting algebra. Therefore, Heyting algebras are characteristic of intuitionistic logic in much the same way as Boolean algebras are characteristic of classical logic (for details, cf. [14, 204]).

(4) As exemplified by intuitionistic logic, study of some types of nonclassical logics leads to various classes of algebraic systems which are distributive lattices. The most popular instances are as follows: an *implicative lattice* or pseudocomplemented lattice, a *topological Boolean algebra* (i.e., a Boolean algebra B with the operation $\mathbf{I} : B \to B$ obeying the *interior axioms*: $\mathbf{I}(x \wedge y) = \mathbf{I}x \wedge \mathbf{I}y$, $x \leq y \to \mathbf{I}x \leq \mathbf{I}y$; $\mathbf{I}^2 = \mathbf{I}$; $\mathbf{I0} = \mathbf{0}$; and $\mathbf{I1} = \mathbf{1}$), a *Post algebra*, etc. (see [14, 69, 204]). A general theory of lattices is an established direction of research which bears a firm and deep relationship with various branches of mathematics.

(5) Origination of all these logics or lattices is associated with "investigation of the laws of thought" in the spirit of the Boole design we have cited.

Analysis of the laws of the microcosm gives rise to a principally different type of logic. The logic of quantum mechanics differs significantly from classical, intuitionistic, and modal logics.

An *ortholattice* is a lattice L with zero, unity, and a unary operation of orthocomplementation $(\,\cdot\,)^\perp : L \to L$ obeying the following conditions:

$$x \wedge x^\perp = \mathbf{0}, \quad x \vee x^\perp = \mathbf{1};$$
$$x^{\perp\perp} := (x^\perp)^\perp = x;$$
$$(x \vee y)^\perp = x^\perp \wedge y^\perp, \quad (x \wedge y)^\perp = x^\perp \vee y^\perp.$$

A distributive ortholattice is a Boolean algebra.

We call two elements x and y *orthogonal* and write $x \perp y$ if $x \leq y^\perp$ or, which is equivalent, $y \leq x^\perp$. An ortholattice L is an *orthomodular lattice* or a *quantum logic* provided that to all $x, y \in L$, $x \leq y$, there is an element $z \in L$ such that $x \perp z$ and $x \vee z = y$, which amounts to the fact that $x \leq y$ implies $y = x \vee (y \wedge x^\perp)$.

The lattice of all closed subspaces of a Hilbert space with orthogonal complementation provides an example of quantum logic.

1.2. Representation of a Boolean Algebra

The Stone Theorem opens up a distinct possibility of representing a Boolean algebra as the Boolean algebra of clopen subsets of a compact space. The basic goal of this section is to prove this theorem and to describe some opportunities it affords.

1.2.1. Let $\mathbf{2} := \mathbb{Z}_2 := \mathscr{P}(\{\varnothing\}) := \{\mathbf{0}, \mathbf{1}\}$ be the underlying set of the two-element Boolean algebra now viewed as a field with the following operations:

$$\mathbf{0}+\mathbf{0} := \mathbf{0}, \quad \mathbf{0}+\mathbf{1} = \mathbf{1}+\mathbf{0} := \mathbf{1}, \quad \mathbf{1}+\mathbf{1} := \mathbf{0},$$
$$\mathbf{0}\cdot\mathbf{1} = \mathbf{1}\cdot\mathbf{0} := \mathbf{0}, \quad \mathbf{0}\cdot\mathbf{0} := \mathbf{0}, \quad \mathbf{1}\cdot\mathbf{1} := \mathbf{1}.$$

Note that every member of $\mathbf{2}$ is idempotent.

Consider an arbitrary set B with the structure of an associative ring whose every element is idempotent: $(\forall b \in B)(b^2 = b)$. In this case B is called a *Boolean ring.* A Boolean ring is commutative and obeys the identity $b = -b$ for $b \in B$. Each Boolean ring is obviously a vector space and, at the same time, a commutative algebra over $\mathbf{2}$. Recall that the unity of an algebra differs from its zero by definition. So, we may and will identify the field $\mathbf{2}$ with the subring of a Boolean ring comprising the zero and unity of the latter. We usually reflect the practice in symbols by letting $\mathbf{0}$ stand for the zero and $\mathbf{1}$, for the unity of whatever ring. This agreement leads clearly to a rather popular notational collision: the addition and multiplication of $\mathbf{2}$ may be redefined on making $\mathbf{0}$ play the role of $\mathbf{1}$ and vice versa.

It is customary to endow a Boolean ring B with some order by the rule:

$$b_1 \leq b_2 \leftrightarrow b_1 b_2 = b_1 \quad (b_1, b_2 \in B).$$

The poset $(B, \leq)$ obviously becomes a distributive lattice with the least element $\mathbf{0}$ and the greatest element $\mathbf{1}$. In this event the lattice and ring operations are connected as follows:

$$x \vee y = x + y + xy, \quad x \wedge y = xy.$$

Moreover, to each element $b \in B$ there is a unique $b^* \in B$, the complement of b, such that

$$b^* \vee b = \mathbf{1}, \quad b^* \wedge b = \mathbf{0}.$$

Obviously, $b^* = \mathbf{1} + b$. Hence, each Boolean ring is a Boolean algebra under the above order.

In turn, we may transform a Boolean algebra B into a ring by putting

$$x + y := x \vartriangle y, \quad xy := x \wedge y \quad (x, y \in B).$$

In this case $(B, +, \cdot, \mathbf{0}, \mathbf{1})$ becomes a unital Boolean ring whose natural order coincides with the initial order on B. Therefore, a Boolean algebra can be viewed as a unital algebra over $\mathbf{2}$ whose every element is idempotent.

1.2.2. Let B be an arbitrary Boolean algebra.

(1) A *character* χ of B is a Boolean homomorphism or, which is the same, a ring homomorphism χ from B to $\mathbf{2}$. Denote by $\mathrm{X}(B)$ the set of all characters of B and make $\mathrm{X}(B)$ into a topological space on furnishing it with the topology of pointwise convergence. To put it more explicitly, the topology on $\mathrm{X}(B)$ is induced by the product topology of $\mathbf{2}^B$, where we consider $\mathbf{2}$ with the unique compact Hausdorff topology on this set, the discrete topology of $\mathbf{2}$. Recall that a topological space X is *connected* whenever the only clopen subsets of X are $\varnothing$ and X. A topological space X is *totally disconnected* provided that each connected subspace of X is at most a singleton. The topological space $\mathbf{2}^B$, called sometimes a *Cantor discontinuum*, is Hausdorff, compact, and totally disconnected. A topological space with all these properties is a *Boolean space*. Evidently, $\mathrm{X}(B)$ is a closed subset of $\mathbf{2}^B$. Therefore, $\mathrm{X}(B)$ itself is a Boolean space. Say that the Boolean space $\mathrm{X}(B)$ is the *character space* of a Boolean algebra B.

(2) Recall that a nonempty subset $\mathscr{F}$ of B is a *filter* on B provided that

$$x \in \mathscr{F} \wedge y \in \mathscr{F} \to x \vee y \in \mathscr{F},$$
$$x \in \mathscr{F} \wedge x \leq y \to y \in \mathscr{F}.$$

A filter other than B is *proper*. A maximal element of the inclusion ordered set of all proper filters on B is an *ultrafilter* on B.

Let $U(B)$ stand for the set of all ultrafilters on B, and denote by $U(b)$ the set of ultrafilters containing b. Introduce in $U(B)$ the topology with base $\{U(b) : b \in B\}$. This definition is sound since it is easy to check that $U(x \wedge y) = U(x) \cap U(y) \quad (x, y \in B)$; i.e., $U(B)$ is closed under finite intersections. The topological space $U(B)$ is often referred to as the *Stone space of* B and is denoted by $\mathrm{St}(B)$.

(3) Denote by $M(B)$ the set of all maximal (proper) ideals of a Boolean algebra B. An ideal here may be understood in accord with 1.1.6 (2) or in the conventional sense of ring theory. Clearly, a set J in B is an ideal of B if and only if $J^* := \{x^* : x \in J\}$ is a filter on B. Moreover, $J \in M(B) \leftrightarrow J^* \in U(B)$. Therefore, the mapping $J \mapsto J^*$ is a bijection between $M(B)$ and $U(B)$. The set $M(B)$ is usually called the *maximal ideal space* of B and is always furnished with the inverse image topology translated from $U(B)$ which makes the mapping $J \mapsto J^*$ a homeomorphism.

1.2.3. Recall the prerequisites we need for applying the Gelfand transform in the case of a Boolean algebra.

(1) *A Boolean ring B is a field if and only if B is the pair of* $\mathbf{0}$ *and* $\mathbf{1}$. *Hence, there is a unique Boolean field to within isomorphism; namely,* $\mathbf{2}$.

◁ Indeed, a nonzero element $x \in B$ is invertible, and so the following implications hold:

$$xx^{-1} = \mathbf{1} \to xxx^{-1} = \mathbf{1} \to xx^{-1} = x \to x = \mathbf{1}. \ \triangleright$$

Given $\chi \in \mathrm{X}(B)$, denote by χ^* the mapping $x \mapsto \chi(x)^*$ ($x \in B$). Obviously, $\ker(\chi) := \{x \in B : \chi(x) = \mathbf{0}\}$ is an ideal, and $\ker(\chi)^*$ is a filter.

(2) *The mappings* $\chi \mapsto \ker(\chi)$ $(\chi \in \mathrm{X}(B))$ *and* $\chi \mapsto \ker(\chi)^*$ $(\chi \in \mathrm{X}(B))$ *are homeomorphisms of* $\mathrm{X}(B)$ *onto* $M(B)$ *and* $U(B)$, *respectively.*

◁ The mapping $\chi \mapsto \ker(\chi)$ is injective. If $J \in M(B)$ then B/J is a field and, by (1), B/J is isomorphic to $\mathbf{2}$. Fix such an isomorphism $\lambda : B/J \to \mathbf{2}$; and put $\chi := \lambda \circ \varphi$, where $\varphi : B \to B/J$ is the factor mapping. Obviously, $\ker(\chi) = J$ and so the mapping under discussion is bijective. The remaining claims are obvious. ▷

(3) *For* x *in* B *to equal zero it is necessary and sufficient that* $\chi(b) = \mathbf{0}$ *for all* $\chi \in \mathrm{X}(B)$.

◁ Assume that $x \neq \mathbf{0}$. Then the principal ideal $\{y \in B : y \leq x^*\}$ is proper, and so it can be extended to a maximal ideal $J \in M(B)$. This claim, known as the *Krull Theorem*, is immediate from the Kuratowski–Zorn Lemma (cf. A.3.9). By (2), $J = \ker(\chi)$ for some $\chi \in \mathrm{X}(B)$. Since $x \notin J$; therefore, $\chi(x) \neq \mathbf{0}$. ▷

1.2.4. *Stone Theorem.* *Each Boolean algebra* B *is isomorphic to the Boolean algebra of clopen sets of a Boolean space unique up to homeomorphism, the Stone space of* B.

◁ Denote by $C(\mathrm{X}(B), \mathbf{2})$ the algebra of continuous $\mathbf{2}$-valued functions on the character space $\mathrm{X}(B)$ of B which is a Boolean space. The Gelfand transform $\mathscr{G}_B$ sends an element $x \in B$ to the $\mathbf{2}$-valued function

$$\widehat{x} : \chi \mapsto \chi(x) \quad (\chi \in \mathrm{X}(B)).$$

Obviously, $\mathscr{G}_B : B \to C(\mathrm{X}(B), \mathbf{2})$ is a injective homomorphism, i.e., a *monomorphism* (cf. 1.2.3 (3)). Take $f \in C(\mathrm{X}(B), \mathbf{2})$ and put $V_f := \{\chi \in \mathrm{X}(B) : f(\chi) = \mathbf{1}\}$. The set V_f is clopen. By the definition of the topology of $\mathrm{X}(B)$, there are $b_1, \dots, b_k \in B$ and $c_1, \dots, c_l \in B$ such that

$$V_f := \{\chi \in \mathrm{X}(B) : \chi(b_n) = \mathbf{1}\ (n \leq k),\ \chi(c_m) = \mathbf{0}\ (m \leq l)\}.$$

Assign $b_0 := b_1 \wedge \dots \wedge b_k, c_0 := c_1 \vee \dots \vee c_l$, and $b := b_0 \wedge c_0^*$. The set V_f can be presented as follows:

$$\begin{aligned} V_f &= \{\chi \in \mathrm{X}(B) : \chi(b_0) = \mathbf{1}, \chi(c_0) = \mathbf{0}\} \\ &= \{\chi \in \mathrm{X}(B) : \chi(b) = \mathbf{1}\} = \{\chi \in \mathrm{X}(B) : \widehat{b}(\chi) = \mathbf{1}\}. \end{aligned}$$

Therefore, $f = \widehat{b}$, and so $\mathcal{G}_B$ is an isomorphism.

Assume now that Q_1 and Q_2 are Boolean spaces such that the mapping $h : C(Q_1, \mathbf{2}) \to C(Q_2, \mathbf{2})$ is an isomorphism of these algebras.

If χ is a character of $C(Q_2, \mathbf{2})$ then $\chi \circ h$ is a character of $C(Q_2, \mathbf{2})$. Hence, $\chi \mapsto \chi \circ h$ is a homeomorphism between the character spaces.

On the other hand, the character space of $C(Q_k, \mathbf{2})$ is homeomorphic to Q_k. The Boolean spaces Q_1 and Q_2 are thus homeomorphic. It suffices to note that the algebra $C(\mathrm{X}(B), \mathbf{2})$ is isomorphic to the algebra of clopen sets of the space $\mathrm{X}(B)$ and so, of the space $U(B)$ as well. ▷

The isomorphism of this theorem between B and $\mathrm{Clop}(\mathrm{St}(B))$ is occasionally called the *Stone transform* of B.

1.2.5. In the sequel we are mostly interested in complete Boolean algebras. The notion of a complete Boolean algebra is closely tied with that of an extremally disconnected compact space. Recall that a Hausdorff topological space X is *extremally disconnected* provided that the closure of each open set in X is open too. Clearly, an extremally disconnected space is totally disconnected.

Ogasawara Theorem. *A Boolean algebra is complete if and only if its Stone space is extremally disconnected.*

◁ Let B be a complete Boolean algebra. Assume further that h is an isomorphism of B onto the algebra of clopen sets of the compact space $Q := \mathrm{St}(B)$. Take an open set $G \subset Q$. Since Q is totally disconnected; therefore, $G = \bigcup \mathcal{U}$, where $\mathcal{U}$ stands for the set of all clopen subsets of G.

Put $\mathcal{U}' := \{h^{-1}(U) : U \in \mathcal{U}\}$ and $b := \bigvee \mathcal{U}'$. The clopen set $h(b)$ is the closure of G. Indeed, $\mathrm{cl}(G) \subset h(b)$ and $h(b) \setminus \mathrm{cl}(G)$ is open. If the last set is nonempty then $h(c) \subset h(b) \setminus \mathrm{cl}(G)$ for some $\mathbf{0} \neq c \in B$. This implies in turn that $h(c) \vee h(u) \leq h(b)$ for all $u \in \mathcal{U}'$, which contradicts the equality $b = \bigvee \mathcal{U}$. Consequently, $\mathrm{cl}(G) = h(b)$ is an open set.

Assume now that the compact space Q is extremally disconnected. Let $\mathcal{G}$ stand for some collection of clopen subsets of Q, and put $G := \bigcup \mathcal{G}$. The set G is open and the closure $\mathrm{cl}(G)$ of G must be open by the hypothesis about Q. Obviously, $\mathrm{cl}(G)$ is the least upper bound of $\mathcal{G}$ in the Boolean algebra of clopen sets $\mathrm{Clop}(Q)$. ▷

1.2.6. Examples.

(1) The Stone space of the Boolean algebra $\{\mathbf{0}, \mathbf{1}\}$ is a singleton. In case a Boolean algebra B is finite, it has $\mathbf{2}^n$ elements for some $n \in \mathbb{N}$, and the Stone space of B consists of n points.

(2) Take a nonempty set X. The Stone space of the boolean $\mathcal{P}(X)$ of X is the *Stone–Čech compactification* $\beta(X)$ of X made into a discrete topological space.

(3) *If Q is a Boolean space then the Stone space of the algebra* Clop(Q) *of all clopen subsets of Q is homeomorphic to Q.*

(4) Assume that B and B' are Boolean algebras, and $h : B \to B'$ is a homomorphism between them. Denote the Stone transforms of B and B' by $\imath : B \to \mathrm{Clop}(\mathrm{St}(B))$ and $\imath' : B' \to \mathrm{Clop}(\mathrm{St}(B'))$. There is a unique continuous mapping $\theta : \mathrm{St}(B') \to \mathrm{St}(B)$ such that

$$h(x) = (\imath')^{-1}\theta^{-1}(\imath(x)) \quad (x \in B).$$

The mapping $h \mapsto \mathrm{St}(h) := \theta$ is a bijection between the sets of all homomorphisms from B to B' and the set of all continuous mappings from $\mathrm{St}(B')$ to $\mathrm{St}(B)$. If B'' is another Boolean algebra and $g : B' \to B''$ is a homomorphism, then $\mathrm{St}(g \circ h) = \mathrm{St}(h) \circ \mathrm{St}(g)$. Moreover, $\mathrm{St}(I_B) = I_{\mathrm{St}(B)}$.

Denote by Boole the category of Boolean algebras and homomorphisms, and let Comp stand for the category of Hausdorff compact spaces and continuous mappings. Then the above may be paraphrased as follows (see A.3):

Theorem. *The mapping $\mathscr{S}$ is a contravariant functor from the category* Boole *to the category* Comp.

Two important particular cases of the situation under consideration are worthy of special attention.

(5) *A Boolean algebra B_0 is isomorphic to a subalgebra of a Boolean algebra B if and only if the Stone space* $\mathrm{St}(B_0)$ *of B_0 is a continuous image of the Stone space* $\mathrm{St}(B)$ *of B.*

(6) *A Boolean algebra B' is the image of a Boolean algebra B under a homomorphism (or B' is isomorphic with a factor algebra of B)* (see 1.1.6 (4)) *if and only if the Stone space* $\mathrm{St}(B')$ *of B' is homomorphic to a closed subset of the Stone space* $\mathrm{St}(B)$ *of B.*

(7) Assume that $B := \prod_{\alpha\in\mathrm{A}} B_\alpha$, with $(B_\alpha)_{\alpha\in\mathrm{A}}$ a nonempty family of Boolean algebras. The Stone space $\mathrm{St}(B)$ of B coincides with the Stone–Čech compactification of the topological sum $\bigcup_{\alpha\in\mathrm{A}} \mathrm{St}(B_\alpha) \times \{\alpha\}$ of the Stone spaces $\mathrm{St}(B_\alpha)$ of B_α.

(8) Let $B := \bigotimes_{\alpha\in\mathrm{A}} B_\alpha$ be the Boolean product of a nonempty family of Boolean algebras (cf. 1.1.6 (6)). Then the Stone space $\mathrm{St}(B)$ of B is homeomorphic to the product $\prod_{\alpha\in\mathrm{A}} \mathrm{St}(B_\alpha)$.

(9) An *absolute* of a compact space X is a compact set aX meeting the following conditions:

(a) X is a continuous *irreducible* image of aX; i.e., there is a continuous surjection of aX onto X whereas X is not a continuous image of any proper closed subset of aX;

(b) every continuous irreducible inverse image of X is homeomorphic to aX.

If oB is the completion of a Boolean algebra B then $\mathrm{St}(oB) = a\,\mathrm{St}(B)$; *i.e., an absolute of the Stone space of B is homeomorphic to the Stone space of the completion oB of B.*

1.2.7. An *atom* of a Boolean algebra B is a nonzero element a of B such that $\{x \in B : \mathbf{0} \leq x \leq a\} = \{\mathbf{0}, a\}$. In other words, $a \neq \mathbf{0}$ is an atom of B if $a \leq x$ or $a \leq x^*$ for whatever $x \in B$.

An algebra B is *atomic* if to each nonzero element $x \in B$ there exists an atom $a \leq x$. A Boolean algebra is *atomless* if it contains no atom.

Say that a Boolean algebra B is *completely distributive* if the following *complete distributive laws* hold

$$\bigwedge_{m \in M} \bigvee_{n \in N} x_{m,n} = \bigvee_{f \in N^M} \bigwedge_{m \in M} x_{m,f(m)}$$

for $x_{m,n} \in B$, with m and n ranging over arbitrary sets M and N. As usual, N^M is the set of all mappings $f : M \to N$. Assuming M and N countable, say that B is *σ-distributive* or *countably distributive* (see 5.2.15 (6) below).

Theorem. *Let B be a complete Boolean algebra. The following are equivalent:*

(1) *B is isomorphic to the boolean $\mathscr{P}(A)$ of a nonempty set A;*

(2) *B is completely distributive;*

(3) *B is atomic.*

◁ (1) → (2) It suffices to note that the set-theoretic union and intersection obey the complete distributive laws.

(2) → (3) Consider a double family $\{x_{b,t} \in B : b \in B, t \in \mathbf{2}\}$, where $\mathbf{2} := \{\mathbf{0}, \mathbf{1}\}$, $x_{b,\mathbf{0}} := b^*$, and $x_{b,\mathbf{1}} := b$. In this case

$$\mathbf{1} = \bigwedge_{b \in B} x_{b,\mathbf{0}} \vee x_{b,\mathbf{1}} = \bigwedge_{b \in B} \bigvee_{t \in \mathbf{2}} x_{b,t}.$$

Since B is a completely distributive Boolean algebra; therefore,

$$\mathbf{1} = \bigvee \{c(f) : f \text{ is a function from } B \text{ to } \mathbf{2}\},$$

where $c(f) := \bigvee \{x_{b,f(b)} : b \in B\}$. This yields $b = \vee\{b \wedge c(f) : f \in \mathbf{2}^B\}$ for $b \in B$. Hence, to a nonzero $b \in B$ there is some $g \in \mathbf{2}^B$ such that $b \wedge c(g) \neq \mathbf{0}$. On the other hand, for arbitrary $b \in B$ and $f \in \mathbf{2}^B$ only the following two cases are possible

(a) $f(b) = \mathbf{0} \to x_{b,f(b)} = b^* \to c(f) \leq b^* \leftrightarrow b \wedge c(f) = \mathbf{0}$,

(b) $f(b) = \mathbf{1} \to x_{b,f(b)} = b \to c(f) \leq b$.

Therefore, if $b \neq \mathbf{0}$ then either $b \wedge c(f) = \mathbf{0}$ or $c(f) \leq b$; i.e., $c(f)$ is an atom of B provided that $c(f) \neq \mathbf{0}$. However, there are sufficiently many nonzero $c(f)$, and so B is atomic.

(3) $\to$ (1) Denote by A the set of all atoms of B. Given $x \in B$, denote by $h(x)$ the set of all atoms $a \in B$ such that $a \leq x$. The mapping $h : B \to \mathscr{P}(A)$ is clearly an isomorphism. ▷

1.2.8. COMMENTS.

(1) The Stone Theorem shows that every Boolean algebra is perfectly determined from its Stone space. In more detail, each property of a Boolean algebra B translates into the topological language, becoming a property of the Stone space $\mathrm{St}(B)$ of B. This way of studying Boolean algebras is the *representation method.*

(2) The basic idea behind the Stone Theorem remains workable in the case of distributive lattices. For a distributive lattice L the role of the Stone space $\mathrm{St}(L)$ of L is played by the set of all prime ideals (or filters) which is equipped with a topology in a special way. Recall that a proper ideal $J \subset L$ is *prime* whenever

$$x \wedge y \in J \to x \in J \vee y \in J.$$

The Stone spaces of distributive lattices may be used for constructing new lattices and finding the topological meaning of lattice-theoretic properties (the representation method), cf. [14, 69, 204].

1.3. Von Neumann–Gödel–Bernays Theory

The axiom of replacement ZF_4^{φ} of Zermelo–Fraenkel set theory ZFC is in fact an axiom-schema embracing infinitely many axioms because of arbitrariness in the choice of a formula φ. It stands to reason to introduce some primitive object that is determined from each formula φ participating in ZF_4^{φ}. With these objects available, we may paraphrase the content of the axiom-schema ZF_4^{φ} as a single axiom about new objects. To this end, we need the axioms that guarantee existence for the objects determined from a set-theoretic formula.

Since all formulas are constructed by a unique procedure in finitely many steps, we find highly plausible the possibility of achieving our goal with finitely many axioms. It is this basic idea stemming from von Neumann that became a cornerstone of the axiomatics of set theory which was elaborated by Gödel and Bernays and is commonly designated by NGB.

The initial undefinable object of NGB is a class. A *set* is a *class* that is a member of some class. A class other than any set is a *proper* class. Objectivization of classes constitutes the basic difference between NGB and ZFC, with the metalanguage of the latter treating "class" and "property" as synonyms.

Axiomatic presentation of NGB uses as a rule one of the two available modifications of the language of ZFC. The first consists in adding a new unary predicate symbol M to the language of ZFC, with $M(X)$ implying semantically that X is a set. The second modification uses two different types of variables for sets and classes. It worth observing that these tricks are not obligatory for describing NGB and reside routinely for the sake of convenience.

1.3.1. The system NGB is a first-order theory. Strictly speaking, the language of NGB does not differ at all from that of ZFC. However, the upper case Latin letters $X, Y, Z, \dots$, possible with indices, are commonly used for variables, while the lower case Latin letters are left for the argo resulting from introducing the abbreviations that are absent in the language of NGB.

Let $M(X)$ stand for the formula $(\exists Y)(X \in Y)$. We read $M(X)$ as "X is a set."

Introduce the lower case Latin letters x, y, z, ... (with indices) for the *bound* variables ranging over sets. To be more exact, the formulas $(\forall x)\varphi(x)$ and $(\exists x)\varphi(x)$, called *generalization* and *instantiation* of φ by x, are abbreviations of the formulas $(\forall X)(M(X) \to \varphi(X))$ and $(\exists X)(M(X) \wedge \varphi(X))$, respectively. Semantically these formulas imply: "φ holds for every set" and "there is a set for which φ is true." In this event the variable X must not occur in φ nor in the formulas comprising the above abbreviations.

The rules for using upper case and lower case letters will however be observed only within the present section. On convincing ourselves that the theory of classes may be formalized in principle, we will gradually return to the cozy and liberal realm of common mathematical parlance. For instance, abstracting the set-theoretic concept of function to the new universe of discourse, we customarily speak about a *class-function* F implying that F might be other than a set but still obeys the conventional properties of a function. This is a sacrosanct privilege of the working mathematician.

We now proceed with stating the special axioms of NGB.

1.3.2. *Axiom of Extensionality* NGB$_1$. *Two classes coincide if and only if they consist of the same elements*:

$$(\forall X)(\forall Y)(X = Y \leftrightarrow (\forall Z)(Z \in X \leftrightarrow Z \in Y)).$$

1.3.3. We now list the axioms for sets:

(1) *Axiom of Pairing* NGB$_2$:

$$(\forall x)(\forall y)(\exists z)(\forall u)(u \in z \leftrightarrow u = x \vee u = y);$$

(2) ***Axiom of Union*** **NGB₃:**

$$(\forall\, x)(\exists\, y)(z \in y \leftrightarrow (\exists\, u)(u \in x \wedge z \in u));$$

(3) ***Axiom of Powerset*** **NGB_4:**

$$(\forall\, x)(\exists\, y)(\forall\, z)(z \in y \leftrightarrow z \subset x);$$

(4) ***Axiom of Infinity*** **NGB_5:**

$$(\exists\, x)(\varnothing \in x \wedge (\forall\, y)(y \in x \leftrightarrow y \cup \{y\} \in x)).$$

These axioms coincide obviously with their counterparts in ZFC, cf. A.2.3, A.2.4, A.2.7, and A.2.8. However, we should always bear in mind that the verbal formulations of NGB_1–NGB_5 presume a "set" to be merely a member of another class. Recall also that the lower case Latin letters symbolize abbreviations (cf. 1.3.1). By way of illustration, we remark that, in partially expanded form, the axiom of powerset NGB_4 looks like

$$(\forall X)(M(X) \rightarrow (\exists Y)(M(Y) \wedge (\forall Z)(M(Z) \rightarrow (Z \in Y \leftrightarrow Z \subset X)))).$$

The record of the axiom of infinity uses the following abbreviation

$$\varnothing \in x := (\exists\, y)(y \in x \wedge (\forall\, u)(u \notin y)).$$

Existence of the *empty set* is a theorem rather than a postulate in NGB in much the same way as in ZFC. Nevertheless, it is common to enlist the existence of the empty set in NGB as a special axiom:

(5) ***Axiom of the Empty Set:***

$$(\exists\, y)(\forall\, u)(u \notin y).$$

1.3.4. ***Axiom of Replacement*** **NGB_6.** *If X is a single-valued class then, for each set y, the class of the second components of those pairs of X whose first components belong to y, is a set:*

$$(\forall\, X)(\mathrm{Un}\,(X) \rightarrow (\forall\, y)(\exists\, z)(\forall\, u)(u \in z \leftrightarrow (\exists\, v)((v, u) \in X \wedge v \in y))),$$

where $U_n(X) := (\forall\, u)(\forall\, v)(\forall\, w)((u, v) \in X \wedge (u, w) \in X \rightarrow v = w).$

As was intended, the axiom-schema of replacement ZF_4^{φ} turns into a single axiom. Note that the axiom-schema of comprehension in ZFC (see A.2.5) also transforms into a single axiom, the *axiom of comprehension*. This latter reads that, to each set x and each class Y, there is a set consisting of the common members of x and Y;

$$(\forall x)(\forall Y)(\exists z)(\forall u)(u \in z \leftrightarrow u \in x \wedge u \in Y).$$

The axiom of comprehension is weaker than the axiom of replacement since the former ensues from NGB_6 and Theorem 1.3.14 below. However, comprehension is often convenient for practical purposes.

The collection of axioms to follow, NGB_7–NGB_{13}, relates to the formation of classes. These axioms state that, given some properties expressible by formulas, we may deal with the classes of the sets possessing the requested properties. As usual, uniqueness in these cases results from the axiom of extensionality for classes NGB_1.

1.3.5. *Axiom of Membership* $\mathbf{NGB_7}$. *There is a class comprising every ordered pair of sets whose first component is a member of the second*:

$$(\exists X)(\forall y)(\forall z)((y, z) \in X \leftrightarrow y \in z).$$

1.3.6. *Axiom of Intersection* $\mathbf{NGB_8}$. *There is a class comprising the common members of every two classes*:

$$(\forall X)(\forall Y)(\exists Z)(\forall u)(u \in Z \leftrightarrow u \in X \wedge u \in Y).$$

1.3.7. *Axiom of Complement* $\mathbf{NGB_9}$. *To each class X there is a class comprising the nonmembers of X*:

$$(\forall X)(\exists Y)(\forall u)(u \in Y \leftrightarrow u \notin X).$$

This implies the existence of the *universal class* $\mathbf{U} := \overline{\varnothing}$ which is the complement of the *empty class* $\varnothing$.

1.3.8. *Axiom of Domain* $\mathbf{NGB_{10}}$. *To each class X of ordered pairs there is a class $Y := \operatorname{dom} X$ comprising the first components of the members of X*:

$$(\forall X)(\exists Y)(\forall u)(u \in Y \leftrightarrow (\exists v)((u, v) \in X)).$$

1.3.9. *Axiom of Product* $\mathbf{NGB_{11}}$. *To each class X there is a class $Y := X \times \mathbf{U}$ comprising the ordered pairs whose first components are members of X*:

$$(\forall X)(\exists Y)(\forall u)(\forall v)((u, v) \in Y \leftrightarrow u \in X).$$

1.3.10. Axioms of Permutation $\mathbf{NGB_{12}}$ and $\mathbf{NGB_{13}}$. Assume that $\sigma := (\imath_1, \imath_2, \imath_3)$ is a permutation of $\{1, 2, 3\}$. A class Y is a *σ-permutation* of a class X provided that $(x_1, x_2, x_3) \in Y$ whenever $(x_{\imath_1}, x_{\imath_2}, x_{\imath_3}) \in X$.

To each class X, there are $(2,3,1)$- and $(1,3,2)$-permutations of X:

$$(\forall X)(\exists Y)(\forall u)(\forall v)(\forall w)((u,v,w) \in Y \leftrightarrow (v,w,u) \in X);$$
$$(\forall X)(\exists Y)(\forall u)(\forall v)(\forall w)((u,v,w) \in Y \leftrightarrow (u,w,v) \in X).$$

The above axioms of class formation proclaim existence of unique classes, as was mentioned above. It is so in common parlance to speak about *the* complement of a class, *the* intersection of classes, etc.

1.3.11. *Axiom of Regularity* $\mathbf{NGB_{14}}$. *Each nonempty class X has a member having no common elements with X:*

$$(\forall X)(X \neq \varnothing \rightarrow (\exists y)(y \in X \wedge y \cap X = \varnothing)).$$

1.3.12. *Axiom of Choice* $\mathbf{NGB_{15}}$. *To each class X there is a choice class-function on X; i.e., a single-valued class assigning an element of X to each nonempty member of X:*

$$(\forall X)(\exists Y)(\forall u)(u \neq \varnothing \wedge u \in X \rightarrow (\exists! v)(v \in u \wedge (u,v) \in Y)).$$

This is a very strong form of the axiom of choice which amounts to a possibility of a simultaneous choice of an element from each nonempty set.

The above axiom makes the list of the special axioms of NGB complete. A moment's inspection shows that NGB, unlike ZFC, has finitely many axioms. Another convenient feature of NGB is the opportunity to treat sets and properties of sets as formal objects, thus implementing the objectivization that is absolutely inaccessible to the expressive means of ZFC.

1.3.13. We now derive a few consequences of the axioms of class formation which are needed in the sequel.

(1) *To each class X there corresponds the $(2,1)$-permutation of X:*

$$(\forall X)(\exists Z)(\forall u)(\forall v)((u,v) \in Z \leftrightarrow (v,u) \in X).$$

$\lhd$ The axiom of product guarantees existence for the class $X \times \mathbf{U}$. Consecutively applying the axioms of the $(2,3,1)$-permutation and $(1,3,2)$-permutation to the $X \times \mathbf{U}$, arrive at the class Y of 3-tuples (alternatively, *triples*) (v,u,w) such that $(v,u) \in X$. Appealing to the axiom of domain, conclude that $Z := \mathrm{dom}(Y)$ is the sought class. $\rhd$

(2) *To each pair of classes there corresponds their product:*

$$(\forall X)(\forall Y)(\exists Z)(\forall w)$$
$$(w \in Z \leftrightarrow (\exists u \in X)(\exists v \in Y)(w = (u, v))).$$

◁ To prove the claim, apply consecutively the axiom of product, (1), and the axiom of intersection to arrange $Z := (\mathbf{U} \times Y) \cap (X \times \mathbf{U})$. ▷

Given $n \geq 2$, we may define the class $\mathbf{U}^n$ of all ordered n-tuples by virtue of 1.3.13 (2).

(3) *To each class X there corresponds the class $Z := (\mathbf{U}^n \times \mathbf{U}^m) \cap (X \times \mathbf{U}^m)$:*

$$(\forall X)(\exists Z)(\forall x_1)\dots(\forall x_n)(\forall y_1)\dots(\forall y_m)$$
$$((x_1, \dots, x_n, y_1, \dots, y_m) \in Z \leftrightarrow (x_1, \dots, x_n) \in X).$$

(4) *To each class X there corresponds the class $Z := (\mathbf{U}^m \times \mathbf{U}^n) \cap (\mathbf{U}^m \times X)$:*

$$(\forall X)(\exists Z)(\forall x_1)\dots(\forall x_n)(\forall y_1)\dots(\forall y_m)$$
$$((y_1, \dots y_m, x_1, \dots, x_n) \in Z \leftrightarrow (x_1, \dots, x_n) \in X).$$

◁ To demonstrate (3) and (4), apply the axiom of product and the axiom of intersection. ▷

(5) *To each class X there corresponds the class Z satisfying*

$$(\forall x_1)\dots(\forall x_n)(\forall y_1)\dots(\forall y_m)$$
$$((x_1, \dots, x_{n-1}, y_1, \dots, y_m, x_n) \in Z \leftrightarrow (x_1, \dots, x_n) \in X).$$

◁ Appeal to the axioms of permutation and the axiom of product. ▷

1.3.14. *Theorem.* *Let φ be a formula whose variables are among $X_1, \dots, X_n$, $Y_1, \dots, Y_m$ and which is **predicative**; i.e., all bound variables of φ range over sets. Then the following is provable in* NGB:

$$(\forall Y_1)\dots(\forall Y_m)(\exists Z)(\forall x_1)\dots(\forall x_n)$$
$$((x_1, \dots, x_n) \in Z \leftrightarrow \varphi(x_1, \dots, x_n, Y_1, \dots, Y_m)).$$

◁ Assume that φ is written so that the only bound variables of φ are those for sets. It suffices to consider only φ containing no subformulas of the shape

$Y \in W$ and $X \in X$, since the latter might be rewritten in equivalent form as $(\exists x)(x = Y \wedge x \in W)$ and $(\exists u)(u = X \wedge u \in X)$. Moreover, the symbol of equality may be eliminated from φ on substituting for $X = Y$ the expression $(\forall u)(u \in X \leftrightarrow u \in Y)$, which is sound by the axiom of extensionality. The proof proceeds by induction on the *complexity* or *length* k of φ; i.e., by the number k of propositional connectives and quantifiers occurring in φ.

In case $k = 0$ the formula φ is atomic and has the form $x_\imath \in x_\jmath$, or $x_\jmath \in x_\imath$, or $x_\imath \in Y_l$ ($\imath < \jmath \leq n$, $l \leq m$). If $\varphi := x_\imath \in x_\jmath$ then, by the axiom of membership, there is a class W_1 satisfying

$$(\forall x_\imath)(\forall x_\jmath)((x_\imath, x_\jmath) \in W_1 \leftrightarrow x_\imath \in x_\jmath).$$

If $\varphi := x_\jmath \in x_\imath$ then, using the axiom of membership again, we find a class W_2 with the property

$$(\forall x_\imath)(\forall x_\jmath)((x_\jmath, x_\imath) \in W_2 \leftrightarrow x_\jmath \in x_\imath),$$

and apply 1.3.13 (1). In result, we obtain a class W_3 such that

$$(\forall x_\imath)(\forall x_\jmath)((x_\imath, x_\jmath) \in W_3 \leftrightarrow x_\jmath \in x_\imath).$$

Hence, in each of these two cases there is a class W satisfying the following formula:

$$\Phi := (\forall x_\imath)(\forall x_\jmath)((x_\imath, x_\jmath) \in W \leftrightarrow \varphi(x_1, \dots, x_n, Y_1, \dots, Y_m)).$$

By 1.3.13 (4), we may replace the subformula $(x_\imath, x_\jmath) \in W$ of Φ with the containment $(x_1, \dots, x_{\imath-1}, x_\imath) \in Z_1$ for some other class Z_1 and insert the quantifiers $(\forall x_1) \dots (\forall x_{\imath-1})$ in the prefix of Φ.

Let Ψ be the so-obtained formula. By 1.3.13 (5), there is some class Z_2 for Ψ so that it is possible to write $(x_1, \dots, x_{\imath-1}, x_\imath, x_\jmath) \in Z_1$ instead of the subformula $(x_1, \dots, x_\imath, x_{\imath+1}, \dots, x_\jmath) \in Z_2$ and to insert the quantifiers $(\forall x_{\imath+1}) \dots (\forall x_{\jmath-1})$ in the prefix of Ψ. Finally, on applying 1.3.13 (3) to Z_2, find a class Z satisfying the following formula:

$$(\forall x_1) \dots (\forall x_n)((x_1, \dots, x_n) \in Z \leftrightarrow \varphi(x_1, \dots, x_n, Y_1, \dots, Y_m)).$$

In the remaining case of $x_\imath \in Y_l$, the claim follows from existence of the products $W := \mathbf{U}^{\imath-1} \times Y_l$ and $Z := W \times \mathbf{U}^{n-\imath}$. This completes the proof of the theorem for $k = 0$.

Assume now that the claim of the theorem is demonstrated for all $k < p$ and the formula φ has p propositional connectives and quantifiers. It suffices to consider the cases in which φ results from some other formulas by negation, implication, and generalization.

Suppose that $\varphi := \neg\psi$. By the induction hypothesis, there is a class V such that

$$(\forall x_1)\dots(\forall x_n)((x_1,\dots,x_n) \in V \leftrightarrow \psi(x_1,\dots,x_n,Y_1,\dots,Y_m)).$$

By the axiom of complement, the class $Z := \mathbf{U} - V := \mathbf{U}\backslash V$ meets the required conditions.

Suppose that $\varphi := \psi \to \theta$. Again, by the induction hypothesis, there are classes V and W making the claim holding for V and ψ and such that

$$(\forall x_1)\dots(\forall x_n)((x_1,\dots,x_n) \in W \leftrightarrow \theta(x_1,\dots,x_n,Y_1,\dots,Y_m)).$$

The sought class $Z := \mathbf{U} - (V \cap (\mathbf{U} - W))$ exists by the axioms of intersection and complement.

Suppose that $\varphi := (\forall x)\psi$, and let V and ψ be the same as above. Applying the axiom of domain to the class $X := \mathbf{U} - V$, obtain the class Z_1 such that

$$(\forall x_1)\dots(\forall x_n)((x_1,\dots,x_n) \in Z_1 \leftrightarrow (\exists x)\neg\psi(x_1,\dots,x_n,Y_1,\dots,Y_m)).$$

The class $Z := \mathbf{U} - Z_1$ exists by the axiom of complement and is the one we seek since the formula $(\forall x)\,\psi$ amounts to $\neg(\exists x)(\neg\psi)$. $\rhd$

1.3.15. Each of the axioms of class formation NGB_7–NGB_{13} is a corollary to Theorem 1.3.14 provided that the formula φ is duly chosen. On the other hand, the theorem itself, as shown by inspection of its proof, ensues from the axioms of class formation. It is remarkable that we are done on using finitely many axioms NGB_7–NGB_{13} rather than infinitely many assertions of Theorem 1.3.14.

Theorem 1.3.14 allows us to prove the existence of various classes. For instance, to each class Y there corresponds the class $\mathscr{P}(Y)$ of all subsets of Y, as well as the union $\bigcup Y$ of all elements of Y. These two classes are defined by the conventional formulas:

$$(\forall u)(u \in \mathscr{P}(Y) \leftrightarrow u \subset Y),$$
$$(\forall u)(u \in \bigcup(Y) \leftrightarrow (\exists v)\ (v \in Y \wedge u \in v)).$$

The above claims of existence are easy on putting $\varphi(X,Y) := X \subset Y$ and $\varphi(X,Y) := (\exists V)(X \in V \wedge V \in Y)$. Analogous arguments corroborate the definitions of Z^{-1}, $\mathrm{im}(Z)$, $Z \restriction Y$, $Z``Y$, $X \cup Y$, etc., with X, Y, and Z arbitrary classes.

1.3.16. *Theorem.* *Each theorem of* ZFC *is a theorem of* NGB.

$\lhd$ Each axiom of ZFC is a theorem of NGB. The only nonobvious part of the claim concerns the axiom of replacement ZF_4^{φ} which we will proof.

Assume that y is not free in φ , and let $\{x, t, z_1, \dots, z_m\}$ stand for the complete list of variables in the construction of φ. Assume further that, for all x, u, v, $z_1, \dots, z_m$, the following formula holds:

$$\varphi(x, u, z_1, \dots, z_m) \wedge \varphi(x, v, z_1, \dots, z_m) \to u = v.$$

The formula is predicative since each bound variable of φ ranges over sets. By Theorem 1.3.14, there is a class Z such that

$$(\forall\, x)(\forall\, u)((x, u) \in Z \leftrightarrow \varphi(x, u, z_1, \dots, z_m)).$$

This property of φ shows that the class Z is single-valued; i.e., $\mathrm{Un}\,(Z)$ is provable in NGB. By the axiom of replacement NGB_6, there is a set y satisfying

$$(\forall\, v)(v \in y \leftrightarrow (\exists\, u)((u, v) \in Z \wedge u \in x)).$$

Obviously, y satisfies the desired formula

$$(\forall\, z_1) \dots (\forall\, z_m)(\forall\, v)(v \in y \leftrightarrow (\exists\, u \in x)\, \varphi(u, v, z_1, \dots, z_m)). \ \rhd$$

1.3.17. *Theorem.* *Each formula of* ZFC *that is a theorem of* NGB *is a theorem of* ZFC.

$\lhd$ The proof may be found, for instance, in [30]. It uses some general facts of model theory which lie beyond the framework of the present book. $\rhd$

Theorems 1.3.16 and 1.3.17 are often paraphrased as follows.

1.3.18. *Theorem.* *Von Neumann–Gödel–Bernays set theory is conservative over Zermelo–Fraenkel set theory.*

1.3.19. COMMENTS.

(1) Expositions of set theory are in plenty. We mention a few: [18, 26, 30, 32, 48, 55, 60, 73, 77, 83, 88, 94, 153, 166, 168, 208, 241, 254].

The formal theory NGB, as well as ZFC, is one of the most convenient and simple axiomatic set theories. To survey other axiomatics, see [18, 55, 218, 254].

(2) Among the other axiomatic set theories, we mention the so-called Bernays–Morse theory that extends NGB. Bernays–Morse set theory assumes the special axioms NGB_1–NGB_5, NGB_{14} and the following axiom-schema of comprehension:

$$(\exists\, X)(\forall\, Y)(Y \in X \leftrightarrow M(Y) \wedge \varphi(Y, X_1, \dots, X_n)),$$

with φ an arbitrary formula without free occurrences of X.

(3) Theorem 3.1.17 belongs to A. Mostowski. It implies in particular that ZF is consistent if and only if so is NGB. The latter fact was established by I. Novak and J. Shoenfield (cf. [217, 254]).

It is immediate from 1.3.14 that if the quantifiers of φ range over sets then the axiom-schema of comprehension is a theorem of NGB. The Bernays–Morse set theory allows quantification over arbitrary classes in the axiom-schema of comprehension. This theory may be enriched with the axiom of choice NGB_{15}.

1.4. Ordinals

The concept of ordinal is a key to studying infinite sets. It is designed for transfinite iteration of various mathematical constructions and arguments as well as for measuring cardinality. The topic of the present section is to explain how this is done.

1.4.1. Consider classes X and Y. Say that X is an *order relation* or, simply, an *order* on Y provided that X is an antisymmetric, reflexive, and transitive relation on Y.

The antisymmetry, reflexivity, and transitivity properties of a relation within NGB are written in much the same way as in the language of ZFC (cf. A.1.10). An order of X on Y is *total* or *linear* if $Y \times Y \subset X \cup X^{-1}$.

A relation X *well orders* Y or is a *well-ordering* on Y, or Y is a *well ordered class* provided that X is an order on Y and each nonempty subclass of Y has a least element with respect to X.

Classes X_1 and X_2, furnished with some order relations R_1 and R_2, are *similar* or *equivalent* if there is exists a bijection h from X_1 on X_2 such that $(x, y) \in R_1 \leftrightarrow (h(x), h(y)) \in R_2$ for all $x, y \in X_1$.

1.4.2. By definition we let

$$(x, y) \in E \leftrightarrow (x \in y \vee x = y).$$

The class E exists by the axiom of membership NGB_7 and Theorem 1.3.14. A moment's thought shows that E is an order on the universal class $\mathbf{U}$.

A class X is *transitive* (not to be confused with a transitive relation) if each member of X is also a subset of X:

$$\mathrm{Tr}\,(X) := (\forall\, y)(y \in X \to y \subset X).$$

An *ordinal class* is a transitive class well ordered by the membership relation. The record $\mathrm{Ord}\,(X)$ means that X is ordinal. If x is a set and $\mathrm{Ord}\,(X)$ then we call X an *ordinal.* The terms "*ordinal number*" or "*transfinite number*" are also in common parlance. Denote by On the class of all ordinals. We usually let lower case Greek letters stand for ordinals. Moreover, we use the following abbreviations:

$$\alpha < \beta := \alpha \in \beta, \quad \alpha \leq \beta := \alpha \in \beta \vee \alpha = \beta, \quad \alpha + 1 := \alpha \cup \{\alpha\}.$$

If $\alpha < \beta$ then we say that α precedes β and β succeeds α. Using the axiom of regularity NGB_{14}, we may easily prove the following:

1.4.3. *A class X is an ordinal if and only if X is a transitive class well ordered by membership.*

$\lhd$ Assume that a transitive class X is well ordered by membership. Choose a nonempty subclass $Y \subset X$ and show that Y has a least element. There is at least one element $y \in Y$. If $y = 0$ then y is the sought least element in Y. If $y \neq 0$ then, by the axiom of regularity, there is an element $x \in y$ such that $x \cap y = 0$.

In this case x is the least element of y because y is well ordered. Since the class Y is well ordered by membership, x is the least element in the class Y as well. Hence, X is an ordinal class. Sufficiency of the hypothesis is thus proven, while necessity is obvious. $\rhd$

Therefore, NGB and ZFC allow us to use a simpler definition of ordinal as follows:

$$\mathrm{Ord}\,(X) \leftrightarrow \mathrm{Tr}\,(X) \wedge (\forall u \in X)(\forall v \in X)(u \in v \vee u = v \vee v \in u).$$

It is worth observing that the equivalence of the above definitions of ordinal can be established without the axiom of choice. Most of the properties of ordinals below may be deduced without the axiom of regularity, using only the initial definition of ordinal. This peculiarity, important as regards proof of consistency of the axiom of regularity with the remaining axioms of ZF, is immaterial to our further aims.

1.4.4. In the sequel we use some auxiliary facts about ordinals which are listed now.

Assume that X and Y are arbitrary classes.

(1) *If X is an ordinal class, Y is a transitive class, and $X \neq Y$; then the formulas $Y \subset X$ and $Y \in X$ are equivalent.*

$\lhd$ If $Y \in X$ then the class Y is a set and $Y \subset X$ since X is transitive.

Conversely, assume that $Y \subset X$. Since $X \neq Y$; therefore, $Z := X - Y \neq \varnothing$. The class Z has the least element $x \in Z$ with respect to the order by membership. This implies that $x \cap Z = \varnothing$ or $x \subset Y$. Moreover, $x \subset X$ since $x \in X$ and X is a transitive class.

Take $y \in Y$. Since X is totally ordered; therefore, $x \in y$, or $x = y$, or, finally, $y \in x$. By transitivity of Y, the first two relations yield $x \in Y$, which contradicts the membership $x \in Z$. Hence, $y \in x$ and so $Y \subset x$. Considering inclusion $x \subset Y$ proven above, conclude that $x = Y$ and, finally, $x = Y \wedge x \in X \to Y \in X$. $\rhd$

(2) *The intersection of every two ordinal classes is an ordinal class.*

$\lhd$ This is obvious. $\rhd$

(3) *If X and Y are ordinal classes then*

$$X \in Y \vee X = Y \vee Y \in X.$$

$\lhd$ Let the intersection $X \cap Y = Z$ coincide with none of the classes X and Y. Then, according to (1) and (2), $Z \in X$ and $Z \in Y$; i.e., $Z \in X \cap Y = Z$. For $Z \in X$, however, the relation $Z \in Z$ is impossible. Hence, either $Z = X$ and $Y \subset X$, or $Z = Y$ and $X \subset Y$. We are left with appealing to (1). $\rhd$

1.4.5. *Theorem.* *The following hold:*

(1) *Each member of an ordinal class is an ordinal;*
(2) *The class* On *is the only ordinal class that is not an ordinal;*
(3) *For every α, the set $\alpha + 1$ is the least of all ordinals succeeding α;*
(4) *The union of a nonempty class of ordinals $X \subset$ On is an ordinal class. If X is a set then the union $\bigcup X$ is an upper bound of the set X in the ordered class* On.

$\lhd$ (1) Take an ordinal class X and $x \in X$. Since X is transitive, $x \subset X$ and so x is totally ordered by membership. Prove $\mathrm{Tr}(x)$. If $z \in y \in x$ then $z \in X$ since X is transitive.

Of the three possibilities: $z = x$, $x \in z$, and $z \in x$, the first two result in the cycles, $z \in y \in z$ and $z \in y \in x \in z$, each contradicting the axiom of regularity. Therefore, $z \in x$ and so $z \in y \to z \in x$; i.e., $y \subset x$, which proves $\mathrm{Tr}(x)$ and, at the same time, $\mathrm{Ord}(x)$.

(2) By 1.4.4 (3), the class On is totally ordered; by (1), it is transitive. Hence, Ord (On). If On were a set then On would be an ordinal, which leads to the contradiction On $\in$ On.

Hence, On is an ordinal class but not an ordinal. For an arbitrary ordinal class X, the formula $X \notin$ On yields $X =$ On. Indeed, 1.4.4 (3) opens the sole possibility: On $\in X$, which contradicts the fact that On is a proper class.

(3) If α is an ordinal, then, obviously, the set $\alpha + 1 := \alpha \cup \{\alpha\}$ is totally ordered. Given $x \in \alpha + 1$, we obtain either $x \in \alpha$ or $x = \alpha$, and in both cases $x \subset \alpha$. However, $\alpha \subset \alpha + 1$. Hence, $x \subset \alpha + 1$, which proves that $\alpha + 1$ is transitive. All in all, $\alpha + 1$ is an ordinal and $\alpha < \alpha + 1$. If $\alpha < \beta$ for some β then $\alpha \in \beta$ and $\alpha \subset \beta$, i.e., $\alpha \cup \{\alpha\} \subset \beta$. By 1.4.4 (1), either $\alpha \cup \{\alpha\} \in \beta$ or $\alpha \cup \{\alpha\} = \beta$. Therefore, $\alpha + 1 \leq \beta$.

(4) Assume that $X \subset$ On. Take $y \in Y := \bigcup X$ and choose $x \in X$ so that $y \in x$. Since x is an ordinal; therefore, $y \subset x$ and, moreover, $y \subset Y$. The class On is transitive (see (2)), and so $x \in X$ yields $x \subset$ On. Consequently, $Y \subset$ On.

Thus, Y is a transitive subclass On, and so Y is an ordinal. If $\alpha \in X$ then $\alpha \subset Y$ and, by 1.4.4 (1), $\alpha \leq Y$. While if β is an ordinal and $\beta \geq \alpha$ for all $\alpha \in X$ then $Y \subset \beta$ and $Y \leq \beta$ by 1.4.4 (1). Hence, $Y = \sup(X)$. $\rhd$

1.4.6. The least upper bound of a set of ordinals x is usually denoted by $\lim(x)$. An ordinal α is a *limit ordinal* if $\alpha \neq \varnothing$ and $\lim(\alpha) = \alpha$.

In other words, α is a limit ordinal provided that α cannot be written down as $\alpha = \beta + 1$ with $\beta \in \mathrm{On}$. Let K_{II} stand for the class of all limit ordinals. The class of nonlimit ordinals K_{I} is the complement of K_{II}; i.e., $K_{\mathrm{I}} := \mathrm{On} - K_{\mathrm{II}} = \{\alpha \in \mathrm{On} : (\exists \beta \in \mathrm{On})\ (\alpha = \beta + 1)\}$. Denote by ω the least limit ordinal whose existence is ensured by Theorem 1.4.5 and the axiom of infinity. It is an easy matter to show that ω coincides with the class of nonlimit ordinals α such that each predecessor of α is also a nonlimit ordinal:

$$\omega = \{\alpha \in \mathrm{On} : \alpha \cup \{\alpha\} \in K_{\mathrm{I}}\}.$$

The members of ω are *finite ordinals*, or *positive integers*, or *natural numbers*, or simply *naturals*. This is why ω is called the *naturals* in common parlance.

The least ordinal, the zero set $0 := \varnothing$, belongs to ω. The successor $1 := 0 + 1 = 0 \cup \{0\} = \{\varnothing\}$ contains the only element 0. Furthermore, $2 := 1 \cup \{1\} = \{0\} \cup \{1\} = \{0, 1\} = \{0, \{0\}\}$, $3 := 2 \cup \{2\} = \{0, \{0\}, \{\{0, \{0\}\}\}$, etc. Thus,

$$\omega := \{0, \{0\}, \{0, \{0\}\}, \dots\} = \{0, 1, 2, \dots\}.$$

The following notation is also used:

$$\mathbb{N} := \omega - \{0\} = \{1, 2, \dots\}.$$

Recall that it is a mathematical tradition of long standing to apply the term "natural" only to the members of $\mathbb{N}$. Historically, zero is "less" natural if not "unnatural."

The next theorem displays the basic properties of the naturals ω which are known as *Peano's axioms*.

1.4.7. Theorem. *The following hold:*

(1) *Zero belongs to* ω;

(2) *The successor* $\alpha + 1$ *of a natural* α *is a natural too;*

(3) $0 \neq \alpha + 1$ *for all* $\alpha \in \omega$;

(4) *If* α *and* β *in* ω *and* $\alpha + 1 = \beta + 1$ *then* $\alpha = \beta$;

(5) *If a class* X *contains the empty set and the successor of each member of* X *then* $\omega \subset X$.

1.4.8. *Theorem* (the principle of transfinite induction). *Let* X *be a class with the following properties:*

(1) $0 \in X$;

(2) *If* α *is an ordinal and* $\alpha \in X$ *then* $\alpha + 1 \in X$;

(3) *If* x *is a set of ordinals contained in* X *then* $\lim(x) \in X$.

Then $\mathrm{On} \subset X$.

◁ Assume to the contrary that $\mathrm{On} \not\subset X$. Then the nonempty subclass $\mathrm{On} - X$ of the well ordered class On has the least element $\alpha \in \mathrm{On} - X$, which means that $\alpha \cap (\mathrm{On} - X) = 0$ or $\alpha \subset X$ and $\alpha \neq 0$ by (1). If $\alpha \in K_{\mathrm{I}}$, i.e., $\alpha = \beta + 1$ for some $\beta \in \mathrm{On}$; then $\beta \in \alpha \subset X \to \beta \in X$ and, by (2), $\alpha = \beta + 1 \in X$. In turn, if $\alpha \in K_{\mathrm{II}}$ then from (3) we deduce $\alpha = \lim(\alpha) \in X$. In both cases $\alpha \in X$, which contradicts the membership $\alpha \in \mathrm{On} - X$. ▷

1.4.9. *Theorem* (the principle of transfinite recursion). *Let G be some class-function. Then there is a unique function F satisfying*

(1) $\mathrm{dom}(F) = \mathrm{On}$;

(2) $F(\alpha) = G(F \restriction \alpha)$ *for all* $\alpha \in \mathrm{On}$, *where* $F \restriction \alpha := F \cap (\alpha \times \mathbf{U})$ *is the restriction of F to α.*

◁ Define the class Y by the formula

$$f \in Y \leftrightarrow \mathrm{Fnc}(f) \wedge \mathrm{dom}(f) \in \mathrm{On} \wedge (\forall \alpha \in \mathrm{dom}(f))\ (f(\alpha) = G(f \restriction \alpha)).$$

If $f,\ g \in Y$ then either $f \subset g$ or $g \subset f$.

Indeed, if $\beta := \mathrm{dom}(f)$ and $\gamma := \mathrm{dom}(g)$ then either $\beta \leq \gamma$ or $\gamma \leq \beta$. Assuming for instance that $\gamma < \beta$, put $z := \{\alpha \in \mathrm{On} : \alpha < \gamma \wedge f(\alpha) \neq g(\alpha)\}$. If $z \neq 0$ then z contains the least element δ.

In this case for all $\alpha < \delta$ we obtain $f(\alpha) = g(\alpha)$; i.e., $f \restriction \delta = g \restriction \delta$. By the definition of Y, we however have $f(\delta) = G(f \restriction \delta)$ and $g(\delta) = G(g \restriction \delta)$. Hence, $f(\delta) = g(\delta)$ and $\delta \notin z$.

This contradicts the choice of δ. So, $z = 0$; i.e., $f(\alpha) = g(\alpha)$ for all $\alpha < \gamma$, which yields the required inclusion $g \subset f$. Put $F := \bigcup Y$. Obviously, F is a function, $\mathrm{dom}(F) \subset \mathrm{On}$, and $F(\alpha) = G(F \restriction \alpha)$ for all $\alpha \in \mathrm{dom}(F)$.

If $\alpha \in \mathrm{dom}(F)$ then $(\alpha, G(F \restriction \alpha)) \in f$ for some $f \in Y$. Then $\alpha \in \beta := \mathrm{dom}(f) \subset \mathrm{dom}(F)$. Since β is transitive, we obtain $\alpha \subset \mathrm{dom}(F)$. Therefore, the class $\mathrm{dom}(F)$ is transitive and, by 1.4.4 (1), either $\mathrm{dom}(F) = \mathrm{On}$ or $\mathrm{dom}(F) \in \mathrm{On}$. However, the latter containment is impossible. Indeed, it follows from $\delta := \mathrm{dom}(F) \in \mathrm{On}$ that the function $f := F \cup \{(\delta, G(F))\}$ belongs to Y. Hence, $f \subset F$, which leads to a contradiction as follows: $f \subset F \to \mathrm{dom}(f) \subset \mathrm{dom}(F) \to \delta \in \mathrm{dom}(F) = \delta$. ▷

1.4.10. A binary relation R is *well founded* if the class $R^{-1}(x)$ is a set for all $x \in \mathbf{U}$ and to each nonempty $x \in \mathbf{U}$ there is an element $y \in x$ such that $x \cap R^{-1}(y) = 0$.

The last condition (on assuming the axiom of choice) amounts to the fact that there is no infinite sequence (x_n) with the property $x_n \in R(x_{n+1})$ for all $n \in \omega$. The membership $\in$ provides an example of a well founded relation. It is often more convenient to apply the principles of transfinite induction and recursion in the following form:

1.4.11. Theorem. *Let R be a well founded relation. The following hold:*

(1) (induction on R) *If a class X is such that for all $x \in \mathbf{U}$ the formula $R^{-1}(x) \subset X$ implies $x \in X$, then $X = \mathbf{U}$;*

(2) (recursion on R) *To each function $G : \mathbf{U} \to \mathbf{U}$ there is a function F such that* $\mathrm{dom}(F) = \mathbf{U}$ *and $F(x) = G(F \restriction R^{-1}(x))$ for all $x \in \mathbf{U}$.*

1.4.12. Two sets are *equipollent*, or *equipotent*, or *of the same cardinality* if there is a bijection of one of them onto the other. An ordinal that is equipotent to no preceding ordinal is a *cardinal.* Every natural is a cardinal.

A cardinal other than a natural is an *infinite* cardinal. Therefore, ω is the least infinite cardinal.

Given an ordinal α, we denote by ω_α an infinite cardinal such that the ordered set of all infinite cardinals less than ω_α is similar to α. If such a cardinal exists then it is unique.

1.4.13. Theorem (the principle of cardinal comparability). *The following hold:*

(1) *Infinite cardinals form a well ordered proper class;*

(2) *To each ordinal α there is a cardinal ω_α so that the mapping $\alpha \mapsto \omega_\alpha$ is a similarity between the class of ordinals and the class of infinite cardinals;*

(3) *There is a mapping $|\cdot|$ from the universal class $\mathbf{U}$ onto the class of all cardinals such that the sets x and $|x|$ are equipollent for all $x \in \mathbf{U}$.*

$\lhd$ The proof may be found for instance in [168]. $\rhd$

The cardinal $|x|$ is called the *cardinality* or the *cardinal number* of a set x. Hence, any set is equipollent to a unique cardinal which is its cardinality.

A set x is *countable* provided that $|x| = \omega_0 := \omega$, and x is *at most countable* provided that $|x| \leq \omega_0$.

1.4.14. Given an ordinal α, we denote by $\mathbf{2}^{\omega_\alpha}$ the cardinality of $\mathscr{P}(\omega_\alpha)$; i.e., $\mathbf{2}^{\omega_\alpha} := |\mathscr{P}(\omega_\alpha)|$. This denotation is justified by the fact that $\mathbf{2}^x$ and $\mathscr{P}(X)$ are equipollent for all x, with $\mathbf{2}^x$ standing for the class of all mappings from x to $\mathbf{2}$.

A theorem, proven by G. Cantor, states that $|x| < |\mathbf{2}^x|$ for whatever set x. In particular, $\omega_\alpha < \mathbf{2}^{\omega_\alpha}$ for each ordinal α. In this case, appealing to Theorem 1.4.13, we obtain $\omega_{\alpha+1} \leq \mathbf{2}^{\omega_\alpha}$.

The *generalized problem of the continuum* asks whether or not there are intermediate cardinals between $\omega_{\alpha+1}$ and $\mathbf{2}^{\omega_\alpha}$; i.e., whether or not the equality $\omega_{\alpha+1} = \mathbf{2}^{\omega_\alpha}$ holds. For $\alpha = 0$ this is the classical *problem of the continuum.*

The *continuum hypothesis* CH is the equality $\omega_1 = \mathbf{2}^\omega$. Similarly, the *generalized continuum hypothesis* GCH is the equality $\omega_{\alpha+1} = \mathbf{2}^{\omega_\alpha}$ for all $\alpha \in \mathrm{On}$.

1.4.15. Furnish the class On × On with some order that will be called *canonical.* To this end, take α_1, α_2, β_1, $\beta_2 \in$ On. Agree to assume that $(\alpha_1, \alpha_2) \leq (\beta_1, \beta_2)$ if one of the following conditions is fulfilled:

(1) $\alpha_1 = \beta_1$ and $\alpha_2 = \beta_2$;

(2) $\sup\{\alpha_1, \alpha_2\} < \sup\{\beta_1, \beta_2\}$;

(3) $\sup\{\alpha_1, \alpha_2\} = \sup\{\beta_1, \beta_2\}$ and $\alpha_1 < \beta_1$;

(4) $\sup\{\alpha_1, \alpha_2\} = \sup\{\beta_1, \beta_2\}$ and $\alpha_1 = \beta_1$ and $\alpha_2 < \beta_2$.

Therefore, the pairs (α, β) are compared by using $\sup\{\alpha, \beta\}$. Also, the set of ordered pairs (α, β) with the same $\sup\{\alpha, \beta\}$ has the lexicographic order. We may easily prove that On × On with the canonical order is a well ordered class. By analogy, we may define the canonical well-ordering on the class On × On × On and so on.

1.4.16. COMMENTS.

(1) The idea of transfinite ranks among the most profound and original discoveries by G. Cantor. Using this idea, he created a powerful method for qualitative analysis of infinity and penetrated deeply into its essence.

The notion of infinity can be traced in religious and philosophical doctrines since the ancient times. The whole totality of the views of the infinite had however been a primarily humanitarian subject prior to G. Cantor who made the very concept of infinity a topic of mathematical research.

Invoked and inspired by the Infinite, "Mathematics is the Science of Infinity." So reads one of the most popular definitions of the present-day mathematics, witnessing the grandeur of the G. Cantor idea.

(2) The problem of the continuum stems from G. Cantor and is the first in the epoch-making report by D. Hilbert at the turn of the twentieth century [20]. Remaining unsolved for decades, this problem gave rise to in-depth foundational studies of set theory. In 1939 K. Gödel established consistency of the generalized continuum hypothesis with ZFC [59]. In 1963 P. J. Cohen proved that the negation of the generalized continuum hypothesis is also consistent with ZFC. Each of these results has brought about new ideas, methods, and problems.

(3) By G. Cantor, an ordinal is the *order type* of some well ordered set x; i.e., the class of all ordered sets similar to x. Each order type, with the exception of the empty set, is a proper class however. This peculiarity prevents us from developing the theory of order types within NGB since it is impossible to consider the classes of order types. The definition of ordinal in 1.4.2 leans on choosing a canonical representative in each order type. This definition belongs to J. von Neumann.

(4) In this section we present only the basic facts on ordinals; details, and further information may be found in [115, 168].

1.5. Hierarchies of Sets

Recursive definitions, basing on Theorem 1.4.9 or its modifications, bring about, in particular, decreasingly (or increasingly) nested transfinite sequences of sets which are known as cumulative hierarchies. Of a profound interest for our tasks are the hierarchies appearing in the models of set theory.

1.5.1. Consider a set x_0 and two single-valued classes Q and R. Starting with them, we construct a new single-valued class G. To begin with, put $G(0) := x_0$. Further, if x is a function and $\mathrm{dom}(x) = \alpha + 1$ for some $\alpha \in \mathrm{On}$ then we let $G(x) := Q(x(\alpha))$. Whereas if $\mathrm{dom}(x) = \alpha$ is a limit ordinal then, to obtain $G(x)$, we "collect" the values of $x(\beta)$ for $\beta < \alpha$ and apply R to the whole collection; i.e., $G(x) := R(\bigcup \mathrm{im}(x))$. In every remaining case we assume that $G(x) = 0$. By Theorem 1.4.9 of transfinite recursion, there exists a single-valued class F satisfying the conditions:

$$F(0) = x_0,$$
$$F(\alpha + 1) = Q(F(\alpha)),$$
$$F(\alpha) = R\Big(\bigcup_{\beta<\alpha} F(\beta)\Big) \quad (\alpha \in K_{\mathrm{II}}).$$

Each $F(\alpha)$ is a *floor* of F, while F itself is a *cumulative hierarchy*. The union of the class $\mathrm{im}(F)$, i.e. the class

$$\bigcup_{\alpha\in\mathrm{On}} F(\alpha) := \bigcup \mathrm{im}(F),$$

is the *limit* of the cumulative hierarchy $(F(\alpha))_{\alpha\in\mathrm{On}}$.

1.5.2. In the sequel, we are interested only in the particular case in which x_0 is the empty set, R is the identity mapping of the universal class $\mathbf{U}$, and Q is a class-function with $\mathrm{dom}(Q) = \mathbf{U}$. In this case the cumulative hierarchies are constructed inductively, starting with the empty set, by successively applying the operation Q. Varying Q, we arrive at different cumulative hierarchies.

The least ordinal α for which $x \in F(\alpha + 1)$ is called the (*ordinal*) *rank* of x in the hierarchy $(F(\alpha))_{\alpha\in\mathrm{On}}$ and is denoted by $\mathrm{rank}(x)$. This definition is justified by Theorem 1.3.14 claiming that we may find a unique class rank obeying the condition

$$(\forall\, x)(\forall\, y)((x, y) \in \mathrm{rank} \leftrightarrow \varphi(x, y, F, \mathrm{On})),$$

with φ standing for the predicative formula

$$(\exists\, \alpha \in \mathrm{On})(y = \alpha \wedge x \in F(\alpha + 1) \wedge (\forall\, \beta \in \mathrm{On})(x \in F(\beta + 1) \to \alpha \leq \beta)).$$

In this event, Un(rank), dom(rank) $= \bigcup \operatorname{im} F$, and im(rank) $\subset$ On; i.e., rank is a function from $\bigcup \operatorname{im}(F)$ to On. We abstain from inserting F in the notation of the rank of a set in F since the context always prompts us the hierarchy F in what follows.

1.5.3. As the simplest example, consider the case in which $x_0 = 0$, $R = I_{\mathbf{U}}$, and $Q := \mathscr{P}_{\mathrm{tr}}$, with $\mathscr{P}_{\mathrm{tr}}$ sending $x \in \mathbf{U}$ to the class $\mathscr{P}_{\mathrm{tr}}(x)$ of all transitive subsets of x. Since a transitive subset of an ordinal is an ordinal; therefore, $Q(\alpha) = \alpha \cup \{\alpha\} = \alpha + 1$ and $F(\alpha + 1) = \alpha + 1$ for every ordinal α. If α is a limit ordinal then

$$F(\alpha) = \bigcup_{\beta<\alpha} F(\beta) = \bigcup_{\beta+1<\alpha} F(\beta+1) = \bigcup_{\beta+1<\alpha} \beta+1 = \alpha.$$

Therefore, the limit of our increasingly nested cumulative hierarchy is the class of ordinals On.

1.5.4. Assigning the rôle of Q to the powerset operation $\mathscr{P}$ and taking $x_0 = 0$ and $R = I_{\mathbf{U}}$, we come to the familiar cumulative hierarchy (cf. the Appendix):

$$\begin{aligned} V_0 &:= 0, \\ V_{\alpha+1} &:= \mathscr{P}(V_\alpha) \quad (\alpha \in \mathrm{On}), \\ V_\alpha &:= \bigcup_{\beta<\alpha} V_\beta \quad (\alpha \in K_{\mathrm{II}}). \end{aligned}$$

The class $\mathbf{V} := \bigcup_{\alpha\in\mathrm{On}} V_\alpha$ is the *von Neumann universe*. Note that the lower floors of $\mathbf{V}$ are as follows: $V_1 = \mathscr{P}(0) = \{0\} = 1$, $V_2 = \mathscr{P}(1) = \{0, \{0\}\} = 2$, $V_3 = \mathscr{P}(V_2) = \{0, \{0\}, \{\{0\}\}, \{0, \{0\}\}\} \neq 3$, etc.

1.5.5. *The following hold:*

(1) V_α *is a transitive set for all* $\alpha \in$ On;

(2) $V_\beta \in V_\alpha$ *and* $V_\beta \subset V_\alpha$ *for all* α, $\beta \in$ On, $\beta < \alpha$;

(3) *If* $x \in y \in \mathbf{V}$ *then* rank$(x) <$ rank(y);

(4) *The class of ordinals* On *is included in* $\mathbf{V}$;

(5) rank$(\alpha) = \alpha$ *for all* $\alpha \in$ On;

(6) *If* x *is a set and* $x \subset \mathbf{V}$ *then* $x \in \mathbf{V}$.

$\lhd$ (1) Proceed by transfinite induction. For $\alpha = 0$, the class $V_0 = 0$ is a transitive set. Assume proven that V_α is a transitive set. Since $V_{\alpha+1} = \mathscr{P}(V_\alpha)$, note that $V_{\alpha+1}$ is a set and, for all x and y, it follows from $x \in y \in V_{\alpha+1}$ that $y \subset V_\alpha$ and $x \in V_\alpha$. By the induction hypothesis, either $x \subset V_\alpha$ or $x \in V_{\alpha+1}$, and so $y \subset V_{\alpha+1}$. If $\alpha \in K_{\mathrm{II}}$ and V_β is a transitive set for all $\beta < \alpha$; then, for all $x \in V_\alpha$, we have

$$(\exists\, \beta < \alpha)(x \in V_\beta) \to (\exists\, \beta < \alpha)(x \subset V_\beta) \to x \subset V_\alpha.$$

Moreover, V_α is a set as the union of a set of sets.

(2) Transitivity of V_α is shown in (1). We are thus left with demonstrating that $V_\beta \in V_\alpha$ $(\beta < \alpha)$. Proceed by transfinite induction on α.

In case $\alpha = 1$, nothing is left to proof. Let $\alpha > 1$ and $V_\beta \in V_\alpha$ for all $\beta < \alpha$. The inequality $\beta < \alpha + 1$ holds only if $\alpha = \beta$ or $\beta < \alpha$. If $\alpha = \beta$ then

$$V_\beta = V_\alpha \in \mathscr{P}(V_\alpha) = V_{\alpha+1}.$$

If $\beta < \alpha$ then, by the induction hypothesis, $V_\beta \in V_\alpha$ and, by (1), $V_\alpha \subset V_{\alpha+1}$. Hence, $V_\beta \in V_{\alpha+1}$. Given a limit ordinal $\alpha \in K_{\text{II}}$, it suffices to note that $V_\beta \in V_\alpha$ for $\beta < \alpha$ since

$$V_\beta \in V_{\beta+1} \subset \bigcup_{\gamma<\alpha} V_\gamma = V_\alpha.$$

(3) A moment's thought shows that $\alpha = \operatorname{rank}(x)$ if and only if $x \in V_{\alpha+1}$ and $x \notin V_\alpha$. Hence, if $x \in y$ then $y \not\subset V_\alpha$ and so $y \notin V_{\alpha+1}$. By definition, $\operatorname{rank}(y) > \alpha$.

(4), (5) Proceed again by transfinite induction.

In case $\alpha = 0$ note that $0 \in V_0 \subset \mathbf{V}$ and $\operatorname{rank}(0) = 0$ since $0 \notin V_0$.

Take $\alpha \in \mathbf{V}$ with $\operatorname{rank}(\alpha) \in \alpha$. Then $\alpha + 1 = \alpha \cup \{\alpha\} \subset V_{\alpha+1}$, or $\alpha + 1 \in \mathscr{P}(V_{\alpha+1}) = V_{\alpha+2}$. On the other hand, if $\alpha + 1 \in V_{\alpha+1}$ then $\alpha \cup \{\alpha\} \subset V_\alpha$, yielding $\alpha \in V_\alpha$, which is a contradiction. Therefore, $\alpha+1 \notin V_{\alpha+1}$ and so $\operatorname{rank}(\alpha+1) = \alpha+1$.

Assume now that $\alpha \in K_{\text{II}}$ and, for all $\beta < \alpha$, it is established that $\beta \in \mathbf{V}$ and $\operatorname{rank}(\beta) = \beta$. In this event

$$\alpha = \{\beta \in \mathrm{On} : \beta < \alpha\} \subset \bigcup_{\beta<\alpha} V_{\beta+1} \subset V_\alpha,$$

whence $\alpha \in V_{\alpha+1}$. Moreover, the membership $\alpha \in V_\alpha$ implies that $\alpha \in V_\beta$ for some $\beta < \alpha$. Applying (3) and the induction hypothesis, we immediately arrive at a contradiction: $\beta = \operatorname{rank}(\beta) < \operatorname{rank}(\alpha) < \beta$.

(6) Put $\alpha := \sup\{\operatorname{rank}(y) : y \in x\}$. Obviously, $x \subset V_{\alpha+1}$ and $x \in V_{\alpha+2} \subset \mathbf{V}$. $\rhd$

1.5.6. *Theorem.* *The axiom of regularity* NGB_{14} *amounts to the equality* $\mathbf{U} = \mathbf{V}$, *i.e., to the coincidence of the universal class and the von Neumann universe.*

$\lhd$ Suppose that $\mathbf{U} = \mathbf{V}$ and take a nonempty class X. There is an element $x \in X$ with the least rank α; i.e., $\operatorname{rank}(x) = \alpha$ and $\operatorname{rank}(x) \le \operatorname{rank}(y)$ for all $y \in X$. If $u \in x \cap X$ then, by 1.5.5 (3), $\operatorname{rank}(u) < \alpha = \operatorname{rank}(x)$, which contradicts the definition of α. Hence, $x \cap X = 0$.

Demonstrate now that the supposition $\mathbf{V} \ne \mathbf{U}$ contradicts the axiom of regularity. To this end, apply this axiom to the nonempty class $\mathbf{U} - \mathbf{V}$ and find a set $y \in \mathbf{U} - \mathbf{V}$ satisfying $y \cap (\mathbf{U} - \mathbf{V}) = 0$. The last equality yields $y \subset \mathbf{V}$, whereas from 1.5.5 (6) we deduce $y \in \mathbf{V}$, which contradicts the choice of y. $\rhd$

1.5.7. Theorem. *The following hold:*

(1) induction on membership: *If a class X has the property that $x \subset X$ implies $x \in X$ for every set x, then $X = \mathbf{V}$;*

(2) recursion on membership: *If G is a single-valued class then there is a unique function F with domain $\mathbf{V}$ satisfying $F(x) = G(\operatorname{im}(F \upharpoonright x))$ for $x \in \mathbf{V}$;*

(3) induction on rank: *If a class X has the property that the inclusion $\{y \in \mathbf{V} : \operatorname{rank}(y) < \operatorname{rank}(x)\} \subset X$ implies the membership $x \in X$ for every set x, then $X = \mathbf{V}$.*

$\lhd$ As shown in 1.5.6, the universe $\mathbf{V}$ coincides with the universal class $\mathbf{U}$. Therefore, all claims are immediate from 1.4.11 provided that the relations $\in := \{(x,y) \in \mathbf{V}^2 : x \in y\}$ and $R := \{(x,y) \in \mathbf{V}^2 : \operatorname{rank}(x) \le \operatorname{rank}(y)\}$ are well founded. For the membership relation $\in$, this follows from the axiom of regularity (cf. 1.4.10).

As regards R, proceed by way of contradiction. Take a sequence $(x_n)_{n\in\omega}$ with $x_n \in \mathbf{V}$ such that $x_{n+1} \in R(x_n)$ for all $n \in \omega$. Then the sequence of the ordinals $\alpha_n := \operatorname{rank}(x_n)$ would obey the condition $\alpha_{n+1} < \alpha_n$ $(n \in \omega)$ (cf. 1.5.5 (3)). This would contradict the fact that On is well ordered. Hence, R is well founded. $\rhd$

1.5.8. Let $\sim$ be an equivalence on a class W. The collection of all members of W which are equivalent to some element of W is a proper class in general, which is an obstacle to combining these equivalence classes into a unique factor class. We may obviate the obstacle by using the ordinal rank.

Frege–Russel–Scott Theorem. *There is a function $F : W \to \mathbf{V}$ such that, for all $x, y \in W$, the following holds:*

$$F(x) = F(y) \leftrightarrow x \sim y.$$

$\lhd$ By Theorem 1.3.14, there is a class F such that, for all $x, y \in W$, we have

$$(x,y) \in F \leftrightarrow \varphi(x, y, W, \sim, \operatorname{rank}),$$

where the predicative formula φ is as follows

$$(\forall z)(z \in y \leftrightarrow z \in W \wedge x \sim z \wedge (\forall u)\,(x \sim u \to \operatorname{rank}(z) \le \operatorname{rank}(u))).$$

Therefore, F is a function, and $F(x)$ stands for the class of sets z equivalent to x and having the least possible rank.

If $\alpha = \operatorname{rank}(x)$ then $F(x) \subset W \cap V_{\alpha+1}$. Hence, $F(x)$ is a set. Moreover, $\operatorname{dom}(F) = W$, and for all $x, y \in W$ we have $x \sim y \leftrightarrow F(x) = F(y)$. Indeed, if $F(x) = F(y)$ then there is an element w in W satisfying $x \sim w$ and $y \sim w$; i.e., $x \sim y$. The reverse implication is obvious. $\rhd$

It follows from the axiom of domain NGB_{10} and 1.3.13 (1) that to F there corresponds the class $\operatorname{im}(F) := \{F(x) : x \in W\}$. Call $\operatorname{im}(F)$ the *factor class* of W by $\sim$ and denote it by $W/{\sim}$; i.e., $W/{\sim} := \operatorname{im}(F)$. In this event we say that F is the *factor mapping* or the *canonical projection* from W to $W/{\sim}$.

1.5.9. Let B be a set with at least two members. Put $Q := \mathscr{P}^{(B)} : x \mapsto B^x$ $(x \in \mathbf{V})$, where B^x stands as usual for the set of all mappings from x to B. The cumulative hierarchy arising in this case (cf. 1.5.1, with $x_0 = 0$ and $R = I_{\mathbf{V}}$) is denoted by $(V_\alpha^{(B)})_{\alpha\in\mathrm{On}}$. The resultant *B-valued universe*

$$\mathbf{V}^{(B)} := \bigcup_{\alpha\in\mathbf{On}} V_\alpha^{(B)}$$

is a subclass of $\mathbf{V}$ consisting of B-valued functions given on the sets of B-valued functions. The conventional interpretation for the membership sign $\in$ in $\mathbf{V}^{(B)}$ yields nothing of interest, since the set-theoretic membership $u \in v$ for B-valued functions u and v holds only in trivial cases.

Happily, the hierarchies (V_α) and $(V_\alpha^{(B)})$ differ significantly which circumstance gives grounds for nonstandard interpretations of set theory in the universe $\mathbf{V}^{(B)}$. This is elaborated in Chapter 2.

1.5.10. For the sake of completeness, we mention one more cumulative hierarchy.

The following operations over sets are the *Gödel operations* (they are eight in total): pairing; (set-theoretic) difference; (Cartesian) product; the $(2,3,1)$-, $(3,2,1)$-, and $(1,3,2)$-permutations (see 1.3.10); restricted membership $x \mapsto x^2 \cap \in$; and domain $x \mapsto \mathrm{dom}(x)$.

Given some set or a set of sets x, define the *Gödel closure* $\mathrm{cl}_G(x)$ of x as the least set containing X and closed under the Gödel operations. Assign $Q(x) := \mathscr{P}(X) \cap \mathrm{cl}_G(x \cup \{x\})$. The resultant cumulative hierarchy $(L_\alpha)_{\alpha\in\mathrm{On}}$ is the *constructible hierarchy*. The *constructible universe* is the class $\mathbf{L} := \bigcup_{\alpha\in\mathrm{On}} L_\alpha$; the elements of $\mathbf{L}$ are *constructible* sets (for details see [83, 172]).

1.5.11. Comments.

(1) It was J. von Neumann who first considered the cumulative hierarchy $(V_\alpha)_{\alpha\in\mathrm{On}}$ now named after him. The relativization of the axiom of regularity to the von Neumann universe $\mathbf{V}$ is provable in the theory $\mathrm{NGB} \setminus \mathrm{NGB}_{14}$, which implies that NGB_{14} is consistent with the rest of the axioms of NGB. Another technique shows that the negation $\neg\,\mathrm{NGB}_{14}$ is consistent with the axioms of NGB; i.e., NGB_{14} is an independent axiom.

(2) If B is a complete Heyting lattice (cf. 1.1.8 (3)) then the universe $\mathbf{V}^{(B)}$ may be transformed into a model of intuitionistic set theory by using the structure of B and the hierarchy $(V_\alpha^{(B)})$. In particular, if B is a complete Boolean algebra then we arrive at a Boolean valued model of set theory (more details will appear in 2.1.10 (3)).

(3) If $B := [0,1]$ is the interval of the real axis then the class $\mathbf{V}^{(B)}$ is called the *universe of Zadeh fuzzy sets* or, briefly, *fuzzy universe* [157, 260–262]. This universe can provide a model for some set theory with an appropriate many-valued logic, which may be of use for studying fuzzy sets.

(4) The constructible universe $\mathbf{L}$ is the least transitive model of ZFC containing all ordinals which is called the *Gödel model*. This universe satisfies the axiom of choice and the generalized continuum hypothesis. Therefore, AC and GCH are consistent with ZF.

The *axiom of constructibility* reads: "every set is constructible"; in symbols, $\mathbf{V} = \mathbf{L}$. The relativization of the formula $\mathbf{V} = \mathbf{L}$ to $\mathbf{L}$ is provable in ZF. Hence, $\mathbf{V} = \mathbf{L}$ is consistent with ZF. All these results, as well as the definition of constructible set, belong to K. Gödel [59] (see also [83, 172]). The corresponding assertions of consistency of the axiom of choice and GCH also hold for NGB (cf. [30, 83, 168, 172]).

(5) It is proven in [234] that if B is a quantum logic (see 1.1.8 (5)) then the universe $\mathbf{V}^{(B)}$ serves as a model for some quantum set theory in the sense analogous to that of Section 2.4 to follow. Treating quantum theories as logical systems, constructing quantum set theory and developing the corresponding quantum mathematics is an intriguing and actual field of research, slow progress wherein notwithstanding. Apparently, the adequate mathematical means and opportunities, together with sound reference points, are traceable within the theory of von Neumann algebras proliferating numerous "noncommutative" branches (noncommutative probability theory, noncommutative integration, etc.).

Chapter 2

Boolean Valued Universes

It is the use of various rather unconventional models of set theory that unifies the available nonstandard methods of analysis. In particular, the technique of Boolean valued analysis bases on the properties of a certain cumulative hierarchy $\mathbf{V}^{(B)}$ whose every successive floor comprises all functions with domain on the preceding floors and range in a complete Boolean algebra B fixed in advance.

Our main topic in the present section is the construction and study of the hierarchy, i.e., the Boolean valued universe $\mathbf{V}^{(B)}$. The idea behind the construction of $\mathbf{V}^{(B)}$ is very simple. We first observe that the characteristic function of a set is a good substitute for the set itself. Rising in the hierarchy whose limit is the von Neumann universe and carrying out the successive substitutions, we arrive at another representation of the von Neumann universe which consists only of two-valued functions. Replacing **2** with an arbitrary Boolean algebra B and repeating the above construction, we arrive at the desired $\mathbf{V}^{(B)}$.

The subtlest aspects, deserving special attention, relate to elaboration of the sense in which we may treat $\mathbf{V}^{(B)}$ as a model of set theory. We set forth the basic technique that lay grounds for Boolean valued analysis, i.e. the transfer, mixing, and maximum principles.

Considerations of logical rigor and expositional independence have requested an ample room for constructing a separated universe and interpreting NGB inside $\mathbf{V}^{(B)}$. The reader, interested only in solid applications to analysis, may just cast a casual glance at these rather sophisticated fragments of exposition while getting first acquaintance with the content of the present book.

2.1. The Universe over a Boolean Algebra

In this section we define a Boolean valued universe and the corresponding Boolean truth values for set-theoretic formulas. We also present the simplest relevant facts and details.

2.1.1. We start with informal heuristic considerations facilitating acquaintance with some features of Boolean valued universes and truth values.

Recall that $\mathbf{2} := \{\mathbf{0}, \mathbf{1}\}$ stands for the two-element Boolean algebra (as usual, we refuse to distinguish between various representations of this simplest Boolean algebra). Take an arbitrary set x, a member of the von Neumann universe $\mathbf{V}$, and associate with x a certain (characteristic) function χ_x ranging in $\mathbf{2}$ and determined (nonuniquely, in general) by the conditions that $x \subset \operatorname{dom}(\chi_x)$ and $\chi_x(t) = \mathbf{1}$ whenever $t \in x$. Clearly, there are many sound reasons to identify x with any of these functions χ_x.

To ensure that the members of the domain $\operatorname{dom}(\chi_x)$ of a two-valued function χ_x are also interpretable as two-valued functions, we surely have to substitute the appropriate characteristic function for each element on the floor V_β, $\beta < \operatorname{rank}(x)$, which includes $\operatorname{dom}(\chi_x)$. Intending to serve so to the whole world of sets, i.e., the universe $\mathbf{V}$, we must start from the zero floor which is $\varnothing$.

Formalizing these observations, we come to the notion of the $\mathbf{2}$-*valued universe*:

$$\mathbf{V}^{(\mathbf{2})} := \{x : (\exists\, \alpha \in \mathrm{On})\ (x \in V_\alpha^{(\mathbf{2})})\},$$

where $V_0^{(\mathbf{2})} := \varnothing$, $V_1^{(\mathbf{2})} := \{\varnothing\}$, $V_2^{(\mathbf{2})} := \{\{\varnothing\}, (\{\varnothing\}, \mathbf{1})\}$, etc. In more detail, acting by analogy with $\mathbf{V}$ and using recursion on membership, we define the cumulative hierarchy

$$V_\alpha^{(\mathbf{2})} := \{x : \operatorname{Fnc}(x) \wedge \operatorname{im}(x) \subset \mathbf{2} \wedge (\exists\, \beta < \alpha)(\operatorname{dom}(x) \in V_\beta^{(\mathbf{2})})\}.$$

Obviously, $\mathbf{V}^{(\mathbf{2})}$ consists of two-valued functions, in which case we associate with each element $x \in \mathbf{V}^{(\mathbf{2})}$ the unique set $\bar{x} := \{y \in \mathbf{V}^{(\mathbf{2})} : x(y) = \mathbf{1}\}$. It worth observing that distinct elements of $\mathbf{V}^{(\mathbf{2})}$ may be assigned to the same set. For this reason, we identify functions x and $y \in \mathbf{V}^{(\mathbf{2})}$ such that $\bar{x} = \bar{y}$, neglecting formal thorns and obstacles which are inevitable on this way.

Take arbitrary $x, y \in \mathbf{V}^{(\mathbf{2})}$. By the identification agreement, the equality $x = y$ holds if and only if $\bar{x} = \bar{y}$. Furthermore, we naturally assume the formula $x \in y$ holding whenever $x \in \bar{y}$. Putting $[\![x = y]\!] := \mathbf{1}$ and $[\![x \in y]\!] := \mathbf{1}$ when $x = y$ and $x \in y$ hold, we let $[\![x = y]\!] := \mathbf{0}$ and $[\![x \in y]\!] := \mathbf{0}$ otherwise. We then have the following presentations:

$$[\![x \in y]\!] = \bigvee_{t \in \operatorname{dom}(y)} y(t) \wedge [\![t \in x]\!],$$

$$[\![x = y]\!] = \bigvee_{t \in \operatorname{dom}(x)} x(t) \Rightarrow [\![t \in y]\!] \wedge \bigvee_{t \in \operatorname{dom}(y)} y(t) \Rightarrow [\![t \in x]\!].$$

It stands to reason to compare these formulas with the following propositions of set theory:

$$u \in v \leftrightarrow (\exists\, w)(w \in v \wedge w = u),$$
$$u = v \leftrightarrow (\forall\, w)(w \in u \to w \in v) \wedge (w \in v \to w \in u).$$

2.1.2. Fix a complete Boolean algebra B which is of course an element of the von Neumann universe $\mathbf{V}$. The *Boolean valued universe* $\mathbf{V}^{(B)}$ over B arises as the limit of the cumulative hierarchy, cf. (1.5.1), provided that $x_0 := 0$ and $R := I_{\mathbf{V}}$, while Q is determined from the formula

$$y \in Q(x) \leftrightarrow \mathrm{Fnc}\,(y) \wedge \mathrm{dom}(y) \subset x \wedge \mathrm{im}(y) \subset B.$$

Therefore, the hierarchy $(V_\alpha^{(B)})_{\alpha\in\mathrm{On}}$ has the form

$$V_0^{(B)} := 0,$$
$$V_{\alpha+1}^{(B)} := \{y : \mathrm{Fnc}\,(y) \wedge \mathrm{dom}(y) \subset V_\alpha^{(B)} \wedge \mathrm{im}(y) \subset B\},$$
$$V_\alpha^{(B)} := \bigcup\{V_\beta^{(B)} : \beta < \alpha\} \quad (\alpha \in K_{\mathrm{II}}).$$

By definition, we assign

$$\mathbf{V}^{(B)} := \bigcup_{\alpha\in\mathrm{On}} V_\alpha^{(B)}.$$

Since the empty set is the function whose domain is void, we easily see that the first and the second floors of the Boolean valued universe are $V_1^{(B)} = \{0\}$ and $V_2^{(B)} = \{0\} \cup \{(0,b) : b \in B\}$. The ordinal rank of $x \in \mathbf{V}^{(B)}$ is further denoted by $\rho(x)$.

2.1.3. Since the membership relation $y \in \mathrm{dom}(x)$ is well founded, the following *induction principle* for $\mathbf{V}^{(B)}$ ensues from 1.4.11 (1):

$$(\forall\, x \in \mathbf{V}^{(B)})((\forall\, y \in \mathrm{dom}(x))\varphi(y) \to \varphi(x)) \to (\forall\, x \in \mathbf{V}^{(B)})\varphi(x),$$

with φ standing for an arbitrary formula of ZFC.

2.1.4. Our nearest aim is to ascribe some truth value to each formula of ZFC whose free variables are replaced with elements of $\mathbf{V}^{(B)}$. The Boolean truth value as a "metafunction" must act to B in such a way that every theorem of ZFC holds "true" inside $\mathbf{V}^{(B)}$; i.e., it acquires the greatest Boolean truth value, the unity of B denoted by $\mathbf{1}$.

To start, we define the Boolean truth value at the atomic formulas $x \in y$ and $x = y$. This is done with the two class functions, $[\![\cdot \in \cdot]\!]$ and $[\![\cdot = \cdot]\!]$, each acting from $\mathbf{V}^{(B)} \times \mathbf{V}^{(B)}$ to B.

Given $x, y \in \mathbf{V}^{(B)}$, we put

(1) $[\![x \in y]\!] := \bigvee_{z \in \mathrm{dom}(y)} y(z) \wedge [\![z = x]\!]$,

(2) $[\![x = y]\!] := \bigwedge_{z \in \mathrm{dom}(y)} y(z) \Rightarrow [\![z \in x]\!] \wedge \bigwedge_{z \in \mathrm{dom}(x)} x(z) \Rightarrow [\![z \in y]\!]$.

Recursion on $(\rho(x), \rho(y))$ and the above formulas lead to the functions $[\![\cdot \in \cdot]\!]$ and $[\![\cdot = \cdot]\!]$ on assuming that $\mathrm{On} \times \mathrm{On}$ is canonically well ordered (see 1.4.15). Indeed, considering the zero floor with $(\rho(x), \rho(y)) = (0, 0)$, obtain (cf. 1.1.1)

$$[\![0 \in 0]\!] = \bigvee \varnothing = \mathbf{0}_B, \quad [\![0 = 0]\!] = \bigwedge \varnothing = \mathbf{1}_B.$$

Now, given $z \in \mathrm{dom}(y)$ or $z \in \mathrm{dom}(x)$, observe that $(\rho(x), \rho(z)) < (\rho(x), \rho(y))$ or $(\rho(z), \rho(y)) < (\rho(x), \rho(y))$, respectively.

It is also possible to choose another road and to proceed by transfinite recursion 1.4.9. Namely, granted the Boolean truth values of $[\![u \in v]\!]$ and $[\![u = v]\!]$ for all $u, v \in V_\alpha^{(B)}$, take $x, y \in V_{\alpha+1}^{(B)}$ and put

$$[\![x = y]\!] = \bigwedge_{u \in \mathrm{dom}(x)} \left(x(u) \Rightarrow \bigvee_{v \in \mathrm{dom}(y)} y(v) \wedge [\![u = v]\!] \right)$$
$$\wedge \bigwedge_{v \in \mathrm{dom}(y)} \left(y(v) \Rightarrow \bigvee_{u \in \mathrm{dom}(x)} x(u) \wedge [\![u = v]\!] \right),$$

since $\mathrm{dom}(x) \subset V_\alpha^{(B)}$ and $\mathrm{dom}(y) \subset V_\alpha^{(B)}$. Now the Boolean truth value $[\![x = z]\!]$ is available for every $z \in \mathrm{dom}(y)$ and so we may calculate

$$[\![x \in y]\!] = \bigvee_{z \in \mathrm{dom}(y)} y(z) \wedge [\![z = x]\!].$$

The case in which α is a limit ordinal causes no problem.

2.1.5. To elaborate the above recursive definition 2.1.4, we now inspect it in more detail.

Choosing $k := 1, 2, 3, 4$, put

$$\pi_x^k(u, v)$$
$$:= \bigvee \{ b \in B : (\exists\, c_1, c_2, c_3, c_4 \in B)((u, v, c_1, c_2, c_3, c_4) \in x \wedge c_k = b) \}.$$

Denote by π_1 and π_2 the functions that send each ordered 6-tuple (alternatively, *hexad*) $(u, v, c_1, c_2, c_3, c_4)$ to the first and second components, i.e., to u and v. With this notation, we describe some single-valued class Q. Given an arbitrary $x \in \mathbf{V}$, let the set $Q(x)$ consist of all 6-tuples $(u, v, c_1, c_2, c_3, c_4)$ satisfying the following conditions:

$$\begin{gathered}
\operatorname{Fnc}(u), \quad \operatorname{Fnc}(v), \quad \operatorname{im}(u) \cup \operatorname{im}(v) \subset B, \\
\operatorname{dom}(u) \subset \pi_1``x, \quad \operatorname{dom}(v) \subset \pi_2``x; \\
b_1 = \bigvee_{z \in \operatorname{dom}(v)} v(z) \wedge \pi_x^3(u, z), \\
b_2 = \bigvee_{z \in \operatorname{dom}(u)} u(z) \wedge \pi_x^4(v, z), \\
b_3 = b_4 = \bigwedge_{z \in \operatorname{dom}(u)} u(z) \Rightarrow \pi_x^1(z, v) \wedge \bigwedge_{z \in \operatorname{dom}(v)} v(z) \Rightarrow \pi_x^2(u, z).
\end{gathered}$$

By 1.5.1, we may find the cumulative hierarchy $(F(\alpha))_{\alpha \in \mathrm{On}}$ satisfying

$$\begin{gathered}
F(0) = (0, 0, \mathbf{0}_B, \mathbf{0}_B, \mathbf{1}_B, \mathbf{1}_B), \\
F(\alpha + 1) = Q(F(\alpha)) \quad (\alpha \in \mathrm{On}), \\
F(\alpha) = \bigcup_{\beta < \alpha} F(\beta) \quad (\alpha \in K_{\mathrm{II}}).
\end{gathered}$$

The class $X := \operatorname{im}(F)$ is obviously a function with $\operatorname{im}(X) \subset B^4$ and $\operatorname{dom}(X) = \mathbf{V}^{(B)} \times \mathbf{V}^{(B)}$.

If $P_k : B^4 \to B$ is the kth projection then we define

$$[\![\cdot \in \cdot]\!] := P_1 \circ X, \quad [\![\cdot = \cdot]\!] := P_3 \circ X.$$

2.1.6. We now describe the way of treating every formula of set theory as a proposition concerning the elements of a Boolean valued universe. In other words, granted B, we intend to interpret the classical set theory in $\mathbf{V}^{(B)}$ by using the functions $[\![\cdot \in \cdot]\!]$ and $[\![\cdot = \cdot]\!]$ of 2.1.4.

To this end, we first define the *interpretation class* I to be the class of all mappings from the set of the symbols of variables in the language of set theory to the universe $\mathbf{V}^{(B)}$.

By the *interpretation of a variable* x we mean the valuation that assigns to each $\nu \in I$ the element $\bar{x}(\nu) := \nu(x)$.

As interpretations of the formulas $x \in y$ and $x = y$ we choose the following functions:

$$\nu \mapsto [\![\bar{x}(\nu) \in \bar{y}(\nu)]\!], \quad \nu \mapsto [\![\bar{x}(\nu) = \bar{y}(\nu)]\!] \quad (\nu \in I).$$

Given a formula $\varphi(x_1,\dots,x_n)$ with n free variables, we now define the interpretation $\nu \mapsto [\![\varphi(\bar{x}_1(\nu),\dots,\bar{x}_n(\nu))]\!]$ by inducting on the length of φ in accord with the following rules

$$
\begin{aligned}
[\![\varphi(x)\wedge\psi(y)]\!] &: \nu\mapsto [\![\varphi(\bar{x}(\nu))]\!]\wedge[\![\psi(\bar{y}(\nu))]\!],\\
[\![\varphi(x)\vee\psi(y)]\!] &: \nu\mapsto [\![\varphi(\bar{x}(\nu))]\!]\vee[\![\psi(\bar{y}(\nu))]\!],\\
[\![\neg\varphi(x)]\!] &: \nu\mapsto [\![\varphi(\bar{x}(\nu))]\!]^*,\\
[\![\varphi(x)\to\psi(y)]\!] &: \nu\mapsto [\![\varphi(\bar{x}(\nu))]\!]\Rightarrow[\![\psi(\bar{y}(\nu))]\!],\\
[\![(\forall\, t)\varphi(t,x)]\!] &: \nu\mapsto \bigwedge\{[\![\varphi(\bar{t}(\nu'),\bar{x}(\nu'))]\!] : \nu'\in I_\nu(x)\},\\
[\![(\exists\, t)\varphi(t,x)]\!] &: \nu\mapsto \bigvee\{[\![\varphi(\bar{t}(\nu'),\bar{x}(\nu'))]\!] : \nu'\in I_\nu(x)\},
\end{aligned}
$$

where $x := (x_1,\dots,x_n)$, $y := (y_1,\dots,y_m)$, $\bar{x}(\nu) := (\bar{x}_1(\nu),\dots,\bar{x}_n(\nu))$, $\bar{y}(\nu) := (\bar{y}_1(\nu),\dots,\bar{y}_m(\nu))$, $I_\nu(x) := \{\nu'\in I : \nu(x) = \nu'(x)\}$, and all free variables of the formulas φ and ψ are listed within $t,x_1,\dots,x_n$ and $t,y_1,\dots,y_m$, respectively.

Note that $[\![\varphi(\bar{x}(\nu))]\!]$ depends only on $\bar{x}_k(\nu)=\nu(x_k)$ $(k:=1,\dots,n)$. Therefore, we write $[\![\varphi(u_1,\dots,u_n)]\!]$ rather than $[\![\varphi(\bar{x}(\nu))]\!] = [\![\varphi(\bar{x}_1(\nu),\dots,\ \bar{x}_n(\nu))]\!]$ provided that $u_k := \bar{x}_k(\nu)\in \mathbf{V}^{(B)}$ $(k:=1,\dots,n)$.

Given a formula $\varphi(x_1,\dots,x_n)$ of ZFC and members $u_1,\dots,u_n$ of $\mathbf{V}^{(B)}$, we call $[\![\varphi(u_1,\dots\ u_n)]\!]$ the Boolean *truth value* of $\varphi(u_1,\dots,u_n)$. We also agree that the record $\mathbf{V}^{(B)}\models\varphi(u_1,\dots,u_n)$ stands for the equality $[\![\varphi(u_1,\dots,u_n)]\!] = \mathbf{1}_B$. In this event we say that φ is *satisfied inside* $\mathbf{V}^{(B)}$ by the assignment of $u_1,\dots,u_n$ to $x_1,\dots,x_n$ or simply that $\varphi(u_1,\dots,u_n)$ *holds* inside $\mathbf{V}^{(B)}$. Sometimes, we use a formula φ that is expressed in the natural language; to mark this, we apply the quotes: $\mathbf{V}^{(B)}\models$ "φ."

Observe also that the *satisfaction sign* $\models$ occasionally inspires the use of model-theoretic expressions like "$\mathbf{V}^{(B)}$ is a *Boolean valued model* for φ" instead of $\mathbf{V}^{(B)}\models\varphi$, etc.

2.1.7. The above concept of interpretation makes it possible to judge the elements of $\mathbf{V}^{(B)}$. More convenient for this purpose appears however to be a somewhat richer language than the original language of set theory.

Namely, the alphabet of the new language, the *B-language* for short, contains an extra constant for each member of $\mathbf{V}^{(B)}$. As usual, the elements of $\mathbf{V}^{(B)}$ are identified with the corresponding symbols of constants. We call the formulas and sentences of the B-language *B-formulas* and *B-sentences*. In this event each B-formula (B-sentence) results from inserting values of $\mathbf{V}^{(B)}$ in place of some (respectively, all) free variables in a formula of set theory.

We now inspect the simplifications due to the B-language in the definitions of Boolean truth values in 2.1.6. Namely, the Boolean truth value of a B-sentence may be obtained by putting

$$[\![\sigma \wedge \tau]\!] := [\![\sigma]\!] \wedge [\![\tau]\!],$$
$$[\![\sigma \vee \tau]\!] := [\![\sigma]\!] \vee [\![\tau]\!],$$
$$[\![\neg\sigma]\!] := [\![\sigma]\!]^*,$$
$$[\![\sigma \to \tau]\!] := [\![\sigma]\!] \Rightarrow [\![\tau]\!],$$
$$[\![(\forall\, x)\varphi(x)]\!] := \bigwedge\{[\![\varphi(u)]\!] : u \in \mathbf{V}^{(B)}\},$$
$$[\![(\exists\, x)\varphi(x)]\!] := \bigvee\{[\![\varphi(u)]\!] : u \in \mathbf{V}^{(B)}\},$$

where σ and τ are arbitrary B-sentences, while φ is a B-formula with a single free variable x.

A B-sentence σ is *true* inside $\mathbf{V}^{(B)}$ if $[\![\sigma]\!] = \mathbf{1}_B$, and we write $\mathbf{V}^{(B)} \models \sigma$. Without further specification, we apply both linguistic meanings of *verity* in $\mathbf{V}^{(B)}$, i.e., those given in 2.1.6 and 2.1.7. We also use the same letters for denoting the variables of B-language and the members of $\mathbf{V}^{(B)}$.

If several Boolean algebras, $B, C, \dots$, are considered simultaneously and there is a necessity to distinguish between their details then, alongside with $[\![\varphi]\!]$, we write $[\![\varphi]\!]^B$, $[\![\varphi]\!]^C$, etc.

2.1.8. Theorem. *If a $\varphi(u_1, \dots, u_n)$ is provable in predicate calculus then $\mathbf{V}^{(B)} \models \varphi(x_1, \dots, x_n)$ for all $x_1, \dots, x_n \in \mathbf{V}^{(B)}$. In particular, for $x, y, z \in \mathbf{V}^{(B)}$, the following hold:*

(1) $[\![x = x]\!] = \mathbf{1}$;
(2) $x(y) \le [\![y \in x]\!]$ *for all* $y \in \mathrm{dom}(x)$;
(3) $[\![x = y]\!] = [\![y = x]\!]$;
(4) $[\![x = y]\!] \wedge [\![y = z]\!] \le [\![x = z]\!]$;
(5) $[\![x \in y]\!] \wedge [\![x = z]\!] \le [\![z \in y]\!]$;
(6) $[\![y \in x]\!] \wedge [\![x = z]\!] \le [\![y \in z]\!]$;
(7) $[\![x = y]\!] \wedge [\![\varphi(x)]\!] \le [\![\varphi(y)]\!]$ *for every B-formula φ.*

$\lhd$ It is an easy matter to show that the axioms of predicate calculus hold inside $\mathbf{V}^{(B)}$ and the rules of inference preserve satisfaction. Strictly speaking, if some formulas $\varphi_1, \dots, \varphi_n$ imply φ in predicate calculus then $[\![\varphi_1]\!] \wedge \dots \wedge [\![\varphi_n]\!] \le [\![\varphi]\!]$.

We now prove (1)–(7).

(1) This is established by induction on the well founded relation $y \in \mathrm{dom}(x)$.

By way of induction, assume that $[\![y = y]\!] = \mathbf{1}$ for all $y \in \mathrm{dom}(x)$. Using 2.1.4 (1), obtain

$$[\![y \in x]\!] = \bigvee_{t \in \mathrm{dom}(x)} x(t) \wedge [\![t = y]\!] \ge x(y) \wedge [\![y = y]\!] \ge x(y),$$

and so, by 1.1.4 (4),

$$[\![x = x]\!] = \bigwedge_{y \in \mathrm{dom}(x)} x(y) \Rightarrow [\![y \in x]\!] = \mathbf{1}.$$

(2) Considering 2.1.4 (1) and (1), for $y \in \mathrm{dom}(x)$ we find

$$[\![y \in x]\!] \geq x(y) \wedge [\![y = y]\!] = x(y).$$

(3) This is immediate from the definitions because of the symmetry of the formula 2.1.4 (2) defining the Boolean truth value of equality.

Items (4)–(6) are demonstrated by simultaneous induction.

Denote by $\rho(x, y, z) := (\alpha, \beta, \gamma) \in \mathrm{On}^3$ the permutation of the 3-tuple of ordinals $\rho(x)$, $\rho(y)$, and $\rho(z)$ such that $\alpha \geq \beta \geq \gamma$. (The class On^3 is furnished with the canonical well-ordering of 1.4.15.)

Take $x, y, z \in \mathbf{V}^{(B)}$ and assume that inequalities (4)–(6) are true for all $u, v, w \in \mathbf{V}^{(B)}$ if $\rho(u, v, w) < \rho(x, y, z)$.

We justify the induction step by cases.

(4) Consider $t \in \mathrm{dom}(x)$. Since $[\![x = y]\!] \leq x(t) \Rightarrow [\![t \in y]\!]$, from 1.1.4 (3) it follows that

$$x(t) \wedge [\![x = y]\!] \leq [\![t \in y]\!],$$
$$x(t) \wedge [\![x = y]\!] \wedge [\![y = z]\!] \leq [\![t \in y]\!] \wedge [\![y = z]\!].$$

On observing that $\rho(t, y, z) < \rho(x, y, z)$ and applying the induction hypothesis for (6), find

$$[\![t \in y]\!] \wedge [\![y = z]\!] \leq [\![t \in z]\!],$$
$$x(t) \wedge [\![y = x]\!] \wedge [\![y = z]\!] \leq [\![t \in z]\!].$$

Again use 1.1.4 (3) to obtain

$$[\![x = y]\!] \wedge [\![y = z]\!] \leq x(t) \Rightarrow [\![t \in z]\!],$$

implying

$$[\![x = y]\!] \wedge [\![y = z]\!] \leq \bigwedge_{t \in \mathrm{dom}(x)} x(t) \Rightarrow [\![t \in z]\!].$$

Analogously,

$$[\![x = y]\!] \wedge [\![y = z]\!] \leq \bigwedge_{t \in \mathrm{dom}(z)} z(t) \Rightarrow [\![t \in x]\!].$$

By 2.1.4 (2), conclude that $[\![x = y]\!] \wedge [\![y = z]\!] \leq [\![x = z]\!]$.

(5) Take $t \in \operatorname{dom}(y)$. Clearly, $\rho(t, x, z) < \rho(x, y, z)$ and so, by the induction hypothesis for (4), infer

$$y(t) \wedge [\![t = x]\!] \wedge [\![x = z]\!] \leq y(t) \wedge [\![t = z]\!] \leq [\![z \in y]\!].$$

By 1.1.5 (2), this gives

$$[\![x = z]\!] \wedge \bigvee_{t \in \operatorname{dom}(y)} y(t) \wedge [\![t = x]\!] \leq [\![z \in y]\!],$$

or $[\![x = z]\!] \wedge [\![x \in y]\!] \leq [\![z \in y]\!]$.

(6) Take $t \in \operatorname{dom}(x)$ again to obtain

$$x(t) \wedge [\![x = z]\!] \leq [\![t \in z]\!],$$
$$[\![t = y]\!] \wedge x(t) \wedge [\![x = z]\!] \leq [\![t = y]\!] \wedge [\![t \in z]\!].$$

Since $\rho(t, y, z) < \rho(x, y, z)$, we may use the induction hypothesis for (5) and 1.1.5 (2) to derive

$$x(t) \wedge [\![x = z]\!] \wedge [\![t = y]\!] \leq [\![y \in z]\!],$$
$$[\![x = z]\!] \wedge \bigvee_{t \in \operatorname{dom}(x)} x(t) \wedge [\![t = y]\!] \leq [\![y \in z]\!].$$

Therefore, by 2.1.4 (1), $[\![x = z]\!] \wedge [\![y \in x]\!] \leq [\![y \in z]\!]$.

(7) This is proved by induction on the length of a formula on using the already-established properties. $\rhd$

As a corollary to Theorem 2.1.8 appear the following rules for calculating the Boolean truth values of bounded formulas.

2.1.9. *For every B-formula φ with a single free variable x and for every $u \in \mathbf{V}^{(B)}$ the following hold:*

$$[\![(\exists\, x \in u)\varphi(x)]\!] = \bigvee_{v \in \operatorname{dom}(u)} u(v) \wedge [\![\varphi(v)]\!],$$

$$[\![(\forall\, x \in u)\varphi(x)]\!] = \bigwedge_{v \in \operatorname{dom}(u)} u(v) \Rightarrow [\![\varphi(v)]\!].$$

$\lhd$ These claims are mutually dual. In other words, replacing φ with $\neg\varphi$ and applying the De Morgan laws, we transform one of the sought formula into the other. So, it suffices to prove either of the claims, say, the first.

By 2.1.8 (2),

$$[\![(\exists x \in u)\varphi(x)]\!] \geq \bigvee_{v \in \operatorname{dom}(u)} u(v) \wedge [\![\varphi(v)]\!].$$

On the other hand, by 2.1.4 (1) and 2.1.8 (7),

$$\begin{aligned}[\![(\exists x \in u)\varphi(x)]\!] &= \bigvee_{t \in \mathbf{V}^{(B)}} \bigvee_{v \in \operatorname{dom}(u)} u(v) \wedge [\![t = v]\!] \wedge [\![\varphi(t)]\!] \\ &\leq \bigvee_{v \in \operatorname{dom}(u)} u(v) \wedge [\![\varphi(v)]\!],\end{aligned}$$

which completes the proof. ▷

2.1.10. COMMENTS.

(1) Given a particular formula φ of set theory, $u_1, \dots, u_n \in \mathbf{V}^{(B)}$, and $b \in B$, we see that the expression $[\![\varphi(u_1, \dots, u_n)]\!] = b$ is again a formula of set theory.

In ZFC, however, the mapping $\varphi \mapsto [\![\varphi]\!]$ is not a definable class, admitting only a metalinguistic definition. That is why we call it a "metafunction."

(2) Boolean valued universes are used in proving relative consistency of set-theoretic propositions as follows.

Suppose that some theories $\mathscr{T}$ and $\mathscr{T}'$ are extensions of ZF such that consistency of ZF implies that of $\mathscr{T}'$. Assume further that we may define a Boolean algebra B so that $\mathscr{T}' \models$ "B is a complete Boolean algebra" and $\mathscr{T}' \models [\![\varphi]\!]^B = \mathbf{1}$ for every axiom φ of the theory $\mathscr{T}$. In this case the consistency of ZF implies that of $\mathscr{T}$ (see [11, 84, 209, 241]).

(3) Let Ω be a complete Heyting lattice (see 1.1.8 (3)). Define the *pseudocomplement* b^* of an element $b \in \Omega$ by the formula $x^* := x \Rightarrow \mathbf{0}$, with $\Rightarrow$ standing for the relative pseudocomplementation of Ω. Slightly changing the formulas of 2.1.4, define the truth values $[\![\cdot \in \cdot]\!]^{\Omega}$ and $[\![\cdot = \cdot]\!]^{\Omega}$ which act from $\mathbf{V}^{(\Omega)} \times \mathbf{V}^{(\Omega)}$ to Ω. Understanding verity in $\mathbf{V}^{(\Omega)}$ in the same manner as in 2.1.6, it is possible to show that in this event all theorems of intuitionistic predicate calculus hold inside $\mathbf{V}^{(\Omega)}$ (cf. [54, 70, 238, 239]).

2.2. Transformations of a Boolean Valued Universe

Each homomorphism of a Boolean algebra B induces a transformation of the Boolean valued universe $\mathbf{V}^{(B)}$. The topic to be discussed in this section is the behavior of these transformations and, in particular, the manner in which they change the Boolean truth value of a formula.

2.2.1. Assume that π is a homomorphism of B to a complete Boolean algebra C. By recursion on a well founded relation $y \in \operatorname{dom}(x)$, we define the mapping $\pi^* : \mathbf{V}^{(B)} \to \mathbf{V}^{(C)}$ using the formulas $\operatorname{dom}(\pi^* x) := \{\pi^* y : y \in \operatorname{dom}(x)\}$ and

$$\pi^* x : v \mapsto \bigvee \{\pi(x(z)) : z \in \operatorname{dom}(x),\ \pi^* z = v\}.$$

If π is injective then π^* is also injective. Moreover,

$$\pi^* x : \pi^* y \mapsto \pi(x(y)) \quad (y \in \operatorname{dom}(x)).$$

$\triangleleft$ It is fully enough to establish that the restriction of π^* to $V_\alpha^{(B)}$ is injective for an arbitrary ordinal α. To this end, assume that the claim holds for all $\beta < \alpha$. Let $x, y \in V_\alpha^{(B)}$ be such that $\pi^* x = \pi^* y$. In this event $\pi^* x : \pi^* z \mapsto \pi(x(z))$ $(z \in \operatorname{dom}(x))$ and $\pi^* y : \pi^* z \mapsto \pi(y(z))$ $(z \in \operatorname{dom}(y))$. Therefore, we come to the equality

$$\{(\pi^* z, \pi(x(z))) : z \in \operatorname{dom}(x)\} = \{(\pi^* u, \pi(y(u))) : u \in \operatorname{dom}(y)\}.$$

Since the sets $\operatorname{dom}(x)$ and $\operatorname{dom}(y)$ lie in $V_\beta^{(B)}$ for some $\beta < \alpha$; therefore, π^* is injective on each of these sets. Since π is injective, obtain

$$\{(z, x(z)) : z \in \operatorname{dom}(x)\} = \{(u, y(u)) : u \in \operatorname{dom}(y)\},$$

or, which is the same, $x = y$. $\triangleright$

A homomorphism $\pi : B \to C$ is *complete* if $\pi\left(\bigvee M\right) = \bigvee \pi(M)$ for every set $M \subset B$.

Throughout this section π is a complete homomorphism from B to a complete Boolean algebra C.

2.2.2. Theorem. *The following hold:*

(1) *If ρ is a complete homomorphism of C to a complete Boolean algebra D then $(\rho \circ \pi)^* = \rho^* \circ \pi^*$;*

(2) *If a homomorphism π is injective (surjective) then the mapping π^* is injective (respectively, surjective);*

(3) *$[\![\pi^* x = \pi^* y]\!]^C = \pi([\![x = y]\!]^B)$ and $[\![\pi^* x \in \pi^* y]\!]^C = \pi([\![x \in y]\!]^B)$ for all x and $y \in \mathbf{V}^{(B)}$;*

(4) *$[\![t \in \pi^* x]\!]^C = \bigvee_{u \in \mathbf{V}^{(B)}} \pi([\![u = x]\!]^B) \wedge [\![t \in \pi^* u]\!]^C$ for all $x \in \mathbf{V}^{(B)}$ and $t \in \mathbf{V}^{(C)}$.*

$\triangleleft$ (1) Assume that $(\rho \circ \pi)^* y = (\rho^* \circ \pi^*) y$ for all $y \in \operatorname{dom}(x)$. Then, for $u := (\rho \circ \pi)^* y$ with $y \in \operatorname{dom}(x)$, deduce (cf. 1.1.5 (9)):

$$
\begin{aligned}
&((\rho \circ \pi)^* x)u \\
&= \bigvee \{(\rho \circ \pi)(x(z)) : z \in \operatorname{dom}(x) \wedge (\rho^* \circ \pi^*)z = (\rho^* \circ \pi^*)y\} \\
&= \bigvee \{\rho \left(\bigvee \{\pi(x(z)) : z \in \operatorname{dom}(x),\ \pi^* z = v\} \right) : v \in \operatorname{dom}(\pi^* x),\ \rho^* v = (\rho^* \circ \pi^*)y\} \\
&= \bigvee \{\rho((\pi^* x)(v)) : v \in \operatorname{dom}(\pi^* x),\ \rho^* v = \rho^*(\pi^* y)\} \\
&= (\rho^*(\pi^* x))(\rho^*(\pi^* y)) = ((\rho^* \circ \pi^*)x)u.
\end{aligned}
$$

Therefore, $(\rho \circ \pi)^* x = \rho^*(\pi^* x)$, and the sought result follows from 2.1.3.

(2) The case of an injective π was settled in 2.2.1. Assume now that π is a surjective mapping. In this case there are a principal ideal B_0 of a Boolean algebra B and a surjection $\rho : C \to B_0$ such that ρ^{-1} coincides with the restriction π_0 of π to B_0. If $x \in \mathbf{V}^{(C)}$ then, by (1), $x = I_C^* x = (\pi_0 \circ \rho)^* x = \pi_0^*(\rho^* x) \in \operatorname{im}(\pi_0^*)$. Hence, π_0^* sends $\mathbf{V}^{(B_0)}$ onto $\mathbf{V}^{(C)}$. Note also that $\mathbf{V}^{(B_0)} \subset \mathbf{V}^{(B)}$ and the restriction of π^* to $\mathbf{V}^{(B_0)}$ coincides with π_0^*.

(3) Proceed by induction on $(\rho(x), \rho(y))$, assuming that the class $\mathrm{On} \times \mathrm{On}$ is canonically well ordered (cf. 1.4.15).

Suppose that the formulas in question are fulfilled for all $u, v \in \mathbf{V}^{(B)}$ provided that $(\rho(u), \rho(v)) < (\rho(x), \rho(y))$.

Obviously, $\max\{(\rho(z), \rho(x)), (\rho(z), \rho(y))\} < (\rho(x), \rho(y))$ if $z \in \operatorname{dom}(x)$ or $z \in \operatorname{dom}(y)$. Hence, the following hold (cf. 1.1.5 (2, 9)):

$$
\begin{aligned}
&[\![\pi^* x \in \pi^* y]\!] \\
&= \bigvee_{t \in \operatorname{dom}(\pi^* y)} (\pi^* y)(t) \wedge [\![t = \pi^* x]\!] = \bigvee_{z \in \operatorname{dom}(y)} (\pi^* y)(\pi^* z) \wedge [\![\pi^* z = \pi^* x]\!] \\
&= \bigvee_{z \in \operatorname{dom}(y)} \left(\bigvee \{\pi(y(u)) : u \in \operatorname{dom}(y),\ \pi^* u = \pi^* z\} \right) \wedge [\![\pi^* z = \pi^* x]\!] \\
&= \bigvee_{z \in \operatorname{dom}(y)} \bigvee \{\pi(y(u)) \wedge [\![\pi^* z = \pi^* x]\!] : u \in \operatorname{dom}(y),\ \pi^* u = \pi^* z\} \\
&= \bigvee_{u \in \operatorname{dom}(y)} \pi(y(u)) \wedge \pi([\![u = x]\!]) = \pi\left(\bigvee_{u \in \operatorname{dom}(y)} y(u) \wedge [\![u = x]\!] \right) \\
&= \pi([\![x \in y]\!]).
\end{aligned}
$$

Note that analogous calculations come through in the case of the Boolean truth value of equality (on successively applying 2.1.4 (1), 2.2.1, 1.1.5 (10), and 2.1.4 (2)):

$$
[\![\pi^* x = \pi^* y]\!]
$$

$$= \bigwedge_{t\in\mathrm{dom}(\pi^* y)} (\pi^* y)(t) \Rightarrow [\![t \in \pi^* x]\!] \wedge \bigwedge_{t\in\mathrm{dom}(\pi^* x)} (\pi^* x)(t) \Rightarrow [\![t \in \pi^* y]\!]$$

$$= \bigwedge_{z\in\mathrm{dom}(y)} (\pi^* y)(\pi^* z) \Rightarrow [\![\pi^* z \in \pi^* x]\!]$$

$$\wedge \bigwedge_{z\in\mathrm{dom}(x)} (\pi^* x)(\pi^* z) \Rightarrow [\![\pi^* z \in \pi^* y]\!]$$

$$= \bigwedge_{z\in\mathrm{dom}(y)} \bigwedge\{\pi(y(u)) \Rightarrow \pi([\![u \in x]\!]) : u \in \mathrm{dom}(y),\ \pi^* u = \pi^* z\}$$

$$\wedge \bigwedge_{z\in\mathrm{dom}(x)} \bigwedge\{\pi(x(u)) \Rightarrow \pi([\![u \in y]\!]) : u \in \mathrm{dom}(x),\ \pi^* u = \pi^* z\}$$

$$= \bigwedge_{u\in\mathrm{dom}(x)} \pi(x(u) \Rightarrow [\![u \in y]\!]) \wedge \bigwedge_{u\in\mathrm{dom}(y)} \pi(y(u) \Rightarrow [\![u \in x]\!]) = \pi([\![x = y]\!]).$$

(4) By (3) and 2.1.8 (4), the following estimates hold for $x \in \mathbf{V}^{(B)}$ and $t \in \mathbf{V}^{(C)}$:

$$[\![t \in \pi^* x]\!]$$

$$= \bigvee_{s\in\mathrm{dom}(\pi^* x)} (\pi^* x)(s) \wedge [\![s = t]\!] = \bigvee_{u\in\mathrm{dom}(x)} (\pi^* x)(\pi^* u) \wedge [\![\pi^* u = t]\!]$$

$$\leq \bigvee_{u\in\mathbf{V}^{(B)}} \pi([\![u = x]\!]) \wedge [\![\pi^* u = t]\!]$$

$$= \bigvee_{u\in\mathbf{V}^{(B)}} [\![\pi^* u = \pi^* x]\!] \wedge [\![\pi^* u = t]\!] \leq [\![t \in \pi^* x]\!],$$

which completes the proof. $\triangleright$

2.2.3. Theorem. *Let $\varphi(x_1, \dots, x_n)$ be a formula of ZFC, $u_1, \dots, u_n \in \mathbf{V}^{(B)}$, and π be a complete homomorphism from B to C. Then the following hold:*

(1) *If φ is a formula of class Σ_1 and π is arbitrary then*

$$\pi([\![\varphi(u_1, \dots, u_n)]\!]^B) \leq [\![\varphi(\pi^* u_1, \dots, \pi^* u_n]\!]^C;$$

(2) *If φ is a bounded formula and π is arbitrary, or π is an epimorphism and φ is arbitrary; then*

$$\pi([\![\varphi(u_1, \dots, u_n)]\!]^B) = [\![\varphi(\pi^* u_1, \dots, \pi^* u_n)]\!]^C.$$

$\triangleleft$ The claim for atomic formulas ensues from 2.2.2.

We settle the general case by induction on the length of φ. A nontrivial step occurs only when φ has the form $(\exists\, x)\varphi_0$ or $(\forall\, x)\varphi_0$. It is in this case that we need the additional hypotheses about φ and π.

(1) Assume that we encounter a bounded universal quantifier in the induction step; i.e., if φ has the form $(\forall x \in u)\varphi_0(x, u_1, \dots, u_n)$. Then, recalling 1.1.5 (3, 10), proceed with the following chain of equalities:

$$\begin{aligned}
&[\![\varphi(\pi^* u, \pi^* u_1, \dots, \pi^* u_n)]\!] \\
&= \bigwedge_{x \in \mathrm{dom}(\pi^* u)} (\pi^* u)(x) \Rightarrow [\![\varphi_0(x, \pi^* u_1, \dots, \pi^* u_n)]\!] \\
&= \bigwedge_{x \in \mathrm{dom}(u)} (\pi^* u)(\pi^* x) \Rightarrow [\![\varphi_0(\pi^* x, \pi^* u_1, \dots, \pi^* u_n)]\!] \\
&= \bigwedge_{x \in \mathrm{dom}(u)} \bigwedge \{\pi(u(z)) \Rightarrow [\![\varphi_0(\pi^* x, \pi^* u_1, \dots, \pi^* u_n)]\!] : z \in \mathrm{dom}(u),\ \pi^* z = \pi^* x\} \\
&= \bigwedge_{x \in \mathrm{dom}(u)} \pi(u(x) \Rightarrow [\![\varphi_0(x, u_1, \dots, u_n)]\!]) \\
&= \pi[\![(\forall\, x \in u)\varphi_0(x, u_1, \dots, u_n)]\!] = \pi[\![\varphi(u, u_1, \dots, u_n)]\!].
\end{aligned}$$

Furthermore, in the case of an unbounded existential quantifier we immediately deduce from definitions that

$$\begin{aligned}
&[\![(\exists\, x)\varphi_0(x, \pi^* u_1, \dots, \pi^* u_n)]\!] \\
&\geq \bigvee \{[\![\varphi_0(x, \pi^* u_1, \dots, \pi^* u_n)]\!] : x \in \mathrm{im}(\pi^*)\} \\
&= \bigvee \{[\![\varphi_0(\pi^* u, \pi^* u_1, \dots, \pi^* u_n)]\!] : u \in \mathbf{V}^{(B)}\} \\
&= \bigvee \{\pi([\![\varphi_0(u, u_1, \dots, u_n)]\!]) : u \in \mathbf{V}^{(B)}\} = \pi([\![(\exists\, x)\varphi_0(x, u_1, \dots, u_n)]\!]).
\end{aligned}$$

(2) Note first of all that if π is a surjection then π^* is a surjection too; i.e., $\mathrm{im}(\pi^*) = \mathbf{V}^{(C)}$ (cf. 2.2.2 (2)). Therefore, considering the formula $\varphi := (\exists\, x)\varphi_0$, obtain

$$\begin{aligned}
&[\![\varphi(\pi^* u_1, \dots, \pi^* u_n)]\!] \\
&= \vee \{[\![\varphi_0(x, \pi^* u_1, \dots, \pi^* u_n)]\!] : x \in \mathbf{V}^{(C)} = \mathrm{im}(\pi^*)\} \\
&= \vee \{[\![\varphi_0(\pi^* u, \pi^* u_1, \dots, \pi^* u_n)]\!] : u \in \mathbf{V}^{(B)}\} \\
&= \vee \{\pi([\![\varphi_0(u, u_1, \dots, u_n)]\!]) : u \in \mathbf{V}^{(B)}\} = \pi([\![\varphi(u_1, \dots, u_n)]\!]).
\end{aligned}$$

The same arguments apply to each formula φ of type $(\forall\, x)\varphi_0(x, u_1, \dots, u_n)$.

If the existential quantifier under study is bounded, i.e., if $\varphi(u_1,\dots,u_n)$ has the form $(\exists\, x \in u)\varphi_0(x,u_1,\dots,u_n)$ with $u,u_1,\dots,u_n \in \mathbf{V}^{(B)}$; then (see the definitions and 1.1.5 (2, 9)) we may proceed as follows:

$$\begin{aligned}
&[\![\varphi(\pi^*u,\pi^*u_1,\dots,\pi^*u_n)]\!]\\
&= \bigvee_{x\in\mathrm{dom}(\pi^*u)} (\pi^*u)(x)\wedge[\![\varphi_0(x,\pi^*u_1,\dots,\pi^*u_n)]\!]\\
&= \bigvee_{x\in\mathrm{dom}(u)} (\pi^*u)(\pi^*x)\wedge[\![\varphi_0(\pi^*x,\pi^*u_1,\dots,\pi^*u_n)]\!]\\
&= \bigvee_{x\in\mathrm{dom}(u)} \bigvee\{\pi(u(z))\wedge[\![\varphi_0(\pi^*x,\pi^*u_1,\dots,\pi^*u_n)]\!] : z\in\mathrm{dom}(u),\ \pi^*z=\pi^*x\}\\
&= \bigvee_{z\in\mathrm{dom}(u)} \pi\,(u(z)\wedge[\![\varphi_0(z,u_1,\dots,u_n)]\!]) = \pi([\![\varphi(u,u_1,\dots,u_n)]\!]).
\end{aligned}$$

The case of a bounded universal quantifier was settled earlier. $\rhd$

2.2.4. Corollary. *Assume that π, φ, and $u_1,\dots,u_n$ are the same as in 2.2.3. Assume further that one of the following conditions is fulfilled:*

(1) *$\varphi(x_1,\dots,x_n)$ is a formula of class Σ_1 and π is arbitrary;*

(2) *π is an epimorphism and $\varphi(x_1,\dots,x_n)$ is arbitrary.*

Then

$$\mathbf{V}^{(B)} \models \varphi(u_1,\dots,u_n) \to \mathbf{V}^{(C)} \models \varphi(\pi^*u_1,\dots,\pi^*u_n).$$

2.2.5. Corollary. *Assume that π, φ, and $u_1,\dots,u_n$ are the same as in 2.2.3. Assume further that one of the following conditions is fulfilled:*

(1) *φ is bounded and π is a monomorphism;*

(2) *π is an isomorphism and φ is arbitrary.*

Then

$$\mathbf{V}^{(B)} \models \varphi(u_1,\dots,u_n) \leftrightarrow \mathbf{V}^{(C)} \models \varphi(\pi^*u_1,\dots,\pi^*u_n).$$

2.2.6. We now consider an important particular case of the situation under study.

Let B_0 be a regular subalgebra of a complete Boolean algebra B. This implies that B_0 is a complete subalgebra, and the supremum and infimum of every subset of B_0 are the same irrespectively of whether they are calculated in B_0 or B. In these circumstances $\mathbf{V}^{(B_0)} \subset \mathbf{V}^{(B)}$. Moreover, denoting by $\imath$ the identical embedding of B_0 in B, we then see that $\imath^*$ is an embedding of $\mathbf{V}^{(B_0)}$ to $\mathbf{V}^{(B)}$.

If $\varphi(x_1,\dots,x_n)$ is a bounded formula and $u_1,\dots,u_n \in \mathbf{V}^{(B_0)}$ then it follows from 2.2.5 (1) that

$$\mathbf{V}^{(B_0)} \models \varphi(u_1,\dots,u_n) \leftrightarrow \mathbf{V}^{(B)} \models \varphi(u_1,\dots,u_n).$$

Since the two-valued algebra $\mathbf{2} := \{\mathbf{0}, \mathbf{1}\}$ may be viewed as a regular subalgebra of the Boolean algebra B; therefore, the above applies to the universe $\mathbf{V}^{(\mathbf{2})}$.

We shall see in the sequel that $\mathbf{V}^{(\mathbf{2})}$ is naturally isomorphic to the von Neumann universe $\mathbf{V}$.

2.2.7. Given an arbitrary $x \in \mathbf{V}$, define the element $x^\wedge \in \mathbf{V}^{(\mathbf{2})} \subset \mathbf{V}^{(B)}$ by recursion on the well founded relation $y \in x$. To this end, put

$$\operatorname{dom}(x^\wedge) := \{y^\wedge : y \in x\}, \quad \operatorname{im}(x^\wedge) := \{\mathbf{1}_B\}.$$

From 2.2.2 (3) it follows for $x, y \in V$ that

$$[\![x^\wedge \in y^\wedge]\!]^B \in \mathbf{2}, \quad [\![x^\wedge = y^\wedge]\!]^B \in \mathbf{2}.$$

The mapping $x \mapsto x^\wedge$ $(x \in \mathbf{V})$ is the *canonical embedding* of $\mathbf{V}$ into the Boolean valued universe $\mathbf{V}^{(B)}$. The elements of $\mathbf{V}^{(B)}$ of the form $x^\wedge$ with $x \in \mathbf{V}$ are *standard*. Sometimes we call $x^\wedge$ the *standard name* of x in $\mathbf{V}^{(B)}$.

2.2.8. Theorem. *The following hold:*

(1) *If $x \in \mathbf{V}$ and $y \in \mathbf{V}^{(B)}$ then*

$$[\![y \in x^\wedge]\!] = \bigvee \{[\![y = u^\wedge]\!] : u \in x\};$$

(2) *If $x, y \in \mathbf{V}$ then*

$$x \in y \leftrightarrow \mathbf{V}^{(B)} \models x^\wedge \in y^\wedge, \quad x = y \leftrightarrow \mathbf{V}^{(B)} \models x^\wedge = y^\wedge;$$

(3) *The mapping $x \mapsto x^\wedge$ is injective;*

(4) *To each $y \in \mathbf{V}^{(\mathbf{2})}$ there is a unique element $x \in \mathbf{V}$ such that $\mathbf{V}^{(B)} \models x^\wedge = y$;*

(5) *If π is a complete homomorphism from B to C then $\pi^* x^\wedge = x^{\overset{\wedge}{\wedge}}$, where $x \in \mathbf{V}$ and $(\,\cdot\,)^{\overset{\wedge}{\wedge}}$ is the canonical embedding of $\mathbf{V}$ to $\mathbf{V}^{(C)}$.*

$\lhd$ (1) Straightforward calculation, together with the definitions of 2.1.4 and 2.2.7, gives

$$[\![y \in x^\wedge]\!] = \bigvee_{t \in \operatorname{dom}(x^\wedge)} x^\wedge(t) \wedge [\![t = y]\!]$$

$$= \bigvee_{t \in x} x^\wedge(t^\wedge) \wedge [\![t^\wedge = y]\!] = \bigvee_{t \in x} [\![t^\wedge = y]\!].$$

(2) Assume that, for all $z \in \mathbf{V}$ with $\operatorname{rank}(z) < \operatorname{rank}(y)$, the following hold:

$$(\forall\, x)(x \in z \leftrightarrow [\![x^\wedge \in z^\wedge]\!] = \mathbf{1}),$$
$$(\forall\, x)(x = z \leftrightarrow [\![x^\wedge = z^\wedge]\!] = \mathbf{1}),$$
$$(\forall\, x)(z \in x \leftrightarrow [\![z^\wedge \in x^\wedge]\!] = \mathbf{1}).$$

By (1), $[\![x^\wedge \in y^\wedge]\!] = \bigvee\{[\![t^\wedge = x^\wedge]\!] : t \in y\}$. Since $\operatorname{rank}(t) < \operatorname{rank}(y)$ for $t \in y$, by the induction hypothesis we conclude that $[\![x^\wedge \in y^\wedge]\!] = \mathbf{1}$ if and only if $[\![t^\wedge = x^\wedge]\!] = \mathbf{1}$ or $t = x$ for some $t \in y$.

By definition, we then have

$$[\![x^\wedge = y^\wedge]\!] = \bigwedge_{t \in x} [\![t^\wedge \in y^\wedge]\!] \wedge \bigwedge_{s \in y} [\![s^\wedge \in x^\wedge]\!]$$

and $\operatorname{rank}(s) < \operatorname{rank}(y)$ for $s \in y$. Therefore, from the above and the induction hypothesis, deduce that the right-hand side of the last equality is equal to unity if and only if $t \in y$ for all $t \in x$ and $s \in x$ for all $s \in y$; i.e., if $x = y$. Appealing to (1) again, obtain

$$[\![y^\wedge \in x^\wedge]\!] = \bigvee\{[\![y^\wedge = t^\wedge]\!] : t \in x\}.$$

Hence, $[\![y^\wedge \in x^\wedge]\!] = \mathbf{1}$ holds only if $[\![y^\wedge = t^\wedge]\!] = \mathbf{1}$ for some $t \in x$. The last proposition amounts, by virtue of the above, to the proposition $(\exists\, t \in x)(t = y)$ stating the membership $y \in x$.

(3) This ensues from (2).

(4) Assume that $y \in \mathbf{V}^{(\mathbf{2})}$ and to each $t \in \operatorname{dom}(y)$ there is an element u in $\mathbf{V}$ such that $[\![t = u^\wedge]\!] = \mathbf{1}$. Define $x \in \mathbf{V}$ by the equality

$$x := \{u \in \mathbf{V} : (\exists\, t \in \operatorname{dom}(y))(y(t) = \mathbf{1} \wedge [\![u^\wedge = t]\!] = \mathbf{1})\}.$$

Granted $u \in x$, obtain

$$[\![u^\wedge \in y]\!] = \bigvee_{t \in \operatorname{dom}(y)} y(t) \wedge [\![t = u^\wedge]\!] = \mathbf{1}.$$

Moreover, using the induction hypothesis, deduce for $t \in \operatorname{dom}(y)$ that

$$y(t) \leq [\![t \in x^\wedge]\!] = \bigvee_{u \in x} [\![t = u^\wedge]\!].$$

Summarizing the above, conclude

$$[\![x^\wedge = y]\!] = \bigwedge_{t \in \operatorname{dom}(y)} y(t) \Rightarrow [\![t \in x^\wedge]\!] \wedge \bigwedge_{u \in x} [\![u^\wedge \in y]\!] = \mathbf{1}.$$

(5) Proceed by induction on the well founded relation $y \in x$.

Assume that $(\forall y \in x)(\pi^* y^\wedge = y^{\hat{\wedge}})$. Then

$$\mathrm{dom}(\pi^* x^\wedge) = \{y^{\hat{\wedge}} : y \in x\} = \mathrm{dom}(x^{\hat{\wedge}}).$$

Therefore, granted $y \in x$, observe

$$(\pi^* x^\wedge)(y^{\hat{\wedge}}) = (\pi^* x^\wedge)(\pi^* y^\wedge)$$
$$= \bigvee \{\pi(x^\wedge(y^\wedge)) : z \in \mathrm{dom}(x^\wedge) \wedge \pi^* z = \pi^* y^\wedge\}$$
$$\geq \pi(x^\wedge(y^\wedge)) = \mathbf{1}_B = x^{\hat{\wedge}}(y^{\hat{\wedge}}).$$

Finally, $\pi^* x^\wedge = x^{\hat{\wedge}}$, which justifies the induction step. $\triangleright$

2.2.9. *Suppose that* $u_1, \dots, u_n \in \mathbf{V}$, *and* $\varphi(x_1, \dots, x_n)$ *is a formula of* ZFC. *Then*

(1) $\varphi(u_1, \dots, u_n) \leftrightarrow \mathbf{V}^{(2)} \models \varphi(u_1^\wedge, \dots, u_n^\wedge)$;

(2) *If* φ *is a bounded formula then*

$$\varphi(u_1, \dots, u_n) \leftrightarrow \mathbf{V}^{(B)} \models \varphi(u_1^\wedge, \dots, u_n^\wedge);$$

(3) *If* φ *is a formula of class* Σ_1 *then*

$$\varphi(u_1, \dots, u_n) \rightarrow \mathbf{V}^{(B)} \models \varphi(u_1^\wedge, \dots, u_n^\wedge).$$

$\triangleleft$ Note that only (1) has to be proven, since both (2) and (3) ensue from (1), 2.2.4 (1) and 2.2.5 (1).

Considering atomic formulas, find that (1) is a consequence of 2.2.8 (2). Induction on the length of φ is nontrivial only when we encounter a new existential quantifier. We thus assume that φ has the form $(\exists\, x)\psi(x, u_1, \dots, u_n)$ and $[\![\varphi(u_1^\wedge, \dots, u_n^\wedge)]\!] = \mathbf{1}$, with (1) holding for ψ. Then

$$\mathbf{1} = \bigvee \{[\![\psi(u, u_1^\wedge, \dots, u_n^\wedge)]\!]^{\mathbf{2}} : u \in \mathbf{V}^{(2)}\}.$$

Therefore, $[\![\psi(v, u_1^\wedge, \dots, u_n^\wedge)]\!] = \mathbf{1}$ for some $v \in \mathbf{V}^{(2)}$. By 2.2.8 (4), there is an element u_0 in $\mathbf{V}$ such that $[\![u_0^\wedge = u]\!] = \mathbf{1}$. Hence, from 2.1.8 (7) we obtain

$$\mathbf{1} = [\![\psi(v, u_1^\wedge, \dots, u_n^\wedge)]\!] \wedge [\![v = u_0^\wedge]\!] \leq [\![\psi(u_0^\wedge, \dots, u_n^\wedge)]\!].$$

By the induction hypothesis, $\psi(u_0, \dots, u_n)$. Consequently, $\varphi(u_1, \dots, u_n)$ holds too.

Conversely, if $\varphi(u_1, \dots, u_n)$ then $\psi(u_0, u_1, \dots, u_n)$ for some $u_0 \in \mathbf{V}$. By the induction hypothesis, $[\![\psi(u_0^\wedge, u_1^\wedge, \dots, u_n^\wedge)]\!] = \mathbf{1}$. Since $[\![(\exists\, x)\psi(x, u_1^\wedge, \dots, u_n^\wedge)]\!] \geq [\![\psi(u_0^\wedge, u_1^\wedge, \dots, u_n^\wedge)]\!]$ by definition; therefore, $[\![\psi(u_1^\wedge, \dots, u_n^\wedge)]\!] = \mathbf{1}$. $\triangleright$

2.2.10. Comments.

(1) Let $\mathfrak{U}$ be an ultrafilter in a Boolean algebra B. Denote by $\mathfrak{U}'$ the ideal dual of $\mathfrak{U}$; i.e., $\mathfrak{U}' := \{b^* : b \in \mathfrak{U}\}$. Then the factor algebra $B/\mathfrak{U}'$ has two elements and we may identify it with the Boolean algebra $\mathbf{2} := \{\mathbf{0}, \mathbf{1}\}$.

The factor homomorphism $\pi : B \to \mathbf{2}$ is not complete in general, and so we cannot use 2.2.4 and 2.2.5 for revealing relationship between the truth values in $\mathbf{V}^{(B)}$ and $\mathbf{V}^{(\mathbf{2})}$. If, however, π is complete (i.e., $\mathfrak{U}$ is a principal ultrafilter) then from 2.2.5 it is evident that for every formula $\varphi(x_1, \dots, x_n)$ and all $u_1, \dots, u_n \in \mathbf{V}^{(B)}$ we have

$$\mathbf{V}^{(\mathbf{2})} \models \varphi(\pi^* u_1, \dots, \pi^* u_n) \leftrightarrow [\![\varphi(u_1, \dots, u_n)]\!] \in \mathfrak{U},$$

since for $b \in B$ the equality $\pi(b) = \mathbf{1}$ and the membership $b \in \mathfrak{U}$ are equivalent.

(2) Using factorization, we may arrange a model other than $\mathbf{V}^{(\mathbf{2})}$ given the universe $\mathbf{V}^{(B)}$ and an ultrafilter $\mathfrak{U}$. Indeed, equip $\mathbf{V}^{(B)}$ with the relation $\sim_{\mathfrak{U}}$ by the formula

$$\sim_{\mathfrak{U}} := \{(x, y) \in \mathbf{V}^{(B)} \times \mathbf{V}^{(B)} : [\![x = y]\!] \in \mathfrak{U}\}.$$

Obviously, $\sim_{\mathfrak{U}}$ is an equivalence on $\mathbf{V}^{(B)}$. Let $\mathbf{V}^{(B)}/\mathfrak{U}$ stand for the factor class (see 1.5.8) of $\mathbf{V}^{(B)}$ by $\sim_{\mathfrak{U}}$. We also endow $\mathbf{V}^{(B)}/\mathfrak{U}$ with the binary relation

$$\in_{\mathfrak{U}} := \{(\tilde{x}, \tilde{y}) : x, y \in \mathbf{V}^{(B)} \wedge [\![x \in y]\!] \in \mathfrak{U}\},$$

where $x \mapsto \tilde{x}$ is the factor mapping from $\mathbf{V}^{(B)}$ to $\mathbf{V}^{(B)}/\mathfrak{U}$. It is possible to demonstrate that

$$\mathbf{V}^{(B)}/\mathfrak{U} \models \varphi(\tilde{x}_1, \dots, \tilde{x}_n) \leftrightarrow [\![\varphi(x_1, \dots, x_n)]\!] \in \mathfrak{U}$$

for all $x_2, \dots, x_n \in \mathbf{V}^{(B)}$ and every formula φ.

The reader familiar with the theory of ultraproducts will recognize in (2) the celebrated Łoš Theorem (cf. [10, 27, 48, 83]). Other in-depth ties with the classical model-theoretic constructions may also be revealed.

In (3) and (4) we arrange ultraproducts by factorizing an appropriate Boolean valued universe.

(3) Let T be a nonempty set consisting of some (not necessarily all) principal ultrafilters on a Boolean algebra B. As usual, denote by $\mathbf{V}^T$ the class of all mappings from T to $\mathbf{V}$. By 2.2.8 (4), to each $x \in \mathbf{V}^{(\mathbf{2})}$ there is a unique element $x^\vee \in \mathbf{V}$ such that $[\![(x^\vee)^\wedge = x]\!] = \mathbf{1}$. We now define a mapping $h : \mathbf{V}^{(B)} \to \mathbf{V}^T$ by putting

$$h(x) := \{(t, \pi_t^* x) : t \in T\} \quad (x \in \mathbf{V}^{(B)}),$$

where π_t is the complete homomorphism from B to $\mathbf{2}$ determined by the ultrafilter t; i.e., $\pi_t(b) = \mathbf{1}$ if $b \in t$ and $\pi_t(b) = \mathbf{0}$ if $b \in t'$. It is possible to demonstrate that h is a surjective mapping. On the other hand, h is injective if and only if each element

$b \in B$ belongs to some ultrafilter $t \in T$; i.e., $(\forall b \in B)(\exists t \in T)(b \in t)$ (which means that T defines a dense subset of the Stone space of the algebra B, or that B is atomic, or that B is isomorphic to the boolean $\mathscr{P}(T)$ of T).

The claim of injection is in fact the above mentioned Łoš Theorem. In this case for all $u_1, \dots, u_n \in \mathbf{V}^{(B)}$ and every formula $\varphi(x_1, \dots, x_n)$ we have

$$[\![\varphi(u_1, \dots, u_n)]\!] \leq b \leftrightarrow (\forall t \in T)([\![\varphi(\pi_t^* u_1, \dots, \pi_t^* u_n)]\!] = \mathbf{1} \to b \in t).$$

(4) Assume that T is a set and $\mathfrak{U}$ is an ultrafilter on the boolean $\mathscr{P}(T)$ of T. Let $\mathbf{V}^{(B)}/\mathfrak{U}$ be the conventional *ultrapower* of the class $\mathbf{V}$ over $\mathfrak{U}$ with the factor mapping $g : \mathbf{V}^T \to \mathbf{V}^T/\mathfrak{U}$ (cf. 1.5.7). Put $\lambda(\tilde{x}) := g \circ h(x)$, with h as in (3) and $x \mapsto \tilde{x}$ the same as in (3). We have so defined some bijection λ between $\mathbf{V}^{(\mathscr{P}(T))}/\mathfrak{U}$ and $\mathbf{V}^T/\mathfrak{U}$. In this event, given a formula $\varphi(x_1, \dots, x_n)$ and functions $u_1, \dots, u_n \in \mathbf{V}^T$, note

$$\mathbf{V}^T/\mathfrak{U} \models \varphi(\tilde{u}_1, \dots, \tilde{u}_n) \leftrightarrow \{t \in T : \varphi(u_1(t), \dots, u_n(t))\} \in \mathfrak{U}.$$

(5) It is worthwhile to compare 2.2.4 and 2.2.5 with the following proposition.

Let M be a *transitive model* of ZFC; i.e., M is a transitive class that is a model of ZFC. Assume further that $\varphi(x_1, \dots, x_n)$ is a bounded formula, $\psi(x_1, \dots, x_n)$ is a formula of class Σ_1, and $u_1, \dots, u_n \in M$. Then

$$(M \models \varphi(u_1, \dots, u_n)) \leftrightarrow \varphi(u_1, \dots, u_n),$$
$$(M \models \psi(u_1, \dots, u_n)) \to \psi(u_1, \dots, u_n).$$

2.3. Mixing and the Maximum Principle

Consider a family of functions $(f_\xi)_{\xi \in \Xi}$ with domain A. If $(A_\xi)_{\xi \in \Xi}$ is a disjoint family of subsets of A then we may define on A the function f whose restriction to A_ξ coincides with the restriction of f_ξ to A_ξ for all $\xi \in \Xi$. This function is naturally called the *disjoint mixing* of $(f_\xi)_{\xi \in \Xi}$ by $(A_\xi)_{\xi \in \Xi}$.

Every Boolean valued universe is complete in the sense that it contains all disjoint mixings of families of its elements. This peculiarity allows us to construct various special elements inside $\mathbf{V}^{(B)}$. We will now elaborate details.

2.3.1. A disjoint subset of a Boolean algebra is an *antichain*. In other words, a subset A of B is an antichain provided that $a_1 \wedge a_2 = \mathbf{0}$ for all distinct $a_1, a_2 \in A$. Accordingly, a family $(a_\xi)_{\xi \in \Xi}$ is called an *antichain* if $a_\xi \wedge a_\eta = \mathbf{0}$ whenever $\xi \neq \eta$ for $\xi, \eta \in \Xi$.

An antichain A in B is a *partition of an element* $b \in B$ (and so a *partition of unity* when b is the unity of B) provided that $b = \bigvee A$.

Take an antichain $(b_\xi)_{\xi\in\Xi}$ in a Boolean algebra B and a family $(x_\xi)_{\xi\in\Xi}$ in the universe $\mathbf{V}^{(B)}$. The *disjoint mixing* or simply the *mixing* of (x_ξ) by (b_ξ) (or with respect to (b_ξ) or even with probabilities (b_ξ)) is an element x in $\mathbf{V}^{(B)}$ meeting the conditions

$$\operatorname{dom}(x) := \bigcup\{\operatorname{dom}(x_\xi) : \xi \in \Xi\},$$
$$x(t) := \bigvee\{b_\xi \wedge x_\xi(t) : \xi \in \Xi\} \quad (t \in \operatorname{dom}(x)).$$

The last equality implies that $x_\xi(t) = \mathbf{0}$ for $t \in \operatorname{dom}(x) - \operatorname{dom}(x_\xi)$. Since $\alpha := \sup_{\xi\in\Xi} \rho(x_\xi) \in \mathrm{On}$; therefore, $\operatorname{dom}(x) \subset \mathbf{V}^{(B)}_{\alpha+1}$. Hence, the above formula indeed determines a certain element $x \in \mathbf{V}^{(B)}$. The following symbolic notation is common: $\operatorname{mix}_{\xi\in\Xi}(b_\xi x_\xi) := \operatorname{mix}\{b_\xi x_\xi : \xi \in \Xi\} := x$.

In order to study the basic properties of mixing, we start with proving an auxiliary fact.

2.3.2. *Take $x \in \mathbf{V}^{(B)}$ and $b \in B$. Define the function bx by the rules:*

$$\operatorname{dom}(bx) := \operatorname{dom}(x), \quad bx : t \mapsto b \wedge x(t) \quad (t \in \operatorname{dom}(x)).$$

Then $bx \in \mathbf{V}^{(B)}$; and, for all $x, y \in \mathbf{V}^{(B)}$, the following hold:

$$[\![x \in by]\!] = b \wedge [\![x \in y]\!], \quad [\![bx = by]\!] = b \Rightarrow [\![x = y]\!].$$

$\lhd$ The first equality follows from straightforward calculation of Boolean truth values on using the infinite distributive law 1.1.5 (2).

Indeed,

$$[\![x \in by]\!] = \bigvee_{t\in\operatorname{dom}(by)} (by)(t) \wedge [\![t = x]\!]$$
$$= b \wedge \bigvee_{t\in\operatorname{dom}(y)} y(t) \wedge [\![t = x]\!] = b \wedge [\![x \in y]\!].$$

Use the preceding equality and successively apply 1.1.4 (2), 1.1.5 (6), 1.1.4 (4), 1.1.4 (2), and 1.1.5 (6) to derive the next chain of equalities

$$[\![bx = by]\!]$$
$$= \bigwedge_{t\in\operatorname{dom}(by)} (by)(t) \Rightarrow [\![t \in bx]\!] \wedge \bigwedge_{t\in\operatorname{dom}(bx)} (bx)(t) \Rightarrow [\![t \in by]\!]$$

$$
\begin{aligned}
&= \bigwedge_{t\in\mathrm{dom}(y)} (b\wedge y(t))\Rightarrow(b\wedge [\![t\in x]\!])\\
&\wedge \bigwedge_{t\in\mathrm{dom}(x)} (b\wedge x(t))\Rightarrow(b\wedge [\![t\in y]\!])\\
&= \bigwedge_{t\in\mathrm{dom}(y)} ((b\wedge y(t))\Rightarrow b)\wedge(b\wedge y(t))\Rightarrow [\![t\in x]\!])\\
&\wedge \bigwedge_{t\in\mathrm{dom}(x)} ((b\wedge x(t))\Rightarrow b)\wedge((b\wedge x(t))\Rightarrow [\![t\in y]\!])\\
&= \bigwedge_{t\in\mathrm{dom}(y)} b\Rightarrow(y(t)\Rightarrow [\![t\in x]\!])\wedge \bigwedge_{t\in\mathrm{dom}(x)} b\Rightarrow(x(t)\Rightarrow [\![t\in y]\!])\\
&= b\Rightarrow [\![x=y]\!],
\end{aligned}
$$

which completes the proof. ▷

2.3.3. *Theorem* (the mixing principle). *Assume that* $(b_\xi)_{\xi\in\Xi}$ *is an antichain in* B *and* $(x_\xi)_{\xi\in\Xi}$ *is a family in* $\mathbf{V}^{(B)}$. *Put* $x := \mathrm{mix}_{\xi\in\Xi}(b_\xi x_\xi)$. *Then*

$$[\![x = x_\xi]\!] \geq b_\xi \quad (\xi\in\Xi).$$

Moreover, if $(b_\xi)_{\xi\in\Xi}$ *is a partition of unity and an element* $y\in\mathbf{V}^{(B)}$ *obeys the condition* $[\![y=x_\xi]\!]\geq b_\xi$ *for all* $\xi\in\Xi$ *then* $[\![x=y]\!]=\mathbf{1}$.

◁ By the definition of mixing, $b_\xi x = b_\xi x_\xi$ for all $\xi\in\Xi$. Applying 2.3.2, deduce

$$\mathbf{1} = [\![b_\xi x = b_\xi x_\xi]\!] = b_\xi \Rightarrow [\![x_\xi = x]\!].$$

Therefore, $[\![x=x_\xi]\!]\geq b_\xi$ for all $\xi\in\Xi$ by 1.1.4 (4).

Assume now that (b_ξ) is a partition of unity and $[\![y=x_\xi]\!]\geq b_\xi$ $(\xi\in\Xi)$. By 2.1.8 (4), note then that

$$b_\xi \leq [\![x=x_\xi]\!]\wedge[\![x_\xi=y]\!]\leq[\![x=y]\!] \quad (\xi\in\Xi).$$

Hence, observe

$$\mathbf{1} = \bigvee\{b_\xi : \xi\in\Xi\}\leq[\![x=y]\!]\leq\mathbf{1},$$

so completing the proof. ▷

2.3.4. *Let* $x\in\mathbf{V}^{(B)}$. *Define* $\bar{x}\in\mathbf{V}^{(B)}$ *by the rule*

$$\mathrm{dom}(\bar{x}) := \mathrm{dom}(x),\quad \bar{x}(t) := [\![t\in x]\!] \quad (t\in\mathrm{dom}(x)).$$

Then $\mathbf{V}^{(B)}\models x=\bar{x}$.

◁ The aim can be achieved by performing the following simple calculations which use the definitions of 2.1.4 as well as 1.1.4 (4) and 2.1.8 (2):

$$[\![x = \bar{x}]\!]$$

$$= \bigwedge_{t\in\mathrm{dom}(x)} x(t) \Rightarrow [\![t \in \bar{x}]\!] \wedge \bigwedge_{t\in\mathrm{dom}(\bar{x})} [\![t \in x]\!] \Rightarrow [\![t \in x]\!]$$

$$= \bigwedge_{t\in\mathrm{dom}(x)} x(t) \Rightarrow \Big(\bigvee_{u\in\mathrm{dom}(\bar{x})} \bar{x}(u) \wedge [\![u = t]\!]\Big)$$

$$\geq \bigwedge_{t\in\mathrm{dom}(x)} x(t) \Rightarrow [\![t \in x]\!] = \mathbf{1}.$$

This ends the proof. ▷

2.3.5. *Assume given a partition of unity $(b_\xi)_{\xi\in\Xi} \subset B$ and a family $(x_\xi)_{\xi\in\Xi} \subset \mathbf{V}^{(B)}$. Put $x := \mathrm{mix}_{\xi\in\Xi}(b_\xi x_\xi)$. Then the following hold:*

(1) *If $(x'_\xi)_{\xi\in\Xi} \subset \mathbf{V}^{(B)}$ and $\mathbf{V}^{(B)} \models x_\xi = x'_\xi$ $(\xi \in \Xi)$ then*

$$\mathbf{V}^{(B)} \models x = \mathop{\mathrm{mix}}_{\xi\in\Xi}(b_\xi x'_\xi);$$

(2) *If an element $y \in \mathbf{V}^{(B)}$ is such that $\mathrm{dom}(y) = \mathrm{dom}(x)$ and*

$$y(t) := \bigvee_{\xi\in\Xi} b_\xi \wedge [\![t \in x_\xi]\!] \quad (t \in \mathrm{dom}(y)),$$

then $\mathbf{V}^{(B)} \models x = y$.

◁ Put $x' := \mathrm{mix}_{\xi\in\Xi}(b_\xi x'_\xi)$. By hypothesis,

$$b_\xi \leq [\![x_\xi = x'_\xi]\!] \wedge [\![x_\xi = x]\!] \wedge [\![x'_\xi = x']\!] \leq [\![x = x']\!],$$

and so $[\![x = x']\!] = \mathbf{1}$. Claim (2) follows from (1) and 2.3.4. ▷

2.3.6. *If $b \in B$ and $x \in \mathbf{V}^{(B)}$ then*

$$[\![bx = x]\!] = b \vee [\![x = \varnothing]\!], \quad [\![bx = \varnothing]\!] = b^* \vee [\![x = \varnothing]\!].$$

In particular,

$$\mathbf{V}^{(B)} \models bx = \mathrm{mix}\{bx, b^*\varnothing\}.$$

◁ Note that $[\![t \in bx \to t \in x]\!] = \mathbf{1}$ since, by virtue of 2.3.2, $[\![t \in bx]\!] = b \wedge [\![t \in x]\!] \leq [\![t \in x]\!]$. Therefore, $[\![bx = x \leftrightarrow (\forall t)(t \in x \to t \in bx)]\!] = \mathbf{1}$. Using this equality, proceed with the calculation

$$
\begin{aligned}
[\![bx = x]\!] &= \bigwedge_{t\in\mathbf{V}^{(B)}} [\![t \in x]\!] \Rightarrow [\![t \in bx]\!] \\
&= \bigwedge_{t\in\mathbf{V}^{(B)}} [\![t \in x]\!]^* \vee (b \wedge [\![t \in x]\!]) \\
&= \bigwedge_{t\in\mathbf{V}^{(B)}} (b \vee [\![t \in x]\!]^*) \wedge ([\![t \in x]\!]^* \vee [\![t \in x]\!]) \\
&= \bigwedge_{t\in\mathbf{V}^{(B)}} b \vee [\![t \in x]\!]^* = b \vee \bigwedge_{t\in\mathbf{V}^{(B)}} [\![t \in x]\!]^* \\
&= b \vee [\![(\forall\, t)(t \notin x)]\!] = b \vee [\![x = \varnothing]\!].
\end{aligned}
$$

On the other hand, appealing to 2.3.2 again and using the equality $b\varnothing = \varnothing$, conclude that

$$
b^* \vee [\![x = \varnothing]\!] = b \Rightarrow [\![x = \varnothing]\!] = [\![bx = b\varnothing]\!] = [\![bx = \varnothing]\!]. \ \triangleright
$$

2.3.7. *Assume that (b_ξ) is a partition of unity in B and let a family $(x_\xi) \subset \mathbf{V}^{(B)}$ be such that $\mathbf{V}^{(B)} \models x_\xi \neq x_\eta$ for all $\xi \neq \eta$. Then there is an element x in $\mathbf{V}^{(B)}$ satisfying $[\![x = x_\xi]\!] = b_\xi$ for all ξ.*

$\triangleleft$ Put $x := \operatorname{mix}(b_\xi x_\xi)$ and $a_\xi := [\![x = x_\xi]\!]$. By hypothesis,

$$
a_\xi \wedge a_\eta = [\![x = x_\xi]\!] \wedge [\![x_\eta = x]\!] \leq [\![x_\xi \neq x_\eta]\!]^* = \mathbf{0}
$$

for $\xi \neq \eta$. Moreover, by the properties of mixing, $b_\xi \leq a_\xi$ for all ξ. Hence, (a_ξ) is also a partition of unity in B.

On the other hand,

$$
b_\xi^* = \bigvee_{\eta\neq\xi} b_\eta \leq \bigvee_{\eta\neq\xi} a_\eta = a_\xi^*,
$$

and so $b_\xi^* \leq a_\xi^* \to b_\xi \geq a_\xi$. Therefore, the partitions of unity (b_ξ) and (a_ξ) coincide. $\triangleright$

The following fact whose proof rests on mixing a two-element set often makes it possible to diminish bulky calculations.

2.3.8. *Consider B-formulas $\varphi(x)$ and $\psi(x)$. Assume that $[\![\varphi(u_0)]\!] = \mathbf{1}$ for some $u_0 \in \mathbf{V}^{(B)}$. Then*

$$
[\![(\forall\, x)(\varphi(x) \to \psi(x))]\!] = \bigwedge\{[\![\psi(u)]\!] : u \in \mathbf{V}^{(B)}, [\![\varphi(u)]\!] = \mathbf{1}\},
$$
$$
[\![(\exists\, x)(\varphi(x) \wedge \psi(x))]\!] = \bigwedge\{[\![\psi(u)]\!] : u \in \mathbf{V}^{(B)}, [\![\varphi(u)]\!] = \mathbf{1}\}.
$$

$\triangleleft$ Prove the first equality.

To begin with, it is evident (cf. 2.1.7) that

$$c := [\![(\forall x)(\varphi(x) \to \psi(x))]\!] = \bigwedge_{t \in \mathbf{V}^{(B)}} [\![\varphi(t)]\!] \Rightarrow [\![\psi(t)]\!]$$

$$\leq \bigwedge_{t \in \mathbf{V}^{(B)},\, [\![\varphi(t)]\!]=\mathbf{1}} [\![\varphi(t)]\!]^* \vee [\![\psi(t)]\!] = \bigvee_{t \in \mathbf{V}^{(B)},\, [\![\varphi(t)]\!]=\mathbf{1}} [\![\psi(t)]\!] =: d.$$

To show the reverse inequality $d \leq c$, choose an arbitrary element $t \in \mathbf{V}^{(B)}$ and put $u := \operatorname{mix}\{bt, b^* u_0\}$, where $b := [\![\varphi(t)]\!]$. Using 2.1.8 (7) and 2.3.3, proceed with estimation

$$b \leq [\![\varphi(t)]\!] \wedge [\![t = u]\!] \leq [\![\varphi(u)]\!],$$
$$b^* \leq [\![\varphi(u_0)]\!] \wedge [\![u = u_0]\!] \leq [\![\varphi(u)]\!].$$

Therefore, $[\![\varphi(u)]\!] = \mathbf{1}$.

Furthermore, by the same considerations,

$$b \wedge [\![\psi(u)]\!] \leq [\![u = t]\!] \wedge [\![\psi(u)]\!] \leq [\![\psi(t)]\!].$$

Hence, the following estimates hold:

$$[\![\psi(u)]\!] \leq b^* \vee (b \wedge [\![\psi(u)]\!]) \leq b^* \vee [\![\psi(t)]\!]$$
$$= b \Rightarrow [\![\psi(t)]\!] = [\![\varphi(t)]\!] \Rightarrow [\![\psi(t)]\!].$$

Since $d \leq [\![\psi(u)]\!]$; therefore, $d \leq [\![\varphi(t)]\!] \Rightarrow [\![\psi(t)]\!]$ $(t \in \mathbf{V}^{(B)})$.

Passing to the infimum over t on the right-hand side of the last inequality, find $d \leq c$.

The second equality under proof is dual to the first and so it is easy to check on applying the De Morgan laws (cf. 1.1.2). $\triangleright$

2.3.9. We intend now to establish the key result of the present section, the maximum principle, which asserts that the least upper bound is attained at some $u_0 \in \mathbf{V}^{(B)}$ in the formula

$$[\![(\exists x)\varphi(x)]\!] = \bigvee \{[\![\varphi(u)]\!] : u \in \mathbf{V}^{(B)}\}.$$

To begin with, recall a certain fundamental property of complete Boolean algebras.

Let B be a complete Boolean algebra. A subset E of B *minorizes* a subset B_0 of B or is a *minorant* for B_0 if to each $0 < b \in B_0$ there is some x in E such that $0 < x \leq b$. It is also in common parlance to call E a *minorizing*, or *minorant*, or *coinitial* set to B_0.

(1) Theorem (the exhaustion principle). *Let M be a nonempty subset of a complete Boolean algebra B. Assume given a subset E of B that minorizes the band B_0 of B generated by M. Then some antichain E_0 in E exists such that $\bigvee E_0 = \bigvee M$ and to each $x \in E_0$ there is an element y in M satisfying $x \leq y$.*

$\lhd$ Consider the set $\mathfrak{A}$ of all antichains A obeying the following conditions: (a) $A \subset E$; (b) to each $x \in A$ there is some $y \in M$ satisfying $x \leq y$.

If $\mathbf{0} \neq y \in M$ then, by hypothesis, $y \geq x$ for some $\mathbf{0} \neq x \in E$. Hence, $\{x\} \in \mathfrak{A}$ and $\mathfrak{A}$ is nonempty. The inclusion ordered set $\mathfrak{A}$ clearly obeys the hypotheses of the Kuratowski–Zorn Lemma. Therefore, there is a maximal element $E_0 \in \mathfrak{A}$.

Show that the elements $b_0 := \bigvee E_0$ and $b := \bigvee M$ coincide. It follows from the definition of $\mathfrak{A}$ that $b_0 \leq b$. If $b_0 \neq b$, then there are elements $\mathbf{0} \neq x_0 \in B$ and $x \in M$ such that $x_0 \wedge b_0 = \mathbf{0}$ and $x_0 \leq x$. By hypothesis, $\mathbf{0} < y \leq x$ for some $y \in E$.

The set $E_0 \cup \{y\}$ belongs to $\mathfrak{A}$ and has essentially more elements than E_0. This contradicts the fact that E_0 is minimal, and so $b_0 = b$. $\rhd$

(2) Corollary. *To each nonempty set $M \subset B$, there is an antichain $A \subset B$ with the following properties: $\bigvee A = \bigvee M$ and, given $x \in A$, we may find y in M such that $x \leq y$.*

$\lhd$ Choose $E := \bigcup_{y \in M} [\mathbf{0}, y]$ as a minorant for M and appeal to (1). $\rhd$

2.3.10. Theorem (the maximum principle). *Assume given $u_1, \dots, u_n \in \mathbf{V}^{(B)}$ and a formula $\varphi(x, x_1, \dots, x_n)$ of* ZFC. *Then there is an element $u_0 \in \mathbf{V}^{(B)}$ such that*

$$[\![(\exists\, x)\varphi(x, u_1, \dots, u_n)]\!] = [\![\varphi(u_0, u_1, \dots, u_n)]\!].$$

In particular, if $\mathbf{V}^{(B)} \models (\exists\, x)\varphi(x, u_1, \dots, u_n)$ then $\mathbf{V}^{(B)} \models \varphi(u_0, u_1, \dots, u_n)$ for some $u_0 \in \mathbf{V}^{(B)}$.

$\lhd$ By definition,

$$b := [\![(\exists\, x)\varphi(x, u_1, \dots, u_n)]\!] = \bigvee_{u \in \mathbf{V}^{(B)}} [\![\varphi(u, u_1, \dots, u_n)]\!].$$

The class $A := \{[\![\varphi(u, u_1, \dots, u_n)]\!] : u \in \mathbf{V}^{(B)}\}$ is a subset of the Boolean algebra B. By 2.3.9 (2), there are a partition $(b_\xi)_{\xi \in \Xi}$ of b and a family $(u_\xi)_{\xi \in \Xi}$ in $\mathbf{V}^{(B)}$ obeying the following conditions:

$$b_\xi \leq [\![\varphi(u_\xi, u_1, \dots, u_n)]\!] \quad (\xi \in \Xi), \ b = \bigvee \{[\![\varphi(u_\xi, u_1, \dots, u_n)]\!] : \xi \in \Xi\}.$$

Put $u_0 := \mathrm{mix}_{\xi \in \Xi}(b_\xi u_\xi)$ and recall that $b_\xi \leq [\![u_0 = u_\xi]\!]$ for all $\xi \in \Xi$ by 2.3.3.

Obviously,

$$[\![\varphi(u_0, u_1, \dots, u_n)]\!] \leq b.$$

On the other hand, applying 2.1.8 (7), obtain

$$b_\xi \le [\![u_0 = u_\xi]\!] \wedge [\![\varphi(u_\xi, u_1, \dots, u_n)]\!] \le [\![\varphi(u_0, \dots, u_n)]\!].$$

Therefore,

$$[\![\varphi(u_0, \dots, u_n)]\!] \ge \bigvee_{\xi \in \Xi} b_\xi = b.$$

The second claim of the theorem is an immediate consequence of the first. ▷

2.4. The Transfer Principle

In this section we show that the universe $\mathbf{V}^{(B)}$ over a complete Boolean algebra B, together with the Boolean truth values $[\![\cdot \in \cdot]\!]$ and $[\![\cdot = \cdot]\!]$, serves as a Boolean valued model of ZFC. In other words, we prove the following theorem.

2.4.1. ***Theorem*** (the transfer principle). *Every theorem of* ZFC *holds inside* $\mathbf{V}^{(B)}$; *in symbols,* $\mathbf{V}^{(B)} \models \mathrm{ZFC}$.

The demonstration of this theorem consists in proving the formulas $\mathbf{V}^{(B)} \models \mathrm{ZF}_k$ for $k := 1, 2, \dots, 6$ and, finally, $\mathbf{V}^{(B)} \models \mathrm{AC}$. Most effort is put into the routine calculation of Boolean truth values which we give in full detail for the sake of rigor and completeness.

2.4.2. *The axiom of extensionality* ZF_1 *holds inside* $\mathbf{V}^{(B)}$:

$$\mathbf{V}^{(B)} \models (\forall\, x)(\forall\, y)(x = y \leftrightarrow (\forall\, z)(z \in x \leftrightarrow z \in y)).$$

◁ The proof is immediate from 2.1.9 and the definition of the Boolean truth value of equality 2.1.4 (2).

Indeed, given x and y in $\mathbf{V}^{(B)}$, put

$$c := c(x, y) := [\![(\forall\, z \in x)(z \in y)]\!] = \bigwedge_{z \in \mathrm{dom}(x)} x(z) \Rightarrow [\![z \in y]\!].$$

Obviously, $c(x, y) \wedge c(y, x) = [\![x = y]\!]$. On the other hand,

$$c(x, y) \wedge c(y, x) = [\![(\forall\, z)(z \in x \leftrightarrow z \in y)]\!].$$

Hence, by 1.1.4 (5),

$$[\![x = y \leftrightarrow (\forall\, z)(z \in x \leftrightarrow z \in y)]\!] = \mathbf{1} \quad (x, y \in \mathbf{V}^{(B)}).$$

Taking infima over x and y, complete the proof. ▷

2.4.3. *The axiom of union* ZF_2 *holds inside* $\mathbf{V}^{(B)}$:

$$\mathbf{V}^{(B)} \models (\forall\, x)(\exists\, y)(\forall\, z)(z \in y \leftrightarrow (\exists\, u \in x)(z \in u)).$$

$\lhd$ Given $x \in \mathbf{V}^{(B)}$, define $y \in \mathbf{V}^{(B)}$ by the formulas

$$\mathrm{dom}(y) := \bigcup \{\mathrm{dom}(u) : u \in \mathrm{dom}(x)\},$$
$$y(t) := [\![(\exists\, u \in x)(t \in u)]\!] \quad (t \in \mathrm{dom}(y)).$$

It suffices to show that $[\![y = \bigcup x]\!] = \mathbf{1}$. By 2.1.9,

$$[\![y \subset \bigcup x]\!] = [\![(\forall\, t \in y)(\exists\, u \in x)(t \in u)]\!]$$
$$= \bigwedge_{t \in \mathrm{dom}(y)} [\![(\exists\, u \in x)(t \in u)]\!] \Rightarrow [\![(\exists\, u \in x)(t \in u)]\!] = \mathbf{1}.$$

Granted $u \in \mathrm{dom}(x)$ and $z \in \mathrm{dom}(u)$ and recalling 2.1.8 (2) and 2.1.9, observe that

$$x(u) \wedge u(z) \leq x(u) \wedge [\![z \in u]\!] \leq \bigvee_{u \in \mathrm{dom}(x)} x(u) \wedge [\![z \in u]\!]$$
$$= [\![(\exists\, u \in x)(z \in u)]\!] = y(z) \leq [\![z \in y]\!].$$

The above formula yields $x(u) \Rightarrow (u(z) \Rightarrow [\![z \in y]\!]) = \mathbf{1}$ (cf. 1.1.4 (2–4)). Using this equality together with 1.1.5 (6) and 2.1.9, proceed with the calculation

$$[\![\bigcup x \subset y]\!] = [\![(\forall\, u \in x)(\forall\, z \in u)(z \in y)]\!]$$
$$= \bigwedge_{u \in \mathrm{dom}(x)} x(u) \Rightarrow \Big(\bigwedge_{z \in \mathrm{dom}(u)} u(z) \Rightarrow [\![z \in y]\!] \Big)$$
$$= \bigwedge_{u \in \mathrm{dom}(x)} \bigwedge_{z \in \mathrm{dom}(u)} x(u) \Rightarrow (u(z) \Rightarrow [\![z \in y]\!]) = \mathbf{1}.$$

Therefore, $[\![y = \bigcup x]\!] = \mathbf{1}$, and so

$$[\![(\exists\, u)(u = \bigcup x)]\!] = \bigvee_{u \in \mathbf{V}^{(B)}} [\![u = \bigcup x]\!] \geq [\![y = \bigcup x]\!] = \mathbf{1}.$$

Passing to the infimum over $x \in \mathbf{V}^{(B)}$, find the desired result:

$$[\![(\forall\, x)(\exists\, y)(y = \bigcup x)]\!] = \bigwedge_{x \in \mathbf{V}^{(B)}} [\![(\exists\, y)(y = \bigcup x)]\!] = \mathbf{1}. \ \rhd$$

2.4.4. *The axiom of powerset* ZF_3 *holds inside* $\mathbf{V}^{(B)}$:

$$\mathbf{V}^{(B)} \models (\forall\, x)(\exists\, y)(\forall\, z)(z \in y \leftrightarrow z \subset x).$$

$\lhd$ Given $x \in \mathbf{V}^{(B)}$, define $y \in \mathbf{V}^{(B)}$ as follows:

$$\mathrm{dom}(y) := B^{\mathrm{dom}(x)},$$
$$y(z) := [\![z \subset x]\!] \quad (z \in \mathrm{dom}(y)).$$

It suffices to show that $[\![z \in y \leftrightarrow z \subset x]\!] = \mathbf{1}$ for every $z \in \mathbf{V}^{(B)}$.

Clearly,

$$[\![z \in y]\!] = \bigvee_{t \in \mathrm{dom}(y)} y(t) \wedge [\![t = z]\!] = \bigvee_{t \in \mathrm{dom}(y)} [\![t \subset x]\!] \wedge [\![t = z]\!] \le [\![z \subset x]\!].$$

Hence, $[\![z \in y \rightarrow z \subset x]\!] = \mathbf{1}$ by 1.1.4 (4).

We now must justify the equality $[\![z \subset x \rightarrow z \in y]\!] = \mathbf{1}$. To this end, modify z slightly; i.e., consider the element $z' \in \mathrm{dom}(y)$ defined as follows: $\mathrm{dom}(z') := \mathrm{dom}(x)$ and $z'(t) := [\![t \in z]\!]$ $(t \in \mathrm{dom}(z'))$.

Granted $t \in \mathbf{V}^{(B)}$, obtain

$$[\![t \in z']\!] = \bigvee_{u \in \mathrm{dom}(z')} z'(u) \wedge [\![t = u]\!] \quad \bigvee_{u \in \mathrm{dom}(z')} [\![u \in z]\!] \wedge [\![u = t]\!] \le [\![t \in z]\!].$$

Consequently, $[\![z' \subset z]\!] = \mathbf{1}$.

On the other hand, from 2.1.8 (5) and 2.1.9 deduce

$$[\![t \in z \cap x]\!] = \bigvee_{u \in \mathrm{dom}(x)} x(u) \wedge [\![t = u]\!] \wedge [\![t \in z]\!] \le \bigvee_{u \in \mathrm{dom}(x)} z'(u) \wedge [\![t = u]\!] = [\![t \in z']\!].$$

Hence, $[\![z \cap x \subset z']\!] = \mathbf{1}$ on appealing to 1.1.4 (4) once again.

Moreover,

$$[\![z \subset x]\!] = \bigwedge_{t \in \mathbf{V}^{(B)}} [\![t \in z]\!] \Rightarrow [\![t \in x]\!] \le \bigwedge_{t \in \mathrm{dom}(z')} z'(t) \Rightarrow [\![t \in x]\!]$$
$$= [\![(\forall\, t \in z')(t \in x)]\!] = [\![z' \subset x]\!] = y(z') \le [\![z' \in y]\!].$$

Summarizing all that was said about z and z', infer

$$[\![z \subset x]\!] \le [\![x \cap z \subset z']\!] \wedge [\![z' \subset z]\!] \wedge [\![z \subset x]\!] \le [\![z = z']\!],$$
$$[\![z \subset x]\!] \le [\![z' \in y]\!].$$

The last two formulas immediately yield

$$[\![z \subset x]\!] = [\![z \subset x]\!] \wedge [\![z = z']\!] \le [\![z' \in y]\!] \wedge [\![z = z']\!] \le [\![z \in y]\!];$$

i.e., $[\![z \subset x]\!] \le [\![z \in y]\!]$, which amounts to the sought result by 1.1.4 (4). $\rhd$

2.4.5. *The axiom of replacement* ZF_4^{φ} *holds inside* $\mathbf{V}^{(B)}$:

$$\mathbf{V}^{(B)} \models (\forall\, u)(\forall\, v_1)(\forall\, v_2)\ (\varphi(u,v_1) \wedge \varphi(u,v_2) \to v_1 = v_2) \\ \to ((\forall\, x)(\exists\, y)(\forall\, t)(\forall\, s \in x)(\varphi(s,t) \leftrightarrow t \in y)).$$

$\triangleleft$ The axiom of replacement is deducible from the axiom of separation (cf. A.2.5) and the formula

$$\Phi := (\forall\, x)((\forall\, t \in x)(\exists\, u)\varphi(t,u) \to (\exists\, y)(\forall\, t \in x)(\exists\, u \in y)\varphi(t,u))$$

(y is not a free variable in φ); i.e., $\Phi \wedge \Psi \to \mathrm{ZF}_4^{\varphi}$, where Ψ is the axiom of separation. Therefore, it suffices to show that $\mathbf{V}^{(B)} \models \Phi$ and $\mathbf{V}^{(B)} \models \Psi$.

(1) $\mathbf{V}^{(B)} \models \Psi := (\forall\, x)(\exists\, y)(\forall\, t)(t \in y \leftrightarrow t \in x \wedge \psi(t))$.

Take $x \in \mathbf{V}^{(B)}$ and consider the function $y \in \mathbf{V}^{(B)}$ defined the formulas

$$\mathrm{dom}(y) := \mathrm{dom}(x), \\ y(t) := x(t) \wedge [\![\psi(t)]\!] \quad (t \in \mathrm{dom}(y)).$$

Then $[\![(\forall\, t)(t \in y \leftrightarrow t \in x \wedge \psi(t))]\!] = a \wedge b$ where

$$a := [\![(\forall\, t \in y)(t \in x \wedge \psi(t))]\!], \quad b := [\![(\forall\, t \in x)(\psi(t) \to t \in y)]\!].$$

From 2.1.8 (2) and 2.1.9, it is however immediate that $a = b = \mathbf{1}$. Indeed,

$$a = \bigwedge_{t \in \mathrm{dom}(y)} y(t) \Rightarrow [\![t \in x \wedge \psi(t)]\!] \\ = \bigwedge_{t \in \mathrm{dom}(y)} x(t) \wedge [\![\psi(t)]\!] \Rightarrow [\![t \in x]\!] \wedge [\![\psi(t)]\!] = \mathbf{1}.$$

By analogy,

$$b = \bigwedge_{t \in \mathrm{dom}(x)} x(t) \Rightarrow ([\![\psi(t)]\!] \Rightarrow [\![t \in y]\!]) \\ = \bigwedge_{t \in \mathrm{dom}(x)} x(t) \wedge [\![\psi(t)]\!] \Rightarrow [\![t \in x]\!] \wedge [\![\psi(t)]\!] = \mathbf{1}.$$

(2) $\mathbf{V}^{(B)} \models \Phi$.

Let x be an arbitrary element of $\mathbf{V}^{(B)}$. Since B is a set; therefore, for every fixed $t \in \mathrm{dom}(x)$, so is the class

$$K := \{[\![\varphi(t,u)]\!] : u \in \mathbf{V}^{(B)}\} \subset B.$$

The axiom of replacement for sets (i.e., in $\mathbf{V}$) implies that there is an ordinal $\alpha(t)$ such that

$$\{[\![\varphi(t,u)]\!] : u \in V^{(B)}_{\alpha(t)}\} = K.$$

Put $\alpha := \sup\{\alpha(t) : t \in \operatorname{dom}(x)\}$ and define $y \in \mathbf{V}^{(B)}$ by the formulas

$$\operatorname{dom}(y) := \mathbf{V}^{(B)}_{\alpha}, \quad \operatorname{im}(y) = \{\mathbf{1}\}.$$

Note now that y is a sought element, as follows from the easy calculations:

$$[\![(\forall\, t \in x)(\exists\, u)\varphi(t,u)]\!] = \bigwedge_{t\in\operatorname{dom}(x)} x(t) \Rightarrow \Big(\bigvee_{u\in\mathbf{V}^{(B)}} [\![\varphi(t,u)]\!]\Big)$$
$$= \bigwedge_{t\in\operatorname{dom}(x)} x(t) \Rightarrow \Big(\bigvee_{u\in\mathbf{V}^{(B)}_{\alpha(t)}} [\![\varphi(t,u)]\!]\Big)$$
$$\leq \bigwedge_{t\in\operatorname{dom}(x)} x(t) \Rightarrow \Big(\bigvee_{u\in\mathbf{V}^{(B)}_{\alpha}} [\![\varphi(t,u)]\!]\Big)$$
$$= \bigwedge_{t\in\operatorname{dom}(x)} x(t) \Rightarrow [\![(\exists\, u \in y)\varphi(t,u)]\!] = [\![(\forall\, t \in x)(\exists\, u \in y)\varphi(t,u)]\!].$$

The proof is complete. $\triangleright$

2.4.6. *The axiom of infinity* ZF_5 *holds inside* $\mathbf{V}^{(B)}$:

$$\mathbf{V}^{(B)} \models (\exists\, x)(0 \in x \wedge (\forall\, t)(t \in x \to t \cup \{t\} \in x)).$$

$\triangleleft$ To satisfy this axiom, assign $x := \omega^{\wedge}$ (cf. 2.2.7).

To demonstrate, note at first that $[\![0^{\wedge} \in \omega^{\wedge}]\!] = \mathbf{1}$ since $0^{\wedge} \in \operatorname{dom}(\omega^{\wedge})$.

Granted $t \in \mathbf{V}$ and $u := t \cup \{t\}$, observe now that $[\![u^{\wedge} = t^{\wedge} \cup \{t^{\wedge}\}]\!] = \mathbf{1}$. Indeed, by 2.2.8 (1),

$$[\![v \in u^{\wedge}]\!] = \bigvee_{s\in u} [\![s^{\wedge} = v]\!] = [\![t^{\wedge} = v]\!] \vee \bigvee_{s\in t} [\![s^{\wedge} = v]\!]$$
$$= [\![t^{\wedge} = v]\!] \vee [\![v \in t^{\wedge}]\!] = [\![t^{\wedge} = v \vee v \in t^{\wedge}]\!] = [\![v \in t^{\wedge} \cup \{t^{\wedge}\}]\!].$$

Using this together with 2.1.9 and 2.2.8 (2), proceed with easy calculation of the Boolean truth values

$$[\![(\forall\, t \in \omega^{\wedge})(t \cup \{t\}) \in \omega^{\wedge}]\!] = \bigwedge_{t\in\omega} [\![t^{\wedge} \cup \{t^{\wedge}\} \in \omega^{\wedge}]\!]$$
$$= \bigwedge_{t\in\omega} [\![(t \cup \{t\})^{\wedge} \in \omega^{\wedge}]\!] = \mathbf{1},$$

so completing the proof. $\triangleright$

2.4.7. *The axiom of regularity* ZF_6 *holds in* $\mathbf{V}^{(B)}$:

$$\mathbf{V}^{(B)} \models (\forall\, x)(\exists\, y)(x = 0 \vee (y \in x \wedge y \cap x = 0)).$$

◁ Take $x \in \mathbf{V}^{(B)}$. Show that

$$b := [\![x \neq 0 \wedge (\forall\, y \in x)(y \cap x \neq 0)]\!] = \mathbf{0}_B.$$

Assume to the contrary that $b \neq \mathbf{0}_B$. Since $b \leq [\![(\exists\, u)(u \in x)]\!]$, there is an element y_0 in $\mathbf{V}^{(B)}$ such that $[\![y_0 \in x]\!] \wedge b \neq \mathbf{0}$ and $\rho(y_0) \leq \rho(y)$ for $[\![y \in x]\!] \wedge b \neq \mathbf{0}$ $(y \in \mathbf{V}^{(B)})$.

Furthermore, given $y \in \mathbf{V}^{(B)}$, note the estimate

$$[\![y \in x]\!] \wedge b \leq [\![y \cap x \neq 0]\!] = \bigvee_{z \in \mathrm{dom}(y)} y(z) \wedge [\![z \in x]\!].$$

Hence, $[\![z \in x]\!] \wedge [\![y_0 \in x]\!] \wedge b \neq \mathbf{0}$ for some $z \in \mathrm{dom}(y_0)$. However, $\rho(z) < \rho(y_0)$, which contradicts the choice of y_0.

Therefore, $b = \mathbf{0}_B$ implying that

$$\begin{aligned} \mathbf{1}_B = b^* &= [\![\neg(x \neq 0 \wedge (\forall\, y \in x)(y \cap x \neq 0))]\!] \\ &= [\![(\exists\, y)(x = 0 \vee (y \in x \wedge y \cap x = 0))]\!]. \end{aligned}$$

The proof is completed by passing to the infimum over $x \in \mathbf{V}^{(B)}$. ▷

2.4.8. We are left with checking the axiom of choice inside $\mathbf{V}^{(B)}$. To this end, we need a few auxiliary constructions more.

Take $x, y \in \mathbf{V}^{(B)}$ arbitrarily. Define the *singleton* $\{x\}^B$, the *pair* or *unordered pair* $\{x, y\}^B$, and the *ordered pair* $(x, y)^B$ inside $\mathbf{V}^{(B)}$ by the formulas

$$\begin{gathered} \mathrm{dom}(\{x\}^B) := \{x\}, \quad \mathrm{im}(\{x\}^B) := \{\mathbf{1}\}; \\ \mathrm{dom}(\{x, y\}^B) := \{x, y\}, \quad \mathrm{im}(\{x, y\}^B) := \{\mathbf{1}\}; \\ (x, y)^B := \{\{x\}^B, \{x, y\}^B\}^B. \end{gathered}$$

The elements $\{x\}^B$, $\{x, y\}^B$, and $(x, y) \in \mathbf{V}^{(B)}$ answer to their names:

Theorem. *The following hold:*

$$\begin{gathered} \mathbf{V}^{(B)} \models (\forall\, t)(t \in \{x\}^B \leftrightarrow t = x), \\ \mathbf{V}^{(B)} \models (\forall\, t)(t \in \{x, y\}^B \leftrightarrow t = x \vee t = y), \end{gathered}$$

$$\mathbf{V}^{(B)} \models \text{“}(x,y)^B \textit{ is the ordered pair of } x \textit{ and } y,\text{”}$$

or, in brief,

$$[\![\{x\}^B = \{x\}]\!] = [\![\{x,y\}^B = \{x,y\}]\!] = [\![(x,y)^B = (x,y)]\!] = \mathbf{1}.$$

$\lhd$ By way of example, check the claim about an unordered pair.

Given $t \in \mathbf{V}^{(B)}$, note

$$[\![t \in \{x,y\}^B]\!] = \bigvee\{[\![t = s]\!] : s \in \operatorname{dom}(\{x,y\}^B)\}$$
$$= [\![t = x]\!] \vee [\![t = y]\!] = [\![t = x \vee t = y]\!].$$

Hence,

$$[\![(\forall t)(t \in \{x,y\}^B \leftrightarrow t = x \vee t = y)]\!] = \mathbf{1}. \ \rhd$$

2.4.9. The notions of the preceding subsection about pairs may be easily abstracted to n-tuples for $n > 2$.

Take $x : n \to \mathbf{V}^{(B)}$. By definition, $s := (x(0),\dots,x(n-1))^B \in \mathbf{V}^{(B)}$ provided that there is a mapping $y : n \mapsto \mathbf{V}^{(B)}$ satisfying

$$y(0) = x(0), \quad y(n-1) = s,$$
$$y(k) = (y(k-1), x(k))^B \quad (0 < k \le n-1).$$

Obviously, this defines a function from $(\mathbf{V}^{(B)})^n$ to $\mathbf{V}^{(B)}$ as follows:

$$(x_0,\dots,x_{n-1}) \mapsto (x_0,\dots,x_{n-1})^B \quad (x_0,\dots,x_{n-1} \in \mathbf{V}^{(B)}).$$

We note an important property of this function, confining exposition to the case of $n = 2$ for simplicity.

Recall that for all x, y, x', $y' \in \mathbf{V}$ the equivalence holds:

$$(x,y) = (x',y') \leftrightarrow x = x' \wedge y = y'.$$

This proposition is a theorem of ZF and so it remains true in $\mathbf{V}^{(B)}$ (by 2.4.2–2.4.7). In consequence, given x, y, x', $y' \in \mathbf{V}^{(B)}$, infer

$$[\![(x,y) = (x',y')]\!] = [\![x = x']\!] \wedge [\![y = y']\!].$$

Since $(x,y)^B$ is an ordered pair inside $\mathbf{V}^{(B)}$; therefore,

$$[\![(x,y)^B = (x',y')^B]\!] = [\![x = x']\!] \wedge [\![y = y']\!].$$

In particular,

$$\mathbf{V}^{(B)} \models (x,y)^B = (x',y')^B \leftrightarrow \mathbf{V}^{(B)} \models x = x' \wedge y = y';$$

i.e., “$(\,\cdot\,,\,\cdot\,)^B$ is an injective function in the internal sense.” It goes without saying that this function is also injective in the sense of $\mathbf{V}$; i.e., if $(x,y)^B$ and $(x',y')^B$ coincide as elements of $\mathbf{V}$ then $x = x'$ and $y = y'$. But still these two are different properties.

2.4.10. Recall that by Theorem 1.4.3 an ordinal may be defined as a transitive set totally ordered by the membership relation. In symbols,

$$\mathrm{Ord}\,(x) \leftrightarrow ((\forall\, u \in x)(\forall\, v \in u)(v \in x)$$
$$\wedge(\forall\, u \in x)(\forall\, v \in x)(u \in v \vee u = v \vee v \in u)).$$

Thus, $\mathrm{Ord}\,(x)$ is a bounded formula, and so

$$\alpha \in \mathrm{On} \leftrightarrow \mathbf{V}^{(B)} \models \mathrm{Ord}\,(\alpha^{\wedge})$$

by 2.2.9 (2).

Moreover, as established in 2.2.8 (2),

$$[\![\alpha^{\wedge} = \beta^{\wedge}]\!] = \mathbf{1} \leftrightarrow \alpha = \beta \quad (\alpha, \beta \in \mathrm{On}).$$

2.4.11. *The axiom of choice* AC *holds inside* $\mathbf{V}^{(B)}$:

$$\mathbf{V}^{(B)} \models (\forall\, x)(\exists\, y)(y \textit{ is a choice function on } x).$$

$\lhd$ We may prove in ZF that there is a choice function for a set x whenever we may find an ordinal α and a function f such that $\alpha = \mathrm{dom}(f)$ and $\mathrm{im}(f) \supset u := \bigcup x$.

Indeed, we may define a choice function y by the formula

$$(t, s) \in y \leftrightarrow s \in t \wedge t \in x \wedge (\exists\, \alpha_0 \in \alpha)(f(\alpha_0) = s)$$
$$\wedge(\forall\, \beta \in \alpha)(f(\beta) \in t \to \alpha_0 \le \beta).$$

Thus, $y(t) = f(\alpha_0)$, where α_0 is the least element of the set of ordinals $\{\beta \in \alpha : f(\beta) \in t\}$.

By 2.4.2–2.4.7, the same proposition holds inside $\mathbf{V}^{(B)}$, and so it suffices to show that

$$\mathbf{V}^{(B)} \models (\forall\, u)(\exists\, \alpha)(\exists\, f)(\mathrm{Ord}\,(\alpha) \wedge \mathrm{Fnc}\,(f) \wedge \mathrm{dom}(f) = \alpha \wedge \mathrm{im}(f) \supset u).$$

Take $u \in \mathbf{V}^{(B)}$ and, using the axiom of choice for sets, find an ordinal α and a function g so that $\mathrm{dom}(g) = \alpha$ and $\mathrm{dom}(u) \subset \mathrm{im}(g) \subset \mathbf{V}^{(B)}$.

Define $f \in \mathbf{V}^{(B)}$ by the formula

$$f := \{(\beta^{\wedge}, g(\beta))^{B} : \beta < \alpha\} \times \{\mathbf{1}_B\}.$$

Show that f obeys all conditions we require:

(1) $\mathbf{V}^{(B)} \models$ "f is a binary relation."

Indeed, granted an arbitrary $f \in \mathbf{V}^{(B)}$, observe

$$[\![t \in f]\!] = \bigvee_{\beta<\alpha} [\![t = (\beta^\wedge, g(\beta))^B]\!]$$

$$\leq \bigvee\{[\![t = (x,y)^B]\!] : x,y \in \mathbf{V}^{(B)}\} = [\![(\exists\, x)(\exists\, y)(t = (x,y))]\!].$$

(2) $\mathbf{V}^{(B)} \models \mathrm{Fnc}\,(f)$.

In view of (1), we have only to show that f is single-valued inside $\mathbf{V}^{(B)}$. To this end, take arbitrary $t, s_1, s_2 \in \mathbf{V}^{(B)}$ and proceed with applying 2.1.4 (1), 2.4.9, 2.1.8 (4), and 2.2.8 (2) successively to obtain:

$$[\![(t,s_1) \in f \wedge (t,s_2) \in f]\!] = [\![(t,s_1)^B \in f]\!] \wedge [\![(t,s_2)^B \in f]\!]$$

$$= \bigvee_{\beta<\alpha} \bigvee_{\gamma<\alpha} [\![(t,s_1)^B = (\beta^\wedge, g(\beta))^B]\!] \wedge [\![(t,s_2)^B = (\gamma^\wedge, g(\gamma))^B]\!]$$

$$= \bigvee_{\beta<\alpha} \bigvee_{\gamma<\alpha} [\![t = \beta^\wedge]\!] \wedge [\![t = \gamma^\wedge]\!] \wedge [\![s_1 = g(\beta)]\!] \wedge [\![s_2 = g(\gamma)]\!]$$

$$\leq \bigvee_{\beta<\alpha} \bigvee_{\gamma<\alpha} [\![\beta^\wedge = \gamma^\wedge]\!] \wedge [\![s_1 = g(\beta)]\!] \wedge [\![s_2 = g(\gamma)]\!]$$

$$= \bigwedge_{\beta<\alpha} [\![s_1 = g(\beta)]\!] \wedge [\![s_2 = g(\beta)]\!] \leq [\![s_1 = s_2]\!].$$

(3) $\mathbf{V}^{(B)} \models \mathrm{Ord}\,(\alpha^\wedge) \wedge \mathrm{dom}(f) = \alpha^\wedge$.

The formula $\mathbf{V}^{(B)} \models \mathrm{Ord}\,(\alpha^\wedge)$ was discussed in 2.4.10. Furthermore, given $t \in \mathbf{V}^{(B)}$, infer

$$[\![t \in \mathrm{dom}(f)]\!] = [\![(\exists\, s)(t,s) \in f]\!] = \bigvee_{s\in\mathbf{V}^{(B)}} [\![(t,s) \in f]\!]$$

$$= \bigvee_{s\in\mathbf{V}^{(B)}} \bigvee_{\beta<\alpha} [\![(t,s) = (\beta^\wedge, g(\beta))]\!]$$

$$= \bigvee_{\beta<\alpha} \bigwedge_{s\in\mathbf{V}^{(B)}} [\![t = \beta^\wedge]\!] \wedge [\![s = g(\beta)]\!]$$

$$= \bigvee_{\beta<\alpha} [\![t = \beta^\wedge]\!] = \bigvee_{\beta\in\mathrm{dom}(\alpha^\wedge)} [\![t = \beta]\!] = [\![t \in \alpha^\wedge]\!].$$

(4) $\mathbf{V}^{(B)} \models \mathrm{im}(f) \supset u$.

Take $s \in \mathbf{V}^{(B)}$ and carry out the following calculations:

$$
\begin{aligned}
[\![s \in u]\!] = \bigvee_{v\in \mathrm{dom}(u)} u(v) \wedge [\![s = v]\!] \le \bigvee_{\beta\le\alpha} [\![s = g(\beta)]\!] \\
= \bigvee_{\beta<\alpha}\Big([\![s = g(\beta)]\!] \wedge \bigvee_{t\in \mathbf{V}^{(B)}} [\![\beta^{\wedge} = t]\!]\Big) \\
= \bigvee_{\beta<\alpha}\bigvee_{t\in \mathbf{V}^{(B)}} [\![(t,s) = (\beta^{\wedge}, g(\beta))]\!] \\
= \bigvee_{t\in \mathbf{V}^{(B)}} [\![(t,s)\in f]\!] = [\![(\exists\, t)(t,s)\in f]\!] = [\![s \in \mathrm{im}(f)]\!].
\end{aligned}
$$

The proof of Theorem 2.4.1 is complete. ▷

2.4.12. Comments.

(1) Substituting the laws of intuitionistic logic (see 2.1.10 (3)) for the logical part of the language of ZF, we come to intuitionistic set theory $\mathrm{ZF_I}$. The models of $\mathrm{ZF_I}$ may also be constructed as above. Namely, if Ω is a complete Heyting lattice then the universe $\mathbf{V}^{(\Omega)}$ becomes a Heyting valued model of $\mathrm{ZF_I}$ provided that the corresponding truth values $[\![\cdot \in \cdot]\!]$ and $[\![\cdot = \cdot]\!]$, acting from $\mathbf{V}^{(\Omega)} \times \mathbf{V}^{(\Omega)}$ to $\mathbf{V}^{(\Omega)}$, are defined. For details, consult [54, 70, 238].

(2) Let B be a quantum logic (see 1.5.11 (5)). If the truth values $[\![\cdot \in \cdot]\!]$ and $[\![\cdot = \cdot]\!]$ are defined as in 2.1.4 and the truth values of formulas are given as in 2.1.7; then the axioms ZF_2–ZF_6 and AC hold inside the universe $\mathbf{V}^{(B)}$. Therefore, we may develop the corresponding set theory inside $\mathbf{V}^{(B)}$. In particular, the reals inside $\mathbf{V}^{(B)}$ will correspond to the observables in the mathematical model of a quantum-mechanical system (cf. [234]).

2.5. Separated Boolean Valued Universes

In this section, we construct a separated Boolean valued universe and interpret NGB therein (cf. [155]).

2.5.1. Given elements x and y of the universe $\mathbf{V}^{(B)}$ which satisfy the condition $\mathbf{V}^{(B)} \models x = y$, we cannot assert in general that x and y are equal as sets, i.e., as elements of $\mathbf{V}$. Indeed, take an ordinal α and define $x_\alpha \in \mathbf{V}^{(B)}$ by the formulas $\mathrm{dom}(x_\alpha) = V_\alpha^{(B)}$ and $\mathrm{im}(x_\alpha) := \{\mathbf{0}\}$. Then, it is evident that $[\![x_\alpha = 0]\!] = \mathbf{1}$ for all α. Therefore, every element of the class $\{x_\alpha : \alpha \in \mathrm{On}\}$ depicts the empty set inside $\mathbf{V}^{(B)}$.

It can be shown that to each $x \in \mathbf{V}^{(B)}$ there corresponds the proper class of all $y \in \mathbf{V}^{(B)}$ satisfying $[\![x = y]\!] = \mathbf{1}$. This peculiarity causes considerable technical inconveniences and, in particular, hampers translations from the language of $\mathbf{V}^{(B)}$, i.e. the B-language, into the language of $\mathbf{V}$, i.e. the conventional language of ZFC. This deficiency of $\mathbf{V}^{(B)}$ is eliminated by a proper factorization (cf. 1.5.8).

2.5.2. Furnish the universe $\mathbf{V}^{(B)}$ over a Boolean algebra B with the equivalence

$$\sim := \{(x, y) \in \mathbf{V}^{(B)} \times \mathbf{V}^{(B)} : [\![x = y]\!] = \mathbf{1}_B\}.$$

Consider the factor class $\widetilde{\mathbf{V}}^{(B)} := \mathbf{V}^{(B)}/\!\sim$ and let $\pi : \mathbf{V}^{(B)} \to \widetilde{\mathbf{V}}^{(B)}$ stand for the factor mapping.

The class $\widetilde{\mathbf{V}}^{(B)}$ is the *separated Boolean valued universe* over B. Define the Boolean truth values for the equality $[\![\cdot = \cdot]\!]_s$ and the membership $[\![\cdot \in \cdot]\!]_s$ in $\widetilde{\mathbf{V}}^{(B)}$ on using the quotients of the corresponding Boolean truth values $[\![\cdot = \cdot]\!]$ and $[\![\cdot \in \cdot]\!]$ by $\sim$:

$$[\![\cdot = \cdot]\!]_s := [\![\cdot = \cdot]\!] \circ (\pi^{-1} \times \pi^{-1}),$$
$$[\![\cdot \in \cdot]\!]_s := [\![\cdot \in \cdot]\!] \circ (\pi^{-1} \times \pi^{-1}).$$

Given a formula $\varphi(u_1, \dots, u_n)$ and $\tilde{x}, \dots, \tilde{x}_n \in \widetilde{\mathbf{V}}^{(B)}$, define $[\![\varphi(\widetilde{x}_1, \dots, \widetilde{x}_n)]\!] \in B$ in exactly the same way as in 2.1.7 to obtain

$$[\![\varphi(x_1, \dots, x_n)]\!] = [\![\varphi(\pi x_1, \dots, \pi x_n)]\!]_s \quad (x_1, \dots, x_n \in \mathbf{V}^{(B)}).$$

Define the truth of formulas in $\widetilde{\mathbf{V}}^{(B)}$ as in 2.1.6:

$$\widetilde{\mathbf{V}}^{(B)} \models \varphi(\widetilde{x}_1, \dots, \widetilde{x}_n) \leftrightarrow [\![\varphi(\widetilde{x}_1, \dots, \widetilde{x}_n)]\!]_s = \mathbf{1}_B.$$

The soundness of the above definitions is obvious since, by 2.1.8 (7),

$$\mathbf{1} = [\![x = y]\!] \to [\![\varphi(x)]\!] = [\![\varphi(y)]\!] \quad (x, y \in \mathbf{V}^{(B)})$$

for every formula φ of ZFC. Therefore, calculating Boolean truth values in a separated Boolean valued universe, we may take arbitrary representatives of the equivalence classes under study. From this observation it is obvious in particular that Theorem 2.1.8 remains true with $\widetilde{\mathbf{V}}^{(B)}$ in place of $\mathbf{V}^{(B)}$ and the Boolean truth values decorated with the index s.

As a somewhat unexpected example, consider the following definition: Given $\widetilde{x} \in \widetilde{\mathbf{V}}^{(B)}$, denote by $\vee\widetilde{x}$ the *level* of $\widetilde{x}$, i.e., the element of B defined as follows:

$$\vee\widetilde{x} := \bigvee_{t \in \operatorname{dom}(x)} x(t),$$

where $x \in \mathbf{V}^{(B)}$ is the equivalence class of $\widetilde{x} \in \mathbf{V}^{(B)}$.

At first sight, this definition seems illegitimate since the domains of elements equal inside $\mathbf{V}^{(B)}$ may differ. However,

$$[\![(\exists\, y \in \widetilde{x})]\!]_s = [\![(\exists\, y \in \widetilde{x}) y = y]\!]_s$$
$$= \bigvee_{t \in \operatorname{dom}(x)} x(t) \wedge [\![t = t]\!] = \bigvee_{t \in \operatorname{dom}(x)} x(t) = \vee \widetilde{x}.$$

Obviously, $\vee \widetilde{x} = [\![x \neq \varnothing]\!]_s$, and so the definition of level is sound.

By analogy, given $\widetilde{x}$ in $\widetilde{\mathbf{V}}^{(B)}$ and b in B, we may correctly define the element $\widetilde{bx} : t \mapsto b \wedge x(t)$ $(t \in \operatorname{dom}(x))$. Indeed, if $[\![x_1 = x_2]\!] = \mathbf{1}$ then, by 2.3.2, $[\![bx_1 = bx_2]\!] = b \Rightarrow [\![x_1 = x_2]\!] = \mathbf{1}$.

In view of this it is customary to use the designation $0 = \varnothing$, which implies in particular that $0\varnothing = \varnothing = 0\widetilde{x}$ for every $x \in \widetilde{\mathbf{V}}^{(B)}$.

2.5.3. Note that the facts of 2.2–2.4 hold true in $\widetilde{\mathbf{V}}^{(B)}$ on assuming obvious specification and clarification.

For instance, $\widetilde{\mathbf{V}}^{(B)}$ is a model of ZFC in the sense of 2.4. Similarly, if ρ is a complete homomorphism of Boolean algebras then ρ^* keeps invariant every equivalence class. Hence, ρ^* induces a unique mapping of the corresponding separated universes which is also denoted by ρ^*, proving that an analog of 2.2.2 holds, etc.

Assume that $(x_\xi) \subset \mathbf{V}^{(B)}$ and (b_ξ) is a disjoint family in B. Put $x = \operatorname{mix}(b_\xi x_\xi)$. We will continue to use the name "mixing" for calling the element $\widetilde{x} := \pi x$ and preserve the notation $\widetilde{x} = \operatorname{mix}(b_\xi \widetilde{x}_\xi)$ $(\widetilde{x}_\xi = \pi x_\xi)$. This definition of mixing in $\widetilde{\mathbf{V}}^{(B)}$ is clearly correct (cf. 2.3.5 (1)). Therefore, if $\widetilde{x} \in \widetilde{\mathbf{V}}^{(B)}$ and $(\widetilde{x}_\xi) \subset \widetilde{\mathbf{V}}^{(B)}$ then the record $\widetilde{x} = \operatorname{mix}(b_\xi \widetilde{x}_\xi)$ means that

$$b_\xi \leq [\![\widetilde{x} = \widetilde{x}_\xi]\!]_s \quad (\xi \in \Xi).$$

Note that if (b_ξ) is a partition of unity then the mixing $\operatorname{mix}(b_\xi x_\xi)$ is unique due to separation (cf. 2.3.3).

The equality (cf. 2.4.9)

$$[\![(x, y)^B = (x', y')^B]\!] = [\![x = x']\!] \wedge [\![y = y']\!]$$

shows that the mapping $(\,\cdot\,,\,\cdot\,)^B$ to be stable under the equivalence relation of 2.5.2. Hence, there is an injective embedding $\widetilde{\mathbf{V}}^{(B)} \times \widetilde{\mathbf{V}}^{(B)} \to \widetilde{\mathbf{V}}^{(B)}$ denoted by the same symbol $(\,\cdot\,,\,\cdot\,)^B$ and satisfying $(\pi x, \pi y)^B = \pi((x, y)^B)$. In this event

$$[\![(\widetilde{x}, \widetilde{y})^B = (\widetilde{x}, \widetilde{y})]\!]_s = \mathbf{1} \quad (\widetilde{x}, \widetilde{y} \in \mathbf{V}^{(B)}).$$

The maximum principle is still true and admits the following clarification.

2.5.4. *Assume that $\varphi(u, u_1, \dots, u_n)$ is a formula, $\widetilde{x}_1, \dots, \widetilde{x}_n \in \widetilde{\mathbf{V}}^{(B)}$, and $\widetilde{\mathbf{V}}^{(B)} \models (\exists! u) \varphi(u, \widetilde{x}_1, \dots, \widetilde{x}_n)$. Then there is a unique element $\widetilde{x}_0 \in \widetilde{\mathbf{V}}^{(B)}$ such that $\widetilde{\mathbf{V}}^{(B)} \models \varphi(\widetilde{x}_0, \widetilde{x}_1, \dots, \widetilde{x}_n)$.*

◁ Put $\widetilde{x}_k := \pi(x_k)$, where $x_k \in \mathbf{V}^{(B)}$ $(k := 1, \dots, n)$. Note then that $\mathbf{V}^{(B)} \models (\exists! u)\varphi(u, x_1, \dots, x_n)$. By the transfer principle, there is an element $x_0 \in \mathbf{V}^{(B)}$, such that $\mathbf{V}^{(B)} \models \varphi(x_0, x_1, \dots, x_n)$. Assign $\widetilde{x}_0 := \pi(x_0)$. Obviously, $\widetilde{\mathbf{V}}^{(B)} \models \varphi(\widetilde{x}_0, \widetilde{x}_1, \dots, \widetilde{x}_n)$. If $\widetilde{\mathbf{V}}^{(B)} \models \varphi(z, \widetilde{x}_1, \dots, \widetilde{x}_n)$ holds for $z \in \widetilde{\mathbf{V}}^{(B)}$ then $\widetilde{\mathbf{V}}^{(B)} \models \varphi(\widetilde{x}_0, \dots, \widetilde{x}_n) \wedge \varphi(z, \widetilde{x}_1, \dots, \widetilde{x}_n)$. By hypothesis, $\widetilde{\mathbf{V}}^{(B)} \models z = \widetilde{x}_0$, which implies $z = \widetilde{x}_0$ since $\widetilde{\mathbf{V}}^{(B)}$ is separated. ▷

2.5.5. Given b and $c \in B$, put (cf. 1.1.4)

$$[\![b = c]\!] := b \Leftrightarrow c := (b \,\Delta\, c)^* = (b \wedge c) \vee (b^* \wedge c^*).$$

Note that, by 1.1.4 (3), $a \leq [\![b = c]\!]$ if and only if $a \wedge b = a \wedge c$.

Consider a function $f : \operatorname{dom}(f) \to B$ whose domain $\operatorname{dom}(f)$ is contained in $\widetilde{\mathbf{V}}^{(B)}$. Say that f is *extensional*

$$[\![x = y]\!]_s \leq [\![f(x) = f(y)]\!] \quad (x, y \in \operatorname{dom}(f)).$$

The extensionality of f amounts clearly to the formula

$$f(x) \wedge [\![x = y]\!]_s \leq f(y) \quad (x, y \in \operatorname{dom}(f)).$$

If $u : \operatorname{dom}(u) \to B$ is an arbitrary function and $\operatorname{dom}(u) \subset \widetilde{\mathbf{V}}^{(B)}$ then we may related to u the extensional function $\bar{u} : \widetilde{\mathbf{V}}^{(B)} \to B$ by the formula

$$\bar{u} : x \mapsto \bigvee_{t \in \operatorname{dom}(u)} u(t) \wedge [\![t = x]\!]_s \quad (x \in \widetilde{\mathbf{V}}^{(B)}).$$

Another class of extensional functions arises as follows. Let φ be a B-formula. Then the following function is extensional

$$\bar{\varphi} : x \mapsto [\![\varphi(x)]\!]_s \quad (x \in \widetilde{\mathbf{V}}^{(B)}).$$

2.5.6. Theorem. *If $u : \operatorname{dom}(u) \to B$ is a function with $\operatorname{dom}(u) \subset \widetilde{\mathbf{V}}^{(B)}$ and $\operatorname{dom}(u) \in \mathbf{V}$ then there is a unique $x \in \widetilde{\mathbf{V}}^{(B)}$ such that $\bar{u}(t) = [\![t \in x]\!]_s$ for all $t \in \widetilde{\mathbf{V}}^{(B)}$.*

Conversely, if $x \in \widetilde{\mathbf{V}}^{(B)}$ then there is a function $u : \operatorname{dom}(u) \to B$ such that $\operatorname{dom}(u) \subset \widetilde{\mathbf{V}}^{(B)}$, $\operatorname{dom}(u) \in \mathbf{V}$, and $\bar{u}(t) = [\![t \in x]\!]_s$ $(t \in \widetilde{\mathbf{V}}^{(B)})$.

◁ Denote by D the subset of the unseparated Boolean valued universe $\mathbf{V}^{(B)}$ whose image under the factor mapping π is $\operatorname{dom}(u)$. Define an element $x' \in \mathbf{V}^{(B)}$ by the formula

$$\operatorname{dom}(x') := D, \quad x'(t) := u(\pi t) \quad (t \in D).$$

Finally, put $x := \pi(x')$.

Given $t \in \widetilde{\mathbf{V}}^{(B)}$, find then that

$$[\![t \in x]\!]_s = \bigvee_{y \in D} x'(y) \wedge [\![t = \pi y]\!]_s = \bigvee_{y \in \mathrm{dom}(u)} x(y) \wedge [\![y = t]\!] = \bar{u}(t).$$

If another element $z \in \widetilde{\mathbf{V}}^{(B)}$ has the same properties then $[\![t \in x]\!]_s = [\![t \in z]\!]_s$ for all $t \in \widetilde{\mathbf{V}}^{(B)}$. Hence,

$$\widetilde{\mathbf{V}}^{(B)} \models (\forall t)\ (t \in x \leftrightarrow t \in z).$$

By the axiom of extensionality, arguing inside $\widetilde{\mathbf{V}}^{(B)}$, note that $[\![x = z]\!]_s = \mathbf{1}$. As $\widetilde{\mathbf{V}}^{(B)}$ is separated, $x = z$.

Conversely, take $x \in \widetilde{\mathbf{V}}^{(B)}$, and let x' be an element of the separated universe such that $x = \pi(x')$. Put $\mathrm{dom}(u) := \pi``(\mathrm{dom}(x'))$ and define $u : \mathrm{dom}(u) \to B$ so that $u(\pi t) = x'(t)$ $(t \in \mathrm{dom}(x'))$. In this event, granted $t \in \widetilde{\mathbf{V}}^{(B)}$, observe

$$\begin{aligned} [\![t \in x]\!]_s &= \bigvee_{y \in \mathrm{dom}(x')} x'(y) \wedge [\![t = \pi y]\!]_s \\ &= \bigvee_{y \in \mathrm{dom}(u)} u(y) \wedge [\![y = t]\!]_s = \bar{u}(t), \end{aligned}$$

so completing the proof. ▷

2.5.7. Throughout the sequel we as a rule deal with a separated Boolean valued universe $\widetilde{\mathbf{V}}^{(B)}$. Moreover, calculating Boolean truth values, we often replace elements of $\widetilde{\mathbf{V}}^{(B)}$ with their representatives in $\mathbf{V}^{(B)}$ without further specification (recall a similar practice of analysis of handling the spaces of cosets of measurable functions).

Furthermore, starting with the sentence to follow, we will omit the sign $\sim$ and index s and simply write $\mathbf{V}^{(B)}$, $[\![\cdot = \cdot]\!]$, and $[\![\cdot \in \cdot]\!]$ instead of $\widetilde{\mathbf{V}}^{(B)}$, $[\![\cdot = \cdot]\!]_s$, and $[\![\cdot \in \cdot]\!]_s$. We also carry out all analogous simplifications since this leads to no confusion.

As seen from 2.5.6, each member of $\mathbf{V}^{(B)}$ defines some extensional mapping from $\mathbf{V}^{(B)}$ to B. However, only part of extensional mappings from $\mathbf{V}^{(B)}$ in B are determined by elements in $\mathbf{V}^{(B)}$. This peculiarity motivates the following definition:

2.5.8. A *class inside* $\mathbf{V}^{(B)}$ or the $\mathbf{V}^{(B)}$*-class* is an extensional mapping $X : \mathbf{V}^{(B)} \to B$ that is a class in the conventional set-theoretic sense; i.e., in the sense of $\mathbf{V}$.

To each element $x \in \mathbf{V}^{(B)}$ we assign the $\mathbf{V}^{(B)}$-class

$$\langle x \rangle := [\![\cdot \in x]\!] : t \mapsto [\![t \in x]\!] \quad (t \in \mathbf{V}^{(B)}).$$

This correspondence is obviously injective.

Given $\mathbf{V}^{(B)}$-classes X and Y and an element $z \in \mathbf{V}^{(B)}$, we now introduce Boolean truth values by putting

$$[\![\langle z\rangle \in X]\!] := X(z),$$

$$[\![X = Y]\!] := \bigwedge_{u \in \mathbf{V}^{(B)}} [\![\langle u\rangle \in X]\!] \Leftrightarrow [\![\langle u\rangle \in Y]\!],$$

$$[\![X \in Y]\!] := \bigvee_{u \in \mathbf{V}^{(B)}} [\![\langle u\rangle = X]\!] \wedge [\![\langle u\rangle \in Y]\!].$$

The first and third formulas are consistent, since the fact that X is extensional implies

$$[\![\langle z\rangle \in X]\!] = \bigvee_{u \in \mathbf{V}^{(B)}} X(u) \wedge [\![u = z]\!];$$

and, moreover, $[\![\langle z\rangle = \langle u\rangle]\!] = [\![z = u]\!]$ for all $u, z \in \mathbf{V}^{(B)}$. It follows from the definitions that $[\![X = Y]\!] = \mathbf{1}$ implies $X = Y$.

The function $\mathbf{U}_B : x \mapsto \mathbf{1}_B$ $(x \in \mathbf{V}^{(B)})$ is the *universal class* inside $\mathbf{V}^{(B)}$. The empty $\mathbf{V}^{(B)}$-class is the identically zero function over $\mathbf{V}^{(B)}$.

2.5.9. Recall that a (set-theoretic) formula φ is *predicative* if each bound variable of φ ranges over sets (cf. 1.3.1 and 1.3.14).

(1) We define the Boolean truth value for a predicative formula by induction on length (cf. 2.1.6).

Dealing with propositional connectives, we proceed in much the same way as in 2.1.7. We are thus left with elaborating the case of quantifiers by variables ranging over sets. Moreover, we may consider only the formulas having no subformulas of the type $X_1 \in X_2$, since the latter formula is equivalent to the formula $(\exists\, x)(x = X_1 \wedge x \in X_2)$.

So, assume that φ is a predicative formula with free variables $X, X_1, \dots, X_n$ and $Y_1, \dots, Y_n$ are some $\mathbf{V}^{(B)}$-classes.

By definition, put

$$[\![(\forall\, x)\varphi(x, Y_1, \dots, Y_n)]\!] = \bigwedge_{y \in \mathbf{V}^{(B)}} [\![\varphi(y, Y_1, \dots, Y_n)]\!],$$

$$[\![(\exists\, x)\varphi(x, Y_1, \dots, Y_n)]\!] = \bigvee_{y \in \mathbf{V}^{(B)}} [\![\varphi(y, Y_1, \dots, Y_n)]\!].$$

We say that a predicative formula $\varphi(X_1, \dots, X_n)$ *holds* or is *satisfied* inside $\mathbf{V}^{(B)}$ by the assignment of $Y_1, \dots, Y_n$ to the variables $X_1, \dots, X_n$ if $[\![\varphi(Y_1, \dots, Y_n)]\!] = \mathbf{1}$.

As in 2.1.6, in this event we write

$$V^{(B)} \models \varphi(Y_1,\dots,Y_n) \leftrightarrow [\![\varphi(Y_1,\dots,Y_n)]\!] = \mathbf{1}.$$

(2) The notion of satisfaction in $\mathbf{V}^{(B)}$ extends to nonpredicative formulas as follows:

If $\varphi(X, X_1,\dots,X_n)$ is a nonpredicative formula, then we put

$$\mathbf{V}^{(B)} \models (\forall X)\varphi(X, Y_1,\dots,Y_n) \quad (\mathbf{V}^{(B)} \models (\exists X)\varphi(X, Y_1,\dots,Y_n))$$

if and only if $[\![\varphi(Y, Y_1,\dots,Y_n)]\!] = \mathbf{1}$ for every $\mathbf{V}^{(B)}$-class Y (respectively, there is some $\mathbf{V}^{(B)}$-class Y such that $[\![\varphi(Y, Y_1,\dots, Y_n)]\!] = \mathbf{1}$).

A $\mathbf{V}^{(B)}$-class Y is a $\mathbf{V}^{(B)}$-*set* provided that $\mathbf{V}^{(B)} \models M(Y)$, where $M(X) := (\exists Z)(X \in Z)$ (cf. 1.3.1).

It would simpler to use the term "B-set" instead of "$\mathbf{V}^{(B)}$-set." However, the former is reserved for another special mission (cf. 3.4).

2.5.10. *For every $x \in \mathbf{V}^{(B)}$, the $\mathbf{V}^{(B)}$-class $\langle x\rangle$ is a $\mathbf{V}^{(B)}$-set. Conversely, if a $\mathbf{V}^{(B)}$-class X is a $\mathbf{V}^{(B)}$-set then $X = \langle x\rangle$ for some $x \in \mathbf{V}^{(B)}$.*

$\lhd$ Granted an arbitrary element $x \in \mathbf{V}^{(B)}$, observe

$$[\![\langle x\rangle \in \langle\{x\}^B\rangle]\!] = [\![x \in \{x\}^B]\!] = \mathbf{1},$$

and so $\mathbf{V}^{(B)} \models M(\langle x\rangle)$. Assume that $\mathbf{V}^{(B)} \models M(X)$ for a $\mathbf{V}^{(B)}$-class X. Then, by definition (cf. 2.5.9 (2)), there is a $\mathbf{V}^{(B)}$-class Z such that

$$\bigvee_{t\in\mathbf{V}^{(B)}} Z(t) \wedge [\![\langle t\rangle = X]\!] = \mathbf{1}.$$

Hence, using the exhaustion principle, we may choose a partition of unity $(b_\xi)_{\xi\in\Xi}$ and a family $(x_\xi)_{\xi\in\Xi} \subset \mathbf{V}^{(B)}$ such that

$$[\![\langle x_\xi\rangle = X]\!] \geq b_\xi \quad (\xi \in \Xi).$$

If $x := \mathrm{mix}(b_\xi x_\xi)$ then

$$[\![\langle x\rangle = X]\!] \geq [\![\langle x\rangle = \langle x_\xi\rangle]\!] \wedge [\![\langle x_\xi\rangle = X]\!] \geq b_\xi,$$

and so $[\![\langle x\rangle = X]\!] = \mathbf{1}$ or $\langle x\rangle = X$. $\rhd$

This fact enables us to identify an element $x \in \mathbf{V}^{(B)}$ and the respective $\mathbf{V}^{(B)}$-set $\langle x\rangle$ in the sequel.

2.5.11. Assume that C is another complete Boolean algebra and $\pi : B \to C$ is a complete homomorphism of B to C. Consider a $\mathbf{V}^{(B)}$-class X and define

$$(x, b) \in \pi^* X \leftrightarrow b = \bigvee_{t \in \mathbf{V}^{(B)}} (\pi \circ X)(t) \wedge [\![x = \pi^* t]\!]^C.$$

Then $\pi^* X$ is a class inside $\mathbf{V}^{(B)}$.

Indeed, $\pi^* X$ is a subclass of $\mathbf{V}$ by Theorem 1.3.14, since

$$\pi^* X = \{(x, b) : \varphi(x, b, B, C, X, \pi^*, \ [\![\cdot = \cdot]\!], \mathbf{V}^{(B)})\}$$

for the predicative formula

$$\varphi(Y, Z, B, \dots) :\, Z = \bigvee_{t \in \mathbf{V}^{(B)}} (\pi \circ X)(t) \wedge [\![Y = \pi^* t]\!].$$

In addition, $\pi^* X$ is an extensional function:

$$(\pi^* X)(x) \wedge [\![x = y]\!] = \bigvee_{t \in \mathbf{V}^{(B)}} (\pi \circ X)(t) \wedge [\![x = \pi^* t]\!]$$

$$\wedge [\![x = y]\!] \leq \bigvee_{t \in \mathbf{V}^{(B)}} (\pi \circ X)(t) \wedge [\![y = \pi^* t]\!] = (\pi^* X)(y).$$

It is easy that 2.2.2 (1) holds for classes; i.e., if ρ is a complete homomorphism then

$$(\rho \circ \pi)^* X = (\rho^* \circ \pi^*) X.$$

Furthermore, if $\mathbf{V}^{(B)} \models M(X)$ then $\mathbf{V}^{(C)} \models M(\pi^* X)$. Indeed, if $X = \langle x \rangle$, $x \in \mathbf{V}^{(B)}$ then, by 2.2.2 (4),

$$(\pi^* x)(t) = \bigvee_{u \in \mathbf{V}^{(B)}} \pi([\![u = x]\!]) \wedge [\![t = \pi^* u]\!]$$

$$= \bigvee_{u \in \mathbf{V}^{(B)}} (\pi \circ \langle x \rangle)(u) \wedge [\![t = \pi^* u]\!] = (\pi^* \langle x \rangle)(t).$$

Therefore, $\langle \pi^* x \rangle = \pi^* \langle x \rangle = \pi^* X$.

The converse proposition is also true provided that π is injective.

Note finally that the definition above agrees with 2.2.1 because of 2.2.2 (4).

2.5.12. *For every* $\mathbf{V}^{(B)}$*-class* X *and every predicative* B*-formula* φ *with a single free variable, the following hold:*

$$[\![(\forall x \in \pi^* X)\varphi(x)]\!]^C = \bigwedge_{t \in \mathbf{V}^{(B)}} \pi \circ X(t) \Rightarrow [\![\varphi(\pi^* t)]\!]^C,$$

$$[\![(\exists x \in \pi^* X)\varphi(x)]\!]^C = \bigvee_{t \in \mathbf{V}^{(B)}} \pi \circ X(t) \wedge [\![\varphi(\pi^* t)]\!]^C.$$

$\triangleleft$ It suffices to prove either of the these formulas, say, the first. The needed calculations follow (on using 1.1.5 (3), 2.1.8 (7), and $(a \wedge b) \Rightarrow (c \wedge b) = (a \wedge b) \Rightarrow c$):

$$[\![(\forall x \in \pi^* X)\varphi(x)]\!] = \bigwedge_{x \in \mathbf{V}^{(C)}} [\![x \in \pi^* X]\!] \Rightarrow [\![\varphi(x)]\!]$$

$$= \bigwedge_{x \in \mathbf{V}^{(C)}} \Bigg(\bigvee_{t \in \mathbf{V}^{(B)}} \pi \circ X(t) \wedge [\![x = \pi^* t]\!] \Bigg) \Rightarrow [\![\varphi(x)]\!]$$

$$= \bigwedge_{t \in \mathbf{V}^{(B)}} \bigwedge_{x \in \mathbf{V}^{(C)}} (\pi \circ X(t) \wedge [\![x = \pi^* t]\!]) \Rightarrow [\![\varphi(x)]\!]$$

$$\leq \bigwedge_{t \in \mathbf{V}^{(B)}} \pi \circ X(t) \Rightarrow [\![\varphi(\pi^* t)]\!]$$

$$= \bigwedge_{t \in \mathbf{V}^{(B)}} \Bigg(\bigwedge_{x \in \mathbf{V}^{(C)}} (\pi \circ X(t))^* \vee [\![x = \pi^* t]\!]^* \vee [\![\varphi(\pi^* t)]\!] \Bigg)$$

$$= \bigwedge_{t \in \mathbf{V}^{(B)}} \bigwedge_{x \in \mathbf{V}^{(C)}} (\pi \circ X(t) \wedge [\![x = \pi^* t]\!]) \Rightarrow ([\![\varphi(\pi^* t)]\!] \wedge [\![x = \pi^* t]\!])$$

$$\leq \bigwedge_{t \in \mathbf{V}^{(B)}} \bigwedge_{x \in \mathbf{V}^{(C)}} (\pi \circ X(t) \wedge [\![x = \pi^* t]\!]) \Rightarrow [\![\varphi(x)]\!]$$

$$= \bigwedge_{x \in \mathbf{V}^{(C)}} \Bigg(\bigvee_{t \in \mathbf{V}^{(B)}} \pi \circ X(t) \wedge [\![x = \pi^* t]\!] \Bigg) \Rightarrow [\![\varphi(x)]\!]$$

$$= \bigwedge_{x \in \mathbf{V}^{(C)}} [\![x \in \pi^* X]\!] \Rightarrow [\![\varphi(x)]\!] = [\![(\forall x \in \pi^* X)\varphi(x)]\!].$$

The proof is complete. $\triangleright$

2.5.13. *For all* $\mathbf{V}^{(B)}$*-classes* X *and* Y*, the following hold:*

$$[\![\pi^* X = \pi^* Y]\!]^C = \pi[\![X = Y]\!]^B, \quad [\![\pi^* X \in \pi^* Y]\!]^C = \pi[\![X \in Y]\!]^B.$$

◁ Note first that $\pi \circ Y(t) = (\pi^*Y)(\pi^*t)$ or $\pi[\![t \in Y]\!]^B = [\![\pi^*t \in \pi^*Y]\!]^C$ for $t \in \mathbf{V}^{(B)}$ (this follows from 2.5.8 and 2.5.11 by 2.2.2 (3)). Then, using the first formula of 2.5.12, deduce

$$
\begin{aligned}
[\![\pi^*X \subset \pi^*Y]\!]^C &= [\![(\forall x \in \pi^*X)(x \in \pi^*Y)]\!]^C \\
&= \bigwedge_{t \in \mathbf{V}^{(B)}} \pi \circ X(t) \Rightarrow [\![\pi^*t \in \pi^*Y]\!]^C \\
&= \bigwedge_{t \in \mathbf{V}^{(B)}} \pi([\![t \in X]\!]^B \Rightarrow [\![t \in Y]\!]^B) = \pi[\![X \subset Y]\!]^B.
\end{aligned}
$$

Whence

$$
[\![\pi^*X = \pi^*Y]\!]^C = [\![\pi^*X \subset \pi^*Y]\!]^C \wedge [\![\pi^*Y \subset \pi^*X]\!]^C = \pi[\![X = Y]\!]^B.
$$

Finally, using the above and the second formula of 2.5.12, obtain

$$
\begin{aligned}
[\![\pi^*X \in \pi^*Y]\!]^C &= [\![(\exists t \in \pi^*Y)\ (t = \pi^*X)]\!]^C \\
&= \bigvee_{t \in \mathbf{V}^{(B)}} \pi \circ Y(t) \wedge [\![\pi^*t = \pi^*X]\!]^C \\
&= \bigvee_{t \in \mathbf{V}^{(B)}} \pi\Big(Y(t) \wedge [\![t = X]\!]^B\Big) = \pi[\![X \in Y]\!]^B,
\end{aligned}
$$

which completes the proof. ▷

2.5.14. The above facts allow us to translate some results of Section 2.2 to a new environment. We list only a few:

(1) *If $\varphi(Y_1, \dots, Y_n)$ is a bounded predicative formula then*

$$
\pi[\![\varphi(X_1, \dots, X_n)]\!]^B = [\![\varphi(\pi^*X_1, \dots, \pi^*X_n)]\!]^C
$$

for all $\mathbf{V}^{(B)}$-classes $X_1, \dots, X_n$. In particular, if π is a monomorphism then

$$
\mathbf{V}^{(B)} \models \varphi(X_1, \dots, X_n) \leftrightarrow \mathbf{V}^{(C)} \models \varphi(\pi^*X_1, \dots, \pi^*X_n).
$$

(2) *If φ is a predicative formula of class Σ_1 then*

$$
\pi[\![\varphi(X_1, \dots, X_n)]\!]^B \le [\![\varphi(\pi^*X_1, \dots, \pi^*X_n)]\!]^C,
$$

with $X_1, \dots, X_n$ the same as before. In particular, the following implication holds:

$$
\mathbf{V}^{(B)} \models \varphi(X_1, \dots, X_n) \to \mathbf{V}^{(C)} \models \varphi(\pi^*X_1, \dots, \pi^*X_n).
$$

◁ The proof is carried out along the lines of 2.2.3. By way of example, consider the case of a bounded universal quantifier: $\varphi := (\forall\, x \in Y)\psi$.

By 2.5.12 and 2.5.13, granted $\mathbf{V}^{(B)}$-classes $Y, X_1, \dots, X_n$, observe

$$
\begin{aligned}
&[\![\varphi(\pi^*Y, \pi^*X_1, \dots, \pi^*X_n)]\!] \\
&= \bigwedge_{x \in \mathbf{V}^{(B)}} [\![\pi^*x \in \pi^*Y]\!] \Rightarrow [\![\psi(\pi^*x, \pi^*X_1, \dots, \pi^*X_n)]\!] \\
&= \bigwedge_{x \in \mathbf{V}^{(B)}} \pi[\![x \in Y]\!] \Rightarrow \pi[\![\psi(x, X_1, \dots, X_n)]\!] \\
&= \pi\Bigg(\bigwedge_{x \in \mathbf{V}^{(B)}} [\![x \in Y]\!] \Rightarrow [\![\psi(x, X_1, \dots, X_n)]\!]\Bigg) \\
&= \pi[\![(\forall\, x \in Y)\psi(x, X_1, \dots, X_n)]\!] = \pi[\![(\varphi(Y, X_1, \dots, X_n)]\!],
\end{aligned}
$$

so completing the proof. ▷

2.5.15. Using the canonical embedding $(\,\cdot\,)^\wedge : \mathbf{V} \to \mathbf{V}^{(B)}$, to each class $X \subset \mathbf{V}$ we assign the $\mathbf{V}^2$-class X' by the formula:

$$
X'(t) := \begin{cases} \mathbf{1_2}, & \text{if } (\exists\, x \in X)(t = x^\wedge), \\ \mathbf{0_2}, & \text{otherwise.} \end{cases}
$$

It is trivial from 2.1.8 (4) that X' is extensional.

We further put $X^\wedge := \imath^*``X'$, where $\imath$ is the identical embedding of $\mathbf{2}$ into B. Hence, $X^\wedge$ is a $\mathbf{V}^{(B)}$-class such that

$$
X^\wedge(t) = \bigvee \{[\![t = x^\wedge]\!] : x \in X\} \quad (t \in \mathbf{V}^{(B)}).
$$

Observe that since $\mathrm{Ord}\,(X)$ is a bounded predicative formula; therefore, by 2.2.8 (4), 2.2.9 (1), and 2.5.14, $\mathrm{On}^\wedge$ is an ordinal class inside $\mathbf{V}^{(B)}$; i.e., $\mathbf{V}^{(B)} \models \mathrm{Ord}\,(\mathrm{On}^\wedge)$. Also, the formulas of 2.5.12 are simplified:

$$
\begin{aligned}
[\![(\forall\, x \in Y^\wedge)\varphi(x)]\!] &= \bigwedge \{[\![\varphi(x^\wedge)]\!] : x \in Y\}, \\
[\![(\exists\, x \in Y^\wedge)\varphi(x)]\!] &= \bigvee \{[\![\varphi(x^\wedge)]\!] : x \in Y\}.
\end{aligned}
$$

2.5.16. *Let φ and ψ be predicative formulas with free variables $X, X_1, \dots, X_n$. Given some $\mathbf{V}^{(B)}$-classes $Y_1, \dots, Y_n$, assume that $[\![\varphi(x_0, Y_1, \dots, Y_n)]\!] = \mathbf{1}$ for some $x_0 \in \mathbf{V}^{(B)}$. Then*

$$\begin{aligned}&[\![(\exists\, x)(\varphi(x,Y_1,\dots,Y_n)\to\psi(x,Y_1,\dots,Y_n))]\!]\\ &=\bigvee\{[\![\psi(x,Y_1,\dots,Y_n)]\!] : x\in \mathbf{V}^{(B)}\wedge [\![\varphi(x,Y_1,\dots,Y_n)]\!]=\mathbf{1}\},\\ &[\![(\forall\, x)(\varphi(x,Y_1,\dots,Y_n)\to\psi(x,Y_1,\dots,Y_n))]\!]\\ &=\bigwedge\{[\![\psi(x,Y_1,\dots,Y_n)]\!] : x\in \mathbf{V}^{(B)}\wedge [\![\varphi(x,Y_1,\dots,Y_n)]\!]=\mathbf{1}\}.\end{aligned}$$

$\lhd$ The proof proceeds along the lines of 2.3.8. $\rhd$

2.5.17. *Theorem* (the maximum principle). *Let $\varphi(x)$ be a predicative B-formula with a single free variable (which implies that φ may contain constants that are $\mathbf{V}^{(B)}$-classes or $\mathbf{V}^{(B)}$-sets). Then the following hold:*

(1) *There is an element x_0 in $\mathbf{V}^{(B)}$ such that $[\![(\exists\, x)\varphi(x)]\!]=[\![\varphi(x_0)]\!]$;*

(2) *If $\mathbf{V}^{(B)}\models(\exists\, x)\varphi(x)$ then there is an element x_0 in $\mathbf{V}^{(B)}$ such that $\mathbf{V}^{(B)}\models\varphi(x_0)$;*

(3) *If $\mathbf{V}^{(B)}\models(\exists!x)\varphi(x)$ then there is a unique element x_0 in $\mathbf{V}^{(B)}$ such that $\mathbf{V}^{(B)}\models\varphi(x_0)$.*

$\lhd$ The proof, basing on the mixing principle (cf. 2.5.3), does not differ from the arguments of 2.3.10 and 2.5.4. $\rhd$

2.5.18. *Theorem* (the transfer principle). *Every theorem of* NGB *holds in* $\mathbf{V}^{(B)}$.

$\lhd$ It suffices to show that the axioms of NGB are satisfied inside $\mathbf{V}^{(B)}$.

(1) The axiom of extensionality for classes inside $\mathbf{V}^{(B)}$ holds, which is immediate from the definitions of 2.5.8 and 2.5.9. $\mathrm{NGB}_2,\dots,\mathrm{NGB}_5$ are true inside $\mathbf{V}^{(B)}$ as shown in Section 2.4.

(2) $\mathbf{V}^{(B)}\models\mathrm{NGB}_6$. The proof proceeds as in 2.4.5. We only need substitute $(t,u)\in X$ for $\varphi(t,u)$ throughout (cf. 2.4.5 and 1.3.4).

(3) $\mathbf{V}^{(B)}\models\bigwedge_{k=7}^{13}\mathrm{NGB}_k$. It suffices to establish that Theorem 1.3.14 holds inside $\mathbf{V}^{(B)}$ since NGB_7–NGB_{13} are particular cases of 1.3.14.

Assume that a formula $\varphi(X_1,\dots,X_n,\ Y_1,\dots,Y_m)$ obeys all hypotheses of 1.3.14. Consider arbitrary $\mathbf{V}^{(B)}$-classes $Y_1,\dots,Y_m$ and define the $\mathbf{V}^{(B)}$-class Z by the formula

$$Z(t):=[\![(\exists\, x_1,\dots,x_n)(t=(x_1,\dots,x_n)\wedge\varphi(x_1,\dots,x_n,Y_1,\dots,Y_m))]\!].$$

It is easy to show that in this case

$$\mathbf{V}^{(B)}\models(\forall\, x_1,\dots,x_n)(\exists\, t)((t=(x_1,\dots,x_n)\wedge t\in Z\leftrightarrow\varphi(x_1,\dots,x_n,Y_1,\dots,Y_N))).$$

(4) $\mathbf{V}^{(B)}\models\mathrm{NGB}_{14}$. Substituting the upper case X for the lower case Latin letter x, obtain the desired.

(5) $\mathbf{V}^{(B)} \models \mathrm{NGB}_{15}$. Let G be a function from On onto $\mathbf{V}^{(B)}$. Put

$$F(t) := \bigvee\{[\![t = (\alpha^\wedge, G(\alpha))^B]\!] : \alpha \in \mathrm{On}\}.$$

Then F is a $\mathbf{V}^{(B)}$-class and by analogy with 2.4.10, we may proceed with the successive calculations: $[\![\mathrm{Fnc}(F)]\!] = \mathbf{1}$, $[\![\mathrm{Ord}(\mathrm{On}^\wedge) \wedge \mathrm{dom}(F) = \mathrm{On}^\wedge]\!] = \mathbf{1}$, and $[\![\mathrm{im}(F) \supset \mathbf{U}_B]\!] = \mathbf{1}$.

Therefore, the universal class $\mathbf{U}_B$ may be well ordered inside $\mathbf{V}^{(B)}$. Hence, $\mathbf{V}^{(B)} \models$ "there exists a choice function of the class $\mathbf{U}^{(B)}$." $\triangleright$

2.5.19. Theorem 2.5.18 opens an opportunity to deal with classes inside $\mathbf{V}^{(B)}$. As an example, we consider the definition of category inside $\mathbf{V}^{(B)}$.

A *category* $\mathfrak{K}$ inside $\mathbf{V}^{(B)}$ consists of some classes $\mathrm{Ob}\,\mathfrak{K}$, $\mathrm{Mor}\,\mathfrak{K}$, and Com inside $\mathbf{V}^{(B)}$ which are called the *class of objects* of $\mathfrak{K}$, the *class of morphisms* of $\mathfrak{K}$, and the *composition* of $\mathfrak{K}$, respectively and which satisfy the condition $\mathbf{V}^{(B)} \models (\mathfrak{K}1)$–$(\mathfrak{K}3)$ where

($\mathfrak{K}1$) There are mappings D and R from $\mathrm{Mor}\,\mathfrak{K}$ to $\mathrm{Ob}\,\mathfrak{K}$ such that, for all objects a and b, the class $\mathfrak{K}(a,b) := H_{\mathfrak{K}}(a,b) := \{\alpha \in \mathrm{Mor}\,\mathfrak{K} : D(\alpha) = a, R(\alpha) = b\}$ is a set (called the *set of morphisms from a to b*);

($\mathfrak{K}2$) Com is an associative partial binary operation on $\mathrm{Mor}\,\mathfrak{K}$ and

$$\mathrm{dom}(\mathrm{Com}) := \{(\alpha, \beta) \in (\mathrm{Mor}\,\mathfrak{K})^2 : D(\beta) = R(\alpha)\};$$

($\mathfrak{K}3$) To every object $a \in \mathrm{Ob}\,\mathfrak{K}$ there is a morphism 1_a called the *identity morphism of a* such that $D(1_a) = R(1_a) = a$, $\mathrm{Com}(1_a, \alpha) = \alpha$ for $R(\alpha) = a$, and $\mathrm{Com}(\beta, 1_a) = \beta$ for $D(\beta) = a$.

We usually write $\beta\alpha$ or $\beta \circ \alpha$ instead of $\mathrm{Com}(\alpha, \beta)$.

2.5.20. COMMENTS.

(1) The Boolean valued model $\mathbf{V}^{(B)}$ over B may be characterized axiomatically. Namely, there is a class $\mathbf{V}^{(B)}$ unique up to a bijection preserving all Boolean truth values and obeying the following conditions: (a) there are two mappings $[\![\cdot \in \cdot]\!]$, $[\![\cdot = \cdot]\!] : \mathbf{V}^{(B)} \times \mathbf{V}^{(B)} \to B$ such that the conventional axioms of equality hold inside $\mathbf{V}^{(B)}$ (cf. 2.1.7 and 2.1.8); (b) $\mathbf{V}^{(B)}$ is separated; i.e., $[\![x = y]\!] = \mathbf{1}_B$ implies that $x = y$ for $x, y \in \mathbf{V}^{(B)}$; (c) the axioms of extensionality and regularity hold inside $\mathbf{V}^{(B)}$; and (d) Proposition 2.5.6 holds for $\mathbf{V}^{(B)}$.

(2) Let π be a complete homomorphism from a complete Boolean algebra B to another complete Boolean algebra C. Then π^* is a unique mapping from $\mathbf{V}^{(B)}$ to $\mathbf{V}^{(C)}$ such that (a) $[\![\pi^* x = \pi^* y]\!]^C = \pi[\![x = y]\!]^B$ $(x, y \in \mathbf{V}^{(B)})$, and (b) $[\![z \in \pi^* y]\!]^C \le \bigvee_{x \in \mathbf{V}^{(B)}} [\![z = \pi^* x]\!]$ for $y \in \mathbf{V}^{(B)}$ and $z \in \mathbf{V}^{(C)}$.

Chapter 3
Functors of Boolean Valued Analysis

The transfer and maximum principles enable us to carry out various constructions of the conventional mathematical practice inside every Boolean valued universe. Therein we encounter the fields of real and complex numbers, Banach spaces, differential operators, etc. The objects, representing them, may be perceived to some extend as nonstandard representations of the original mathematical entities.

Therefore, viewing $\mathbf{V}^{(B)}$ as a nonstandard presentation of the mathematical universe of discourse and recalling that $\mathbf{V}^{(B)}$ is constructed within the von Neumann universe, we may peek in the Boolean valued world, discovering standard objects in a nonstandard disguise. Skipping from one B to another, a keen researcher sees many hypostaces of a sole mathematical idea embodied in a set-theoretic formula. Comparing observations is a method for studying an intrinsic meaning of the formula. The method shows often that essentially different analytical objects are in fact just distinctive appearances of the same concept. This reveals the esoteric reasons for many vague analogies and dim parallelism as well as opens new opportunities to study familiar objects.

The overall picture reminds us of the celebrated cave of Plato. If a casual escapee decided to inform his fellow detainees on what he saw at large, he might build a few bonfires in the night. Then each entity will cast several shadows on the wall of the cave (rather than a single shadow suggested by Plato). Now the detainees acquired a possibility of finding the essence of unknown things from analyzing the collection of shadows bearing more information than a sole shadow of an entity.

Comparative analysis, using Boolean valued models, proceeds usually in two stages which we may agree to call syntactic and semantic.

At the *syntactic stage*, the mathematical statement under investigation (a definition, a construction, a property, etc.) is transformed into a formal text of the symbolic language of set theory or, to be more precise, into a text in a suitable jargon. In this stage we often have to analyze the complexity of the text; in particular, it matters whether the text or some of its fragments is a bounded formula.

The *semantic stage* consists in interpreting a formal text inside a Boolean valued universe. In this stage we use the terms of the conventional set theory, i.e. the von Neumann universe $\mathbf{V}$, to interpret (decode or translate) some meaningful texts that contain truth about the objects of the Boolean valued universe $\mathbf{V}^{(B)}$. This is done by using especial operations on the elements and subsets of the von Neumann universe.

In the present chapter we consider the basic operations of Boolean valued analysis, i.e., the canonical embedding, descent, ascent, and immersion. The most important properties of these operations are conveniently expressed using the notions of category and functor. The reader may resume acquaintance with the preliminaries to category theory by consulting the Appendix.

3.1. The Canonical Embedding

3.1.1. This section is devoted to the way of the embedding class of sets into a Boolean valued universe.

Theorem. *The following statements hold:*

(1) *If a class $X \subset \mathbf{V}$ and an element $z \in \mathbf{V}^{(B)}$ are such that $\mathbf{V}^{(B)} \models z \in X^\wedge$ then $z = \operatorname{mix}_{x\in X}(b_x x^\wedge)$ for some partition of unity $(b_x)_{x\in X}$ in B;*

(2) *To a $\mathbf{V}^{\mathbf{2}}$-class Y there is a unique class $X \subset \mathbf{V}$ such that $\mathbf{V}^{\mathbf{2}} \models X^\wedge = Y$;*

(3) *For $X \subset \mathbf{V}$ and $Y \subset \mathbf{V}$,*

$$X \in Y \leftrightarrow \mathbf{V}^{(B)} \models X^\wedge \in Y^\wedge, \quad X = Y \leftrightarrow \mathbf{V}^{(B)} \models X^\wedge = Y^\wedge;$$

(4) *If $\pi : B \to C$ is a complete homomorphism then $\pi^* X^\wedge = X^{\hat{\wedge}}$ for every class $X \subset \mathbf{V}$ where $X^{\hat{\wedge}}$ is the standard name of X in $\mathbf{V}^{(C)}$.*

$\triangleleft$ (1) Given $x \in X$, put $b_x := [\![x^\wedge = z]\!]$. Then, by 2.2.8 (2),

$$b_x \wedge b_y \leq [\![x^\wedge = y^\wedge]\!] = \mathbf{0}$$

for $x, y \in X$, $x \neq y$.

On the other hand,

$$\bigvee\{b_x : x \in X\} = X^\wedge(z) = [\![z \in X^\wedge]\!] = \mathbf{1},$$

so that $(b_x)_{x\in X}$ is a partition of unity and $z = \operatorname{mix}_{x\in X}(b_x x^\wedge)$.

(2) The claim follows from 2.2.8. Indeed, if $X' := \{y \in \mathbf{V}^{(\mathbf{2})} : [\![y \in Y]\!] = \mathbf{1_2}\}$ and $X := \{x \in \mathbf{V} : x^\wedge \in X'\}$ then, by 2.2.8 (3, 4), for $t \in \mathbf{V}^{(\mathbf{2})}$ obtain

$$X^\wedge(t) = \bigvee\{[\![t = x^\wedge]\!]^{\mathbf{2}} : x \in X\} = \bigvee\{[\![t = x^\wedge]\!]^{\mathbf{2}} : Y(x) = \mathbf{1_2}\}$$
$$= \bigvee\{Y(x) \wedge [\![t = x^\wedge]\!]^{\mathbf{2}} : x \in \mathbf{V}^{(\mathbf{2})}\} = Y(t).$$

Uniqueness ensues from 2.2.8 (4) and 2.5.15.

(3) To prove, compare 2.5.15 and (2).

(4) If $\imath_1$ and $\imath_2$ are embeddings of the two-element algebra **2** into B and C then $\pi \circ \imath_1 = \imath_2$; and, by 2.5.11,

$$\pi^* X^\wedge = \pi^* \circ \imath_1^*(X^\wedge) = \imath_2^* X^\wedge = X^{\hat{\wedge}}. \ \rhd$$

3.1.2. *If x and y are sets then*

$$\{x\}^\wedge = \{x^\wedge\}^B, \quad \{x, y\}^\wedge = \{x^\wedge, y^\wedge\}^B, \quad (x, y)^\wedge = (x^\wedge, y^\wedge)^B.$$

$\lhd$ All these formulas are bounded. Using 2.2.9, deduce

$$\mathbf{V}^{(B)} \models \{x\}^\wedge = \{x^\wedge\} \wedge \{x, y\}^\wedge = \{x^\wedge, y^\wedge\} \wedge (x, y)^\wedge = (x^\wedge, y^\wedge).$$

It suffices now to recall the appropriate formulas of 2.4.8. $\rhd$

3.1.3. *Assume that a formula φ of class Σ_1 obeys all hypotheses of Theorem 1.3.14. Take some classes $Z_1, \dots, Z_n$, $Y_1, \dots, Y_m$, and define the class Y by the formula*

$$Y := \{(x_1, \dots, x_n) :$$
$$x_1 \in Z_1 \wedge \dots \wedge x_n \in Z_n \wedge \varphi(x_1, \dots, x_n, Y_1, \dots, Y_m)\}.$$

Then the following holds inside $\mathbf{V}^{(B)}$:

$$Y^\wedge = \{(x_1, \dots, x_n) :$$
$$x_1 \in Z_1^\wedge \wedge \dots \wedge x_n \in Z_n^\wedge \wedge \varphi(x_1, \dots, x_n, Y_1^\wedge, \dots, Y_m^\wedge)\}.$$

$\lhd$ By Theorem 1.3.14, Y is the only class obeying the conditions $\Phi(Z_1, \dots, Z_n, Y_1, \dots, Y_m)$ and $\Psi(Z_1, \dots, Z_n, Y_1, \dots, Y_m)$, where Φ and Ψ are as follows

$$\Phi := (\forall u \in Y)(\exists x_1 \in Z_1) \dots (\exists x_n \in Z_n)(u = (x_1, \dots, x_n) \wedge \varphi(x_1, \dots, Y_m)),$$
$$\Psi := (\forall x_1 \in Z_1) \dots (\forall x_n \in Z_n)(\exists u)$$
$$(u = (x_1, \dots, x_n) \wedge \varphi(x_1, \dots, Y_m) \to u \in Y).$$

Obviously, Φ and Ψ are formulas of class Σ_1. Hence, from 2.5.14 we infer

$$\mathbf{V}^{(B)} \models \Phi(Z_1^\wedge, \dots, Y_m^\wedge) \wedge \Psi(Z_1^\wedge, \dots, Y_m^\wedge).$$

This amounts to the claim. $\rhd$

3.1.4. *For $X \subset \mathbf{V}$ and $Y \subset \mathbf{V}$ the following hold:*

(1) $\mathbf{V}^{(B)} \models (X \cup Y)^\wedge = X^\wedge \cup Y^\wedge$;

(2) $\mathbf{V}^{(B)} \models (X \times Y)^\wedge = X^\wedge \times Y^\wedge$;

(3) $\mathbf{V}^{(B)} \models (\bigcup X)^\wedge = \bigcup(X^\wedge)$;

(4) $\mathrm{Rel}\,(X) \to \mathbf{V}^{(B)} \models \mathrm{Rel}\,(X^\wedge)$;

(5) $(F : X \to Y) \to \mathbf{V}^{(B)} \models F^\wedge : X^\wedge \to Y^\wedge$;

(6) $\mathrm{Rel}\,(X) \to \mathbf{V}^{(B)} \models (X``Y)^\wedge = (X^\wedge)``(Y^\wedge)$;

(7) $\mathrm{Rel}\,(X) \to \mathbf{V}^{(B)} \models \mathrm{dom}(X^\wedge) = \mathrm{dom}(X)^\wedge \wedge \mathrm{im}(X^\wedge) = \mathrm{im}(X)^\wedge$.

$\lhd$ The claims of (1)–(5) follow from 3.1.3 (cf. A.1.11 and A.1.12). Unfortunately, (6) and (7) fall beyond the scope of applicability of 3.1.3 and so we deduce them by direct calculations, appealing to 2.4.9, 3.1.1, and 3.1.2.

Start with (6):

$$[\![t \in (X^\wedge)``(Y^\wedge)]\!] = [\![(\exists\, u \in X^\wedge)(\exists\, v \in Y^\wedge)(u = (v, t))]\!]$$
$$= \bigvee_{u \in X} \bigvee_{v \in Y} [\![u^\wedge = (v^\wedge, t)]\!] = \bigvee_{v \in Y} \bigvee_{(z,w) \in X} [\![z^\wedge = v^\wedge]\!] \wedge [\![w^\wedge = t]\!]$$
$$= \bigvee \{[\![w^\wedge = t]\!] : v \in Y, (v, w) \in X\}$$
$$= [\![(\exists\, w \in (X^\wedge)``(Y^\wedge))\,(t = w)]\!] = [\![t \in (X``Y)^\wedge]\!].$$

Proceed with checking (7):

$$[\![t \in \mathrm{dom}(X^\wedge)]\!] = [\![(\exists\, u \in X^\wedge)(\exists\, v)(u = (t, v))]\!]$$
$$= \bigvee_{(z,w) \in X} \bigvee_{v \in \mathbf{V}^{(B)}} [\![z^\wedge = t]\!] \wedge [\![w^\wedge = v]\!]$$
$$= \bigvee \{[\![z^\wedge = t]\!] : z \in \mathrm{dom}(X)\} = [\![t \in \mathrm{dom}(X)^\wedge]\!].$$

The proof is complete. $\rhd$

3.1.5. *Theorem.* *Let X and Y be nonempty sets and $F \subset X \times Y$. Consider the correspondence $\Phi := (F, X, Y)$. Then the element $\Phi^\wedge$ of $\mathbf{V}^{(B)}$ satisfies the following conditions:*

(1) $\mathbf{V}^{(B)} \models \Phi^\wedge$ *is a correspondence from* $X^\wedge$ *to* $Y^\wedge$, *and* $\mathrm{Gr}(\Phi^\wedge) = F^\wedge$;

(2) $\mathbf{V}^{(B)} \models \Phi^\wedge(A^\wedge) = \Phi(A)^\wedge$ *for all* $A \in \mathscr{P}(X)$;

(3) $\mathbf{V}^{(B)} \models (\Psi \circ \Phi)^\wedge = \Psi^\wedge \circ \Phi^\wedge$ *for every correspondence* Ψ;

(4) $\mathbf{V}^{(B)} \models (I_X)^\wedge = I_{X^\wedge}$.

◁ (1) Let the formula $\varphi(X, Y, F, \Phi)$ state that Φ is a correspondence from X to Y and $F = \mathrm{Gr}(\Phi)$. Then φ is a bounded formula and the claim ensues from 2.2.9.

(2) This follows from 3.1.4 (6).

(3), (4) Here we again deal with bounded formulas. Hence, it suffices to refer to 2.2.9. ▷

3.1.6. Corollary. *Let $f : X \to Y$ be a mapping. Then $f^\wedge$ satisfies the conditions:*

(1) $\mathbf{V}^{(B)} \models f^\wedge : X^\wedge \to Y^\wedge$;
(2) $\mathbf{V}^{(B)} \models f^\wedge(x^\wedge) = f(x)^\wedge$ *for all* $x \in X$;
(3) $\mathbf{V}^{(B)} \models (g \circ f)^\wedge = g^\wedge \circ f^\wedge$ *for all* $g : Y \to Z$.

3.1.7. We now define the categories $\mathscr{V}_*$ and $\mathscr{V}_*^{(B)}$ that are associated with the universes $\mathbf{V}$ and $\mathbf{V}^{(B)}$. Note that, without further specification, we agree to presume that the classes of objects and morphisms of any category do not intersect (this can be achieved by using extra indices, cf. A.3.2).

Let $\mathscr{V}_*$ be the category of nonempty sets and correspondences, so that $\mathrm{Ob}\,\mathscr{V}_* := \mathbf{V} \setminus \{\varnothing\}$ and $\mathscr{V}_*(x, y)$ is the set of all nonempty correspondences from x to y, with the composition law the conventional composition of correspondences.

The class of objects of the category $\mathscr{V}_*^{(B)}$ consists of nonempty $\mathbf{V}^{(B)}$-sets:

$$\mathrm{Ob}\,\mathscr{V}_*^{(B)} := \{x \in \mathbf{V}^{(B)} : [\![x \neq \varnothing]\!] = \mathbf{1}\}.$$

The set of morphisms from an object $x \in \mathrm{Ob}\,\mathscr{V}_*^{(B)}$ into an object $y \in \mathrm{Ob}\,\mathscr{V}_*^{(B)}$ is defined by the formula

$$\mathscr{V}_*^{(B)}(x, y)$$
$$:= \{\alpha \in \mathbf{V}^{(B)} : [\![\alpha \text{ is a correspondence from } x \text{ to } y \text{ and } \mathrm{Gr}(\alpha) \neq \varnothing]\!] = \mathbf{1}\}.$$

If α and β are morphisms of the category $\mathscr{V}_*^{(B)}$ such that $[\![D(\beta) = R(\alpha)]\!] = \mathbf{1}$ then, by the maximum principle, there is a unique element $\gamma \in \mathbf{V}^{(B)}$ satisfying $[\![\gamma = \beta \circ \alpha]\!] = \mathbf{1}$. We appoint this element γ as the composition of α and β in the category $\mathbf{V}^{(B)}$.

The subcategories of $\mathscr{V}_*$ and $\mathscr{V}_*^{(B)}$, each preserving the original class of objects but with mappings as morphisms, are denoted by $\mathscr{V}$ and $\mathscr{V}^{(B)}$. Assign to a set $x \in \mathbf{V} \setminus \{\mathbf{0}\}$ and a correspondence α the elements $x^\wedge \in \mathbf{V}^{(B)}$ and $\alpha^\wedge \in \mathbf{V}^{(B)}$. Denote the resultant mapping by $\mathscr{F}^\wedge$. The following theorem is straightforward from 3.1.5 and 3.1.6.

3.1.8. Theorem. *The so-defined (pair of mappings) $\mathscr{F}^\wedge$ is a covariant functor from the category $\mathscr{V}_*$ to the category $\mathscr{V}_*^{(B)}$ (as well as from the category $\mathscr{V}$ to the category $\mathscr{V}^{(B)}$).*

The functor $\mathscr{F}^\wedge$ (as well as its restriction to the subcategory $\mathscr{V}$) is the *canonical embedding functor* or the *standard name functor*.

3.1.9. We now inspect the properties of ordinals inside $\mathbf{V}^{(B)}$.

(1) Recall (cf. 2.4.10) that $\mathrm{Ord}(X)$ is a bounded formula. By definition, $\lim(\alpha) \leq \alpha$ for every ordinal α. Therefore, the formula $\mathrm{Ord}(x) \wedge x = \lim(x)$ may be rewritten as

$$\mathrm{Ord}(x) \wedge (\forall t \in x)(\exists s \in x)(t \in s),$$

and so it is bounded too. Finally, the record

$$\mathrm{Ord}(x) \wedge x = \lim(x) \wedge (\forall t \in x)(t = \lim(t) \to t = 0)$$

shows that the concept of "least limit ordinal" is expressed by a bounded formula. Thus, by 2.2.9, α is the least limit ordinal if and only if $\mathbf{V}^{(B)} \models$ "$\alpha^\wedge$ is the least limit ordinal." Since ω is the least limit ordinal (cf. 1.4.6); therefore, $\mathbf{V}^{(B)} \models$ "$\omega^\wedge$ is the least limit ordinal."

(2) It follows from 1.4.5 (2), 2.5.15, and 2.5.16 that $\mathbf{V}^{(B)} \models$ "$\mathrm{On}^\wedge$ is the only ordinal class failing to be an ordinal." Hence, for every $x \in \mathbf{V}^{(B)}$ the following holds:

$$[\![\mathrm{Ord}(x)]\!] = \bigvee\{[\![x = \alpha^\wedge]\!] : \alpha \in \mathrm{On}\}.$$

(3) *For $x \in \mathbf{V}^{(B)}$, the formula $\mathbf{V}^{(B)} \models \mathrm{Ord}(x)$ holds if and only if there are an ordinal $\beta \in \mathrm{On}$ and a partition of unity $(b_\alpha)_{\alpha\in\beta} \subset B$ such that $x = \mathrm{mix}_{\alpha\in\beta}(b_\alpha \alpha^\wedge)$. In other words, each ordinal inside $\mathbf{V}^{(B)}$ is a mixing of some set of standard ordinals.*

$\lhd$ The claim follows from (2) and 3.1.1 (1). $\rhd$

(4) Using 2.5.16, we come to the rules for quantifying over ordinals:

$$[\![(\forall x)(\mathrm{Ord}(x) \to \psi(x))]\!] = \bigwedge_{\alpha\in\mathrm{On}} [\![\psi(\alpha^\wedge)]\!],$$

$$[\![(\exists x)(\mathrm{Ord}(x) \wedge \psi(x))]\!] = \bigvee_{\alpha\in\mathrm{On}} [\![\psi(\alpha^\wedge)]\!].$$

3.1.10. A class X is *finite* if X coincides with the image of a function on a finite ordinal. In symbols, this is expressed as $\mathrm{Fin}(X)$. Namely,

$$\mathrm{Fin}(X) := (\exists n)(\exists f)(n \in \omega \wedge \mathrm{Fnc}(f) \wedge \mathrm{dom}(f) = n \wedge \mathrm{im}(f) = X).$$

Obviously, the above formula is not bounded. By the axiom of replacement NGB_6, it is clear that $\mathrm{Fin}(X) \to M(X)$, and so we shall speak about finite sets instead of finite classes. Denote by $\mathscr{P}_{\mathrm{fin}}(X)$ the class of all finite subsets of X:

$$\mathscr{P}_{\mathrm{fin}}(X) := \{Y \in \mathscr{P}(X) : \mathrm{Fin}(Y)\}.$$

We now check what happens with finite sets under the canonical embedding of $\mathbf{V}$ in $\mathbf{V}^{(B)}$, thus grasping the class $\mathscr{P}_{\mathrm{fin}}(X)^\wedge$. To this end, we first show that

$$\mathbf{V}^{(B)} \models \mathscr{P}_{\mathrm{fin}}(X)^\wedge \subset \mathscr{P}_{\mathrm{fin}}(X^\wedge).$$

$\lhd$ Note that if f is a mapping of $n \in \omega$ to X then $[\![\mathrm{im}(f^\wedge) \in \mathscr{P}_{\mathrm{fin}}(X^\wedge)]\!] = \mathbf{1}$. Indeed, by 3.1.6, $[\![f^\wedge : n^\wedge \to X^\wedge]\!] = [\![n^\wedge \in \omega^\wedge]\!] = \mathbf{1}$, and so

$$[\![\mathrm{im}(f^\wedge) \in \mathscr{P}(X^\wedge) \wedge \mathrm{Fin}(\mathrm{im}(f^\wedge))]\!] = \mathbf{1}.$$

Given $t \in \mathbf{V}^{(B)}$, proceed with easy calculations (cf. 2.2.8 (1), 3.1.4 (7), 3.1.6):

$$\begin{aligned} &[\![t \in \mathscr{P}_{\mathrm{fin}}(X)^\wedge]\!] \\ &= \bigvee_{u \in \mathscr{P}_{\mathrm{fin}}(X)} [\![t = u^\wedge]\!] = \bigvee_{n \in \omega} \bigvee_{f : n \to X} [\![t = \mathrm{im}(f)^\wedge]\!] \\ &= \bigvee_{n \in \omega} \bigvee_{f : n \to X} [\![t = \mathrm{im}(f^\wedge)]\!] \wedge [\![n^\wedge \in \omega^\wedge]\!] \wedge [\![f^\wedge : n^\wedge \to X^\wedge]\!] \\ &\le [\![t \in \mathscr{P}_{\mathrm{fin}}(X^\wedge)]\!], \end{aligned}$$

so completing the proof. $\rhd$

3.1.11. *The following holds*

$$\mathbf{V}^{(B)} \models \mathscr{P}_{\mathrm{fin}}(X)^\wedge = \mathscr{P}_{\mathrm{fin}}(X^\wedge)$$

for an arbitrary class X.

$\lhd$ Assume that for $t \in \mathbf{V}^{(B)}$ the following holds:

$$[\![t \in \mathscr{P}_{\mathrm{fin}}(X^\wedge)]\!] = [\![(\exists n \in \omega^\wedge)(\exists f)(f : n \leftrightarrow X^\wedge \wedge t = \mathrm{im}(f)]\!] = \mathbf{1}.$$

Then there is a countable partition of unity $(b^{(n)})_{n \in \omega} \subset B$ such that

$$[\![(\exists\, f)(f : n^\wedge \to X^\wedge \wedge t = \mathrm{im}(f)]\!] \ge b^{(n)} \quad (n \in \omega).$$

Given $n \in \omega$ and using the maximum principle, find $f'_n \in \mathbf{V}^{(B)}$ obeying the inequality

$$[\![f'_n : n^\wedge \to X^\wedge]\!] \wedge [\![t = \mathrm{im}(f'_n)]\!] \geq b^{(n)}.$$

By 3.1.6, choose $f''_n \in \mathbf{V}^{(B)}$ so that $[\![f''_n : n^\wedge \to X^\wedge]\!] \geq (b^{(n)})^*$, and assign $f_n := \mathrm{mix}\{b^{(n)} f'_n, (b^{(n)})^* f''_n\}$. Then $[\![f_n : n^\wedge \to X^\wedge]\!] = \mathbf{1}$ and $[\![t = \mathrm{im}(f_n)]\!] \geq b^{(n)}$.

Further, considering $k \in n$, note that $[\![f_n(k^\wedge) \in X^\wedge]\!] = \mathbf{1}$. Hence, $f_n(k) = \mathrm{mix}(b_x^{(k)} x^\wedge)$ for some partition of unity $(b_x^{(k)})_{x \in X}$ (cf. 3.1.1 (1)). Therefore,

$$[\![f_n(k^\wedge) = x^\wedge]\!] \geq b_x^{(k)} \quad (x \in X, k \in n).$$

Let X^n stand as usual for the class of all mappings from n to X. Given $g \in X^n$ and $k \in n$, note that

$$[\![f_n(k^\wedge) = g^\wedge(k^\wedge)]\!] = [\![f_n(k^\wedge) = g(k)^\wedge]\!] \geq b_{g(k)}^{(k)}.$$

Hence, $[\![f_n = g^\wedge]\!] \geq b_{g,n}$, where $b_{g,n} := \bigwedge\{b_{g(k)}^{(k)} : k \in n\}$. In this event however we also see that

$$[\![\mathrm{im}(f) = \mathrm{im}(g^\wedge)]\!] \geq b_{g,n} \quad (g \in X^n).$$

By definition, $\mathrm{im}(g) \in \mathscr{P}_{\mathrm{fin}}(X)$, while by 3.1.4 (7),

$$[\![\mathrm{im}(g^\wedge) \in \mathscr{P}_{\mathrm{fin}}(X)^\wedge]\!] = \mathbf{1}.$$

We thus obtain

$$[\![t \in \mathscr{P}_{\mathrm{fin}}(X)^\wedge]\!] \geq [\![t = \mathrm{im}(f_n)]\!]$$
$$\wedge [\![\mathrm{im}(f_n) = \mathrm{im}(g^\wedge)]\!] \wedge [\![\mathrm{im}(g^\wedge) \in \mathscr{P}_{\mathrm{fin}}(X)^\wedge]\!] \geq b^{(n)} \wedge b_{g,n}.$$

Using the definition of $b_{g,n}$ and the distributive laws 1.1.5 (1, 2), calculate

$$\bigvee\{b^{(n)} \wedge b_{g,n} : n \in \omega, g \in X^n\} = \bigvee_{n \in \omega} b^{(n)} \wedge \Big(\bigvee_{g \in X^n} \bigwedge_{k \in n} b_{g(k)}^{(k)} \Big)$$
$$= \bigvee_{n \in \omega} b^{(n)} \wedge \Big(\bigwedge_{k \in n} \bigvee_{g \in X^n} b_{g(k)}^{(k)} \Big) = \bigvee_{n \in \omega} b^{(n)} \wedge \Big(\bigwedge_{k \in n} \bigvee_{x \in X} b_x^{(k)} \Big) = \bigvee_{n \in \omega} b^{(n)} = \mathbf{1}.$$

Clearly, $[\![t \in \mathscr{P}_{\mathrm{fin}}(X)^\wedge]\!] = \mathbf{1}$. So, applying 2.5.16, deduce $[\![\mathscr{P}_{\mathrm{fin}}(X^\wedge) \subset \mathscr{P}_{\mathrm{fin}}(X)^\wedge]\!] = \mathbf{1}$. The reverse inclusion is established in 3.1.10. ▷

3.1.12. *For a class X and $n \in \omega$, the following hold:*

(1) $\mathbf{V}^{(B)} \models (X^n)^\wedge = (X^\wedge)^{n^\wedge}$;

(2) $\mathbf{V}^{(B)} \models \mathscr{P}(X)^\wedge \subset \mathscr{P}(X^\wedge)$.

$\triangleleft$ (1) Given $t \in \mathbf{V}^{(B)}$, by 3.1.6 we may write

$$[\![t \in (X^n)^\wedge]\!] = \bigvee\{[\![t = u^\wedge]\!] : u \in X^n\}$$
$$= \bigvee\{[\![t = u^\wedge]\!] \wedge [\![u^\wedge : n^\wedge \to X^\wedge]\!] : u \in X^n\}$$
$$\leq \bigvee\{[\![t = u]\!] \wedge [\![u : n^\wedge \to X^\wedge]\!] : u \in \mathbf{V}^{(B)}\}$$
$$= [\![(\exists\, u)(u : n^\wedge \to X^\wedge \wedge t = u)]\!] = [\![t \in (X^\wedge)^{n^\wedge}]\!].$$

Therefore, we have established

$$[\![(X^n)^\wedge \subset (X^\wedge)^{n^\wedge}]\!] = \mathbf{1}.$$

To prove the reverse inclusion, consider $u \in \mathbf{V}^{(B)}$ satisfying $[\![u : n^\wedge \to X^\wedge]\!] = \mathbf{1}$. In this event $[\![u(k^\wedge) \in X^\wedge]\!] = \mathbf{1}$ $(k \in n)$, and so $[\![u(k^\wedge) = \mathrm{mix}(b_x^{(k)} x^\wedge)]\!] = \mathbf{1}$ for some partition of unity $(b_x^{(k)})_{x \in X}$ (cf. 3.1.1 (1)).

By refining partitions, we may, if need be, choose a partition of unity (b_ξ) and families $(x_{k,\xi}) \subset X$ $(k \in n)$ such that $[\![u(k^\wedge) = \mathrm{mix}(b_\xi x_{k,\xi}^\wedge)]\!] = \mathbf{1}$ for all $k \in n$.

Define the functions $u_\xi : n \to X$ as follows $u_\xi(k) := x_{k,\xi}$. Then $[\![u = u_\xi^\wedge]\!] \geq b_\xi$ and $[\![u_\xi^\wedge \in (X^n)^\wedge]\!] = \mathbf{1}$. Hence, $[\![u \in (X^n)^\wedge]\!] = \mathbf{1}$. By 2.5.16, conclude $[\![(X^\wedge)^{n^\wedge} \subset (X^n)^\wedge]\!] = \mathbf{1}$.

(2) This follows from straightforward calculation. $\triangleright$

3.1.13. COMMENTS.

(1) Cardinals inside $\mathbf{V}^{(B)}$ are a greater problem than ordinals (cf. 3.1.9). It is easy to note that $\neg\,\mathrm{Card}(x)$ is a Σ_1-formula and so $[\![\mathrm{Card}(\alpha^\wedge)]\!] = \mathbf{1} \to \mathrm{Card}(\alpha)$. The formula $\mathrm{Card}(x)$ is not however of class Σ_1. Therefore, the opposite implication might fail and an ordinal might lose the property of being a cardinal under the canonical embedding in $\mathbf{V}^{(B)}$. In fact, given infinite cardinals $\lambda < \varkappa$, it is possible to choose a complete Boolean algebra B so that $\mathbf{V}^{(B)} \models |\lambda^\wedge| = |\varkappa^\wedge|$. This effect is called the *cardinal shift* or *cardinal displacement.* We may even choose B so that $\mathbf{V}^{(B)} \models 2^{\omega_\alpha} = \omega_{\beta+1}$ for some $\alpha < \beta$. That is how the consistency of $\neg$ GCH and ZFC is established [11, 83, 241].

(2) In spite of what has been said in (1), cardinals inside $\mathbf{V}^{(B)}$ behave themselves provided that B satisfies the *countable chain condition*; i.e., if every disjoint subset of B is at most countable (in this event B is said to has *countable type* in the literature of the Russian provenance). Granted B, observe

$$\mathbf{V}^{(B)} \models \mathrm{Card}(\alpha^{\wedge}) \leftrightarrow \mathrm{Card}(\alpha),$$
$$\mathbf{V}^{(B)} \models (\omega_\alpha)^{\wedge} = \omega_{\alpha^{\wedge}}.$$

(3) The properties of constructible sets (see 1.5.10) inside $\mathbf{V}^{(B)}$ resemble those of cardinals. Namely, if $L(x)$ is the formula stating that x is a constructible set then

$$[\![L(u)]\!] = \bigvee\{[\![u = v]\!] : v \in \mathbf{L}\} \quad (u \in \mathbf{V}^{(B)})$$

and 3.1.9 (2)–(4) remain true on substituting L for Ord (cf. [11, 83, 241]).

(4) In view of 3.1.11, it might seem that we have equality holding in 3.1.12 (2), i.e., $[\![\mathscr{P}(X^{\wedge}) = \mathscr{P}(X)^{\wedge}]\!] = \mathbf{1}$. However, this is not so. Indeed, let B be the algebra of regular closed subsets of the *Cantor set* (which is the ω*-discontinuum*, i.e., the product of countably many discrete two-element Boolean algebras. Then $[\![\mathscr{P}(\omega^{\wedge}) \neq \mathscr{P}(\omega)^{\wedge}]\!] = \mathbf{1}$.

3.2. The Descent Functor

In this section we set forth the basic technique of translating propositions about the members of a Boolean valued universe $\mathbf{V}^{(B)}$ into statements about ordinary sets. The role of the translator is performed by descent. *We use the word "descent" both for the result and the method of presenting the elements of* $\mathbf{V}^{(B)}$ *in the von Neumann universe* $\mathbf{V}$. Paraphrasing this informally, we may say that the descent acts from $\mathbf{V}^{(B)}$ to $\mathbf{V}$.

3.2.1. Take an arbitrary class X inside $\mathbf{V}^{(B)}$, i.e., an extensional mapping from $\mathbf{V}^{(B)}$ to B, and put

$$X{\downarrow} := \{x \in \mathbf{V}^{(B)} : [\![x \in X]\!] = \mathbf{1}_B\}.$$

This equality defines a certain subclass $X{\downarrow}$ of the von Neumann universe $\mathbf{V}$ which is called the *descent* of X. Let $X_\varphi := \bar{\varphi}$ be the class inside $\mathbf{V}^{(B)}$ *definable* by some B-formula φ (see 2.5.5). Then the descent of X_φ has the form

$$X_\varphi{\downarrow} = \{x \in \mathbf{V}^{(B)} : [\![\varphi(x)]\!] = \mathbf{1}\}.$$

In this case the formula $x \in X_\varphi{\downarrow}$ reads: "x satisfies φ inside $\mathbf{V}^{(B)}$." Thus, for instance, if $f \in \mathbf{V}^{(B)}$ and $[\![\mathrm{Fnc}(f)]\!] = \mathbf{1}$ then we say that f is a *function inside* $\mathbf{V}^{(B)}$. It is obvious that the descent of the universal $\mathbf{V}^{(B)}$-class $\mathbf{U}_B$ coincides with $\mathbf{V}^{(B)}$. Also, observe two useful formulas that are immediate from 2.5.16:

$$[\![X_\varphi \subset X_\psi]\!] = \bigwedge\{[\![\psi(x)]\!] : x \in X_\varphi{\downarrow}\},$$

$$[\![X_\varphi \cap X_\psi \neq \varnothing]\!] = \bigvee\{[\![\psi(x)]\!] : x \in X_\varphi{\downarrow}\},$$

where φ and ψ are arbitrary B-formulas.

In what follows we systematically use the following technique of abbreviations. Let a symbol f be a (conventional) notation for some n-ary function; for instance, $\{\cdot,\cdot\}$, $(\cdot,\cdot)$, $\Phi(\cdot)$, $\pi_\Phi(\cdot)$, etc. Then to all $x_1,\dots,x_n \in \mathbf{V}^{(B)}$ there exists a unique element $x_f \in \mathbf{V}^{(B)}$ such that

$$[\![x_f = f(x_1,\dots,x_n)]\!] = [\![(\exists\, x)(x_1,\dots,x_n,x) \in f]\!].$$

In this event we simply write $f(x_1,\dots,x_n){\downarrow}$ instead of $x_f{\downarrow}$. For instance, $\Phi(A){\downarrow}$ is the class determined by the rule

$$y \in \Phi(A) \leftrightarrow ([\![(\exists\, x \in A)(y \in \Phi(x))]\!] = \mathbf{1}).$$

3.2.2. Let X be a subclass of $\mathbf{V}^{(B)}$, i.e., $X \subset \mathbf{V}^{(B)}$ in the sense of $\mathbf{V}$. Say that X is *cyclic* or *universally complete* and write $\mathrm{Cyc}(X)$ provided that X is *closed under mixing*, i.e., if X contains the mixing of its every family by an arbitrary partition of unity. In other words, X is a cyclic class whenever, given a partition of unity $(b_\xi)_{\xi\in\Xi} \subset B$ and a family $(x_\xi)_{\xi\in\Xi} \subset X$, we observe that $\mathrm{mix}_{\xi\in\Xi}(b_\xi x_\xi) \in X$. The intersection of an arbitrary collection of cyclic sets is a cyclic set itself. The least cyclic set, containing a set $M \subset \mathbf{V}^{(B)}$, is the *cyclic hull* or *cyclic completion* or *universal completion* of M. Let $\mathrm{cyc}(M)$ stand for the cyclic hull of M. Obviously, a subset M of $\mathbf{V}^{(B)}$ is cyclic if and only if $M = \mathrm{cyc}(M)$.

3.2.3. *Let X and Y be classes inside $\mathbf{V}^{(B)}$. Then the following hold:*

(1) $[\![X \neq \varnothing]\!] = \mathbf{1} \to X{\downarrow} \neq \varnothing \wedge \mathrm{Cyc}(X{\downarrow})$;

(2) $X \in \mathbf{V}^{(B)} \to X{\downarrow} \in \mathbf{V}$;

(3) $X = Y \leftrightarrow X{\downarrow} = Y{\downarrow}$.

$\lhd$ (1) By the maximum principle, the class $X{\downarrow}$ is nonempty. If $(x_\xi)_{\xi\in\Xi} \subset X{\downarrow}$ and $(b_\xi)_{\xi\in\Xi}$ is a partition of unity then, assigning $x := \mathrm{mix}_{\xi\in\Xi}(b_\xi x_\xi)$, note that

$$[\![x \in X]\!] \geq [\![x = x_\xi]\!] \wedge [\![x_\xi \in X]\!] \geq b_\xi \quad (\xi \in \Xi).$$

Therefore, $[\![x \in X]\!] \geq \bigvee_{\xi\in\Xi} b_\xi = \mathbf{1}$ and $x \in X{\downarrow}$.

(2) Assume that $X \in \mathbf{V}^{(B)}$ and $x \in X{\downarrow}$. Let $u : \mathrm{dom}(u) \to B$ be a function such that $\mathrm{dom}(u) \subset \mathbf{V}^{(B)}$, $\mathrm{dom}(u) \in \mathbf{V}$, and $\bar{u}(\cdot) = [\![\cdot \in X]\!]$ (cf. 2.5.6). Then

$$\bigvee\{u(t) \wedge [\![t = x]\!] : t \in \mathrm{dom}(u)\} = \mathbf{1}.$$

Using the exhaustion principle 2.3.9, find a partition of unity $(b_\xi) \subset B$ and a family $(t_\xi) \subset \operatorname{dom}(u)$ satisfying $u(t_\xi) \wedge [\![x = t_\xi]\!] \geq b_\xi$, which implies the equality $x = \operatorname{mix}(b_\xi t_\xi)$. Denote by $\operatorname{Part}(B)$ the set of all partitions of unity in B and put

$$Y := \bigcup\{(\operatorname{dom}(u))^\theta : \theta \in \operatorname{Part}(B)\}.$$

Consider the function F assigning to each x the set of those ordered pairs (θ, v) for which $\theta \in \operatorname{Part}(B)$; $v : \theta \to \operatorname{dom}(u)$; and if $\theta := (b_\xi)$ then $x = \operatorname{mix}(b_\xi x_\xi)$, with $x_\xi := v(b_\xi)$. Obviously, $\operatorname{dom}(F) \supset X{\downarrow}$, $\operatorname{im}(F) \subset \mathscr{P}(\operatorname{Part}(B) \times Y)$, and $F(x) \cap F(y) = \varnothing$ for $x \neq y$. Therefore, $|X{\downarrow}| \leq |\mathscr{P}(\operatorname{Part}(B) \times Y)|$ and $X{\downarrow} \in \mathbf{V}$.

(3) If $X{\downarrow} = Y{\downarrow}$ then, by 2.5.16,

$$[\![X \subset Y]\!] = \bigwedge_{t \in X\downarrow} [\![t \in Y]\!] = \bigwedge_{t \in Y\downarrow} [\![t \in Y]\!] = \mathbf{1}.$$

Analogously, $[\![Y \subset X]\!] = \mathbf{1}$ and, hence, $[\![X = Y]\!] = \mathbf{1}$. $\triangleright$

3.2.4. Let X and Y be two $\mathbf{V}^{(B)}$-classes. Denote by $X \times_B Y$ their Cartesian product inside $\mathbf{V}^{(B)}$, which exists by virtue of 1.3.13 (2) and 2.5.18.

The mapping

$$(\cdot\,,\cdot)^B : (x, y) \mapsto (x, y)^B \quad (x \in X{\downarrow},\ y \in Y{\downarrow})$$

is a bijection of the class $X{\downarrow} \times Y{\downarrow}$ onto the class $(X \times_B Y){\downarrow}$. Moreover,

$$[\![\operatorname{Pr}_{X\downarrow}(x, y) = \operatorname{Pr}_X(x, y)]\!] = [\![\operatorname{Pr}_{Y\downarrow}(x, y) = \operatorname{Pr}_Y(x, y)]\!] = \mathbf{1}$$
$$(x \in X{\downarrow},\ y \in Y{\downarrow}),$$

where $\operatorname{Pr}_{X\downarrow}$ and $\operatorname{Pr}_{Y\downarrow}$ are the coordinate projections to the factors $X{\downarrow}$ and $Y{\downarrow}$, while Pr_X and Pr_Y stand for the coordinate projections inside $\mathbf{V}^{(B)}$ to X and Y.

(Recall that Pr_X and Pr_Y are classes inside $\mathbf{V}^{(B)}$, whereas $\operatorname{Pr}_{X\downarrow}$ and $\operatorname{Pr}_{Y\downarrow}$ are classes in the sense of $\mathbf{V}$.)

$\triangleleft$ As was mentioned earlier (cf. 2.4.9 and 2.5.3), the function $(\cdot\,,\cdot)^B$ is an injective embedding of $\mathbf{V}^{(B)} \times \mathbf{V}^{(B)}$ into $\mathbf{V}^{(B)}$. Hence, it suffices to establish that $(\cdot\,,\cdot)^B$ sends $X{\downarrow} \times Y{\downarrow} \subset \mathbf{V}^{(B)} \times \mathbf{V}^{(B)}$ to $(X \times_B Y){\downarrow}$. Granted $x \in X{\downarrow}$ and $y \in Y{\downarrow}$, observe

$$[\![(x, y)^B \in X \times Y]\!]$$
$$= [\![(\exists\, u)(\exists\, v)(u \in X \wedge v \in Y \wedge (u, v) = (x, y)^B)]\!]$$

$$= \bigvee_{u \in \mathbf{V}^{(B)}} \bigvee_{v \in \mathbf{V}^{(B)}} [\![u \in X]\!] \wedge [\![v \in Y]\!] \wedge [\![(u,v) = (x,y)^B]\!]$$
$$\geq [\![x \in X]\!] \wedge [\![y \in Y]\!] \wedge [\![(x,y) = (x,y)^B]\!] = \mathbf{1}.$$

Therefore, $(x,y)^B \in (X \times_B Y){\downarrow}$. Now, consider an arbitrary element $z \in (X \times_B Y){\downarrow}$ and note that, by the maximum principle, there are elements x and y of $\mathbf{V}^{(B)}$ satisfying

$$\mathbf{1} = [\![z \in X \times Y]\!] = [\![(\exists\, u \in X)(\exists\, v \in Y)(z = (u,v))]\!]$$
$$= [\![x \in X]\!] \wedge [\![y \in Y]\!] \wedge [\![z = (x,y)]\!].$$

Hence, $x \in X{\downarrow}$, $y \in Y{\downarrow}$, and $z = (x,y)^B$. Finally, given $x \in X{\downarrow}$, $y \in Y{\downarrow}$, and $z \in \mathbf{V}^{(B)}$, infer

$$[\![z = \Pr_X(x,y)]\!] = [\![((x,y),z) \in \Pr_X]\!] = [\![z = x]\!] = [\![z = \Pr_{X\downarrow}(x,y)]\!],$$

which ensures validity of the claimed identity for the projection to X. The situation is analogous with the projection to the second factor. $\triangleright$

3.2.5. Consider a (binary) relation X inside $\mathbf{V}^{(B)}$. This implies that X is a class inside $\mathbf{V}^{(B)}$ and $[\![X \text{ is a relation}]\!] = \mathbf{1}$. By 3.2.4 and the axiom of domain NGB_{10}, there is a class Y satisfying

$$(x,y) \in Y \leftrightarrow (x,y)^B \in X{\downarrow}.$$

Indeed, we may put

$$Y := \operatorname{dom}((\,\cdot\,,\,\cdot\,)^B \cap (\mathbf{V}^{(B)} \times \mathbf{V}^{(B)} \times X{\downarrow})).$$

It is obvious that Y is a relation and that $(\,\cdot\,,\,\cdot\,)^B$ carries out a bijection between Y and $X{\downarrow}$. The class Y is the *descent of* X. We preserve the symbol $X{\downarrow}$ for Y. In much the same way, we define the descent of an n^-ary relation X; namely:

$$X{\downarrow} := \{(x_1, \dots, x_n) \in (\mathbf{V}^{(B)})^n : (x_1, \dots, x_n)^B \in X{\downarrow}\}.$$

Observe that the descent of a class X and the descent of a binary relation X are not the same. Therefore, the common notation $X{\downarrow}$ is just a minor liberty we took for convenience. This particularity is worth remembering to avoid confusion. For instance, the equality $(X \times_B Y){\downarrow} = X{\downarrow} \times Y{\downarrow}$ is simply another record of the first part of 3.2.4. The same remark applies to the descents of correspondences, categories, and their next of kin to appear below.

3.2.6. Theorem. *If X and Y are classes inside $\mathbf{V}^{(B)}$ then the following hold:*

(1) $\mathrm{dom}(X)\!\downarrow = \mathrm{dom}(X\!\downarrow)$, $\mathrm{im}(X)\!\downarrow = \mathrm{im}(X\!\downarrow)$;

(2) $(X \cap Y)\!\downarrow = X\!\downarrow \cap Y\!\downarrow$;

(3) $(X \restriction Y)\!\downarrow = (X\!\downarrow) \restriction (Y\!\downarrow)$;

(4) $(X^{-1})\!\downarrow = (X\!\downarrow)^{-1}$;

(5) $(X \circ Y)\!\downarrow = (X\!\downarrow) \circ (Y\!\downarrow)$;

(6) $(X``Y)\!\downarrow = (X\!\downarrow)``(Y\!\downarrow)$;

(7) $(\mathbf{V}^{(B)} \models \mathrm{Fnc}\,(X)) \leftrightarrow \mathrm{Fnc}\,(X\!\downarrow)$;

(8) $(\mathbf{V}^{(B)} \models X \subset Y) \leftrightarrow X\!\downarrow \subset Y\!\downarrow$;

(9) $[\![x = y]\!] \leq [\![X(x) = X(y)]\!]$ $(x, y \in \mathbf{V}^{(B)})$;

(10) $(X\!\downarrow)^n = (X^{n^\wedge})\!\downarrow$ $(n \in \omega)$.

$\lhd$ (1) By the maximum principle, granted $x \in \mathbf{V}^{(B)}$, note that there is some y in $\mathbf{V}^{(B)}$ satisfying

$$[\![x \in \mathrm{dom}(X)]\!] = [\![(\exists\, u)((x, u) \in X)]\!] = [\![(x, y)^B \in X]\!].$$

Therefore, from $x \in \mathrm{dom}(X)\!\downarrow$ it follows that $x \in \mathrm{dom}(X\!\downarrow)$. Conversely, if $x \in \mathrm{dom}(X\!\downarrow)$ then $[\![(x, y) \in X]\!] = \mathbf{1}$ for some $y \in \mathbf{V}^{(B)}$. Hence,

$$[\![x \in \mathrm{dom}(X)]\!] = \bigvee\{[\![(x, u) \in X]\!] : u \in \mathbf{V}^{(B)}\} \geq [\![(x, y) \in X]\!],$$

and so $x \in \mathrm{dom}(X)\!\downarrow$. The second formula is proven by analogy.

(2) By definition, given $x \in \mathbf{V}^{(B)}$, note that

$$[\![x \in X \cap Y]\!] = [\![x \in X \wedge x \in Y]\!] = [\![x \in X]\!] \wedge [\![x \in Y]\!].$$

Therefore, $x \in (X \cap Y)\!\downarrow$ if and only if $x \in X\!\downarrow$ and $x \in Y\!\downarrow$ simultaneously.

(3) Applying (2), 3.2.4, and the definition of $X \restriction Y$, deduce

$$(X \restriction Y)\!\downarrow = (X \cap (Y \times \mathbf{U}_B))\!\downarrow = X\!\downarrow \cap (Y\!\downarrow \times \mathbf{V}^{(B)}) = (X\!\downarrow) \restriction (Y\!\downarrow).$$

(4) This ensues from the definition of X^{-1}.

(5) Considering a class Z, denote by σZ the σ-permutation of Z, with $\sigma := (\imath_1, \imath_2, \imath_3)$ a permutation of $\{1, 2, 3\}$ (cf. 1.3.10). It is easy to check that $(\sigma Z)\!\downarrow = \sigma(Z\!\downarrow)$. If $Z \in \mathbf{V}^{(B)}$ is such that $\mathbf{V}^{(B)} \models Z = (Y \times \mathbf{U}_B) \cap (\mathbf{U}_B \times X)$ and $\sigma := \{1, 3, 2\}$ then

$$\mathbf{V}^{(B)} \models X \circ Y = \mathrm{dom}(\sigma Z).$$

Now, using (1), (2), and 3.2.4, proceed with the following chain of equalities

$$(X \circ Y)\!\downarrow = \mathrm{dom}(\sigma Z)\!\downarrow = \mathrm{dom}(\sigma(Z\!\downarrow))$$

$$= \mathrm{dom}(\sigma((Y{\downarrow} \times \mathbf{V}^{(B)}) \cap (\mathbf{V}^{(B)} \times X{\downarrow}))) = (X{\downarrow}) \circ (Y{\downarrow}).$$

(6) Successively applying (1) and (3), obtain

$$(X\text{``}Y){\downarrow} = (\mathrm{im}(X \restriction Y)){\downarrow} = \mathrm{im}((X \restriction Y){\downarrow})$$
$$= \mathrm{im}((X{\downarrow}) \restriction (Y{\downarrow})) = (X{\downarrow})\text{``}(Y{\downarrow}).$$

(7) Assume that $[\![\mathrm{Fnc}\,(X)]\!] = \mathbf{1}$. Then $X{\downarrow}$ is a binary relation and, moreover,

$$[\![(x, y) \in X]\!] \wedge [\![(x, z) \in X]\!] \le [\![y = z]\!]$$

for all $x, y, z \in \mathbf{V}^{(B)}$. Hence, granted $(x, y) \in X{\downarrow}$ and $(x, z) \in X{\downarrow}$, infer $[\![y = z]\!] = \mathbf{1}$, i.e., $y = z$. In other words, $\mathrm{Fnc}\,(X{\downarrow})$ is fulfilled. In turn, if $X{\downarrow}$ is a single-valued binary relation then, using 2.5.16, deduce

$$[\![\mathrm{Fnc}\,(X)]\!] = \bigwedge_{x \in \mathbf{V}^{(B)}} \bigwedge \{[\![y = z]\!] : (x, y) \in X{\downarrow}, (x, z) \in X{\downarrow}\} = \mathbf{1}.$$

(8) Applying (2) and 3.2.3 (3), write

$$\mathbf{1} = [\![X \subset Y]\!] \leftrightarrow \mathbf{1} = [\![X \cap Y = X]\!] \leftrightarrow X{\downarrow} \cap Y{\downarrow} = X{\downarrow} \leftrightarrow X{\downarrow} \subset Y{\downarrow}.$$

(9) The formula $(\forall\, x)(\forall\, y)(x = y \to X\text{``}\{x\} = X\text{``}\{y\})$ is a theorem of ZF, and so its Boolean truth value is unity. Expanding the Boolean truth value by the rules for quantification and implication, come to the claim.

(10) If $[\![t : n^\wedge \to X]\!] = \mathbf{1}$ then to every $k \in n$ there is a unique element $x \in X{\downarrow}$ for which $[\![t(k^\wedge) = x]\!] = \mathbf{1}$. Letting $s(k) := x$ for $k \in n$, obtain the mapping $s : n \to X{\downarrow}$ which is also denoted by $t{\downarrow}$. Hence,

$$[\![t{\downarrow}(k) = t(k^\wedge)]\!] = \mathbf{1} \quad (k \in n).$$

Conversely, if $s : n \to X{\downarrow}$ then define $t \in \mathbf{V}^{(B)}$ by the rule

$$t := \{(k^\wedge, s(k))^B : k \in n\} \times \mathbf{1}_B.$$

In this event $[\![t : n^\wedge \to X]\!] = \mathbf{1}$, $[\![t(k^\wedge) = s(k)]\!] = \mathbf{1}$ for $k \in n$ and $t{\downarrow} = s$. Summarizing, conclude that the mapping $t \mapsto t{\downarrow}$ is a bijection between $\{x \in \mathbf{V}^{(B)} : [\![x \in X^{n^\wedge}]\!] = \mathbf{1}\}$ and $(X{\downarrow})^n$.

Proceed with recalling the definition of $s := (x(0), \dots, x(n-1))^B$ (cf. 2.4.9). Let $x : n \to X{\downarrow}$ and $y : n \to X{\downarrow}$ be such that $y(0) = x(0)$, $y(k) = (y(k-1), x(k))^B$ for $0 \ne k \in n$ and $y(n-1) = s$. By the above, there are $p, q \in \mathbf{V}^{(B)}$ satisfying $[\![p, q : n^\wedge \to X]\!] = \mathbf{1}$, in which case $p{\downarrow} = x$ and $q{\downarrow} = y$. It is now easy to check that

$$[\![p(0) = q(0) \wedge (\forall\, k \in n^\wedge)(k \ne 0 \to q(k) = (q(k-1), p(k)))]\!] = \mathbf{1}.$$

Therefore, $[\![q(n^\wedge - 1) = (p(0^\wedge), \dots, p(n^\wedge - 1)) \in X^{n^\wedge}]\!] = \mathbf{1}$. On the other hand, $[\![s = q(n^\wedge - 1)]\!] = \mathbf{1}$, and so $s \in (X^{n^\wedge}){\downarrow}$. Thus, the mapping

$$(x(0), \dots, x(n-1)) \mapsto (x(0), \dots, x(n-1))^B$$

is an injection of $(X{\downarrow})^n$ to $(X^{n^\wedge}){\downarrow}$.

Analogous arguments show that the image of $(X{\downarrow})^n$ is the whole of $(X^{n^\wedge}){\downarrow}$. $\rhd$

3.2.7. The matter with the descents of the complement of a class and the union of a family of classes differs in some respects from the cases settled in 3.2.6.

Consider an arbitrary class $Y \subset \mathbf{V}^{(B)}$. Since the formula $x \in \mathbf{V}^{(B)} \wedge (\forall\, y \in Y)$ $([\![x = y]\!] = 0)$ is predicative, there is a class $Y^{\mathfrak{c}}$ determined from the relation

$$x \in Y^{\mathfrak{c}} \leftrightarrow x \in \mathbf{V}^{(B)} \wedge (\forall\, y \in Y)([\![x = y]\!] = \mathbf{0}).$$

Now, take a class X inside $\mathbf{V}^{(B)}$. Denote by X^c the $\mathbf{V}^{(B)}$-class that is the complement of X inside $\mathbf{V}^{(B)}$; i.e.,

$$\mathbf{V}^{(B)} \models (\forall\, x)(x \in X^c \leftrightarrow x \notin X).$$

The existence of X^c follows from 2.5.18.

Consider the formula

$$\begin{aligned} &\varphi(y, B, Y, \mathbf{V}^{(B)}, [\![\cdot\ =\ \cdot]\!]) \\ &:= (\forall\, a)(\forall\, b)(\forall\, x)(b : a \to Y \wedge \text{“}b \text{ is a partition of unity”} \\ &\wedge x : a \to Y \wedge y = \operatorname*{mix}_{\alpha \in a}(b(\alpha)x(\alpha))), \end{aligned}$$

stating that y is a mixing of a certain family of elements of the class Y. It is easy to see that this formula is predicative, and so there is a class $\operatorname{mix}(Y)$ such that

$$(\forall\, y)(y \in \operatorname{mix}(Y) \leftrightarrow \varphi(y, B, Y, \mathbf{V}^{(B)}, [\![\cdot\ =\ \cdot]\!])).$$

By way of example, granted an arbitrary class $X \subset \mathbf{V}$, observe that $X^\wedge\!\!\downarrow = \operatorname{mix}(X_1)$ where $X_1 := \{x^\wedge : x \in X\}$ and the canonical embedding (cf. 3.1.1 (1)) carries out the injection of X to $\operatorname{mix}(X_1)$.

3.2.8. *If a class Y is a set then*

$$\operatorname{mix}(Y) = \operatorname{cyc}(Y).$$

$\lhd$ We only have to demonstrate that the set $\operatorname{mix}(Y)$ of all possible mixings $\operatorname{mix}_{y \in Y}(b_y y)$ of families of Y is cyclic. To this end, consider a partition of unity $(b_\xi)_{\xi \in \Xi}$ and the elements

$$y_\xi := \operatorname*{mix}_{y \in Y}(b_{\xi,y} y) \quad (\xi \in \Xi)$$

in $\operatorname{mix}(Y)$. Put $y_0 := \operatorname{mix}_{\xi \in \Xi}(b_\xi y_\xi)$ and $b_{(\xi,y)} := b_\xi \wedge b_{\xi,y}$ for $\xi \in \Xi$ and $y \in Y$. If $(\xi, y) \neq (\eta, z)$ then

$$b_{(\xi,y)} \wedge b_{(\eta,z)} = b_\xi \wedge b_\eta \wedge b_{\xi,y} \wedge b_{\eta,z} = \mathbf{0}.$$

Moreover, straightforward calculation gives (cf. 1.1.5 (2))

$$\bigvee_{(\xi,y) \in \Xi \times Y} b_{(\xi,y)} = \bigvee_{\xi \in \Xi} \left(b_\xi \wedge \bigvee_{y \in Y} b_{\xi,y} \right) = \mathbf{1}.$$

Therefore, $(b_{(\xi,y)})$ is a partition of unity. Given $y \in Y$, note that

$$[\![y_0 = y]\!] \geq [\![y_0 = y_\xi]\!] \wedge [\![y_\xi = y]\!] \geq b_\xi \wedge b_{\xi,y}.$$

Whence, $y_0 = \operatorname{mix}(b_{(\xi,y)} y)$, and so $y_0 \in \operatorname{mix}(Y)$; i.e., $\operatorname{mix}(Y)$ is a cyclic set. $\rhd$

3.2.9. *For nonempty classes X and Y inside $\mathbf{V}^{(B)}$, the following hold:*

(1) $X^c{\downarrow} = X{\downarrow}^{\mathfrak{c}}$;

(2) $(X \cup Y){\downarrow} = \operatorname{mix}(X{\downarrow} \cup Y{\downarrow})$.

$\triangleleft$ (1) Using definitions and 2.5.16, derive the following equivalences:

$$x \in X^c{\downarrow} \leftrightarrow [\![x \in X^c]\!] = \mathbf{1}$$
$$\leftrightarrow [\![x \notin X]\!] = \mathbf{1} \leftrightarrow [\![x \in X]\!] = \mathbf{0} \leftrightarrow \bigvee\{[\![x = s]\!] : s \in X{\downarrow}\} = \mathbf{0}$$
$$\leftrightarrow (\forall\, s \in X{\downarrow})([\![s = x]\!] = \mathbf{0}) \leftrightarrow x \in (X{\downarrow})^{\mathfrak{c}}.$$

(2) It is seen from 3.2.6 (8) that $X{\downarrow}\cup Y{\downarrow} \subset (X\cup Y){\downarrow}$. Conversely, if $z \in (X\cup Y){\downarrow}$ then

$$(\exists\, x \in X)(\exists\, y \in Y)(x = z \vee y = z).$$

Using the maximum principle, choose $x_0, y_0 \in \mathbf{V}^{(B)}$ so that $b \vee c = \mathbf{1}$ where $b := [\![x_0 \in X]\!] \wedge [\![x_0 = z]\!]$ and $c := [\![y_0 \in Y]\!] \wedge [\![y_0 = z]\!]$. Choosing $x_1 \in X{\downarrow}$ and $y_1 \in Y{\downarrow}$ arbitrarily, put $x = \operatorname{mix}\{bx_0, b^*x_1\}$ and $y := \operatorname{mix}\{cy_0, c^*y_1\}$. Then $x \in X{\downarrow}$, because

$$b \le [\![x = x_0]\!] \wedge [\![x_0 \in X]\!] \le [\![x \in X]\!],$$
$$b^* \le [\![x_1 = x]\!] \wedge [\![x_1 \in X]\!] \le [\![x \in X]\!].$$

By an analogous reason, $y \in Y{\downarrow}$. Moreover,

$$b \le [\![x = x_0]\!] \wedge [\![x_0 = z]\!] \le [\![x = z]\!],$$
$$b^* \le c \le [\![y = y_0]\!] \wedge [\![y_0 = z]\!] \le [\![y = z]\!];$$

i.e., $z = \operatorname{mix}\{bx, b^*y\}$ and $z \in \operatorname{mix}(X{\downarrow} \cup Y{\downarrow})$. $\triangleright$

It is worth observing in addition that we factually have

(3) $(X \cup Y){\downarrow} = \bigcup_{b\in B} bX{\downarrow} \oplus b^*Y{\downarrow}$, *where* $bX{\downarrow} \oplus b^*Y{\downarrow}$ *is the set of elements of the type* $\operatorname{mix}\{bx, b^*y\}$ $(x \in X{\downarrow},\ y \in Y{\downarrow})$.

3.2.10. Sometimes we are to repeat descending. We now clarify the way this happens.

Let X be a class. Arrange the class-function Y by the formula

$$Y := \{(x, y) : x \in \mathbf{V}^{(B)}, y = x{\downarrow}\}.$$

The *double* or *repeated descent* of X is the class $\bigcup \operatorname{im}(Y{\restriction}(X{\downarrow}))$ denoted by $X{\downarrow\downarrow}$. Therefore,

$$X{\downarrow\downarrow} = \bigcup\{x{\downarrow} : x \in X{\downarrow}\}.$$

Evidently, if $X \in \mathbf{V}^{(B)}$ then $X{\downarrow\downarrow} \in \mathbf{V}$ (cf. 3.2.3 (2)).

3.2.11. *For each nonempty* $\mathbf{V}^{(B)}$*-class* X *the following hold:*

(1) $(\bigcup X)\downarrow = \bigcup(X\downarrow\downarrow)$;

(2) $(\bigcap Y)\downarrow = \bigcap(X\downarrow\downarrow)$;

(3) $\mathscr{P}(X)\downarrow\downarrow \subset \mathscr{P}(X\downarrow)$.

$\triangleleft$ The proof leans on 2.5.16. The due calculations are as follows:

(1) $u \in \bigcup(X\downarrow\downarrow) \leftrightarrow (\exists v \in X\downarrow\downarrow)(u \in v) \leftrightarrow (\exists z \in X\downarrow)(u \in z\downarrow)$
$\leftrightarrow (\exists z \in X\downarrow)([\![u \in z]\!] = \mathbf{1}) \leftrightarrow [\![(\exists z \in X)(u \in z)]\!] = \mathbf{1} \leftrightarrow [\![u \in \bigcup X]\!] = \mathbf{1}$
$\leftrightarrow u \in (\bigcup X)\downarrow$.

(2) $u \in \bigcap(X\downarrow\downarrow) \leftrightarrow (\forall v \in X\downarrow\downarrow)(u \in v) \leftrightarrow (\forall z \in X\downarrow)(u \in z\downarrow)$
$\leftrightarrow (\forall z \in X\downarrow)([\![u \in z]\!] = \mathbf{1}) \leftrightarrow [\![(\forall z \in X)(u \in z)]\!] = \mathbf{1}$
$\leftrightarrow [\![u \in \bigcap X]\!] = \mathbf{1} \leftrightarrow u \in (\bigcap X)\downarrow$.

(3) $u \in \mathscr{P}(X)\downarrow\downarrow \leftrightarrow (\exists z \in \mathscr{P}(X)\downarrow)(u = z\downarrow) \leftrightarrow (\exists z)([\![z \subset X]\!] = \mathbf{1} \wedge u = z\downarrow)$
$\leftrightarrow (\exists z)(z\downarrow \subset X\downarrow \wedge u = z\downarrow) \to u \subset X\downarrow \leftrightarrow u \in \mathscr{P}(X\downarrow)$. $\triangleright$

3.2.12. Theorem. *Let* X, Y, *and* $f \in \mathbf{V}^{(B)}$ *be such that* $[\![X \neq \varnothing]\!] = [\![Y \neq \varnothing]\!] = [\![f : X \to Y]\!] = \mathbf{1}$. *Then there is a unique mapping* $f\downarrow : X\downarrow \to Y\downarrow$, *the descent of* f, *such that*

$$[\![f(x) = f\downarrow(x)]\!] = \mathbf{1} \quad (x \in X\downarrow).$$

The descent $f\downarrow$ *of a mapping* f *inside* $\mathbf{V}^{(B)}$ *has the following properties:*

(1) $f\downarrow$ *is an extensional mapping*, i.e.,

$$[\![x = x']\!] \leq [\![f\downarrow(x) = f\downarrow(x')]\!] \quad (x, x' \in X\downarrow);$$

(2) *If* Z *and* $g \in \mathbf{V}^{(B)}$ *are such that* $[\![Z \neq \varnothing]\!] = [\![g : Y \to Z]\!] = \mathbf{1}$ *then*

$$(g \circ f)\downarrow = g\downarrow \circ f\downarrow;$$

(3) $f\downarrow$ *is surjective, or injective, or bijective if and only if* $[\![f$ *is surjective, or injective, or bijective*$]\!] = \mathbf{1}$.

$\triangleleft$ Let h be the descent of f in the sense of 3.2.5. It follows from 3.2.6 (1, 7) that $h : X\downarrow \to Y\downarrow$. Then, since $(x, h(x))^B \in f\downarrow$ for all $x \in X\downarrow$; therefore,

$$[\![h(x) = f(x)]\!] = [\![(x, h(x)) \in f]\!] = [\![(x, h(x))^B \in f]\!] = \mathbf{1}.$$

The so-defined mapping h is unique. Indeed, if $g : X\downarrow \to Y\downarrow$ has the same property then

$$[\![h(x) = g(x)]\!] \geq [\![g(x) = f(x)]\!] \wedge [\![h(x) = f(x)]\!] = \mathbf{1}.$$

Hence, $h(x) = g(x)$ for every $x \in X\downarrow$ because $\mathbf{V}^{(B)}$ is separated. Using the defining relation of h and 3.2.6 (9), proceed with this calculation

$$[\![x = x']\!] \leq [\![f(x) = f(x')]\!] \wedge [\![f(x) = h(x)]\!]$$
$$\wedge [\![f(x') = h(x')]\!] \leq [\![h(x) = h(x')]\!].$$

We have thus established (1), while (2) follows from 3.2.6 (5).

So, we are left with checking (3). The claim about surjectivity is easy from 3.2.6 (6), while bijectivity is the conjunction of surjectivity and bijectivity. The injectivity of f inside $\mathbf{V}^{(B)}$ is equivalent to the formula

$$[\![x = x']\!] = [\![f(x) = f(x')]\!] = [\![h(x) = h(x')]\!] \quad (x, x' \in X{\downarrow}).$$

Hence, $x = x'$ if and only if $h(x) = h(x')$, which means that the mapping h is injective. ▷

3.2.13. Theorem. *Let $X, Y, F \in \mathbf{V}^{(B)}$ be such that $[\![X \neq \varnothing]\!] = [\![Y \neq \varnothing]\!] = [\![\varnothing \neq F \subset X \times Y]\!] = \mathbf{1}$. Let $\Phi \in \mathbf{V}^{(B)}$ be a correspondence from X to Y with graph F inside $\mathbf{V}^{(B)}$; i.e., $\mathbf{V}^{(B)} \models \Phi = (F, X, Y)$. Then the 3-tuple $\Phi{\downarrow} := (F{\downarrow}, X{\downarrow}, Y{\downarrow})$, the descent of Φ, is a unique correspondence obeying the equality*

$$\Phi{\downarrow}(x) = \Phi(x){\downarrow} \quad (x \in X{\downarrow}).$$

The descent of a correspondence has the following properties:

(1) $\Phi(A){\downarrow} \in \Phi{\downarrow}(A{\downarrow})$ *for every $A \in \mathbf{V}^{(B)}$ satisfying $[\![A \subset X]\!] = \mathbf{1}$;*

(2) $\pi_\Phi(A){\downarrow} = \pi_{\Phi{\downarrow}}(A{\downarrow})$ *for every $A \in \mathbf{V}^{(B)}$ satisfying $[\![A \subset X]\!] = \mathbf{1}$;*

(3) $(\Phi' \circ \Phi){\downarrow} = \Phi'{\downarrow} \circ \Phi{\downarrow}$ *for another correspondence Φ' inside $\mathbf{V}^{(B)}$;*

(4) $(I_X){\downarrow} = I_{X{\downarrow}}$.

◁ All claims but (2) are elementarily deduced from 3.2.6. Note only that the defining relation $\Phi{\downarrow}(x) = \Phi(x){\downarrow}$ $(x \in X{\downarrow})$ must be understood in accord with the remark on 3.2.1.

Indeed, by the maximum principle that there exists a member Ψ in $\mathbf{V}^{(B)}$ such that $[\![\Psi : X \to \mathscr{P}(Y)]\!] = \mathbf{1}$ and $[\![\Phi(x) = \Psi(x)]\!] = \mathbf{1}$ for all $x \in X{\downarrow}$. By 3.2.12, $\Psi{\downarrow} : X{\downarrow} \to \mathscr{P}(Y){\downarrow}$ and $[\![\Phi(x) = \Psi{\downarrow}(x)]\!] = \mathbf{1}$ for $x \in X{\downarrow}$. In this case, however, $\Phi{\downarrow}$ is defined by the relation

$$\Phi{\downarrow}(x) = (\Psi{\downarrow}(x)){\downarrow} = \Psi(x){\downarrow}{\downarrow} \quad (x \in X{\downarrow}).$$

In particular, this yields $\Phi{\downarrow}(A{\downarrow}) = \Psi(A){\downarrow}{\downarrow}$. Using these remarks, turn to proving (2). Note that

$$[\![\pi_\Phi(A) = \bigcap \Psi(A)]\!] = \mathbf{1};$$

i.e., $\pi_\Phi(A) = \bigcap\{\Psi(a) : a \in A\}$ is fulfilled inside $\mathbf{V}^{(B)}$. Whence, using 3.2.11 (2), deduce

$$\pi_\Phi(A){\downarrow} = (\bigcap \Psi(A)){\downarrow} = \bigcap(\Psi(A){\downarrow}{\downarrow})$$
$$= \bigcap\{\Phi{\downarrow}(a) : a \in A{\downarrow}\} = \pi_{\Phi{\downarrow}}(A{\downarrow}),$$

so completing the proof. ▷

3.2.14. We now address families of functions and correspondences inside $\mathbf{V}^{(B)}$.

(1) *Assume that* X *and* Y *are nonempty sets inside* $\mathbf{V}^{(B)}$. *Assume further that a family* $(f_\xi)_{\xi\in\Xi}$ *of members of* $\mathbf{V}^{(B)}$ *is such that*

$$[\![f_\xi : X \to Y]\!] = \mathbf{1} \quad (\xi \in \Xi).$$

Then the mixing $\operatorname{mix}_{\xi\in\Xi}(b_\xi f_\xi)$ *of* $(f_\xi)_{\xi\in\Xi}$ *by each partition of unity* $(b_\xi)_{\xi\in\Xi} \subset B$ *is a function from* X *to* Y *inside* $\mathbf{V}^{(B)}$ *and*

$$\operatorname*{mix}_{\xi\in\Xi}(b_\xi f_\xi){\downarrow}(x) = \operatorname*{mix}_{\xi\in\Xi}(b_\xi f_\xi{\downarrow}(x)) \quad (x \in X{\downarrow}).$$

$\lhd$ Put $g := \operatorname{mix}_{\xi\in\Xi}(b_\xi f_\xi)$. Since

$$b_\xi \le [\![g = f_\xi]\!] \wedge [\![f_\xi : X \to Y]\!] \le [\![g : X \to Y]\!];$$

therefore, $[\![g : X \to Y]\!] = \mathbf{1}$ meaning that g is a function from X to Y inside $\mathbf{V}^{(B)}$. Moreover, by 3.2.12, given $x \in X{\downarrow}$, find

$$\begin{aligned} b_\xi &\le [\![g{\downarrow}(x) = g(x)]\!] \wedge [\![g(x) = f_\xi(x)]\!] \\ &\wedge [\![f_\xi{\downarrow}(x) = f_\xi(x)]\!] \le [\![g{\downarrow}(x) = f_\xi{\downarrow}(x)]\!]. \end{aligned}$$

Hence, $g{\downarrow}(x) = \operatorname{mix}_{\xi\in\Xi}(b_\xi f_\xi{\downarrow}(x))$. $\rhd$

(2) *With* X, Y, *and* (b_ξ) *the same as above, assume that* $(\Phi_\xi)_{\xi\in\Xi}$ *is a family in* $\mathbf{V}^{(B)}$ *consisting of correspondences from* X *to* Y *inside* $\mathbf{V}^{(B)}$. *Then the mixing* $\operatorname{mix}_{\xi\in\Xi}(b_\xi \Phi_\xi)$ *itself is a correspondence from* X *to* Y *inside* $\mathbf{V}^{(B)}$. *Moreover,*

$$\operatorname*{mix}_{\xi\in\Xi}(b_\xi \Phi_\xi){\downarrow}(x) = \operatorname*{mix}_{\xi\in\Xi}(b_\xi \Phi_\xi{\downarrow}(x)) \quad (x \in X{\downarrow}).$$

$\lhd$ The proof is analogous to 3.2.14 (1). $\rhd$

3.2.15. Let $\mathscr{F}^{\downarrow}$ stand for the mapping sending a nonempty $\mathbf{V}^{(B)}$-set X to its descent $X{\downarrow}$ and taking each correspondence Φ inside $\mathbf{V}^{(B)}$ to $\Phi{\downarrow}$.

Theorem. *The mapping* $\mathscr{F}^{\downarrow}$ *is a covariant functor from the category* $\mathscr{V}_*^{(B)}$ *to the category* $\mathscr{V}_*$ *(from the category* $\mathscr{V}^{(B)}$ *to the category* $\mathscr{V}$, *respectively).*

3.2.16. *Theorem.* *Let* $\mathfrak{K}$ *be a category inside* $\mathbf{V}^{(B)}$. *Then there is a unique category* $\mathfrak{K}'$ *(in the sense of* $\mathbf{V}$*) such that* $\operatorname{Ob}\mathfrak{K}' = (\operatorname{Ob}\mathfrak{K}){\downarrow}$, $\operatorname{Mor}\mathfrak{K}' = (\operatorname{Mor}\mathfrak{K}){\downarrow}$, *and* $\mathrm{Com}' = \mathrm{Com}{\downarrow}$, *where* Com' *is the composition of* $\mathfrak{K}'$ *and* $\mathbf{V}^{(B)} \models$ *"*Com *is the composition of the category* $\mathfrak{K}$*."*

◁ It follows from 3.2.6 (7) that Com′ is a partial binary operation on the class $(\operatorname{Mor}\mathfrak{K}){\downarrow}$. Since $[\![\mathrm{Com}(\alpha,\beta) = \mathrm{Com}'(\alpha,\beta)]\!] = \mathbf{1}$ for all α, $\beta \in \operatorname{Mor}\mathfrak{K}'$ and Com is associative inside $\mathbf{V}^{(B)}$; therefore, Com′ is also associative.

Let D and R be the $\mathbf{V}^{(B)}$-classes of the definition of $\mathfrak{K}$ (cf. 2.5.19). Put $D' := D{\downarrow}$ and $R' := R{\downarrow}$. By 3.2.6 (1), (7), D' and R' are mappings from $\operatorname{Mor}\mathfrak{K}'$ to $\operatorname{Ob}\mathfrak{K}'$. Appealing to 3.2.6 (1) again, conclude that for $\alpha, \beta \in \operatorname{Mor}\mathfrak{K}'$ the formulas $(\alpha,\beta) \in \mathrm{dom}(\mathrm{Com}')$ and $[\![(\alpha,\beta) \in \mathrm{dom}(\mathrm{Com})]\!] = \mathbf{1}$ are equivalent. On the other hand, the equality $R'(\alpha) = D'(\beta)$ is fulfilled only if $[\![R(\alpha) = D(\beta)]\!] = \mathbf{1}$. Existence of the identity morphisms in $\mathfrak{K}'$ is obvious. Hence, $\mathfrak{K}$ satisfies all hypotheses of the definition in 2.5.19. ▷

3.2.17. The category $\mathfrak{K}'$ of 3.2.16 is called the *descent* of $\mathfrak{K}$ and denoted by $\mathfrak{K}{\downarrow}$. Let Set_*^B be the category of nonempty sets and correspondences inside $\mathbf{V}^{(B)}$. More explicitly, $\operatorname{Mor}\mathrm{Set}_*^B$, $\operatorname{Ob}\mathrm{Set}_*^B$, and $\mathrm{Com} : \mathbf{V}^{(B)} \to B$ have the form

$$\begin{gathered}\operatorname{Ob}\mathrm{Set}_*^B : x \mapsto [\![x \neq \varnothing]\!],\\ \operatorname{Mor}\mathrm{Set}_*^B : \alpha \mapsto [\![(\exists\, x)(\exists\, y)(\exists\, f)\\ (x \neq \varnothing \wedge y \neq \varnothing \wedge f \neq \varnothing \wedge f \subset x \times y \wedge \alpha = (f, x, y))]\!],\\ \mathrm{Com} : u \mapsto [\![(\exists \alpha)(\exists \beta)(\exists \gamma)\\ (\alpha, \beta, \text{ and } \gamma \text{ are correspondences}) \wedge \gamma = \alpha \circ \beta \wedge u = (\alpha, \beta, \gamma)]\!].\end{gathered}$$

The descent of the category Set_*^B is easily seen to coincide with the category $\mathscr{V}_*^{(B)}$ of 3.1.7. The category Set^B of nonempty sets and mappings inside $\mathbf{V}^{(B)}$ is defined similarly, yielding $\mathscr{V}^{(B)} = \mathrm{Set}^B{\downarrow}$.

3.2.18. COMMENTS.

(1) As was mentioned in 3.2.5, we use the unique symbol ↓ for denoting various operations of the same provenance. Consequently, the record $X{\downarrow}$ is unambiguously understood only if extra information is available on which object X is descending. This runs in a perfect analogy with using the same sign + for denoting many group operations: addition of numbers, vectors, linear operators, etc. The context always prompts the precise meaning.

(2) The double descent of 3.2.10 appears in dealing with other set-theoretic operations. For instance, let $\prod X$ stand for the class of all mappings f from X to $\bigcup X$ such that $f(x) \in x$ for all $x \in X$ and $\sum X := \bigcup\{x \times \{x\} : x \in X\}$. Then to each $X \in \mathbf{V}^{(B)}$ there are natural bijections

$$\Big(\prod X\Big){\downarrow\downarrow} = \prod(X{\downarrow\downarrow}), \quad \Big(\sum X\Big){\downarrow} = \sum(X{\downarrow\downarrow}).$$

The double descent in $(\prod X){\downarrow\downarrow}$ relates to mappings.

(3) The inclusion in 3.2.11 (3) is clearly strict on assuming that $B \neq \mathbf{2}$. Note also that $\mathscr{P}(X)\downarrow$ is an algebraic system of signature $(\vee, \wedge, *, 0, 1)$. It is possible to show that this is a complete Boolean algebra presenting a completion of the inclusion ordered set $\mathscr{P}(X)\downarrow\downarrow$ in the following sense: There is an order preserving injection $\imath : \mathscr{P}(X)\downarrow\downarrow \to \mathscr{P}(X)\downarrow$ satisfying the following condition: given $a \in \mathscr{P}(X)\downarrow$, $a < \mathbf{1}$, we may find $b \in \mathscr{P}(X)\downarrow\downarrow$ so that $a \leq \imath(b) < \mathbf{1}$. This situation is in exact analogy with the construction of the completion of a Boolean algebra (cf. [83, 220]).

(4) Proving 3.2.6 (10), we have established in particular that, for $X \in \mathbf{V}^{(B)}$, the mapping $\downarrow$ is a bijection between the sets $\mathscr{V}(n, X\downarrow)$ and $\mathscr{V}^{(B)}(n^{\wedge}, X)$. This phenomenon is of a rather abstract nature, reflecting deep relationship between the functors $\mathscr{F}^{\wedge}$ and $\mathscr{F}^{\downarrow}$. We elaborate details in Section 3.5.

3.3. The Ascent Functor

In this section we ascend from the von Neumann universe to a Boolean valued universe, considering this as reversal of descent. We define an appropriate functor and study its main properties.

3.3.1. Assume given a subclass X of the class $\mathbf{V}^{(B)}$.

(1) *The formula*

$$Y(t) := \bigvee \{[\![t = x]\!] : x \in X\} \quad (t \in \mathbf{V}^{(B)})$$

defines a $\mathbf{V}^{(B)}$*-class* Y.

$\lhd$ By Theorem 1.3.14, there is a class Y in the sense of $\mathbf{V}$ such that

$$(y, b) \in Y \leftrightarrow y \in \mathbf{V}^{(B)} \wedge b \in B \wedge \left(b = \bigvee_{x \in X} [\![x = y]\!]\right).$$

Clearly, Y is single-valued and $\operatorname{dom}(Y) = \mathbf{V}^{(B)}$; i.e., Y is a mapping from $\mathbf{V}^{(B)}$ to B. Moreover, this mapping is extensional since, by virtue of 2.1.8 (4),

$$\begin{aligned} Y(t) \wedge [\![t = s]\!] &= \bigvee \{[\![t = x]\!] \wedge [\![t = s]\!] : x \in X\} \\ &\leq \bigvee \{[\![s = x]\!] : x \in X\} = Y(s). \end{aligned}$$

Hence, Y is a class inside $\mathbf{V}^{(B)}$. $\rhd$

To each class $X \subset \mathbf{V}^{(B)}$ we have thus assigned the class Y inside $\mathbf{V}^{(B)}$ which is called the *ascent* of X and denoted by $X\uparrow$.

In case X is a set, there is a unique element $y \in \mathbf{V}^{(B)}$ such that $X\uparrow(t) = [\![t \in y]\!]$ for all $t \in \mathbf{V}^{(B)}$ (cf. 2.5.6). This y is the *ascent* of X (cf. 2.5.10). By way of example, note that, for a class $X \subset \mathbf{V}$, the class $X^{\wedge}$ is the ascent of $\{x^{\wedge} : x \in X\}$ (cf. 2.5.15).

(2) Assume now that X is a relation on $\mathbf{V}^{(B)}$; i.e., $X \subset \mathbf{V}^{(B)} \times \mathbf{V}^{(B)}$. In order for X to ascend, we will firstly embed it into $\mathbf{V}^{(B)}$ and, secondly, apply the above procedure. To this end, we use the function $(x, y) \mapsto (x, y)^B$ (cf. 3.2.4). Therefore, we give the following definition of the *ascent* of a binary relation X on $\mathbf{V}^{(B)}$:

$$X\uparrow : t \mapsto \bigvee\{[\![t = (x, y)^B]\!] : (x, y) \in X\}.$$

In particular, if X is the product of some classes $Y \subset \mathbf{V}^{(B)}$ and $Z \subset \mathbf{V}^{(B)}$ then we arrive at the *ascent* of $Y \times Z$:

$$(Y \times Z)\uparrow : t \mapsto \bigvee\{[\![t = (x, y)^B]\!] : y \in Y, z \in Z\}.$$

3.3.2. *Assume that $X \subset \mathbf{V}^{(B)}$ is a nonempty class and φ is a B-formula. Then*

$$[\![(\forall\, u \in X\uparrow)\varphi(u)]\!] = \bigwedge\{[\![\varphi(u)]\!] : u \in X\},$$
$$[\![(\exists\, u \in X\uparrow)\varphi(u)]\!] = \bigvee\{[\![\varphi(u)]\!] : u \in X\}.$$

$\lhd$ We demonstrate only the last formula (cf. 1.1.5 (2, 7)):

$$[\![(\exists\, u \in X\uparrow)\varphi(u)]\!] = [\![(\exists\, u)(u \in X\uparrow \wedge \varphi(u))]\!]$$
$$= \bigvee_{v \in \mathbf{V}^{(B)}} \bigvee_{u \in X} [\![u = v]\!] \wedge [\![\varphi(v)]\!]$$
$$= \bigvee_{u \in X} \Big(\bigvee_{v \in \mathbf{V}^{(B)}} [\![v = u]\!] \wedge [\![\varphi(v)]\!] \Big) = \bigvee\{[\![\varphi(u)]\!] : u \in X\}.$$

The case of a universal quantifier is settled by analogy. $\rhd$

3.3.3. *For an arbitrary class $X \subset \mathbf{V}^{(B)}$ and a nonempty $\mathbf{V}^{(B)}$-class $Y : \mathbf{V}^{(B)} \to B$, the following **arrow cancellation rules** hold:*

(1) $X\uparrow\downarrow = \mathrm{mix}(X)$;

(2) $Y\downarrow\uparrow = Y$.

$\lhd$ (1) If X is empty then the claim is trivial. If $x \in X$ then $[\![x \in X\uparrow]\!] = \mathbf{1}$. Hence, $x \in X\uparrow\downarrow$. This fact, together with 3.2.3, yields $\mathrm{mix}(X) \subset X\uparrow\downarrow$. The reverse inclusion follows from 3.3.2 and the mixing principle.

(2) By 2.5.16, given $y \in \mathbf{V}^{(B)}$, note that

$$[\![y \in Y\downarrow\uparrow]\!] = \bigvee\{[\![y = t]\!] : t \in Y\downarrow\} = [\![(\exists t \in Y)(t = y)]\!] = [\![y \in Y]\!],$$

so completing the proof. $\rhd$

(3) Using the mixing of a family of ordered pairs, we find the following proposition of service:

Assume that $(b_\xi)_{\xi\in\Xi}$ is a partition of unity in B. Assume further that $(x_\xi)_{\xi\in\Xi}$ and $(y_\xi)_{\xi\in\Xi}$ are some families in $\mathbf{V}^{(B)}$. Then

$$\operatorname*{mix}_{\xi\in\Xi} b_\xi (x_\xi, y_\xi)^B = \big(\operatorname*{mix}_{\xi\in\Xi} b_\xi x_\xi, \operatorname*{mix}_{\xi\in\Xi} b_\xi y_\xi\big)^B.$$

◁ Show first that $b(x,y)^B = b(bx,by)^B$ for all $x,y \in \mathbf{V}^{(B)}$ and $b \in B$. To this end, successively apply 2.3.2, 2.4.9, and 2.3.6:

$$\begin{gathered}[\![b(x,y)^B = b(bx,by)^B]\!] = b \to [\![(x,y)^B = (bx,by)^B]\!] = b \\ \to ([\![x = bx]\!] \wedge [\![y = by]\!]) = b \to ((b^* \Rightarrow [\![x = \varnothing]\!]) \\ \wedge (b^* \Rightarrow [\![y = \varnothing]\!])) = b^* \vee ((b \vee [\![x = \varnothing]\!]) \wedge (b \vee [\![y = \varnothing]\!])) \\ = (b^* \vee b \vee [\![x = \varnothing]\!]) \wedge (b^* \vee b \vee [\![y = \varnothing]\!]) = \mathbf{1}.\end{gathered}$$

Now, assign

$$x := \operatorname*{mix}_{\xi\in\Xi} b_\xi x_\xi, \quad y := \operatorname*{mix}_{\xi\in\Xi} b_\xi y_\xi.$$

Summarizing, obtain

$$b_\xi (x_\xi, y_\xi)^B = b_\xi (b_\xi x_\xi, b_\xi y_\xi)^B = b_\xi (b_\xi x, b_\xi y)^B = b_\xi (x,y)^B.$$

To complete the proof, refer to the mixing principle. ▷

This fact allows us to consider mixings in the class $\mathbf{V}^{(B)} \times \mathbf{V}^{(B)}$. Namely, we agree to assign

$$\operatorname*{mix}_{\xi\in\Xi} b_\xi (x_\xi, y_\xi) := \big(\operatorname*{mix}_{\xi\in\Xi} b_\xi x_\xi, \operatorname*{mix}_{\xi\in\Xi} b_\xi y_\xi\big).$$

We are in a position now to assert that the mapping $(x,y) \mapsto (x,y)^B$ preserves mixing.

3.3.4. Theorem. *For all classes $X \subset \mathbf{V}^{(B)}$ and $Y \subset \mathbf{V}^{(B)}$ the following hold:*

(1) $\mathbf{V}^{(B)} \models X\!\uparrow \subset Y\!\uparrow$ *provided that* $X \subset Y$;

(2) $\mathbf{V}^{(B)} \models (X \cup Y)\!\uparrow = X\!\uparrow \cup Y\!\uparrow$;

(3) $\mathbf{V}^{(B)} \models (\mathrm{mix}(X) \cap \mathrm{mix}(Y))\!\uparrow = X\!\uparrow \cap Y\!\uparrow$;

(4) $\mathbf{V}^{(B)} \models (X \times Y)\!\uparrow = X\!\uparrow \times Y\!\uparrow$.

Moreover, if X and Y are relations and Z is a class then

(5) $\mathbf{V}^{(B)} \models \mathrm{dom}(X)\!\uparrow = \mathrm{dom}(X\!\uparrow) \wedge \mathrm{im}(X)\!\uparrow = \mathrm{im}(X\!\uparrow)$;

(6) $\mathbf{V}^{(B)} \models (X^{-1})\!\uparrow = (X\!\uparrow)^{-1}$;

(7) $\mathbf{V}^{(B)} \models (\mathrm{mix}(X)\text{“}\,\mathrm{mix}(Z))\!\uparrow = (X\!\uparrow)\text{“}(Z\!\uparrow)$;

(8) $\mathbf{V}^{(B)} \models (\mathrm{mix}(X) \circ \mathrm{mix}(Y))\!\uparrow = (X\!\uparrow) \circ (Y\!\uparrow)$;

(9) $\mathbf{V}^{(B)} \models (Z^n)\!\uparrow = (Z\!\uparrow)^{n^\wedge}$ *for all* $n \in \mathbb{N}$.

$\triangleleft$ (1) The claim follows from the definition of descent.

(2) This ensues from the following calculation:

$$[\![t \in (X \cup Y)\!\uparrow]\!] = \bigvee\{[\![t = u]\!] : u \in X \cup Y\}$$
$$= \bigvee_{u \in X} [\![t = u]\!] \vee \bigvee_{u \in Y} [\![t = u]\!] = [\![t \in X\!\uparrow \vee\, t \in Y\!\uparrow]\!].$$

(3) Assume proven that the ascent of the intersection of X and Y coincides with the intersection of the ascents $X\!\uparrow$ and $Y\!\uparrow$ inside $\mathbf{V}^{(B)}$. By 3.2.6 (2) and 3.3.3, conclude

$$\mathrm{mix}(X \cap Y) = (X \cap Y)\!\uparrow\downarrow = (X\!\uparrow \cap Y\!\uparrow)\!\downarrow$$
$$= X\!\uparrow\downarrow \cap Y\!\uparrow\downarrow = \mathrm{mix}(X) \cap \mathrm{mix}(Y).$$

Conversely, assume that the cyclic hull of the intersection of X and Y equals the intersection of the cyclic hulls of X and Y. On appealing to 3.2.6 (2) and 3.3.3 again, infer

$$(X \cap Y)\!\uparrow\downarrow = X\!\uparrow\downarrow \cap Y\!\uparrow\downarrow = (X\!\uparrow \cap Y\!\uparrow)\!\downarrow.$$

Hence, $[\![(X \cap Y)\!\uparrow = X\!\uparrow \cap Y\!\uparrow]\!] = \mathbf{1}$ according to 3.2.3 (3).

To complete the proof, apply the above to the classes $\mathrm{mix}(X)$ and $\mathrm{mix}(Y)$ and recall the rules for arrow cancellation of 3.3.3.

(4) Using 3.3.2, proceed with the calculation

$$[\![z \in X\!\uparrow \times Y\!\uparrow]\!] = [\![(\exists\, u \in X\!\uparrow)(\exists\, v \in Y\!\uparrow) z = (u, v)]\!]$$
$$= \bigvee_{u \in X} \bigvee_{v \in Y} [\![z = (u, v)]\!] = \bigvee_{(u,v) \in X \times Y} [\![z = (u, v)^B]\!] = [\![z \in (X \times Y)\!\uparrow]\!].$$

(5) Supposing that X is a binary relation, it is easy to check the following chain of equalities (cf. 1.1.5 (2, 7)):

$$[\![x \in \mathrm{dom}(X\!\uparrow)]\!] = [\![(\exists y)((x, y) \in X\!\uparrow)]\!]$$
$$= \bigvee_{y \in \mathbf{V}^{(B)}} \bigvee_{(s,t) \in X} [\![(x, y)^B = (s, t)^B]\!]$$
$$= \bigvee_{(s,t) \in X} \bigvee_{y \in \mathbf{V}^{(B)}} [\![x = s]\!] \wedge [\![y = t]\!]$$
$$= \bigvee_{s \in \mathrm{dom}(X)} [\![x = s]\!] = [\![x \in \mathrm{dom}(X)\!\uparrow]\!].$$

The claim about im(X) is demonstrated by analogy.

(6) $[\![(x,y) \in (X\uparrow)^{-1}]\!] = [\![(y,x) \in X\uparrow]\!] = \bigvee_{(s,t)\in X} [\![(s,t) = (y,x)]\!]$

$$= \bigvee_{(t,s)\in X^{-1}} [\![(t,s) = (x,y)]\!] = [\![(x,y) \in (X^{-1})\uparrow]\!].$$

(7), (8) It is obvious that

$$\operatorname{mix}(X) \cap (\operatorname{mix}(Z) \times \mathbf{V}^{(B)}) = \operatorname{mix}(X) \cap \operatorname{mix}(Z \times \mathbf{V}^{(B)});$$
$$(\operatorname{mix}(Y) \times \mathbf{V}^{(B)}) \cap (\mathbf{V}^{(B)} \times \operatorname{mix}(X))$$
$$= \operatorname{mix}(Y \times \mathbf{V}^{(B)}) \cap \operatorname{mix}(\mathbf{V}^{(B)} \times X).$$

Proceed further along the lines of 3.2.6 (5, 6), using (3), (4) and the fact that $[\![\mathbf{V}^{(B)}\uparrow = \mathbf{U}_B]\!] = \mathbf{1}$.

(9) Considering 3.3.3 (3), note that $\operatorname{mix}(Z^n) = \operatorname{mix}(Z)^n$. Using 3.2.6 (10) and 3.3.3 (1), conclude

$$((Z\uparrow)^{n^\wedge})\downarrow = (Z\uparrow\downarrow)^n = (Z^n)\uparrow\downarrow.$$

This yields the claim by 3.2.3 (3). ▷

3.3.5. Consider a class X composed of subsets of $\mathbf{V}^{(B)}$; i.e., $X \subset \mathscr{P}(\mathbf{V}^{(B)})$. The *double* or *repeated ascent* of X, denoted by $X\uparrow\uparrow$, is the ascent of the class $\{x\uparrow : x \in X\}$. Hence,

$$[\![t \in X\uparrow\uparrow]\!] = \bigvee\{[\![t = x\uparrow]\!] : x \in X\} \quad (t \in \mathbf{V}^{(B)}).$$

Introduce one more notation:

$$\operatorname{mix} \text{“}X := \{\operatorname{mix}(u) : u \in X\}.$$

Obviously, $[\![X\uparrow\uparrow = (\operatorname{mix} \text{“}X)\uparrow\uparrow]\!] = \mathbf{1}$.

Let $\mathscr{P}_0(X)$ stand for the class of nonempty members of $\mathscr{P}(X)$; i.e.,

$$\mathscr{P}_0(X) := \{z : z \subset X, z \neq \varnothing\}.$$

3.3.6. *Suppose that X is a nonempty $\mathbf{V}^{(B)}$-class and $Y \subset \mathscr{P}(\mathbf{V}^{(B)})$. Then*

(1) $\mathbf{V}^{(B)} \models \bigcup(Y\uparrow\uparrow) = (\bigcup Y)\uparrow$;

(2) $\mathbf{V}^{(B)} \models \bigcap(Y\uparrow\uparrow) = \bigcap(\operatorname{mix} \text{“}(Y\uparrow))$;

(3) $\mathbf{V}^{(B)} \models \bigcup X = (\bigcup(X\downarrow\downarrow))\uparrow$;

(4) $\mathbf{V}^{(B)} \models \mathscr{P}_0(X\downarrow)\uparrow\uparrow = \mathscr{P}_0(X)$.

◁ The proof is left to the reader as an exercise. ▷

3.3.7. We now return to Theorem 3.3.4 and observe by items (1) and (4) of this theorem that the ascent of a correspondence is again a correspondence. For the purposes of analysis, it is desirable that "the images of points and sets," $X(t)$ and X"A, be preserved in ascending. Unfortunately, this is not so as seen from 3.3.4 (7). Moreover, the ascent of a function may fail to be single-valued. This is easy to comprehend on recalling that the consecutive "ascending and descending" provides cyclic hulls (3.3.3 (1)), whereas every descended function is extensional by 3.2.6 (9).

We exhibit an appropriate example. Suppose that $X \subset \mathbf{V}^{(B)}$ is a cyclic set and $f : X \to \{\mathbf{0}^\wedge, \mathbf{1}^\wedge\}$ is a two-valued function. Assume that $f(x) = \mathbf{0}^\wedge$ and $f(y) = \mathbf{1}^\wedge$ for some $x, y \in X$, $x \neq y$, while an element $b \in B$ is other than $\mathbf{0}$ and $\mathbf{1}$.

If f sends $z := \operatorname{mix}\{bx, b^*y\} \in X$ to $\mathbf{0}^\wedge$, then $\mathbf{0} < b^* \leq [\![z = y]\!] \not\leq [\![f(z) = f(y)]\!] = \mathbf{0}$. Analogously, if $f(z) = \mathbf{1}^\wedge$ then $0 < b \leq [\![z = x]\!] \not\leq [\![f(z) = f(x)]\!] = \mathbf{0}$. On the other hand, $[\![z = y]\!] \leq [\![f{\uparrow}(z) = f{\uparrow}(y)]\!]$ by 3.2.6 (9). Therefore, either $[\![f{\uparrow}(y) = f(y)]\!] \neq \mathbf{1}$, or $[\![f{\uparrow}(x) \doteq f(x)]\!] \neq \mathbf{1}$; i.e., the equality $[\![f{\uparrow}(x) = f(x)]\!] = \mathbf{1}$ is fulfilled not for all $x \in X$.

Therefore, we must study in more detail what happens with an ascending correspondence.

3.3.8. *For an arbitrary relation $X \subset \mathbf{V}^{(B)} \times \mathbf{V}^{(B)}$ the following are equivalent:*

(1) *If $b \leq [\![x_1 = x_2]\!]$ for x_1, $x_2 \in \operatorname{dom}(X)$ and $b \in B$ then*

$$\bigvee\{b \wedge [\![y_1 = u]\!] : y_1 \in X(x_1)\} = \bigvee\{b \wedge [\![y_2 = u]\!] : y_2 \in X(x_2)\}$$

for every $u \in \mathbf{V}^{(B)}$;

(2) *If x_1, $x_2 \in \operatorname{dom}(X)$ and $y_1 \in X(x_1)$ then*

$$[\![x_1 = x_2]\!] \leq \bigvee\{[\![y_1 = y_2]\!] : y_2 \in X(x_2)\};$$

(3) $\operatorname{mix}(X(x)) = \operatorname{mix}(X)(x)$ $(x \in \operatorname{dom}(X))$;

(4) $[\![X{\uparrow}(x) = X(x){\uparrow}]\!] = \mathbf{1}$ $(x \in \operatorname{dom}(X))$;

(5) $[\![x_1 = x_2]\!] \leq [\![X(x_1){\uparrow} = X(x_2){\uparrow}]\!]$ $(x_1, x_2 \in \operatorname{dom}(X))$.

$\lhd$ (1) $\to$ (2) Insert $b := [\![x_1 = x_2]\!]$ and $u := y_1$ in (1).

(2) $\to$ (3) The inclusion $\subset$ is obvious. To prove the reverse inclusion, take a partition of unity $(b_\xi) \subset B$ and a family $((x_\xi, y_\xi)) \subset X$ and arrange $(x, y) = \operatorname{mix}(b_\xi(x_\xi, y_\xi))$. The task is to establish that $y \in \operatorname{mix}(X(x))$. It follows from (2) that

$$b_\xi \leq [\![x = x_\xi]\!] \leq \bigvee\{[\![y' = y_\xi]\!] : y' \in X(x)\} = [\![y_\xi \in X(x){\uparrow}]\!].$$

Therefore, $b_\xi \leq [\![y = y_\xi]\!] \wedge [\![y_\xi \in X(x){\uparrow}]\!] \leq [\![y \in X(x){\uparrow}]\!]$, so that $[\![y \in X(x){\uparrow}]\!] = \mathbf{1}$. But then $y \in X(x){\uparrow}{\downarrow} = \operatorname{mix}(X(x))$, which completes the proof.

(3) → (4) Using 3.3.3 (1) and 3.2.6 (6), note that

$$X(x)\uparrow\downarrow = \operatorname{mix}(X(x)) = \operatorname{mix}(X)(x) = (X\uparrow\downarrow)(x) = (X\uparrow(x))\downarrow.$$

Applying 3.3.3 (2), obtain the desired.

(4) → (5) It suffices to apply 3.2.6 (9).

(5) → (1) By 2.3.2, if $b \leq [\![x_1 = x_2]\!]$ and $x_1, x_2 \in \operatorname{dom}(X)$ then $b(X(x_1)\uparrow = b(X(x_2)\uparrow)$. On the other hand, by the definition of descent,

$$[\![u \in b(X(x_k)\uparrow)]\!] = \bigvee\{b \wedge [\![u = y]\!] : y \in X(x_k)\},$$

which yields the claim. ▷

3.3.9. Now return in more diverse circumstances to the notion of extensionality which we have encountered in 3.2.6 (9) and 3.2.12 (1). A binary relation $R \subset \mathbf{V}^{(B)} \times \mathbf{V}^{(B)}$ is *extensional in second coordinate* provided that R obeys one (and hence all) of the equivalent conditions 3.3.8 (1–5). Note that if R is a function then each of the conditions (2) and (5) of 3.3.8 turns into the following formula (cf. 2.5.5)

$$[\![x_1 = x_2]\!] \leq [\![R(x_1) = R(x_2)]\!] \quad (x_1, x_2 \in \operatorname{dom}(R)).$$

Let $X \subset \mathbf{V}^{(B)}$ and $Y \subset \mathbf{V}^{(B)}$ be sets. A correspondence $\Phi := (F, X, Y)$ is *extensional* if the graph F of Φ, viewed as a relation on $\mathbf{V}^{(B)} \times \mathbf{V}^{(B)}$, is extensional in second coordinate.

If, moreover, $\operatorname{dom}(\Phi) = \operatorname{mix}(\operatorname{dom}(\Phi))$ and $\Phi(x) = \operatorname{mix}(\Phi(x))$ for every $x \in \operatorname{dom}\Phi$ then Φ is said to be *fully extensional.* Evidently, if Φ is fully extensional then $F = (X \times Y) \cap \operatorname{mix}(F)$.

Say that some sets A and $C \subset \mathbf{V}^{(B)}$ are *in general position* provided that

$$[\![a = c]\!] \leq \bigvee\{[\![a = b]\!] \wedge [\![b = c]\!] : b \in A \cap C\}$$

for all $a \in A$ and $c \in C$. When this condition is fulfilled, the last inequality becomes an equality since $[\![a = b]\!] \wedge [\![b = c]\!] \leq [\![a = c]\!]$.

The following are equivalent:

(1) $\mathbf{V}^{(B)} \models (A \cap C)\uparrow = A\uparrow \cap C\uparrow$;

(2) $\operatorname{mix}(A \cap C) = \operatorname{mix}(A) \cap \operatorname{mix}(C)$;

(3) *A and C are in general position.*

◁ The equivalence of (1) and (2) results from 3.2.6 (1), 3.3.3 (1), and 3.2.4 (3).

Prove (1) ↔ (3). To this end, note that the inclusion $A\uparrow \cap C\uparrow \subset (A \cap C)\uparrow$ amounts to the formula

$$(\forall\, a \in A\uparrow)(\forall\, c \in C\uparrow)(a = c \rightarrow (\exists\, b \in A \cap C)(a = b \wedge b = c)).$$

The Boolean truth value of this formula is as follows:

$$\bigwedge_{a\in A,c\in C} [\![a=c]\!] \Rightarrow \bigvee_{b\in A\cap C} [\![a=b]\!] \wedge [\![b=c]\!].$$

This implies that (3) is equivalent to the inclusion $A\!\uparrow \cap C\!\uparrow \subset (A\cap C)\!\uparrow$ inside $\mathbf{V}^{(B)}$. The reverse inclusion is always true. $\triangleright$

We thus see that if $A \subset C$ then A and C are in general position for a trivial reason. Also, every two sets of the type $A := \{a^{\wedge} : a \in A'\}$, where $A' \in \mathbf{V}$, are in general position.

The *ascent* of a correspondence $\Phi := (F, X, Y)$ is by definition the element $\Phi\!\uparrow := (F\!\uparrow, X\!\uparrow, Y\!\uparrow)^B \in \mathbf{V}^{(B)}$, where $F\!\uparrow$ is the ascent of the relation F (cf. 3.3.1 (2)).

3.3.10. *Theorem.* *Suppose that X and Y are subsets of $\mathbf{V}^{(B)}$ and Φ is an extensional correspondence from X to Y. The ascent $\Phi\!\uparrow$ is a unique correspondence from $X\!\uparrow$ to $Y\!\uparrow$ inside $\mathbf{V}^{(B)}$ such that*

$$[\![\mathrm{dom}(\Phi\!\uparrow) = (\mathrm{dom}(\Phi))\!\uparrow]\!] = \mathbf{1},$$
$$[\![\Phi\!\uparrow(x) = \Phi(x)\!\uparrow]\!] = \mathbf{1} \quad (x \in \mathrm{dom}(\Phi)).$$

Moreover, the following hold:

(1) *If* $\mathrm{dom}(\Phi)$ *and a set* $A \subset X$ *are in general position then*

$$\mathbf{V}^{(B)} \models \Phi(A)\!\uparrow = \Phi\!\uparrow(A\!\uparrow);$$

(2) *The composition $\Psi \circ \Phi$ of extensional correspondences Φ and Ψ is an extensional correspondence. In this event if* $\mathrm{dom}(\Psi\circ\Phi) = \mathrm{dom}(\Phi)$ *and the sets* $\mathrm{dom}(\Psi)$ *and* $\Phi(x)$ *are in general position for all* $x \in \mathrm{dom}(\Phi)$ *then*

$$\mathbf{V}^{(B)} \models (\Psi\circ\Phi)\!\uparrow = \Psi\!\uparrow \circ \Phi\!\uparrow;$$

(3) $\mathbf{V}^{(B)} \models (I_X)\!\uparrow = I_{X\uparrow}$.

$\triangleleft$ By 3.3.4 and 3.3.8, it suffices to verify uniqueness for $\Phi\!\uparrow$ and the claims (1)–(3). Also, the case of the empty correspondence is obvious and thus omitted.

Let Ψ be another correspondence inside $\mathbf{V}^{(B)}$ obeying the same identities as $\Phi\!\uparrow$; i.e., $[\![\mathrm{dom}(\Psi) = (\mathrm{dom}(\Phi))\!\uparrow]\!] = \mathbf{1}$ and $[\![\Psi(x) = \Phi(x)\!\uparrow]\!] = \mathbf{1}$ $(x \in \mathrm{dom}(\Phi))$. In this case $\mathbf{V}^{(B)} \models \mathrm{dom}(\Psi) = \mathrm{dom}(\Phi\!\uparrow)$ and

$$[\![(\forall\, x \in \mathrm{dom}(\Psi))\Psi(x) = \Phi\!\uparrow(x)]\!]$$

$$= \bigwedge_{x\in\mathrm{dom}(\Phi)} [\![\Psi(x) = \Phi{\uparrow}(x)]\!] = \bigwedge_{x\in\mathrm{dom}(\Phi)} [\![\Psi(x) = \Phi(x){\uparrow}]\!] = \mathbf{1}.$$

(1) Using 3.3.9 (1) and the properties of $\Phi{\uparrow}$ established above, take an arbitrary $y \in \mathbf{V}^{(B)}$ and proceed with the equivalences:

$$\begin{aligned} y \in \Phi{\uparrow}(A{\uparrow}) &\leftrightarrow (\exists\, x)(x \in (\mathrm{dom}(\Phi)){\uparrow} \wedge x \in A{\uparrow} \wedge y \in \Phi{\uparrow}(x)) \\ &\leftrightarrow (\exists\, x)(x \in (A \cap \mathrm{dom}(\Phi)){\uparrow} \wedge y \in \Phi{\uparrow}(x)) \\ &\leftrightarrow (\exists\, x \in (A \cap \mathrm{dom}(\Phi)){\uparrow})\, y \in \Phi(x){\uparrow}. \end{aligned}$$

Hence,

$$[\![y \in \Phi{\uparrow}(A{\uparrow})]\!] = \bigvee_{x\in A\cap\mathrm{dom}(\Phi)} [\![y \in \Phi(x){\uparrow}]\!]$$

$$= \bigvee_{x\in A\cap\mathrm{dom}(\Phi)} \bigvee_{v\in\Phi(x)} [\![y = v]\!] = \bigvee_{v\in\Phi(A)} [\![y = v]\!] = [\![y \in \Phi(A){\uparrow}]\!].$$

(2) Show that the correspondence $\Theta := \Psi \circ \Phi$ is extensional. Take x_1, $x_2 \in \mathrm{dom}(\Theta)$, $y_1 \in \Phi(x_1)$, and $z_1 \in \Psi(y_1)$. By 3.3.8 (2), the following estimates hold:

$$\bigvee_{z_2\in\Theta(x_2)} [\![z_1 = z_2]\!] = \bigvee_{y_2\in\Phi(x_2)} \Bigg(\bigvee_{z_2\in\Psi(y_2)} [\![z_1 = z_2]\!] \Bigg)$$

$$\geq \bigvee_{y_2\in\Phi(x_2)} [\![y_1 = y_2]\!] \geq [\![x_1 = x_2]\!].$$

Using 3.3.8 (2) again, note that Θ is extensional. Therefore, using the above for Θ, infer:

$$[\![\Theta{\uparrow}(x) = \Theta(x){\uparrow}]\!] = \mathbf{1} \quad (x \in \mathrm{dom}(\Theta)).$$

Considering the facts established in (1), proceed inside $\mathbf{V}^{(B)}$ as follows:

$$\begin{aligned} \Theta{\uparrow}(x) = \Theta(x){\uparrow} &= \Psi(\Phi(x)){\uparrow} = \Psi{\uparrow}(\Phi(x){\uparrow}) \\ = \Psi{\uparrow}(\Phi{\uparrow}(x)) &= (\Psi{\uparrow} \circ \Phi{\uparrow})(x) \quad (x \in \mathrm{dom}(\Theta)). \end{aligned}$$

Therefore, from 3.3.2 we derive

$$\mathbf{V}^{(B)} \models (\forall x \in \mathrm{dom}(\Theta{\uparrow}))\,(\Theta{\uparrow}(x) = (\Psi{\uparrow} \circ \Phi{\uparrow})(x)),$$

which amounts to the claim since $\mathrm{dom}(\Psi{\uparrow} \circ \Phi{\uparrow}) = \mathrm{dom}(\Theta{\uparrow})$.

(3) This is obvious. ▷

3.3.11. Theorem. *Let X and Y be subsets of $\mathbf{V}^{(B)}$. Assume further that f is an extensional mapping from X to Y. Then $f\uparrow$ is a unique element of $\mathbf{V}^{(B)}$ satisfying*

$$[\![f\uparrow : X\uparrow \to Y\uparrow]\!] = [\![f\uparrow(x) = f(x)]\!] = \mathbf{1} \quad (x \in X).$$

Moreover, the following hold:

(1) *If Z is a subset of $\mathbf{V}^{(B)}$ and $g : Y \to Z$ is an extensional mapping then $g \circ f$ is also an extensional mapping and*

$$\mathbf{V}^{(B)} \models (g \circ f)\uparrow = g\uparrow \circ f\uparrow;$$

(2) $\mathbf{V}^{(B)} \models f(A)\uparrow = f\uparrow(A\uparrow)$ $(A \subset X)$;

(3) $\mathbf{V}^{(B)} \models$ *"the mapping $f\uparrow$ is injective" if and only if f is injective;*

(4) $\mathbf{V}^{(B)} \models$ *"the mapping $f\uparrow$ is surjective" if and only if* $\operatorname{mix}(\operatorname{im}(f)) = \operatorname{mix}(Y)$.

3.3.12. Proposition 3.3.3 directly yields the arrow cancellation rules for correspondences and mappings.

Let Φ and f be extensional correspondences from X to Y, with f single-valued. Assume further that Ψ is a correspondence inside $\mathbf{V}^{(B)}$. Then the following hold:

(1) $\Phi\uparrow\downarrow(x) = \operatorname{mix}(\Phi(x))$ $(x \in \operatorname{dom}(\Phi))$,

(2) $f\uparrow\downarrow(x) = f(x)$ $(x \in \operatorname{dom}(f))$,

(3) $\Psi\downarrow\uparrow = \Psi$,

(4) $\pi_{\Phi\uparrow\downarrow}(A) = \pi_{\Phi\uparrow}(A\uparrow)\downarrow$ $(A \subset X)$,

(5) $\pi_{\Phi\uparrow\downarrow}(A)\uparrow = \pi_{\Phi\uparrow}(A\uparrow)$ $(A \subset X)$.

Moreover, if Φ is fully extensional and $A \subset \operatorname{dom}(\Phi)$ then

(6) $\pi_{\Phi}(A)\uparrow = \pi_{\Phi\uparrow}(A\uparrow)$.

$\lhd$ (1) Given $x \in \operatorname{dom}(\Phi)$, use 3.2.13, 3.3.10, and 3.3.3 (1) to derive:

$$\Phi\uparrow\downarrow(x) = \Phi\uparrow(x)\downarrow = \Phi(x)\uparrow\downarrow = \operatorname{mix}(\Phi(x)).$$

(2, 3) These are obvious.

(4) Considering $A \subset X$, obtain

$$\begin{aligned} z \in \pi_{\Phi\uparrow}(A\uparrow)\downarrow &\leftrightarrow [\![(\forall\, a \in A\uparrow) z \in \Phi\uparrow(a)]\!] = \mathbf{1} \\ \leftrightarrow \bigwedge_{a \in A} [\![z \in \Phi\uparrow(a)]\!] = \mathbf{1} &\leftrightarrow (\forall\, a \in A)(z \in \Phi\uparrow(a)\downarrow) \\ &\leftrightarrow (\forall\, a \in A) z \in \Phi\uparrow\downarrow(a) \leftrightarrow z \in \pi_{\Phi\uparrow\downarrow}(A). \end{aligned}$$

(5) The sought equality ensues from the above on appealing to 3.3.3 (2).

(6) Granted a fully extensional Φ, use (1) to observe

$$\pi_{\Phi\uparrow\downarrow}(A) = \bigcap_{a\in A} \Phi\uparrow\downarrow(a) = \bigcap_{a\in A} \Phi(a) = \pi_\Phi(A).$$

The claim now ensues from (5). ▷

3.3.13. Consider the category $\mathscr{P}\mathscr{V}_*^{(B)}$ consisting of nonempty subsets of $\mathbf{V}^{(B)}$, extensional correspondences with a nonempty graph and the conventional composition law:

$$\operatorname{Ob} \mathscr{P}\mathscr{V}_*^{(B)} := \mathscr{P}(\mathbf{V}^{(B)}) \setminus \{\varnothing\};$$

$$\mathscr{P}\mathscr{V}_*^{(B)}(X,Y) := \{\Phi : \Phi \text{ is an extensional correspondence from } X \text{ to } Y \text{ and } \operatorname{Gr}(\Phi) \neq \varnothing\},$$

$$\operatorname{Com}(\Phi, \Psi) := \Psi \circ \Phi \quad (\Phi, \Psi \in \operatorname{Mor} \mathscr{P}\mathscr{V}_*^{(B)}).$$

The subcategory of the category $\mathscr{P}\mathscr{V}_*^{(B)}$ which consists of cyclic sets and fully extensional correspondences is denoted by $\mathscr{G}\mathscr{P}\mathscr{V}_*^{(B)}$. Let $\mathscr{P}\mathscr{V}^{(B)}$ and $\mathscr{G}\mathscr{P}\mathscr{V}^{(B)}$ be the respective subcategories of the categories $\mathscr{P}\mathscr{V}_*^{(B)}$ and $\mathscr{G}\mathscr{P}\mathscr{V}_*^{(B)}$ with the same classes of objects but now with extensional mappings as morphisms.

The soundness of this definition is ensured by 3.3.10 and 3.3.11. Consider a mapping $\mathscr{F}^\uparrow$ assigning to every object X and every morphism Φ of the category $\mathscr{P}\mathscr{V}_*^{(B)}$ their ascents $X\uparrow$ and $\Phi\uparrow$. By Theorem 3.3.10, $\mathscr{F}^\uparrow$ acts into the category $\mathscr{V}_*^{(B)}$ (cf. 3.1.7).

3.3.14. *Theorem.* *The mapping $\mathscr{F}^\uparrow$ is a covariant functor from the category $\mathscr{P}\mathscr{V}^{(B)}$ to the category $\mathscr{V}^{(B)}$.*

3.3.15. COMMENTS.

(1) We use the unique symbol $\uparrow$ for denoting various ascents in much the same way as this is done with descents. Therefore, all precautions and agreements of 3.2.5 and 3.2.18 (1) should be taken into account. The terminology of "ascending and descending" was coined by S. S. Kutateladze in [141, 142] in memory of M. C. Escher (cf. [79, 156]).

(2) The functors $\mathscr{F}^\wedge$ and $\mathscr{F}^\uparrow$ act in the same category and resemble one another in many respects (compare, for instance, the definitions 2.5.15 and 3.3.1 (1); the formulas 3.3.2 and 2.5.15; 3.3.3 and 3.1.1 (1); 3.3.4 and 3.1.4; 3.3.10 and 3.1.5; etc.). A deeper analogy is revealed in Section 3.4.

(3) Formulas 3.3.2 and their counterparts of 2.5.15 are particular cases of the following rules. If φ and ψ are predicative formulas in $n+1$ and $m+1$ free variables, while $X_1, \ldots, X_n$ and $Y_1, \ldots, Y_m$ are some $\mathbf{V}^{(B)}$-classes; then

$$[\![(\forall u)(\varphi(u, \bar{X}) \to \psi(u, \bar{Y})]\!] = \bigwedge\{[\![\psi(u, \bar{X})]\!] : x \in A\},$$
$$[\![(\exists u)(\varphi(u, \bar{X}) \wedge \psi(u, \bar{Y})]\!] = \bigvee\{[\![\psi(u, \bar{X})]\!] : x \in A\},$$

where A is any subclass of $\mathbf{V}^{(B)}$ obeying the condition

$$\operatorname{mix}(A) = \{x \in \mathbf{V}^{(B)} : [\![\varphi(x, \bar{X})]\!] = \mathbf{1}\} \neq \varnothing \quad (\bar{X} = (X_1, \ldots, X_n)).$$

(4) Ascending was implicit in Section 2.4. We now explicate this point. Let x be a subset of an unseparated universe. Assume further that $x' \subset \mathbf{V}^{(B)}$ is the image of X under factorization (cf. 2.5.2 and 2.5.7): $x' := \pi``x := \{\pi t : t \in x\}$. Define some element y of the unseparated universe by the formulas: $\operatorname{dom}(y) := x$, $\operatorname{im}(y) := \{\mathbf{1}\}$. Then $[\![\pi y = x'\!\uparrow]\!] = \mathbf{1}$. Indeed,

$$[\![\pi t \in x'\!\uparrow]\!] = \bigvee_{u \in x'} [\![\pi t = u]\!] = \bigvee_{u \in x} [\![\pi t = \pi u]\!]$$
$$= \bigvee_{u \in \operatorname{dom}(y)} y(u) \wedge [\![t = u]\!] = [\![t \in y]\!] = [\![\pi t \in \pi y]\!].$$

Therefore, the element y of 2.4.5 (2), $\{x\}^B$ and $\{x, y\}^B$ of 2.4.8, f of 2.4.11 (1–3) are all ascents in the unseparated universe. Moreover, $X^\wedge$ is the ascent of $\{x^\wedge : x \in X\}$ (cf. 3.3.1 (1)).

(5) The hypothesis of general position is impossible to omit in Theorem 3.3.10. The corresponding counterexamples are easily available on using the following argument: Assume that $A \subset X$ and Φ is a correspondence from X to X with graph $\{(x, x) : x \in M\}$. If $A \subset X$ and $A \cap M = \varnothing$ but $A \cap \operatorname{mix}(M) \neq \varnothing$, then $\Phi(A) = \varnothing$ and $[\![\Phi(A)\!\uparrow = \varnothing]\!] = \mathbf{1}$. On the other hand, $[\![\Phi\!\uparrow\!(A\!\uparrow) \neq \varnothing]\!] = \mathbf{1}$ since $[\![z \in \Phi\!\uparrow\!(A\!\uparrow)]\!] = \mathbf{1}$ for $z \in A \cap \operatorname{mix}(M)$. Observe also that in some of our relevant articles (cf., for instance, [123, 131, 141]) the condition of general position was absent on the implicit presumption that $A \subset \operatorname{dom}(\Phi)$ or $\operatorname{im}(\Phi) \subset \operatorname{dom}(\Psi)$. This inadvertent omission might lead to confusion in the case of general correspondences. However, there is no danger at all in dealing with the correspondences defined everywhere and, in particular, with mappings. The same remarks are appropriate in regard to the rules for calculating polars (cf. 3.3.12 (6)).

3.4. The Immersion Functor

Of utmost import for applications of Boolean valued models to analysis is the following technique: We immerse the object under study, i.e. some set X, in a suitable Boolean valued universe $\mathbf{V}^{(B)}$, making the structure of X less intricate or even transparent. We then continue analysis inside $\mathbf{V}^{(B)}$, which often completes the original task. This immersion procedure proves to be functor-like, enabling us to study not only the inner structure of individual objects but also interplay between them.

3.4.1. A complete Boolean algebra B induces some extra algebraic structure in the descent of every object inside $\mathbf{V}^{(B)}$. Therefore, only those objects may pretend to be immersed into $\mathbf{V}^{(B)}$ that are duly related to the complete Boolean algebra B.

We now introduce the appropriate terminology. Consider an arbitrary set X. A mapping $d : X \times X \to B$ is a *B-semimetric* provided that for all x, y, $z \in X$ the following are fulfilled:

(1) $d(x, x) = \mathbf{0}$;

(2) $d(x, y) = d(y, x)$;

(3) $d(x, y) \le d(x, z) \vee d(z, y)$.

If, moreover, $d(x, y) = \mathbf{0}$ yields $x = y$ then d is called a *B-metric* or *Boolean metric* on X. In this event (X, d) is called a *B-set* or *Boolean set*.

Each set X, lying in $\mathbf{V}^{(B)}$, is equipped with the *canonical B-metric*

$$d(x, y) := [\![x \ne y]\!] = [\![x = y]\!]^* \quad (x, y \in X).$$

The fact that d is a B-metric follows from 2.1.8 (1, 3, 4) and the separation property of $\mathbf{V}^{(B)}$. Considering subsets of $\mathbf{V}^{(B)}$ as B-sets, we always imply that are furnished with the canonical Boolean metric.

Many concepts of Chapter 2 translate naturally to B-sets by dualizing with respect to complementation in B. Thus, we sometimes omit some minor details in introducing new notions.

3.4.2. Let (b_ξ) be a partition of unity in B and let (x_ξ) be a family in a B-set X. The *mixing* of (x_ξ) by (b_ξ) is an element $x \in X$ such that $b_\xi \wedge d(x, x_\xi) = \mathbf{0}$ for all ξ. As before, we write $x = \operatorname{mix}(b_\xi x_\xi)$. This mixing, if existent, is unique. Indeed, if $y \in X$ and $(\forall\, \xi)(b_\xi \wedge d(y, x_\xi) = \mathbf{0})$ then

$$b_\xi \wedge d(x, y) \le b_\xi \wedge (d(x, x_\xi) \vee d(x_\xi, y)) = \mathbf{0}.$$

The infinite distributive law 1.1.5 (2) in B implies

$$d(x, y) = \bigvee \{b_\xi \wedge d(x, y)\} = \mathbf{0},$$

and so $x = y$.

Note that unlike the case of the universe $\mathbf{V}^{(B)}$ (cf. Section 2.3), not all mixings in a B-set may fail to contain each mixing.

3.4.3. Consider a B-set (X, d). Given a subset $A \subset X$, we denote by $\operatorname{mix}(A)$ the set of all mixings of elements of A. If $\operatorname{mix}(A) = A$ then A is a *cyclic subset* of X.

Denote by $\operatorname{cyc}(A)$ the intersection of all cyclic sets containing A. A Boolean set X is *universally complete* or *extended* if X contains the mixing $\operatorname{mix}(b_\xi x_\xi)$ of each family $(x_\xi) \subset X$ by any partition of unity $(b_\xi) \subset B$. In the event when these mixings exist only for finite subsets of X, we call X *finitely complete* (the word "decomposable" prevails in the Russian literature).

In much the same way as in 3.2.8, it is possible to show that if X is a universally complete B-set then $\operatorname{mix}(A) = \operatorname{cyc}(A)$ for all $A \subset X$. A cyclic subset of a B-set is not always a universally complete B-set. Every cyclic subset of $\mathbf{V}^{(B)}$ with the canonical B-metric is a universally complete B-set.

3.4.4. *Let* A *be a set. Assume that to each* $\alpha \in \mathrm{A}$ *there corresponds a* B*-set* (X_α, d_α). *Put* $X := \Pi_{\alpha \in \mathrm{A}} X_\alpha$ *and define the mapping* $d : X \times X \to B$ *as follows:*

$$d(x, y) := \bigvee \{d_\alpha(x(\alpha), y(\alpha)) : \alpha \in \mathrm{A}\}.$$

Then d *is a Boolean metric on* X; *moreover,* (X, d) *is universally complete if and only if* X_α *is universally complete for all* $\alpha \in \mathrm{A}$.

$\triangleleft$ It is easy to prove that the above mapping is a B-metric. Moreover, if (b_ξ) is a partition of unity and (x_ξ) is a family in the product X then $x = \operatorname{mix}(b_\xi x_\xi)$ if and only if $x(\alpha) = \operatorname{mix}(b_\xi x_\xi(\alpha))$ for all $\alpha \in \mathrm{A}$. Whence it follows that X is universally complete. $\triangleright$

In the sequel we always view the product of B-sets as a B-set with the Boolean metric of 3.4.4.

3.4.5. *Let* A *be a subset of a universally complete* B*-set* (X, d). *Then for any* $x \in X$ *the Boolean distance from* x *to* A, *defined as*

$$\operatorname{dist}(x, A) := \bigwedge \{d(x, a) : a \in A\},$$

is attained for some $a \in \operatorname{mix}(A)$. *In other words, to every* $x \in X$ *there is some* $a \in \operatorname{mix}(A)$ *satisfying* $\operatorname{dist}(x, A) = d(x, a)$.

$\triangleleft$ If $b_0 := \operatorname{dist}(x, A)$ then there are a partition (b_ξ) of b_0^* and a family $(a_\xi) \subset A$ such that $b_\xi \wedge d(x, a_\xi) = \mathbf{0}$ for all ξ. Put $a := \operatorname{mix}\{b_0 a_0, b_\xi a_\xi\}$, where a_0 is an arbitrary element of A. Since $(b_\xi) \cup \{b_0\}$ is a partition of unity then $a \in \operatorname{mix}(A)$. Moreover, for every ξ we obtain

$$b_\xi \wedge d(x, a) \leq (b_\xi \wedge d(x, a_\xi)) \vee (b_\xi \wedge d(a_\xi, a)) = \mathbf{0}.$$

Hence, $b_0^* \wedge d(x, a) = \bigvee\{b_\xi \wedge d(x, a)\} = \mathbf{0}$ or $d(x, a) \leq b_0$. The converse claim is immediate. $\triangleright$

3.4.6. Note three useful corollaries to 3.4.5:

(1) *The distance from a point $x \in X$ to a subset A of a universally complete B-set X is equal to zero if and only if $x \in \operatorname{mix}(A)$.*

(2) The *Boolean distance* between $A_1 \subset X$ and $A_2 \subset X$ is defined by the formula

$$\bar{d}(A_1, A_2) := \bigvee_{\alpha \in A_1} \operatorname{dist}(a, A_2) \vee \bigvee_{\alpha \in A_2} \operatorname{dist}(A_1, a).$$

It is easy to check that $\bar{d}$ is a Boolean semimetric on $\mathscr{P}(X)$ but not a metric in general. It would be natural to call $\bar{d}$ the *Hausdorff B-semimetric* associated with d. If X is universally complete then $\bar{d}(A_1, A_2) = \mathbf{0}$ if and only if $\operatorname{mix}(A_1) = \operatorname{mix}(A_2)$.

(3) *Let $\mathscr{P}_{\mathrm{cyc}}(X)$ be the set of all cyclic subsets of a B-set (X, d). Then (X, d) is universally complete if and only if $(\mathscr{P}_{\mathrm{cyc}}(X), \bar{d})$ is a universally complete B-set.*

$\triangleleft$ Indeed, assume that X is universally complete. Then, by (2), $\bar{d}$ is a metric on $\mathscr{P}_{\mathrm{cyc}}(X)$ and we have only to prove that $(\mathscr{P}_{\mathrm{cyc}}(X), \bar{d})$ is universally complete. To this end, consider a partition of unity (b_ξ) and a family (A_ξ) in $\mathscr{P}_{\mathrm{cyc}}(X)$.

Define $A \subset X$ as the union of all mixings of the form $\operatorname{mix}(b_\xi x_\xi)$, where $x_\xi \in A_\xi$ for all ξ. Then, given $x \in A$ and $x' \in A_\xi$ and using 1.1.5 (8), note the equalities

$$b_\xi \wedge \operatorname{dist}(x', A) = \bigwedge\{b_\xi \wedge d(x', a) : a \in A\} = \mathbf{0},$$

$$b_\xi \wedge \operatorname{dist}(x, A_\xi) = \bigwedge\{b_\xi \wedge d(x, a) : a \in A_\xi\} = \mathbf{0}.$$

Finally, by the distributive laws 1.1.5 (1, 2), $b_\xi \wedge \bar{d}(A, A_\xi) = \mathbf{0}$. The last equality holds for all ξ and so $A = \operatorname{mix}(b_\xi A_\xi)$. To prove that A is cyclic, proceed along the lines of 3.2.8.

The converse claim results from the fact that the mapping $x \mapsto \{x\}$ is an injection of X to $\mathscr{P}_{\mathrm{cyc}}(X)$ satisfying $\bar{d}(\{x\}, \{y\}) = d(x, y)$ for all $x, y \in X$. $\triangleright$

3.4.7. Consider B-sets (X, d_X) and (Y, d_Y). A correspondence Φ from X to Y is called *contractive*, or a *contraction correspondence*, or simply a *contraction* provided that

$$\bar{d}_Y(\Phi(x), \Phi(y)) \leq d_X(x, y) \quad (x, y \in \operatorname{dom}(\Phi)),$$

where $\bar{d}_Y$ is the Hausdorff B-semimetric associated with d_Y.

(1) *The contraction property of a correspondence Φ is equivalent to each of the conditions* (*cf.* 3.3.8 (1, 2)):

(a) *If* $d_X(x_1, x_2) \le b$ $(x_1, x_2 \in \operatorname{dom}(\Phi))$ *then*

$$b \vee \operatorname{dist}(y, \Phi(x_1)) = b \vee \operatorname{dist}(y, \Phi(x_2))$$

for all $y \in Y$;

(b) $\operatorname{dist}(y_1, \Phi(x_2)) \le d_X(x_1, x_2)$ *for arbitrary* $x_1, x_2 \in \operatorname{dom}(\Phi)$ *and* $y_1 \in \Phi(x_1)$.

If X and Y are subsets of $\mathbf{V}^{(B)}$ then to denote the same property of a correspondence after introducing the above definition we may use two terms, contraction and extensionality, which are contrasting in common parlance. To avoid ambiguity, recall that extensionality is interpreted with the Boolean truth value of equality $[\![\cdot = \cdot]\!]$, whereas contraction pertains to the B-metric under study.

A correspondence Φ is called *fully contractive* if Φ is contractive and

$$\Phi(x) = \operatorname{mix}(\Phi(x)) \quad (x \in \operatorname{dom}(\Phi)).$$

(2) *The descent of every correspondence is a fully contractive or, which is the same, fully extensional correspondence.*

$\lhd$ The conclusion means that if Ψ is a correspondence inside $\mathbf{V}^{(B)}$ and $\Phi := \Psi{\downarrow}$ then Φ is an extensional correspondence and $\Phi(x)$ is a cyclic set for every $x \in \operatorname{dom}(\Phi)$. By 3.2.6 (9), 3.2.13, and 3.3.8 (5), Φ is extensional; while by 3.2.3 (1) and 3.2.13 (1) $\Phi(x)$ is cyclic. $\rhd$

A mapping $f : X \to Y$ is contractive whenever

$$d_Y(f(x), f(x')) \le d_X(x, x') \quad (x, x' \in X).$$

If the last formula holds with equality then f is a *B-isometry*. A bijective B-isometry is an *isomorphism* of B-sets.

3.4.8. Each set $X \in \mathbf{V}$ becomes a B-set if equipped with the *discrete B-metric*:

$$d(x, y) := \begin{cases} \mathbf{1}_B, & \text{if } x \neq y, \\ \mathbf{0}_B, & \text{if } x = y. \end{cases}$$

In this case the pair (X, d) is called a *discrete B-set*. In a discrete B-set there is no mixing $\operatorname{mix}(b_\xi x_\xi)$ if only the set of elements (x_ξ) has more than one element and the partition of unity (b_ξ) differs from the trivial partition $\{\mathbf{0}_B, \mathbf{1}_B\}$. Every correspondence from a discrete B-set to an arbitrary B-set is contractive.

Discrete and universally complete B-sets are two extreme examples of "B-qualification" offered by the elements of the universes $\mathbf{V}$ and $\mathbf{V}^{(B)}$ (cf. 3.2.3). Compromises are plentiful among the members of $\mathscr{P}(\mathbf{V}^{(B)})$. In analysis, we often encounter B-sets of another provenance.

3.4.9. *Assume that π is a complete monomorphism of B to a Boolean algebra C. Put*

$$d_\pi(x,y) := \bigwedge\{b^* : \pi(b)\wedge x = \pi(b)\wedge y\} \quad (x,y\in C).$$

Then d_π is a B-metric on C, and the Boolean operations on C are contractive.

$\lhd$ If $\pi = I_B$ then $d_\pi(b,b') = (b \Leftrightarrow b')^* = b \vartriangle b'$. Consider one more complete Boolean algebra C' and a complete monomorphism $\pi' : B \to C'$. Then the homomorphism $h : C \to C'$ is a contractive mapping between the B-sets (C, d_π) and $(C', d_{\pi'})$ if and only if $h\circ\pi = \pi'$. Indeed, the fact that h is contractive in the metrics d_π and $d_{\pi'}$ means that $\pi(b)\wedge x = \pi(b)\wedge y$ implies $\pi'(b)\wedge h(x) = \pi'(b)\wedge h(y)$ for all $x,y\in C$ and $b\in B$.

If $\pi' = h\circ\pi$ then, applying h to the equality $\pi(b)\wedge x = \pi(b)\wedge y$, obtain $\pi'(b)\wedge h(x) = \pi'(b)\wedge h(y)$. Conversely, if in the last equality we put $x = \mathbf{1}_C$ and $y := \pi(b)$ then we get either $\pi'(b) = \pi'(b)\wedge h\pi(b)$ or $\pi'(b)\le h\circ\pi(b)$. Since $b\in B$ is arbitrary, deduce $\pi' = h\circ\pi$. $\rhd$

3.4.10. Consider another construction with B-sets analogous to 2.2.10. Let ψ be an ultrafilter on a Boolean algebra D. Consider a Boolean set (X, d_X) with a D-valued B-metric d_X. Equip X with the binary relation $\sim_\psi$ by the formula

$$(x,y)\in\sim_\psi \leftrightarrow d_X(x,y)^*\in\psi.$$

The definition of Boolean metric implies that $\sim_\psi$ is an equivalence. Let $X/{\sim_\psi}$ be the factor set of the set X by $\sim_\psi$. Also, let $\pi_X : X\to X/{\sim_\psi}$ stand for the canonical mapping. If the same is done with the Boolean set $(D,\vartriangle)$ then $D/{\sim_\psi}$ presents the two-element Boolean algebra, so that $D/{\sim_\psi}\simeq\{\mathbf{0}_D,\mathbf{1}_D\}$.

Clearly, there is a unique mapping $\tilde d : X/{\sim_\psi}\to D/{\sim_\psi}$ such that $\tilde d(\pi_X x, \pi_X y) = \pi_D(d(x,y))$ $(x,y\in X)$. Moreover, $\tilde d$ is a discrete Boolean metric on $X/{\sim_\psi}$. If d_X is a discrete metric then $\sim_\psi = I_X$ and $X/{\sim_\psi} = X$. Some set-theoretic operations on X and $X/{\sim_\psi}$ are simply interrelated. If (X_α) is a family in X then $(\bigcup X_\alpha)/{\sim_\psi} = \bigcup(X_\alpha/{\sim_\psi})$.

In the case of powers there is a natural bijection between $X^n/{\sim_\psi}$ and $(X/{\sim_\psi})^n$ given by the formula

$$\pi_{X^n} : (x_1,\dots,x_n)\mapsto(\pi_X x_1,\dots,\pi_X x_n) \quad (x_1,\dots,x_n\in X).$$

Note also that if $A\subset X$ then $A/{\sim_\psi} = \pi_X(A)$ and $\pi_A = \pi_X\restriction A$.

Choose one more B-set (Y,d_Y), and let $F\subset X\times Y$. It is then easily checked that

$$\mathrm{dom}(F/{\sim_\psi}) = \mathrm{dom}(F)/{\sim_\psi}, \quad \mathrm{im}(F/{\sim_\psi}) = \mathrm{im}(F)/{\sim_\psi}.$$

3.4.11. *Assume that ρ is an arbitrary automorphism (homomorphism into itself) of a Boolean algebra B, and ψ_ρ is an element of $\mathbf{V}^{(B)}$ determined by the function $\{(b^\wedge,\rho(b)) : b\in B\}$ in accord with 2.5.6. Then the following hold:*

(1) $\rho(b) = [\![b^\wedge \in \psi_\rho]\!]$ *for all* $b \in B$;

(2) $[\![A^\wedge \subset \psi_\rho \to (\bigwedge A)^\wedge \in \psi_\rho]\!] = \mathbf{1}$ *for* $A \subset B$ *if and only if* $\rho(\bigwedge A) = \bigwedge \rho(A)$;

(3) $[\![\psi_\rho$ *is an ultrafilter on* $B^\wedge]\!] = \mathbf{1}$.

$\lhd$ (1) This is checked by calculation on appealing to 2.2.8 (1, 2).

(2) Using (1) and given $A \subset B$, obtain

$$[\![A^\wedge \subset \psi_\rho]\!] = \bigwedge_{a\in A} [\![a \in \psi_\rho]\!] = \bigwedge_{a\in A} \rho(a) = \bigwedge \rho(A).$$

Since $\rho(\bigwedge A) \le \bigwedge \rho(A)$ by monotonicity of ρ, the inequality $[\![A^\wedge \subset \psi_\rho]\!] \le [\![(\bigwedge A)^\wedge \in \psi_\rho]\!]$ amounts to the equality $\rho(\bigwedge A) = \bigwedge \rho(A)$.

(3) First of all, note that $\mathbf{V}^{(B)} \models \psi_\rho \subset B^\wedge$. Indeed, granted $t \in \mathbf{V}^{(B)}$, observe

$$[\![t \in \psi_\rho]\!] = \bigvee_{b\in B} \rho(b) \wedge [\![t = b^\wedge]\!] \le \bigvee_{b\in B} [\![t = b^\wedge]\!] = [\![t \in B^\wedge]\!].$$

It then follows from (1) that $[\![0^\wedge \notin \psi_\rho]\!] = \mathbf{1}$, while (2) implies that $[\![\psi_\rho$ is a filter base$]\!] = \mathbf{1}$. Moreover, if $b \in B$ then

$$\begin{aligned}[\![(\exists a \in \psi_\rho)(a \le b^\wedge)]\!] &= \bigvee_{a\in B} \rho(a) \wedge [\![a^\wedge \le b^\wedge]\!] = \bigvee_{a\le b} \rho(a)\\ &= \rho(b) = [\![b^\wedge \in \psi_\rho]\!],\end{aligned}$$

so that

$$[\![(\forall\, b \in B^\wedge)((\exists\, a \in \psi_\rho)(a \le b) \to b \in \psi_\rho)]\!] = \mathbf{1}.$$

Therefore, ψ_ρ is a filter on $B^\wedge$ inside $\mathbf{V}^{(B)}$, and we have to show that $\mathbf{V}^{(B)} \models$ "for each $b \in B^\wedge$ either $b \in \psi_\rho$ or $b^* \in \psi_\rho$." This claim is demonstrated by the following formulas:

$$\begin{aligned}&[\![(\forall\, b \in B^\wedge)(b \in \psi_\rho \vee b^* \in \psi_\rho)]\!]\\ &= \bigwedge_{b\in B} [\![b^\wedge \in \psi_\rho]\!] \vee [\![(b^*)^\wedge \in \psi_\rho]\!] = \bigwedge_{b\in B} \rho(b) \vee \rho(b^*)\\ &= \bigwedge\{\rho(b \vee b^*) : b \in B\} = \rho(\mathbf{1}) = \mathbf{1}.\end{aligned}$$

The proof is over. $\rhd$

3.4.12. Let $\psi := \psi_\imath$, where $\imath$ is the identity homomorphism on B. According to 3.4.11, $\mathbf{V}^{(B)} \models$"ψ is an ultrafilter on $B^\wedge$, and $A^\wedge \subset \psi$ implies $\bigwedge(A)^\wedge \in \psi$" for an arbitrary set $A \subset B$.

Take an arbitrary B-set (X, d). It is obvious from 3.1.16 that $(X^\wedge, d^\wedge)$ is a B-set inside $\mathbf{V}^{(B)}$. By 3.4.10, 3.4.11, and the maximum principle, there are $\widetilde{X}$, $\sim := \sim_\psi$, and $\pi_X \in \mathbf{V}^{(B)}$ such that

(1) $\mathbf{V}^{(B)} \models$ "$\sim$ is an equivalence relation on $X^\wedge$";

(2) $\mathbf{V}^{(B)} \models \widetilde{X} := X^\wedge/\sim$;

(3) $\mathbf{V}^{(B)} \models$ "$\pi_X : X \to \widetilde{X}$ is the factor mapping";

(4) $[\![(x^\wedge, y^\wedge)^B \in \sim]\!] = d(x,y)^*$ $(x, y \in X)$.

If we apply the described procedure to a B-set (B, Δ) (cf. 3.4.9) then in place of $\widetilde{B}$ we obtain the two-element Boolean algebra, so that $\mathbf{V}^{(B)} \models \widetilde{B} \simeq \{\mathbf{0}_B^\wedge, \mathbf{1}_B^\wedge\}^B$. Therefore, inside $\mathbf{V}^{(B)}$ there is a unique $\{\mathbf{0}_B^\wedge, \mathbf{1}_B^\wedge\}$-valued Boolean metric $\tilde{d}$ on $\widetilde{X}$ satisfying

$$\mathbf{V}^{(B)} \models (\forall\, x, y \in X^\wedge)\tilde{d}(\pi_X(x), \pi_X(y)) = \pi_B(d^\wedge(x,y)).$$

As seen from 3.4.10, for a discrete B-set (X, d) we have $\sim\, = I_{X^\wedge}$ and $X^\sim = X^\wedge$.

Say that subsets A and C of some B-set (X, d) are in *general position* whenever

$$d(a,c) \geq \bigwedge\{d(a,b) \vee d(b,c) : b \in A \cap C\}$$

for all $a \in A$ and $c \in C$. In much the same way as in 3.3.9, the above inequality is in fact an equality since $d(a,c) \leq d(a,b) \vee d(b,c)$.

(5) *Sets A and C are in general position if and only if*

$$\mathbf{V}^{(B)} \models (A \cap C)^\sim = A^\sim \cap C^\sim.$$

$\lhd$ Note that $(A \cap C)^\sim = \pi_X((A \cap C)^\wedge) = \pi_X(A^\wedge \cap C^\wedge)$ and $A^\sim \cap C^\sim = \pi_X(A^\wedge) \cap \pi_X(C^\wedge)$. Hence, the inclusion $(A \cap C)^\sim \subset A^\sim \cap C^\sim$ holds always, while $A^\sim \cap C^\sim \subset (A \cap C)^\sim$ amounts to the formula

$$(\forall\, a \in A^\wedge)(\forall\, c \in C^\wedge)(a{\sim}c \to (\exists\, b \in (A \cap C)^\wedge)(b{\sim}a \wedge b{\sim}c)).$$

Writing out the Boolean truth value of the last formula and considering the equality $[\![a^\wedge{\sim}c^\wedge]\!] = d(a,c)^*$, obtain

$$\bigwedge_{a \in A, c \in C} d(a,c)^* \Rightarrow \Big(\bigvee_{b \in A \cap C} d(a,b)^* \wedge d(b,c)^*\Big) = \mathbf{1}.$$

It is now evident that $[\![A^\sim \cap C^\sim \subset (A \cap C)^\sim]\!] = \mathbf{1}$ if and only if, for all $a \in A$ and $c \in C$, we have

$$d(a,c)^* \leq \Big(\bigwedge_{b \in A \cap C} d(a,b) \vee d(b,c)\Big)^*.$$

This means that A and C are in general position. $\rhd$

3.4.13. Theorem. *Suppose that (X, d_X) and (Y, d_Y) are some B-sets and Φ is a contractive correspondence from X to Y. Then inside $\mathbf{V}^{(B)}$ there is a unique correspondence $\Phi^\sim$ from $X^\sim$ to $Y^\sim$ such that*

$$\operatorname{dom}(\Phi^\sim) = (\operatorname{dom}\Phi)^\sim,$$
$$[\![\Phi^\sim(\pi_X x^\wedge) = \pi_Y(\Phi(x)^\wedge)]\!] = \mathbf{1} \quad (x \in \operatorname{dom}\Phi).$$

In this event the following hold:

(1) *If $A \subset X$ and $\operatorname{dom}(\Phi)$ are in general position then*

$$\mathbf{V}^{(B)} \models \Phi(A)^\sim = \Phi^\sim(A^\sim);$$

(2) *The composition $\Psi \circ \Phi$ of contractive correspondences Φ and Ψ is contractive, and if $\operatorname{dom}(\Psi \circ \Phi) = \operatorname{dom}(\Phi)$ and the sets $\operatorname{dom}(\Psi)$ and $\Phi(x)$ are in general position for all $x \in \operatorname{dom}(\Phi)$ then*

$$\mathbf{V}^{(B)} \models (\Psi \circ \Phi)^\sim = \Psi^\sim \circ \Phi^\sim;$$

(3) $\mathbf{V}^{(B)} \models (I_X)^\sim = I_{X^\sim}$.

$\lhd$ As follows from 3.1.5, $\mathbf{V}^{(B)} \models$ "$\Phi^\wedge$ is a correspondence from $X^\wedge$ to $Y^\wedge$." Put $\Phi^\sim := \pi_Y \circ \Phi^\wedge \circ \pi_X^{-1}$. It is obvious that $\mathbf{V}^{(B)} \models$ "$\Phi^\sim$ is a correspondence from $X^\sim$ to $Y^\sim$ and $\operatorname{dom}(\Phi^\sim) = \pi_X(\operatorname{dom}(\Phi^\wedge)) = \pi_X(\operatorname{dom}(\Phi)^\wedge) = \operatorname{dom}(\Phi^\sim)$."

Show now that the Boolean truth values $b_1 := [\![y \in \Phi^\sim \circ \pi_X(x^\wedge)]\!]$ and $b_2 := [\![y \in \pi_Y \circ \Phi^\wedge(x^\wedge)]\!]$ coincide for all $x \in Z := \operatorname{dom}(\Phi)$ and $y \in \mathbf{V}^{(B)}$. Indeed,

$$\begin{aligned} b_1 &= [\![(\exists s \in Z^\wedge)(\exists t \in Y^\wedge)(y = \pi_Y(t) \wedge t \in \Phi^\wedge(s) \wedge \pi_X(s) = \pi_X(x^\wedge))]\!] \\ &= \bigvee_{s \in Z} \bigvee_{t \in Y} [\![t^\wedge \in \Phi(s)^\wedge]\!] \wedge [\![y = \pi_Y(t^\wedge)]\!] \wedge [\![\pi_X(s^\wedge) = \pi_X(x^\wedge)]\!] \\ &\geq \bigvee_{t \in Y} [\![y = \pi_Y(t^\wedge)]\!] \wedge [\![t^\wedge \in \Phi(x)^\wedge]\!] \\ &= [\![(\exists t \in Y^\wedge)(y = \pi_Y(t) \wedge t \in \Phi^\wedge(x^\wedge))]\!] = b_2. \end{aligned}$$

On the other hand, using the equalities

$$d_X(s, x)^* = [\![\pi_X(s^\wedge) = \pi_X(x^\wedge)]\!],$$
$$\bar{d}_Y(\Phi(x), \Phi(s))^* = [\![\pi_Y(\Phi(x)^\wedge) = \pi_Y(\Phi(s)^\wedge)]\!]$$

and considering that the correspondence Φ is contractive, infer

$$b_1 \leq \bigvee_{s\in Z}\bigvee_{t\in Y} [\![\pi_Y(\Phi(s)^\wedge) = \pi_Y(\Phi(x)^\wedge)]\!] \wedge [\![t^\wedge \in \Phi(s)^\wedge]\!]$$

$$\wedge [\![y = \pi_Y(t^\wedge)]\!] \leq \bigvee_{s\in Z} [\![y \in \pi_Y(\Phi^\wedge(x^\wedge))]\!] = b_2.$$

Therefore, $b_1 = b_2$, which immediately implies the defining relation $[\![\pi_Y(\Phi(x)^\wedge) = \Phi^\sim(\pi_X(x^\wedge))]\!] = \mathbf{1}$ for all $x \in Z$. Hence, the relation

$$\mathbf{V}^{(B)} \models (\forall x \in (\mathrm{dom}(\Phi))^\wedge)\Phi^\sim(\pi_X x) = \pi_Y \Phi^\wedge(x)$$

holds. Moreover, $\Phi^\sim$ is unique since $\mathrm{dom}(\Phi^\sim) = (\mathrm{dom}(\Phi))^\sim = \pi_X((\mathrm{dom}(\Phi))^\wedge)$.

(1) Using 3.4.12 (5), it is easy to note that

$$\Phi^\sim(A^\sim) = \Phi^\sim(A^\sim \cap \mathrm{dom}(\Phi^\sim)) = \Phi^\sim((A \cap \mathrm{dom}(\Phi))^\sim).$$

On the other hand, $\Phi(A)^\sim = \Phi(A \cap \mathrm{dom}(\Phi))^\sim$ and so there is no loss of generality in assuming that $A \subset \mathrm{dom}(\Phi)$. In this case, however, using the defining property of $\Phi^\sim$, we may write inside $\mathbf{V}^{(B)}$ the following chain of equalities

$$\Phi^\sim(A^\sim) = \bigcup_{a\in A^\sim} \Phi^\sim(a) = \bigcup_{a\in A^\wedge} \Phi^\sim(\pi_X a)$$

$$= \bigcup_{a\in A^\wedge} \pi_Y(\Phi^\wedge(a)) = \pi_Y(\Phi^\wedge(A^\wedge)) = \pi_Y(\Phi(A)^\wedge) = \Phi(A)^\sim.$$

(2) Let Ψ be a contractive correspondence from Y to U. Choose $x_1, x_2 \in Z$, $y_1 \in \Phi(x_1)$ and $u_1 \in \Psi(y_1)$. Then, according to 3.4.7 (1)

$$\mathrm{dist}(u_1, \Psi \circ \Phi(x_2)) \leq \bigwedge\{\mathrm{dist}(u_1, \Psi(y)) : y \in \Phi(x_2)\}$$

$$\leq \bigwedge\{d(y_1, y) : y \in \Phi(x_2)\} = \mathrm{dist}(y_1, \Phi(x_2)) \leq d(x_1, x_2).$$

Since x_1, x_2, y_1, and u_1 are arbitrary; therefore, the correspondence $\Psi \circ \Phi$ is contractive.

Given $x \in Z$, use (1), 3.1.5 (2), and the defining relations of $(\Psi \circ \Phi)^\sim$, $\Psi^\sim$, and $\Phi^\sim$ to obtain

$$(\Psi^\sim \circ \Phi^\sim)(\pi_X x^\wedge) = \Psi^\sim(\Phi(x)^\sim) = \Psi(\Phi(x))^\sim$$

$$= \pi_Y((\Psi \circ \Phi)(x)^\wedge) = \pi_Y((\Psi \circ \Phi)^\wedge(x^\wedge)) = (\Psi \circ \Phi)^\sim(\pi_X x^\wedge).$$

Hence, $[\![(\Psi \circ \Phi)^\sim = \Psi^\sim \circ \Phi^\sim]\!] = \mathbf{1}$ since $Z^\sim = \mathrm{dom}(\Psi^\sim \circ \Phi^\sim)$.

(3) The claim is immediate from 3.1.5 (4). $\triangleright$

3.4.14. Theorem. *To each contractive mapping $f : X \to Y$ there is a unique element $f^\sim \in \mathbf{V}^{(B)}$ such that*

$$[\![f^\sim : X^\sim \to Y^\sim]\!] = [\![f^\sim \circ \pi_X = \pi_Y \circ f^\wedge]\!] = \mathbf{1}.$$

Moreover, the following hold:

(1) $\mathbf{V}^{(B)} \models f(A)^\sim = f^\sim(A^\sim)$ *for* $A \subset X$;

(2) *If* $g : Y \to Z$ *is a contraction then* $g \circ f$ *is a contraction and* $\mathbf{V}^{(B)} \models (g \circ f)^\sim = g^\sim \circ f^\sim$;

(3) $\mathbf{V}^{(B)} \models$ *"$f^\sim$ is injective" if and only if f is a B-isometry*;

(4) $\mathbf{V}^{(B)} \models$ *"$f^\sim$ is surjective" if and only if* $\bigvee\{d(f(x), y) : x \in X\} = \mathbf{1}$ *for every* $y \in Y$.

3.4.15. Consider the categories BSet$_*$ and CBSet$_*$. The objects of these categories are nonempty B-sets and nonempty universally complete B-sets, respectively, while the morphisms are contractive and fully contractive correspondences. As composition of morphisms we take the usual composition of correspondences. The subcategories of the categories BSet$_*$ and CBSet$_*$ consisting of the same objects and of contractive mappings are denoted by BSet and CBSet, respectively. Let $\mathscr{F}^\sim$ be the function assigning to an object X and a morphism Φ of BSet the elements $\mathscr{F}^\sim(X) := X^\sim$ and $\mathscr{F}^\sim(\Phi) := \Phi^\sim$.

3.4.16. *The mapping $\mathscr{F}^\sim$ is a covariant functor from the category* BSet *to the category* $\mathscr{V}^{(B)}$.

3.4.17. Comments.

(1) The concept of a Boolean metric appeared at the beginning of the 1950s in result of studying various "distances" given on abstract sets and taking values in posets (cf. [15, 47, 205]).

Unfortunately, no particularly rich geometry associated with this concept was ever discovered, which fact accounts most likely for the B-metrics being unpopular in the years to follow. The reason of this curiosity is perfectly revealed in Theorems 3.4.13 and 3.5.4.

The geometry of Boolean metrics is rather meaningful and enthralling when combined with topological and functional-analytical structures. In this case the presence of a compatible B-metric opens up a possibility of studying the structure in question by means of Boolean valued analysis.

(2) A mapping $[\![\cdot = \cdot]\!] : X^2 \to B$ is called a *Boolean valued equality* provided that it obeys 2.2.8 (1, 3, 4). These mappings are plentiful in Boolean valued interpretation of first-order theories (cf. [54]).

Clearly, the concept of a Boolean valued equality is just a "reflection" of the idea of a Boolean metric since the conditions of 2.2.8 (1, 3, 4) are met if and only if the mapping $(x,y) \mapsto [\![x = y]\!]^*$ is a Boolean metric. In this context the idea of a Boolean metric proves to be rather fruitful.

(3) Definitions 3.4.1, effective in this section, are motivated by the fact that the algebraic systems typical of analysis often possess a natural B-semimetric, whereas the introduction of some B-valued equality might be artificial.

(4) It is possible to demonstrate that the converse of 3.4.6 is also true. Namely, if ψ is an ultrafilter on $B^\wedge$ inside $\mathbf{V}^{(B)}$; then the mapping $\rho_\psi : B \to B$, defined by the formula $\rho_\psi(b) := [\![b^\wedge \in \psi]\!]$, is an automorphism of $B^\wedge$. Moreover, $\rho_{\psi_\rho} = \rho$ and $[\![\psi_{\rho_\psi} = \psi]\!] = \mathbf{1}$.

(5) Our remarks on 3.3.15 (5) apply fully to the case of 3.4.13 (1, 2).

3.5. Interplay Between the Main Functors

The main functors of the preceding four sections have a fruitful relationship rather productive of applications. This specifies the topic of the present section.

3.5.1. Recall that for an arbitrary $X \in \mathscr{P}(\mathbf{V}^{(B)})$ the set $\operatorname{mix}(X)$ consists of all possible mixings $\operatorname{mix}(b_\xi x_\xi)$ of all families (x_ξ) in X by all partitions of unity (b_ξ) in B (cf. 3.2.7). In this event the operation mix sends X to the cyclic hull of X (cf. 3.2.8). We now abstract mix to extensional correspondences.

Let X and Y be subsets of $\mathbf{V}^{(B)}$. Assume that Φ is an extensional correspondence from X to Y. There is a unique fully extensional correspondence Ψ from $\operatorname{mix}(X)$ *to* $\operatorname{mix}(Y)$ *satisfying*

$$\Psi(x) = \operatorname{mix}(\Phi(x)) \quad (x \in \operatorname{dom}(\Phi)).$$

$\lhd$ To prove, assign $\Psi := \Phi{\uparrow}{\downarrow}$ and use 3.3.12 (1) and 3.4.7 (2). From 3.2.13 and 3.3.3 (1) it follows that $\operatorname{Gr}(\Psi) = \operatorname{mix}(\operatorname{Gr}(\Phi))$. $\rhd$

By definition, $\operatorname{mix}(\Phi) = \Psi$. If Θ is another extensional correspondence and $\operatorname{dom}(\Theta) \subset Y$ then, by 3.2.13 (3) and 3.3.4 (8), we note that $\operatorname{mix}(\Theta \circ \Phi) = \operatorname{mix}(\Theta) \circ \operatorname{mix}(\Phi)$ if and only if $(\Theta \circ \Phi){\uparrow} = \Theta{\uparrow} \circ \Phi{\uparrow}$. Moreover, it is obvious that $\operatorname{mix}(I_X) = I_{\operatorname{mix}(X)}$.

3.5.2. Take a nonempty set X. Denote by $B_0(X)$ the set of all partitions of unity in B of the type $(b_x = b(x))_{x \in X}$:

$$b \in B_0(X) \leftrightarrow (b \in B^X \wedge (\forall x \in X)(\forall y \in X)(x \neq y) \to b(x) \wedge b(y) = 0)).$$

Assign to an element $y \in X$ the partition of unity $\imath y := \imath_X y := (b_x)_{x \in X}$, where $b_x = \mathbf{1}$ for $x = y$ and $b_x = 0$ for $x \neq y$. Evidently, $\imath_X$ is an injection from X to $B_0(X)$. Granted u, $v \in B_0(X)$, define

$$d(u,v) := \bigwedge \{u(x)^* \vee v(x)^* : x \in X\}.$$

It is easy to check that d is a B-metric on $B_0(X)$. Moreover, $(B_0(X), d)$ is a universally complete B-set. The last fact is established by essentially the same arguments as in 3.2.8. Hence, $B_0(\cdot)$ is a mapping from $\mathbf{V}$ to CBSet. We now extend this mapping to correspondences.

Given a correspondence $\Phi := (F, X, Y)$, define $B_0(\Phi) := (G, B_0(X), B_0(Y))$, where

$$\begin{aligned} G := \{&(u, v) \in B_0(X) \times B_0(Y) \\ &\leftrightarrow (\forall\, x \in X)(\forall\, y \in Y)(u(x) \wedge v(y) \neq 0 \rightarrow (x, y) \in F)\}. \end{aligned}$$

If Φ is single-valued then $B_0(\Phi)$ is single-valued too.

By definition,

$$\begin{gathered} B_0(I_X) = I_{B_0(X)}, \\ B_0(\Psi \circ \Phi) = B_0(\Psi) \circ B_0(\Phi), \\ \Phi = \imath_Y^{-1} \circ B_0(\Phi) \circ \imath_X. \end{gathered}$$

Hence, the mapping $B_0(\cdot)$ is a covariant functor from $\mathscr{V}_*$ to CBSet_*.

3.5.3. Some features of interplay between the main operations of Boolean valued analysis have earlier been presented in the form of the arrow cancellation rules. We now paraphrase these rules for functors.

(1) *The descent functor $\mathscr{F}^{\downarrow}$ and the ascent functor $\mathscr{F}^{\uparrow}$ establish an isomorphism between the categories $\mathscr{V}^{(B)}$ and $\mathscr{C}\mathscr{P}\mathscr{V}^{(B)}$. This implies that $\mathscr{F}^{\uparrow} \circ \mathscr{F}^{\downarrow}$ and $\mathscr{F}^{\downarrow} \circ \mathscr{F}^{\uparrow}$ coincide with the identity functors on $\mathscr{V}^{(B)}$ and $\mathscr{C}\mathscr{P}\mathscr{V}^{(B)}$, respectively.*

$\lhd$ The functor $\mathscr{F}^{\uparrow} \circ \mathscr{F}^{\downarrow}$ acts as the identity by the rules for descending and ascending 3.3.3 (2) and 3.3.12 (3). Similarly, the functor $\mathscr{F}^{\downarrow} \circ \mathscr{F}^{\uparrow}$ acts as the identity by the rules for ascending and descending 3.3.3 (1) and 3.3.12 (1). $\rhd$

(2) *The functor* mix : $\mathscr{P}\mathscr{V}^{(B)} \rightarrow \mathscr{C}\mathscr{P}\mathscr{V}^{(B)}$ *coincides with the composition $\mathscr{F}^{\uparrow} \circ \mathscr{F}^{\downarrow}$ and is a $\mathscr{C}\mathscr{P}\mathscr{V}^{(B)}$-reflector of the category $\mathscr{P}\mathscr{V}^{(B)}$. In particular, $\mathscr{C}\mathscr{P}\mathscr{V}^{(B)}$ is a reflective subcategory in $\mathscr{P}\mathscr{V}^{(B)}$.*

$\lhd$ The equality mix $:= \mathscr{F}^{\uparrow} \circ \mathscr{F}^{\downarrow}$ results from 3.3.3 (1) and 3.3.12 (2). Consider nonempty sets A, $C \in \mathscr{P}(\mathbf{V}^{(B)})$, and suppose that C is cyclic. Then each extensional mapping $g : A \rightarrow C$ admits a unique extensional extension $\bar{g} = g{\uparrow\downarrow} :$ mix$(A) \rightarrow C$ (cf. 3.2.12, 3.3.11, and 3.3.12 (2)). Therefore, the restriction mapping $\theta_{A,C} : h \mapsto h \restriction A$ is a bijection of $\mathscr{C}\mathscr{P}\mathscr{V}^{(B)}$ (mix$(A), C$) onto $\mathscr{P}\mathscr{V}^{(B)}(A, C)$. Denote the family of the mappings $\theta_{A,C}$ by θ. Then θ is a adjunction from mix to the functor of the identical embedding of $\mathscr{C}\mathscr{P}\mathscr{V}^{(B)}$ to $\mathscr{P}\mathscr{V}^{(B)}$.

Indeed, if A', $C' \in \mathscr{P}(\mathbf{V}^{(B)})$ and C' is cyclic then, granted extensional mappings $f : \mathrm{mix}(A) \to C$, $g : A' \to A$, and $h : C \to C'$, observe $(f \circ \mathrm{mix}(g)) \upharpoonright A' = (f \upharpoonright A) \circ g$. In turn, this makes obvious the equality

$$(h \circ (f \circ \mathrm{mix}(g))) \upharpoonright A' = h \circ (f \upharpoonright A) \circ g,$$

or, which is the same,

$$\theta_{A',C'}(h \circ f \circ \mathrm{mix}(g)) = h \circ \theta_{A,C}(f) \circ g. \ \triangleright$$

(3) *The composition of the canonical embedding functor and the descent functor is naturally isomorphic to the functor B_0 or, in symbols, $\mathscr{F}{\downarrow} \circ \mathscr{F}^{\wedge} \sim B_0$.*

$\triangleleft$ Given a set X, note that the mapping

$$\theta_X : (b_x)_{x \in X} \mapsto \underset{x \in X}{\mathrm{mix}}(b_x x^{\wedge}) \quad ((b_x)_{x \in X} \in B_0(X))$$

is a bijection of $B_0(X)$ onto $X^{\wedge}{\downarrow}$. The mapping $\theta : X \mapsto \theta_X$ $(X \in \mathrm{Ob}\,\mathscr{V}_*)$ is an isomorphism of the functors B_0 and $\mathscr{F}{\downarrow} \circ \mathscr{F}^{\wedge}$. To see this, it suffices to observe that, for $u \in B_0(X)$, $v \in B_0(Y)$, $a := \theta_X(u)$, and $b := \theta_Y(v)$, we have $(a, b) \in \Phi^{\wedge}{\downarrow}$ if and only if $(x, y) \in \Phi$ whenever $u(x) \wedge v(x) \neq 0$. $\triangleright$

3.5.4. Theorem. *Let (X, d_X) be a B-set and $X' := X^{\sim}{\downarrow}$. Then the following hold:*

(1) *There is an injection $\imath_X : X \to X'$ such that*

$$d_X(x_1, x_2) = [\![\imath_X x_1 \neq \imath_X x_2]\!] \quad (x_1, x_2 \in X);$$

(2) *To each $x' \in X'$ there are a partition of unity (b_ξ) and a family $(x_\xi) \subset X$ such that $x' = \mathrm{mix}(b_\xi \imath x_\xi)$;*

(3) *If Φ is a contractive correspondence from X to a B-set Y, $Y' := Y^{\sim}{\downarrow}$, and $\Phi' := \Phi^{\sim}{\downarrow}$; then Φ' is a unique fully extensional correspondence from X' to Y' satisfying $\mathrm{dom}(\Phi') = \mathrm{mix}(\imath_X(\mathrm{dom}(\Phi)))$ and*

$$\Phi'(\imath_X x) = \mathrm{mix}(\imath_X(\Phi(x))) \quad (x \in \mathrm{dom}(\Phi)).$$

$\triangleleft$ (1) By the definitions of $X^{\sim}$ and π_X (cf. 3.4.12 (1–3)), $[\![\pi_X x^{\wedge} \in X^{\sim}]\!] = \mathbf{1}$ for every $x \in X$. Hence, there is a unique element $x' \in X'$ such that $[\![x' = \pi_X x^{\wedge}]\!] = \mathbf{1}$. Assign $\imath_X x := x'$. This defines the mapping $\imath := \imath_X : X \to X'$ such that $[\![\imath x = \pi_X x^{\wedge}]\!] = \mathbf{1}$ $(x \in X)$. Using the last relation and 3.4.12 (4), for arbitrary x_1, $x_2 \in X$, deduce

$$[\![\imath x_1 \neq \imath x_2]\!] = [\![\pi_X x_1^{\wedge} = \pi_X x_2^{\wedge}]\!]^* = [\![x_1 \sim x_2]\!]^* = d_X(x_1, x_2),$$

which implies in particular that $\imath$ is injective.

(2) Note first that the following formula holds: $[\![t \in \operatorname{im}(\imath)\!\uparrow = \pi_X(X^\wedge)]\!] = \mathbb{1}$. Indeed, given $t \in \mathbf{V}^{(B)}$, by the definition of $\imath$ we have

$$[\![t \in \operatorname{im}(\imath)\!\uparrow]\!] = \bigvee_{x\in X} [\![t = \imath x]\!] = \bigvee_{x\in X} [\![t = \pi_X x^\wedge]\!] = [\![t \in \pi_X(X^\wedge)]\!].$$

Using the arrow cancellation rule 3.3.3 (1), derive

$$X' = \pi_X(X^\wedge)\!\downarrow = \operatorname{im}(\imath)\!\uparrow\downarrow = \imath(X)\!\uparrow\downarrow = \operatorname{mix}(\imath(X)).$$

(3) Since $\Phi^\sim$ is a correspondence from $X^\sim$ to $Y^\sim$ inside $\mathbf{V}^{(B)}$; therefore, Φ' is a fully extensional correspondence from X' to Y' (cf. 3.4.7 (2)). Using the property 3.2.13 (1) of descent and given arbitrary $x \in X$ and $y \in Y$, infer

$$\imath_Y y \in \Phi'(\imath_X x) \leftrightarrow [\![\imath_Y y \in \Phi^\sim(\imath_X x)]\!] = \mathbb{1}.$$

Using the construction of $\imath_X$, substitute $\pi_X x^\wedge$ for $\imath_X x$ on the right-hand side of the above equivalence. Appealing to Theorem 3.4.13, note then that

$$[\![\imath_Y y \in \Phi^\sim(\pi_X x^\wedge)]\!] = [\![\imath_Y y \in \pi_Y(\Phi(x)^\wedge)]\!].$$

All in all, $\imath_Y y \in \Phi'(\imath_X x)$ if and only if $\imath_Y y \in \pi_Y(\Phi(x)^\wedge)\!\downarrow$, which implies the claim. Indeed, using (1) and (2), conclude that $A^\sim\!\downarrow = \pi_Y(A^\wedge)\!\downarrow = \operatorname{mix}(\imath_Y(A))$ for $A \subset Y$. Involving 3.2.13 (1), proceed as follows:

$$\Phi'(\imath_X x) = \Phi^\sim\!\downarrow(\imath_X x) = \Phi^\sim(\pi_X(x^\wedge))\!\downarrow = \pi_Y(\Phi(x)^\wedge) = \operatorname{mix}(\imath_Y(\Phi(x))),$$

where $x \in \operatorname{dom}(\Phi)$. Put $X_1 := \operatorname{im}(\imath_X)$, $Y_1 := \operatorname{im}(\imath_Y)$, and $\Phi_1 := \imath_Y^{-1} \circ \Phi' \circ \imath_X$. Then Φ_1 is an extensional correspondence from X_1 to Y_1, and the following hold:

$$X' = \operatorname{mix}(X_1), \quad Y' = \operatorname{mix}(Y_1), \quad \Phi'(x) = \operatorname{mix}(\Phi_1(x)) \quad (x \in \operatorname{dom}(\Phi_1)).$$

Hence, $\Phi' = \operatorname{mix}(\Phi_1)$, and so Φ' is unique. $\triangleright$

3.5.5. We now describe the modified descents and ascents of correspondences.

(1) Suppose that X is a nonempty B-set, and Y is an arbitrary element $\mathbf{V}^{(B)}$ satisfying $[\![Y \neq \varnothing]\!] = \mathbb{1}$. Consider a member Φ of $\mathbf{V}^{(B)}$ such that $\mathbf{V}^{(B)} \models$ "$\Phi = (F, X^\sim, Y)$ is a correspondence from $X^\sim$ to Y."

By Theorem 3.2.13, $\Phi\!\downarrow$ is a correspondence from $X' := X^\sim\!\downarrow$ to $Y\!\downarrow$. Assign $\Phi\!\rfloor := \Phi\!\downarrow \circ \imath_X$. The correspondence $\Phi\!\rfloor$ is called the *modified descent* of Φ. By virtue of Theorems 3.2.13 and 3.5.4, $\Phi\downarrow$ is a unique fully contractive correspondence from X to $Y\downarrow$ satisfying

$$y \in \Phi\downarrow(x) \leftrightarrow [\![y \in \Phi(\imath_X x)]\!] = \mathbf{1} \quad (x \in X).$$

Note also that $\Phi\downarrow = (F\downarrow^{-}, X, Y\downarrow)$, where

$$F\downarrow^{-} := \{(x, y) \in X \times Y\downarrow : (\imath_X x, y)^B \in F\}.$$

(2) Assume that $\Psi := (F, X, Y\downarrow)$ is a contractive correspondence. The ascent operation of Section 3.3 does not apply directly to Ψ. However, the correspondence $\Psi \circ \imath_X$ is clearly extensional and so it ascends. Assign $\Psi\uparrow := (\Psi \circ \imath_X^{-1})\uparrow$ and call $\Psi\uparrow$ the *modified ascent* of Ψ. By Theorems 3.3.10 and 3.5.4, $\Psi\uparrow$ is a unique correspondence from $X^{\sim}$ to Y inside $\mathbf{V}^{(B)}$ such that

$$[\![\mathrm{dom}(\Psi\uparrow) = (\mathrm{dom}(\Psi))^{\sim}]\!] = \mathbf{1}, \quad [\![\Psi\uparrow(\imath_X x) = \Psi(x)\uparrow]\!] = \mathbf{1} \quad (x \in \mathrm{dom}(\Psi)).$$

Note again that $\Psi\uparrow = (F_{-}\uparrow, X^{\sim}, Y)$, where

$$F_{-} := \{(\imath_X x, y)^B : (x, y) \in F\}.$$

(3) Assume now that X is a discrete B-set. Then $\Phi\downarrow$ is a correspondence from X to $Y\downarrow$ uniquely determined from the formula

$$y \in \Phi\downarrow(x) \leftrightarrow [\![y \in \Phi(x^{\wedge})]\!] = \mathbf{1} \quad (x \in X).$$

On the other hand, in this case each correspondence Ψ from X to $Y\downarrow$ is contractive so that there is a unique correspondence $\Psi\uparrow$ from $X^{\wedge}$ to Y satisfying

$$[\![\Psi\uparrow(x^{\wedge}) = \Psi(x)\uparrow]\!] = \mathbf{1} \quad (x \in X).$$

3.5.6. Theorem. *The modified descent and ascent are inverse to one another, each implementing a bijection between the set of elements $\Phi \in \mathbf{V}^{(B)}$ satisfying $[\![\Phi$ is a correspondence from $X^{\sim}$ to $Y]\!] = \mathbf{1}$ and the set of all fully contractive correspondences from X to $Y\downarrow$.*

$\lhd$ For simplicity, put $\imath := \imath_X$. By 3.5.4 (2) and 3.3.3 (1), $X^{\sim} = \mathrm{im}(\imath)\uparrow$. Hence, in virtue of 3.3.10 (3), note that $I_{X^{\sim}} = (I_{\mathrm{im}(\imath)})\uparrow$. Applying the arrow cancellation rules for correspondences, conclude then that the following holds inside $\mathbf{V}^{(B)}$:

$$\Phi\downarrow\uparrow = ((\Phi\downarrow \circ \imath) \circ \imath^{-1})\uparrow = (\Phi\downarrow \circ I_{\mathrm{im}(\imath)})\uparrow = \Phi\downarrow\uparrow \circ (I_{\mathrm{im}(\imath)})\uparrow = \Phi \circ I_{X^{\sim}} = \Phi.$$

On the other hand, granted a fully contractive Ψ, observe

$$\begin{aligned}\Psi\uparrow\downarrow(x) &= (\Psi \circ \imath^{-1})\uparrow\downarrow(\imath x) = (\mathrm{mix}(\Psi)) \circ \imath^{-1}(\imath x) \\ &= \mathrm{mix}(\Psi)(x) = \Psi(x) \quad (x \in \mathrm{mix}(\mathrm{dom}(\Psi)) = \mathrm{dom}(\Psi)),\end{aligned}$$

which completes the proof. $\rhd$

3.5.7. Theorem. *The descent functor $\mathscr{F}^{\downarrow}$ is right adjoint to the immersion functor $\mathscr{F}^{\sim}$. In this event the modified descent $\downarrow$ is an adjunction, while the modified ascent $\uparrow$ is a coadjunction.*

$\lhd$ Consider the functors $\mathscr{H}^{\sim}$ and $\mathscr{H}^{\downarrow}$ from the category $\mathrm{BSet} \times \mathscr{V}^{(B)}$ to the category $\mathscr{V}$ defined as follows:

$$\mathscr{H}^{\sim}(X,Y) := \mathscr{V}^{(B)}(X^{\sim},Y), \quad \mathscr{H}^{\downarrow}(X,Y) := \mathrm{BSet}_0(X, Y\downarrow);$$
$$\mathscr{H}^{\sim}(\alpha,\beta) := \Phi' \leftrightarrow \mathbf{V}^{(B)} \models \Phi' = \beta \circ \Phi \circ \alpha^{\sim};$$
$$\mathscr{H}^{\downarrow}(\alpha,\beta) := \beta\downarrow \circ \Psi \circ \alpha,$$

where $X \in \operatorname{Ob}\mathrm{BSet}$, $Y \in \operatorname{Ob}\mathscr{V}^{(B)}$, $\alpha \in \mathrm{BSet}(X_1, X)$, $\beta \in \mathscr{V}^{(B)}(Y, Y_1)$, $\Phi \in \mathscr{H}^{\sim}(X,Y)$, and $\Psi \in \mathscr{H}^{\downarrow}(X,Y)$.

The claim is that the modified descent $\downarrow$ is an isomorphism of the functors $\mathscr{H}^{\sim}$ and $\mathscr{H}^{\downarrow}$. By virtue of Theorem 3.5.6, we only have to establish that $\downarrow$ is a functor morphism of the functor $\mathscr{H}^{\sim}$ to the functor $\mathscr{H}^{\downarrow}$ or, in other words, that the following diagram commutes

$$\begin{array}{ccc} \mathscr{H}^{\sim}(X,Y) & \xrightarrow{\ \downarrow\ } & \mathscr{H}^{\downarrow}(X,Y) \\ {\scriptstyle \mathscr{H}^{\sim}(\alpha,\beta)}\downarrow & & \downarrow{\scriptstyle \mathscr{H}^{\downarrow}(\alpha,\beta)} \\ \mathscr{H}^{\sim}(X_1,Y_1) & \xrightarrow[\ \downarrow\ ]{} & \mathscr{H}^{\downarrow}(X_1,Y_1) \end{array}$$

for the above indicated X, X_1, Y, Y_1, α, and β. The commutativity amounts to the fact that the equality $(\mathscr{H}(\alpha,\beta)\Phi)\downarrow = \mathscr{H}^{\downarrow}(\alpha,\beta)(\Phi\downarrow)$ holds for every $\Phi \in \mathscr{H}^{\sim}(X,Y)$ or, in virtue of the definitions of $\mathscr{H}^{\sim}$ and $\mathscr{H}^{\downarrow}$, that the following conditions are compatible:

$$\Psi \in \mathscr{H}^{\downarrow}(X,Y), \quad [\![\Psi = \beta \circ \Phi \circ \alpha^{\sim}]\!] = \mathbf{1},$$
$$(\beta\downarrow) \circ (\Phi\downarrow) \circ \alpha = \Psi\downarrow.$$

These are fulfilled if and only if

$$[\![\beta \circ \Phi \circ \alpha^{\sim} = (\beta\downarrow \circ (\Phi\downarrow) \circ \alpha)\uparrow]\!] = \mathbf{1}.$$

However, the arrow cancellation rules, together with the definitions of modified descent and ascent, imply that the following holds inside $\mathbf{V}^{(B)}$:

$$(\beta\downarrow \circ (\Phi\downarrow) \circ \alpha)\uparrow = (\beta\downarrow \circ (\Phi\downarrow) \circ \imath \circ \alpha \circ \imath^{-1})\uparrow$$
$$= \beta\downarrow\uparrow \circ (\Phi\downarrow\uparrow) \circ (\imath \circ \alpha \circ \imath^{-1})\uparrow = \beta \circ \Phi \circ (\imath \circ \alpha \circ \imath^{-1})\uparrow.$$

To complete the proof, it suffices to note that $[\![(\imath \circ \alpha \circ \imath^{-1})\uparrow = \alpha^{\sim}]\!] = \mathbf{1}$. $\rhd$

3.5.8. We now list some important corollaries to Theorem 3.5.4 with its hypotheses and notation presumed effective.

(1) *If (X, d_X) is a universally complete B-set then $\imath_X$ is a bijection between X and X'.*

$\lhd$ Note that if $x = \mathrm{mix}(b_\xi x_\xi)$ for a partition of unity (b_ξ) and a family $(x_\xi) \subset X$ then $\imath_X x = \mathrm{mix}(b_\xi \imath_X x_\xi)$. $\rhd$

(2) *To each B-set (X, d_X) there is a 3-tuple $(X', d'_X, \imath_X)$ called a B-completion of (X, d_X) and obeying the following conditions:*

(a) *(X', d'_X) is a universally complete B-set, and $\imath_X$ is an isometry of X to X';*

(b) *$X' = \mathrm{mix}(\mathrm{im}(\imath_X))$;*

(c) *to each contractive correspondence Φ from X to a universally complete B-set Y, there is a unique fully contractive correspondence Φ' from X' to Y satisfying $\mathrm{dom}(\Phi') = \mathrm{mix}(\imath(\mathrm{dom}(\Phi)))$ and*

$$\mathrm{mix}(\Phi(x)) = \Phi'(\imath_X x) \quad (x \in \mathrm{dom}(\Phi));$$

(d) *if a 3-tuple $(X'', d''_X, \imath'_X)$ obeys (a)–(c), then there exists some B-isomorphism $\imath$ between X' and X'' satisfying $\imath \circ \imath_X = \imath'_X$.*

$\lhd$ To prove, take some universally complete B-set as Y in 3.5.4 (3) and appeal to (1). $\rhd$

(3) *If $X \in \mathrm{Ob}\,\mathscr{V}^{(B)}$ then there is a member $\jmath_X$ of $\mathbf{V}^{(B)}$ such that $[\![\jmath_X$ is an isomorphism (in the category $\mathscr{V}^{(B)}$) of X onto $X{\downarrow}^{\sim}]\!] = \mathbf{1}$.*

$\lhd$ Indeed, if $Y := X{\downarrow}$ then, letting $\jmath_X := \imath_Y{\uparrow}$, note that $\jmath_X$ is an isomorphism between $Y{\uparrow} = X$ and $Y^{\sim} = X{\downarrow}^{\sim}$, since $\imath_Y$ is an isomorphism between Y and $Y^{\sim}{\downarrow}$. $\rhd$

(4) *If X and Y are universally complete B-sets and Φ is a correspondence from $X^{\sim}$ to $Y^{\sim}$ inside $\mathbf{V}^{(B)}$, then there is a unique fully contractive correspondence Ψ from X to Y such that $\Psi^{\sim} = \Phi$.*

$\lhd$ Indeed, $\Phi' := \Phi{\downarrow}$ is a fully extensional correspondence from $X' := X^{\sim}{\downarrow}$ to $Y' := Y^{\sim}{\downarrow}$. Hence, $\Psi := \imath_Y^{-1} \circ \Phi' \circ \imath_X$ is a fully contractive correspondence from X to Y. If $\Psi' := \Psi^{\sim}{\downarrow}$ then, using 3.5.4 (3), obtain $\imath_Y^{-1} \circ \Psi \circ \imath_X = \imath_Y^{-1} \circ \Psi' \circ \imath_X$. By (1), $\Psi = \Psi'$, and so $\Phi = \Phi'{\uparrow} = \Psi'{\uparrow} = \Psi{\uparrow}$. $\rhd$

(5) *If X and Y are universally complete B-sets then the mapping $\Phi \mapsto \Phi^{\sim}$ is a bijection between the sets of morphisms $\mathrm{CBSet}_*(X, Y)$ and $\mathscr{V}_*^{(B)}(X^{\sim}, Y^{\sim})$.*

3.5.9. *Suppose that X and Y are arbitrary B-sets and Φ is a fully contractive correspondence from X to Y. Then*

$$\mathbf{V}^{(B)} \models \pi_{\Phi}(A)^{\sim} = \pi_{\Phi^{\sim}}(A^{\sim})$$

for every subset A of $\operatorname{dom}(\Phi)$.

$\triangleleft$ Note that the formulas $(\forall\, a \in A^{\wedge})\ (y \in \Phi^{\sim}(\pi_X a))$ and $y \in \pi_{\Phi^{\sim}}(A^{\sim})$ are equivalent since $A^{\sim} = \pi_X(A^{\wedge})$. Using Theorem 3.4.13 and the fact that Φ is fully contractive, take $y \in \imath_Y(Y)$ and proceed with the following equivalences:

$$\begin{aligned} y \in \pi_{\Phi^{\sim}}(A^{\sim}){\downarrow} &\leftrightarrow \bigwedge\{[\![y \in \Phi^{\sim}(\pi_X a^{\wedge})]\!] : a \in A\} = \mathbf{1} \\ \leftrightarrow (\forall\, a \in A)[\![y \in \pi_Y(\Phi(a)^{\wedge})]\!] &= \mathbf{1} \leftrightarrow (\forall\, a \in A)(y \in \Phi(a)^{\sim}{\downarrow}) \\ \leftrightarrow (\forall\, a \in A) y \in \operatorname{mix}(\imath_Y(\Phi(a))) &\leftrightarrow (\forall\, a \in A)\, y \in \imath_Y(\operatorname{mix}(\Phi(a))) \\ \leftrightarrow y \in \bigcap_{a \in A} \imath_A(\Phi(a)) &\leftrightarrow y \in \imath_Y(\pi_{\Phi}(A)). \end{aligned}$$

Hence,

$$\pi_{\Phi^{\sim}}(A^{\sim}) = \imath_Y(\pi_{\Phi}(A)){\uparrow} = \pi_{\Phi}(A)^{\sim}. \ \triangleright$$

3.5.10. Theorem. *The functors $\mathscr{F}^{\sim}$ and $\mathscr{F}^{\downarrow}$ establish equivalence between the categories* CBSet_* *and $\mathscr{V}_*^{(B)}$. In particular, $\mathscr{F}^{\sim}$ and $\mathscr{F}^{\downarrow}$ are mutually adjoint full and faithful functors preserving inductive and projective limits (for the given categories).*

$\triangleleft$ It suffices to demonstrate the following:

(1) the functor $\mathscr{F}^{\downarrow} \circ \mathscr{F}^{\sim}$ is naturally isomorphic to the identity functor on CBSet_*; while the isomorphism is implemented by the mappings $\imath_X : X \mapsto X'$ where $X \in \mathrm{CBSet}_*$;

(2) the functor $\mathscr{F}^{\sim} \circ \mathscr{F}^{\downarrow}$ is naturally isomorphic to the identity functor on $\mathscr{V}_*^{(B)}$; while the isomorphism is accomplished by the mappings $\jmath_X \in \mathscr{V}^{(B)}(X, X{\downarrow}^{\sim})$ where $X \in \mathscr{V}_*^{(B)}$.

To prove (1), involve 3.5.8 (1) and note that, by virtue of 3.5.4 (3), for X, $Y \in \operatorname{Ob}\mathrm{CBSet}_*$ and $\Phi \in \mathrm{CBSet}_*(X, Y)$, the following diagram commutes:

$$\begin{array}{ccc} X & \xrightarrow{\ \imath_X\ } & X^{\sim}{\downarrow} \\ {\scriptstyle \Phi}\big\downarrow & & \big\downarrow{\scriptstyle \Phi^{\sim}\downarrow} \\ Y & \xrightarrow[\ \imath_Y\ ]{} & Y^{\sim}{\downarrow} \end{array}$$

It then ensues from 3.5.8 (3, 4) that, for all X, $Y \in \operatorname{Ob} \mathcal{V}_*^{(B)}$ and $\Phi \in \mathcal{V}_*^{(B)}(X,Y)$, the following diagram commutes:

$$\begin{array}{ccc} X & \xrightarrow{\jmath_X} & X{\downarrow}^{\sim} \\ {\scriptstyle\Phi}\big\downarrow & & \big\downarrow{\scriptstyle\Phi{\downarrow}^{\sim}} \\ Y & \xrightarrow[\jmath_Y]{} & Y{\downarrow}^{\sim} \end{array}$$

This yields (2). ▷

3.5.11. *For all* $X \in \operatorname{Ob}\mathrm{CBSet}_*$ *and* $Y \in \operatorname{Ob} \mathcal{V}_*^{(B)}$, *the following hold:*

$$(\jmath_Y){\downarrow} = \imath_{Y\downarrow}, \quad \mathbf{V}^{(B)} \models (\imath_X)^{\sim} = \jmath_{X^{\sim}}.$$

◁ The first equality is immediate from the definitions: $(\jmath_Y){\downarrow} = (\imath_{Y\downarrow}){\uparrow}{\downarrow} = \imath_{Y\downarrow}$. To prove the second equality, assign

$$b := [\![(\imath_X)^{\sim} = \jmath_{X^{\sim}}]\!], \quad b_x := [\![\imath_{X^{\sim}} \pi_X x^{\wedge} = \jmath_{X^{\sim}} \pi_X x^{\wedge}]\!] \quad (x \in X).$$

Note that $b = \bigwedge\{b_x : x \in X\}$. Hence, we are to check that $b_x = \mathbf{1}$ for every $x \in X$. However, if $x \in X$ then, by 3.4.13 and the definition of $\jmath_X$, obtain $b_x = [\![\pi_{X^{\sim}\downarrow}(\imath_X x)^{\wedge} = (\imath_{X^{\sim}\downarrow}){\uparrow} \circ \pi_X(x^{\wedge})]\!]$. Now, apply the following equalities which hold by the definition of $\imath_X$:

$$[\![\pi_X x^{\wedge} = \imath_X x]\!] = [\![\pi_{X^{\sim}\downarrow} y^{\wedge} = \imath_{X^{\sim}\downarrow} y]\!] = \mathbf{1} \quad (x \in X, y \in Y^{\sim}{\downarrow}).$$

Whence, on letting $y = \imath_X x$ and using 3.5.4 (1), infer

$$b_x = [\![\pi_{X^{\sim}\downarrow}(\imath_X x)^{\wedge} = \imath_{X^{\sim}\downarrow}(\imath_X x)]\!] = \mathbf{1},$$

which completes the proof. ▷

Chapter 4

Boolean Valued Analysis of Algebraic Systems

Every Boolean valued universe has the collection of mathematical objects in full supply: available in plenty are all sets with extra structure: groups, rings, algebras, normed spaces, etc. Applying the descent functor to the established algebraic systems in a Boolean valued model, we distinguish bizarre entities or recognize old acquaintances, which revealing new facts of their life and structure.

This technique of research, known as *direct Boolean valued interpretation*, allows us to produce new theorems or, to be more exact, to extend the semantical content of the available theorems by means of slavish translation. The information we so acquire might fail to be vital, valuable, or intriguing, in which case the direct Boolean valued interpretation turns out to be a leisurely game.

It thus stands to reason to raise the following questions: What structures significant for mathematical practice are obtainable by the Boolean valued interpretation of the most common algebraic systems? What transfer principles hold in this process? Clearly, the answers should imply specific objects whose particular features enable us to deal with their Boolean valued representation which, if understood duly, is impossible to implement for arbitrary algebraic systems.

In the preceding chapter we have shown that an abstract B-set U embeds in the Boolean valued universe $\mathbf{V}^{(B)}$ so that the Boolean distance between the members of U becomes the Boolean truth value of the negation of their equality. The corresponding element of $\mathbf{V}^{(B)}$ is, by definition, the *Boolean valued representation* of U. In case the B-set U has some a priori structure, we may try to equip the Boolean valued representation of U with an analogous structure, intending to apply the technique of ascending and descending to studying the original structure of U. Consequently, the questions we raised above may be treated as instances of the same problem of searching the qualified Boolean valued representation of a B-set furnished with some additional structure.

The present chapter analyzes the problem for the main objects of general algebra. Located at the epicenter of exposition, the notion of an algebraic B-system refers to a nonempty B-set endowed with a few contractive operations and B-predicates, the latter meaning B-valued contractive mappings.

The Boolean valued representation of an algebraic B-system appears to be a conventional two-valued algebraic system of the same type. This means that an appropriate completion of each algebraic B-system coincides with the descent of some two-valued algebraic system inside $\mathbf{V}^{(B)}$. On the other hand, each two-valued algebraic system may be transformed into an algebraic B-system on distinguishing a complete Boolean algebra of congruences of the original system. In this event the task is to find the formulas holding in direct or reverse transition from a B-system to a two-valued system. In other words, we have to seek here some versions of the transfer principle or the identity preservation principle of long standing in some branches of mathematics.

We illustrate the general facts of Boolean valued analysis with particular algebraic systems in which complete Boolean algebras of congruences are connected with the relations of order and disjointness.

4.1. Algebraic B-Systems

We now introduce a class of algebraic systems suitable for the Boolean valued interpretation of first-order languages. These systems arise as B-sets equipped with contractive operations and predicates.

4.1.1. Recall that a *signature* is a 3-tuple $\sigma := (F, P, \mathfrak{a})$, where F and P are some (possibly, empty) sets and $\mathfrak{a}$ is a mapping from $F \cup P$ to ω. If F and P are finite then σ is a *finite signature*. In applications we usually deal with algebraic systems of finite signature.

An *n-ary operation* and an *n-ary-predicate* on a B-set A are contractive mappings $f : A^n \to A$ and $p : A^n \to B$ respectively. By definition, f and p are *contractive mappings* provided that

$$d(f(a_0, \dots, a_{n-1}), f(a'_0, \dots, a'_{n-1})) \leq \bigvee_{k=0}^{n-1} d(a_k, a'_k),$$

$$d_s(p(a_0, \dots, a_{n-1}), p(a'_0, \dots, a'_{n-1})) \leq \bigvee_{k=0}^{n-1} d(a_k, a'_k)$$

for all a_0, $a'_0, \dots, a_{n-1}$, $a'_{n-1} \in A$, where d is the B-metric of A, and d_s is the *symmetric difference* on B; i.e., $d_s(b_1, b_2) := b_1 \mathbin{\triangle} b_2$ (cf. 1.1.4).

Clearly, the above definitions depend on B and it would be cleaner to speak of B-operations, B-predicates, etc. We adhere to a simpler practice whenever this entails no confusion.

An *algebraic B-system* $\mathfrak{A}$ of signature σ is a pair (A, ν), where A is a nonempty B-set, the *underlying set* or *carrier* of $\mathfrak{A}$, and ν is a mapping such that (a) $\operatorname{dom}(\nu) = F \cup P$; (b) $\nu(f)$ is an $\mathfrak{a}(f)$-ary operation on A for all $f \in F$; and (c) $\nu(p)$ is an $\mathfrak{a}(p)$-ary predicate on A for every $p \in P$.

It is in common parlance to call ν the *interpretation* of $\mathfrak{A}$, in which case the notations f^ν and p^ν are familiar substitutes for $\nu(f)$ and $\nu(p)$.

The signature of an algebraic B-system $\mathfrak{A} := (A, \nu)$ is often denoted by $\sigma(\mathfrak{A})$; while the carrier A of $\mathfrak{A}$, by $|\mathfrak{A}|$. Since $A^0 = \{\varnothing\}$, the nullary operations and predicates on A are mappings from $\{\varnothing\}$ to the set A and to the algebra B respectively. We agree to identify a mapping $g : \{\varnothing\} \to A \cup B$ with the element $g(\varnothing)$. Each nullary operation on A thus transforms into a unique member of A. Analogously, the set of all nullary predicates on A turns into the Boolean algebra B. If $F := \{f_1, \dots, f_n\}$ and $P := \{p_1, \dots, p_m\}$ then an algebraic B-system of signature σ is often written down as $(A, \nu(f_1), \dots, \nu(f_n), \nu(p_1), \dots, \nu(p_m))$ or even $(A, f_1, \dots, f_n, p_1, \dots, p_m)$. In this event the expression $\sigma = (f_1, \dots, f_n, p_1, \dots, p_m)$ is substituted for $\sigma = (F, P, \mathfrak{a})$.

4.1.2. If B is the two-element Boolean algebra $\{\mathbf{0}, \mathbf{1}\}$, then instead of algebraic B-system we speak about a *two-valued algebraic system* or simply about an *algebraic system*. In this case an arbitrary set may be treated as a B-set, while an n-ary operation becomes an arbitrary mapping from A^n to A and a predicate P on a B-set transforms into the characteristic function $p : A^n \to \{\mathbf{0}, \mathbf{1}\}$ of $\{x \in A^n : p(x) = \mathbf{1}\}$. Therefore, an *algebraic system* $\mathfrak{A}$ of signature σ is a pair $\mathfrak{A} = (A, \nu)$, where the underlying set A of $\mathfrak{A}$ is nonempty and the interpretation ν of $\mathfrak{A}$ is a function from $\operatorname{dom}(\nu) = F \cup P$ to $\mathbf{V}$ such that

$$\nu(f) : A^{\mathfrak{a}(f)} \to A, \quad \nu(p) \subset A^{\mathfrak{a}(p)} \quad (f \in F,\ p \in P).$$

On the other hand, if (A, ν) is an algebraic system of signature σ and $A \subset \mathbf{V}^{(B)}$ then, considering A as a B-set (with the B-metric $d(a, a') := [\![a = a']\!]^* = [\![a \neq a']\!]$ $(a, a' \in A)$) and given $p \in P$, we may define the n-ary B-predicate $\nu'(p)$ on A with $n := \mathfrak{a}(p)$ by the following formula (cf. 3.4.5)

$$\nu'(p) := (a_0, \dots, a_{n-1}) \mapsto \operatorname{dist}((a_0, \dots, a_{n-1}), \nu(p)).$$

It is obvious that $\nu'(p) : A^n \to B$ is a contractive mapping. Assume further that $\nu(f)$ is a contractive mapping for every $f \in F$. Put $\nu'(f) := \nu(f)$ for all $f \in F$. Then (A, ν') is an algebraic B-system.

Considering a particular algebraic system $\mathfrak{A}$, we describe the ingredients of $\mathfrak{A}$ in a liberal fashion. Rather than solemnly proclaiming the formalities of the signature of $\mathfrak{A}$, we usually indicate only the most significant symbols of operations and predicates and even identify the whole system $\mathfrak{A}$ with its underlying set $|\mathfrak{A}|$. This routine is another sacrosanct privilege of the working mathematician.

4.1.3. An algebraic B-system A is *universally complete* or *finitely complete* provided that A is a universally complete or finitely complete algebraic B-set (cf. 3.4.3). Note that "decomposable" is synonymous with "finitely complete" and "extended" stands for "universally complete" in texts of the Russian provenance.

A B-predicate p on the set A is called *assertive* if there exists an element x in A such that $p(x) = \mathbf{1}$.

(1) *A contractive mapping p from a universally complete B-set A to B is an assertive B-predicate if and only if $\mathbf{1} = \bigvee\{p(x) : x \in A\}$.*

$\lhd$ Indeed, if the proviso is fulfilled then there are a family $(x_\xi) \subset A$ and a partition of unity $(b_\xi) \subset B$ such that $p(x_\xi) \geq b_\xi$. If $x := \operatorname{mix}(b_\xi x_\xi)$ then $p(x) = \mathbf{1}$. $\rhd$

To each algebraic B-system $\mathfrak{A}$ we may relate the algebraic system $\bar{\mathfrak{A}}$ with the same underlying set $|\bar{\mathfrak{A}}| := |\mathfrak{A}|$ and the interpretation $\bar{\nu}$ defined as follows: If f is a function symbol then $\bar{\nu}(f) := \nu(f)$; while if p is a predicate symbol and $n = \mathfrak{a}(p)$, then $\bar{\nu}(p) := \{(x_0, \dots, x_{n-1}) \in A^n : p(x_0, \dots, x_{n-1}) = \mathbf{1}\}$. Clearly, the predicate $\bar{\nu}(p)$ might be empty for some p.

The algebraic system $\bar{\mathfrak{A}}$ is said to be the *purification* or *reduct* of $\mathfrak{A}$. It is in common parlance also to say that $\bar{\mathfrak{A}}$ is obtained from $\mathfrak{A}$ by *purification* or *reduction.*

(2) *If (A, ν) is an algebraic B-system and $(A, \bar{\nu})$ is the purification of (A, ν) then*

$$p^\nu : x \mapsto \operatorname{dist}(x, \bar{\nu}(p))^* \quad (x \in A^{\mathfrak{a}(p)})$$

for every assertive predicate p^ν.

$\lhd$ By corollaries to the theorem on Boolean valued representation of B-sets (cf. 3.5.8), the B-set A has a B-completion $A' \subset \mathbf{V}^{(B)}$, and p^ν admits a unique extension $\nu'(p)$ to some B-predicate on A'.

In this event, $\nu'(p)(x) = \operatorname{dist}(x, \operatorname{mix}(\bar{\nu}(p)))^* = \operatorname{dist}(x, \bar{\nu}(p))^* = [\![x \in p^\nu{\uparrow}]\!]$ $(x \in A^{\mathfrak{a}(p)})$. This yields the desired result since we lose no generality in assuming that $A \subset A'$. $\rhd$

Proposition 4.1.3 (2) makes it possible to identify an algebraic B-system with assertive predicates $\mathfrak{A}$ and some algebraic system, namely, the purification of $\mathfrak{A}$. It is natural to ask the question: What algebraic systems are obtainable by purification of finitely or universally complete algebraic B-systems? The answer to this question is formulated in terms of congruences.

4.1.4. Consider an arbitrary algebraic system $\mathfrak{A} := (A, \nu)$ of signature $\sigma := (F, P, \mathfrak{a})$.

An equivalence ρ on A is a *congruence* of $\mathfrak{A}$ if, for every $f \in F$ and for all $x_0, \dots, x_{n-1}, y_0, \dots, y_{n-1} \in A$, $n = \mathfrak{a}(f)$, from $(x_0, y_0) \in \rho, \dots, (x_{n-1}, y_{n-1}) \in \rho$ it follows that $(f^\nu(x_0, \dots, x_{n-1}), f^\nu(y_0, \dots, y_{n-1})) \in \rho$.

The set of all congruences of $\mathfrak{A}$ is denoted by $\mathrm{Cong}(\mathfrak{A})$. Equip $\mathrm{Cong}(\mathfrak{A})$ with some order by the formula

$$\rho_1 \le \rho_2 \leftrightarrow \rho_1 \subset \rho_2 \quad (\rho_1, \rho_2 \in \mathrm{Cong}(\mathfrak{A})).$$

The *identity congruence* $I_A := \{(x, x) : x \in A\}$ and the trivial, *indiscriminate congruence* $A \times A$ are obviously the least and greatest elements of $\mathrm{Cong}(\mathfrak{A})$.

(1) *Theorem.* *The poset $\mathrm{Cong}(\mathfrak{A})$ is a complete lattice. The greatest lower bound of a subset $\mathscr{P}$ of $\mathrm{Cong}(\mathfrak{A})$ coincides with the intersection $\bigcap\{\rho : \rho \in \mathscr{P}\}$. The least upper bound of a subset $\mathscr{P}$ of $\mathrm{Cong}(\mathfrak{A})$ is the union of all possible compositions $\rho_1 \circ \ldots \circ \rho_n$, where $\{\rho_1, \dots, \rho_n\}$ is an arbitrary finite set in $\mathrm{Cong}(\mathfrak{A})$.*

The poset $\mathrm{Cong}(\mathfrak{A})$ is the *congruence lattice* of $\mathfrak{A}$. The join $\rho_1 \vee \rho_2$ of $\rho_1, \rho_2 \in \mathrm{Cong}(\mathfrak{A})$, as seen from the above theorem, coincides with the union of all possible relations of the form $\rho_1 \circ \rho_2 \circ \rho_1 \circ \ldots \circ \rho_1 \circ \rho_2$. Hence, if ρ_1 and ρ_2 commute, i.e., $\rho_1 \circ \rho_2 = \rho_2 \circ \rho_1$; then $\rho_1 \vee \rho_2 = \rho_1 \circ \rho_2$. Conversely, if $\rho_1 \vee \rho_2 = \rho_1 \circ \rho_2$ then the congruences ρ_1 and ρ_2 commute.

A set of congruences Λ on an algebraic system $\mathfrak{A}$ is *independent* (*finitely independent*) if, to every family (finite family) $(\lambda_\xi)_{\xi \in \Xi}$ in Λ and to every family (finite family) $(a_\xi)_{\xi \in \Xi}$ in A, there is an element a in A satisfying $(a, a_\xi) \in \lambda_\xi$ for all $\xi \in \Xi$.

A set of congruences Λ is *complete* provided that (a) $\inf(\Lambda) := \bigcap(\Lambda) = I_A$ and (b) for all $p \in P$ and an arbitrary n-tuple $(x_0, \dots, x_{n-1}) \in A^n$, $n = \mathfrak{a}(p)$, the formula $(x_0, \dots, x_{n-1}) \notin \nu(p)$ yields the existence of λ in Λ such that $(y_0, \dots, y_{n-1}) \notin \nu(p)$ as soon as $(x_0, y_0) \in \lambda, \dots, (x_{n-1}, y_{n-1}) \in \lambda$ (cf. [164]).

Considering the definition of a complete set of congruences, it is convenient to paraphrase (b) in terms of mixing.

Take a family $(a_\lambda)_{\lambda \in \Lambda}$ in A. If $(a, a_\lambda) \in \lambda$ for some $a \in A$ and all $\lambda \in \Lambda$ then we naturally say that a is the *mixing of* (a_λ) *relative to* Λ. A subset U of A^n is *closed under Λ-mixing* if for each family $((a_\lambda^0, \dots, a_\lambda^{n-1}))_{\lambda \in \Lambda}$ in U we have $(a_0, \dots, a_{n-1}) \in U$, where a_k is the mixing of (a_λ^k) relative to Λ.

(2) *An independent set of congruences Λ of an algebraic system $\mathfrak{A}$ is complete if and only if $\inf(\Lambda) = I_A$ and every predicate $\nu(p)$, $p \in P$, is closed under Λ-mixing.*

$\lhd$ To proof sufficiency, assume that all predicates are closed under Λ-mixing. Assume further that $p \in P$, $n = \mathfrak{a}(p)$, and $(x_0, \dots, x_{n-1}) \notin \nu(p)$ but, nonetheless, to each $\lambda \in \Lambda$ there are $(y_\lambda^0, \dots, y_\lambda^{n-1}) \in \nu(p)$ such that $(x_k, y_\lambda^k) \in \lambda$ $(k = 0, \dots, n-1)$.

Denote by y_k the mixing of the family $(y_{\lambda,k})_{\lambda\in\Lambda}$ relative to Λ. Then $(y_0,\dots,y_{n-1})\in\nu(p)$. At the same time $(x_k,y_k)\in\lambda$ for all $\lambda\in\Lambda$. Hence, $x_k=y_k$ $(k=0,\dots,n-1)$ since $\bigcap\Lambda=I_A$; a contradiction.

Assume conversely that Λ is a complete set of congruences. Take $p\in P$ and a family of n-tuples $(a_{\lambda,0},\dots,a_{\lambda,n-1})$ contained in $\nu(p)$. Let a_k stand for the mixing of $(a_{\lambda,k})_{\lambda\in\Lambda}$ relative to Λ.

If $(a_0,\dots,a_{n-1})\notin\nu(p)$ then, since Λ is complete, there is a congruence $\lambda\in\Lambda$ satisfying $(a_{\lambda,0},\dots,a_{\lambda,n-1})\notin\nu(p)$. This, however, contradicts the choice of $(a_{\lambda,0}$, $\dots$, $a_{\lambda,n-1})$. Hence, $\nu(p)$ is closed under Λ-mixing.

Necessity holds clearly without the assumption that Λ is independent. $\triangleright$

4.1.5. A *Boolean algebra of congruences* is a Boolean algebra $\mathscr{B}\subset\mathrm{Cong}(\mathfrak{A})$ such that the least upper bound of an arbitrary set in $\mathscr{B}$ is inherited from the congruence lattice $\mathrm{Cong}(\mathfrak{A})$ and the least congruence I_A serves as the zero of $\mathscr{B}$.

It is worth observing that the Boolean complement ρ^* of $\rho\in\mathscr{B}$ may fail to be the complement of ρ in the congruence lattice $\mathrm{Cong}(\mathfrak{A})$; i.e., the least upper bound of ρ and ρ^* in $\mathrm{Cong}(\mathfrak{A})$ may be less than $A\times A$.

A *base* for an algebraic system $\mathfrak{A}$ is a complete Boolean algebra of congruences $\mathscr{B}\subset\mathrm{Cong}(\mathfrak{A})$ such that each predicate $\nu(p)$ $(p\in P)$ is closed under Λ^*-mixing for each partition of unity $\Lambda\subset\mathscr{B}$ where $\Lambda^*:=\{b^*:b\in\Lambda\}$.

An algebraic system with base $\mathscr{B}$ is *universally* (*finitely*) *complete* provided that the set of congruences Λ^* is independent where $\Lambda\subset\mathscr{B}$ is an arbitrary (finite) partition of unity.

An algebraic system $\mathfrak{A}$ has a base $\mathscr{B}$ isomorphic to a complete Boolean algebra B if and only if there is an injective mapping $h:B\to\mathrm{Cong}(\mathfrak{A})$ obeying the following conditions:

(1) *h preserves infima and $h(\mathbf{0})=I_A$;*

(2) *every predicate $\nu(p)$ $(p\in P)$ is closed under $h(\Lambda^*)$-mixing for each partition of unity $\Lambda\subset B$.*

In this event $\mathfrak{A}$ is universally (finitely) complete if and only if the set $h(\Lambda^*)$ is independent for every (finite) partition of unity $\Lambda\subset B$.

4.1.6. An algebraic B-system $\mathfrak{A}$ is *full* provided that to each $\mathbf{0}\neq b\in B$ there are elements x, $y\in A$, $x\neq y$, such that $d(x,y)\le b$. It is obvious that a finitely complete B-system is full, but the converse may fail in general.

Theorem. *An algebraic system $\mathfrak{A}$ is the purification of some full algebraic B-system $\mathfrak{A}'$ if and only if $\mathfrak{A}$ has a base isomorphic to B. In this event, $\mathfrak{A}$ and $\mathfrak{A}'$ are universally (finitely) complete or not simultaneously.*

$\triangleleft$ Let $\mathfrak{A}'$ be a full algebraic B-system. Take $b\in B$ and put $h(b):=\{(x,y)\in A^2:d(x,y)\le b\}$.

Since $\nu(f)$ is a contractive mapping for every $f \in F$; therefore, $h(b)$ is a congruence on A. It is obvious that $h(\mathbf{0}) = I_A$ and h preserves infima. Since $\mathfrak{A}$ is full, conclude that h is injective. Assume that the algebraic system $\mathfrak{A}$ is the purification of $\mathfrak{A}'$. Note that every set of the type $\{z \in A : p(z) = \mathbf{1}\}$ is closed under all mixings in the B-set A. Whence, it follows from 4.1.5 that $\mathfrak{A}$ has a base isomorphic to B. Conversely, assume that $\mathfrak{A}$ has a base $\mathscr{B}$ and there exists a Boolean isomorphism h from B to $\mathscr{B}$. Assign

$$d(x,y) := \bigwedge\{b \in B : (x,y) \in h(b)\} \quad (x,y \in A).$$

If $b_1, b_2 \in B$ are such that $(x,z) \in h(b_1)$ and $(z,y) \in h(b_2)$ then $(x,y) \in h(b_2) \circ h(b_1)$. However, $h(b_2) \circ h(b_1) \subset h(b_1 \vee b_2)$ and so $d(x,y) \leq b_1 \vee b_2$.

Taking the infimum over b_1 and b_2, use the distributive law 1.1.5 (1) to conclude that $d(x,y) \leq d(x,z) \vee d(z,y)$. It is now evident that d is a Boolean semimetric on A. Since h preserves infima; therefore,

$$h(d(x,y)) = \bigcap\{h(b) : b \in B, (x,y) \in h(b)\}.$$

Whence we deduce that $d(x,y) \leq b$ if and only if $(x,y) \in h(b)$. In particular, $d(x,y) = \mathbf{0}$ implies that $x = y$; while, given $\mathbf{0} \neq b \in B$, we may find $x, y \in A$ satisfying $x \neq y$ and $d(x,y) \leq b$.

It remains to show that if Λ is a partition of unity in B then for a family $(a_b)_{b\in\Lambda} \subset A$ the mixing relative to $h(\Lambda^*)$ coincides with that in the sense of the B-metric d, i.e., with $\operatorname{mix}_{b\in\Lambda}(ba_b)$. This fact, however, is immediate from the above: $(a, a_b) \in h(b^*) \leftrightarrow d(a, a_b) \leq b^* \leftrightarrow b \wedge d(a, a_b) = \mathbf{0}$. We now define $\mathfrak{A}' := (A', \nu')$ by putting $A' := A$, $\nu'(f) = \nu(f)$, $f \in F$, and

$$\nu'(p) : x \mapsto \operatorname{dist}(x, \nu(p)) \quad (p \in P,\ x \in A^{\mathfrak{a}(p)}).$$

If $f \in F$ and $n = \mathfrak{a}(f)$ then for all $b \in B$ and $x_0, y_0, \dots, x_{n-1}, y_{n-1} \in A$ the containments $(x_k, y_k) \in h(b)$, $k < n$ imply that $(f^\nu(x_0, \dots, x_{n-1}), f^\nu(y_0, \dots, y_{n-1})) \in h(b)$, which gives

$$d(f^\nu(x_0, \dots, x_{n-1}),\ f^\nu(y_0, \dots, y_{n-1})) \leq b.$$

Passing to the infimum over b and observing that

$$\bigwedge\{b : (x_k, y_k) \in h(b), k < n\} = \bigvee_{k=0}^{n-1} d(x_k, y_k),$$

conclude that the mappings $f^\nu = \nu(f)$ are contractions. Choosing $p \in P$, $\mathfrak{a}(p) = m$, take $x := (x_0, \dots, x_{m-1})$ and $y := (y_0, \dots, y_{m-1})$ in A^m. Then

$$d(x,y) \wedge \operatorname{dist}(x, \nu(p)) \leq \operatorname{dist}(y, \nu(p)),$$

which implies that $\nu'(p)$ is a contractive mapping. Moreover, since $\nu(p)$ is closed under mixing (see 4.1.3 (2)), observe that $\nu(p) = \{x \in A^m : \nu'(p)(x) = \mathbf{1}\}$ which makes the contraction property of $\nu'(p)$ obvious. Moreover, since $\nu(p)$ is closed under mixing (cf. 4.1.3 (2)), we see that $\nu(p) = \{x \in A^m : \nu'(p)(x) = \mathbf{1}\}$. Hence, $\mathfrak{A}$ is the purification of the full algebraic B-system $\mathfrak{A}'$. The fact that the systems $\mathfrak{A}$ and $\mathfrak{A}'$ are universally complete implies that Λ^*, where Λ is a partition of unity in $\mathscr{B}$, is an independent set and that (A, d) is closed under any mixings. By similar reasons, the claims about finite completeness of the two systems are also equivalent to each other. ▷

4.1.7. Consider some concrete examples of algebraic B-systems. Recall that an associative ring R is a *Boolean ring* if every element of R is *idempotent*, i.e., if $(\forall\, x \in R)(x^2 = x)$. A unital Boolean ring is a Boolean algebra. Conversely, each Boolean algebra B is a Boolean ring with unity. In this event the zero and unity of a ring coincide with the Boolean zero and unity, respectively (see 1.2.1).

(1) Let B_0 be a Boolean algebra. Assume that X is a unital module over the Boolean ring B_0. Denote by B the completion of B_0 and let $\jmath$ stand for an isomorphism of B_0 onto a dense subalgebra of B. Assign

$$d_\jmath(x, y) := \bigwedge \{\jmath(b) : b^* x = b^* y, b \in B_0\} \quad (x, y \in X).$$

It is easy to see that $d_\jmath$ is a B-semimetric on X. For instance, demonstration of the triangle inequality proceeds as follows: If $b^* x = b^* z$ and $c^* z = c^* y$ then, considering $e := b^* \cdot c^* = b^* \wedge c^* = (b \vee c)^*$, note that $ex = ez$ and $ey = ez$. Therefore, $ex = ey$ and $d_\jmath(x, y) \leq e \leq \jmath(b \vee c) = \jmath(b) \vee \jmath(c)$. Since b and c are arbitrary, obtain $d_\jmath(x, y) \leq d_\jmath(x, z) \vee d_\jmath(z, y)$.

Call X a *laterally faithful* module if for each partition of unity (b_ξ) in B_0 from $(\forall\, \xi)\ (b_\xi x = 0)$ it follows that $x = 0$ for all $x \in X$. It is beyond a doubt that the semimetric $d_\jmath$ is a metric for a laterally faithful unital B_0-module X. By analogy with the triangle inequality for $d_\jmath$, we may show that all module operations are contractive:

$$d_\jmath(x + u, y + v) \leq d_\jmath(x, y) \vee d_\jmath(u, v) \quad (x, y, u, v \in X),$$
$$d_\jmath(bx, cy) \leq d_\jmath(x, y) \vee d_s(b, c) \quad (x, y \in X;\ b, c \in B).$$

The last inequality implies in particular that

$$d_\jmath(bx, by) \leq d_\jmath(x, y) \quad (b \in B;\ x, y \in X).$$

Moreover, $d_\jmath(-x, -y) = d_\jmath(x, y)$. Therefore, the set X, furnished with the operations $+$ and $-$ and the unary operations of multiplication by $b \in B_0$, is an algebraic B-system.

(2) Assume that R is a unital commutative ring. Consider the set $B_0 := \{e \in R : e \cdot e = e\}$ of all idempotents of R. Then B_0 is a Boolean ring with unity and R is a module over B_0. In case B and $\jmath$ are the same as in (1), we notice the B-semimetric $d_\jmath$ on R.

Clearly, R is laterally faithful over B_0. By (1), we deduce that a unital commutative ring R, laterally faithful over the subring B_0 of the idempotents of B, is an algebraic B-system of signature $(+, -, \cdot, \mathbf{1})$.

(3) Assume that C is a Boolean algebra and $\imath$ is a homomorphism from a Boolean algebra B_0 to C. Since $\imath(B_0)$ is a subring of the Boolean ring C, we can readily endow C with the structure of a unital module over B_0. If B and $\jmath$ are the same as in (1) then

$$d_\jmath(x, y) := \bigwedge \{\jmath(b) : \imath(b^*)x = \imath(b^*)y\}.$$

The module C is laterally faithful if $\imath$ is a complete monomorphism. In view of the above mentioned interrelation between Boolean and ring operations, the Boolean algebra C is an algebraic B-system of signature $(\vee, \wedge, *, \mathbf{0}, \mathbf{1})$ in the case when $\imath$ is a complete monomorphism. This system is universally complete if, for instance, B_0 and C are complete Boolean algebras.

4.1.8. We now address the B-valued interpretation of a first-order language.

Consider an algebraic B-system $\mathfrak{A} := (A, \nu)$ of signature $\sigma := \sigma(\mathfrak{A}) := (F, P, \mathfrak{a})$. Let $\varphi(x_0, \dots, x_{n-1})$ be a formula of signature σ with n free variables. Assume given $a_0, \dots, a_{n-1} \in A$. Define the Boolean truth value $|\varphi|^{\mathfrak{A}}(a_0, \dots, a_{n-1}) \in B$ of a formula φ in the system $\mathfrak{A}$ for the given values $a_0, \dots, a_{n-1}$ of the variables $x_0, \dots, x_{n-1}$. The definition proceeds readily by usual recursion on the length of φ: Considering propositional connectives and quantifiers, put

$$
\begin{aligned}
|\varphi \wedge \psi|^{\mathfrak{A}}(a_0, \dots, a_{n-1}) &:= |\varphi|^{\mathfrak{A}}(a_0, \dots, a_{n-1}) \wedge |\psi|^{\mathfrak{A}}(a_0, \dots, a_{n-1});\\
|\varphi \vee \psi|^{\mathfrak{A}}(a_0, \dots, a_{n-1}) &:= |\varphi|^{\mathfrak{A}}(a_0, \dots, a_{n-1}) \vee |\psi|^{\mathfrak{A}}(a_0, \dots, a_{n-1});\\
|\neg\varphi|^{\mathfrak{A}}(a_0, \dots, a_{n-1}) &:= |\varphi|^{\mathfrak{A}}(a_0, \dots, a_{n-1})^*;\\
|(\forall\, x_0)\varphi|^{\mathfrak{A}}(a_1, \dots, a_{n-1}) &:= \bigwedge_{a_0 \in A} |\varphi|^{\mathfrak{A}}(a_0, \dots, a_{n-1});\\
|(\exists\, x_0)\varphi|^{\mathfrak{A}}(a_1, \dots, a_{n-1}) &:= \bigvee_{a_0 \in A} |\varphi|^{\mathfrak{A}}(a_0, \dots, a_{n-1}).
\end{aligned}
$$

Now, the case of atomic formulas is in order. Suppose that $p \in P$ symbolizes an m-ary predicate, $q \in P$ is a nullary predicate, and $t_0, \dots, t_{m-1}$ are terms of signature σ assuming values $b_0, \dots, b_{m-1}$ at the given values $a_0, \dots, a_{n-1}$ of the variables

$x_0, \dots, x_{n-1}$. By definition, we let

$$|\varphi|^{\mathfrak{A}}(a_0, \dots, a_{n-1}) := \nu(q), \text{ if } \varphi = q^\nu;$$
$$|\varphi|^{\mathfrak{A}}(a_0, \dots, a_{n-1}) := d(b_0, b_1)^*, \text{ if } \varphi = (t_0 = t_1);$$
$$|\varphi|^{\mathfrak{A}}(a_0, \dots, a_{n-1}) := p^\nu(b_0, \dots, b_{m-1}), \text{ if } \varphi = p^\nu(t_0, \dots, t_{m-1}),$$

where d is a B-metric on A.

Say that $\varphi(x_0, \dots, x_{n-1})$ is *satisfied* in $\mathfrak{A}$ by the assignment $a_0, \dots, a_{n-1} \in A$ of $x_0, \dots, x_{n-1}$ and write $\mathfrak{A} \models \varphi(a_0, \dots, a_{n-1})$ provided that $|\varphi|^{\mathfrak{A}}(a_0, \dots, a_{n-1}) = \mathbf{1}_B$. Alternative expressions are as follows: $a_0, \dots, a_{n-1} \in A$ *satisfies* $\varphi(x_0, \dots, x_{n-1})$ or $\varphi(a_0, \dots, a_{n-1})$ *holds* in $\mathfrak{A}$. In case $B := \{\mathbf{0}, \mathbf{1}\}$, we arrive at the conventional definition of the satisfaction of a formula in an algebraic system (cf. [48, 164]).

Recall that a closed formula φ of signature σ is a *tautology* or *logically valid* if φ is satisfied in every algebraic $\mathbf{2}$-system of signature σ.

4.1.9. *Theorem.* *Let $\mathfrak{A}$ be an arbitrary algebraic B-system. Then the following hold:*

(1) *Every theorem of predicate calculus holds in $\mathfrak{A}$;*

(2) *Every tautology of signature $\sigma(\mathfrak{A})$ holds in $\mathfrak{A}$.*

$\lhd$ (1) We are to demonstrate that the axioms of predicate calculus are satisfied in $\mathfrak{A}$, and the rules of inference do not destroy satisfaction in $\mathfrak{A}$ (see 2.1.8). To this end, it suffices to inspect the corresponding calculations of Boolean truth values (cf. [11, 48, 123, 131, 240, 241]).

(2) If a closed formula φ fails in $\mathfrak{A}$ then $b := |\varphi|^{\mathfrak{A}} < \mathbf{1}_B$. Let $h : B \to \mathbf{2} := \{\mathbf{0}, \mathbf{1}\}$ be a complete homomorphism satisfying $h(b) = \mathbf{0}$. Such an h exists, since the ideal $[\mathbf{0}, b]$ lies in a maximal ideal that may be taken as $h^{-1}(\mathbf{0})$. If ν is an interpretation of $\mathfrak{A}$ then we put $\nu'(f) := f^\nu$ for function symbols and $\nu'(p) := h \circ p^\nu$ for predicate symbols. Then $\mathfrak{A}' := (|\mathfrak{A}|, \nu')$ is an algebraic $\mathbf{2}$-system and $|\varphi|^{\mathfrak{A}'} = h(b) = \mathbf{0}$; i.e., φ fails in $\mathfrak{A}'$. Hence, φ is not a tautology. $\rhd$

4.1.10. Consider algebraic B-systems $\mathfrak{A} := (A, \nu)$ and $\mathfrak{C} := (C, \mu)$ of the same signature σ. The mapping $h : A \to C$ is a *homomorphism* of $\mathfrak{A}$ to $\mathfrak{C}$ provided that, for all $a_0, \dots, a_{n-1} \in A$, the following hold:

(1) $d_B(h(a_1), h(a_2)) \le d_A(a_1, a_2)$;

(2) $h(f^\nu) = f^\nu$, $\mathfrak{a}(f) = 0$;

(3) $h(f^\nu(a_0, \dots, a_{n-1})) = f^\nu(h(a_0), \dots, h(a_{n-1}))$, $0 \ne n := \mathfrak{a}(f)$;

(4) $p^\nu(a_0, \dots, a_{n-1}) \le p^\mu(h(a_0), \dots, h(a_{n-1}))$, $n := \mathfrak{a}(p)$.

A homomorphism h is called *strong* if

(5) $\mathfrak{a}(p) := n \neq 0$ for all $p \in P$, and the following inequality holds:

$$p^{\mu}(c_0, \dots, c_{n-1})$$
$$\geq \bigvee_{a_0, \dots, a_{n-1} \in A} \{p^{\nu}(a_0, \dots, a_{n-1}) \wedge d_C(c_0, h(a_0)) \wedge \dots \wedge d_C(c_{n-1}, h(a_{n-1}))\}$$

for all $c_0, \dots, c_{n-1} \in C$.

If a homomorphism h is injective and (1) and (4) are equalities then h is said to be an *isomorphism from* $\mathfrak{A}$ *to* $\mathfrak{C}$. Undoubtedly, each surjective isomorphism h and, in particular, the identity mapping $I_A : A \to A$ are strong homomorphisms. The composition of (strong) homomorphisms is a (strong) homomorphism. Clearly, if h is a homomorphism and h^{-1} is a homomorphism too then h is an isomorphism.

Note again that in the case of the two-element Boolean algebra $B := \{\mathbf{0}, \mathbf{1}\}$ we come to the conventional notions of homomorphism, strong homomorphism, and isomorphism (cf. [48, 164]).

4.1.11. Consider some set Φ of formulas of the same fixed signature σ. Define the category B-AS(Φ) as follows: The class Ob B-AS(Φ) consists of all algebraic B-systems of signature σ each of which satisfies all formulas of Φ. The class Mor B-AS(Φ) is the class of all homomorphisms of algebraic B-systems of Ob B-AS(Φ) with the conventional composition of mappings as composition of morphisms. An isomorphism in the category B-AS(Φ) is obviously a B-isometric strong homomorphism. Denote by B-CAS(Φ) the full subcategory of the category B-AS(Φ) whose objects are universally complete algebraic B-systems.

4.1.12. According to 4.1.5 and 4.1.6, the structure of an algebraic B-system $\mathfrak{A}$ may be reconstructed from the complete Boolean algebra of congruences Cong($\mathfrak{A}$). On the other hand, one of the most general methods for obtaining complete Boolean algebras is associated with the abstract concept of disjointness. We now dwell for a while on essential relationship between these notions, starting with some relevant facts to be recalled.

Consider sets X and Y. Assume that Φ is a correspondence from X to Y. Denote by $\pi_{\Phi}(A)$ and $\pi_{\Phi}^{-1}(C)$ the *polar* of $A \subset X$ and the *inverse polar* of $C \subset Y$ with respect to Φ (see A.3.10):

$$\pi_{\Phi}(A) := \bigcap_{x \in A} \Phi(x), \quad \pi_{\Phi}^{-1}(C) := \bigcap_{y \in C} \Phi^{-1}(y).$$

A set $K \subset Y$ is a *Φ-band* or simply a *band* of Φ when the context prompts Φ provided that $K = \pi_{\Phi}(\pi_{\Phi}^{-1}(K))$ or, which is equivalent, $K = \pi_{\Phi}(A)$ for some $A \subset X$. Denote by $\mathfrak{K}_{\Phi}(Y)$ the set of all Φ-bands. Let $[C]$ stand for the least band that includes a subset C of Y; i.e., $[C] = \pi_{\Phi}(\pi_{\Phi}^{-1}(C))$.

(1) Theorem. *The inclusion ordered set $\mathfrak{K}_{\Phi}(Y)$ is a complete lattice. The supremum and infimum of a family $(K_{\xi})_{\xi\in\Xi} \subset \mathfrak{K}_{\Phi}(Y)$ are calculated by the formulas*

$$\bigwedge_{\xi\in\Xi} K_{\xi} = \bigcap_{\xi\in\Xi} K_{\xi}, \quad \bigvee_{\xi\in\Xi} K_{\xi} = \Big[\bigcup_{\xi\in\Xi} K_{\xi}\Big].$$

The inverse polar mapping $K \mapsto \pi_{\Phi}^{-1}(K)$ is an antitonic bijection of $\mathfrak{K}_{\Phi}(Y)$ on $\mathfrak{K}_{\Phi^{-1}}(X)$.

(2) A relation Δ on a set X is a *disjointness relation* or *disjointness* (on X) provided that the following conditions are met:

(a) $\Delta = \Delta^{-1}$; i.e., Δ is symmetric;

(b) $\Delta \cap I_X \subset \Theta \times \Theta$,

with $\Theta := \pi_{\Delta}(X)$ signifying the least Δ-band;

(c) $[x] \cap [y] \subset \Theta \to (x, y) \in \Delta$.

A disjointness Δ is called *simple* if Δ obeys the additional requirement

(d) $(x, y) \in \Delta \to x \in \Theta \vee y \in \Theta$.

Since Δ is symmetric, the lattices $\mathfrak{K}_{\Delta}(X)$ and $\mathfrak{K}_{\Delta^{-1}}(X)$ coincide. If $A \subset X$ then the polar $\pi_{\Delta}(A)$ is called the *disjoint complement* of A in which case we also denote $\pi_{\Delta}(A)$ by $A^{\perp}$. The relations $x \in \pi_{\Delta}(A)$ and $C \subset \pi_{\Delta}(A)$ are rewritten as $x \perp A$ and $C \perp A$. Note also that $A^{\perp\perp} := (A^{\perp})^{\perp} = [A]$.

(3) Theorem. *The inclusion ordered set $\mathfrak{K}_{\Delta}(X)$ of all bands of a disjointness Δ is a complete Boolean algebra. The Boolean complement of a band coincides with its disjoint complement.*

$\lhd$ As mentioned in (1), $\mathfrak{K}_{\Delta}(X)$ is a complete lattice. The zero and unity of this lattice are Θ and X. Applying elementary rules for operations on polars from A.3.10 and using the distributive laws for the set-theoretic operations on arbitrary bands K, L, and M, we may write the following chain of equalities:

$$\begin{aligned}(K \vee L) \wedge M &= ((K \vee L)^{\perp} \cup M^{\perp})^{\perp} = ((K^{\perp} \cap L^{\perp}) \cup M^{\perp})^{\perp} \\ &= ((K^{\perp} \cup M^{\perp}) \cap (L^{\perp} \cup M^{\perp}))^{\perp} = [(K^{\perp} \cup M^{\perp})^{\perp} \cup (L^{\perp} \cup M^{\perp})^{\perp}] \\ &= (K^{\perp\perp} \cap M^{\perp\perp}) \vee (L^{\perp\perp} \cap M^{\perp\perp}) = (K \wedge M) \vee (L \wedge M).\end{aligned}$$

Hence, the lattice $\mathfrak{K}_{\Delta}(X)$ is distributive. Obviously, $K \cap K^{\perp} = \Theta$. On the other hand,

$$K \vee K^{\perp} = [K \cup K^{\perp}] = (K^{\perp} \wedge K)^{\perp} = \Theta^{\perp} = X;$$

i.e., $K^{\perp}$ is the complement of K in $\mathfrak{K}_{\Delta}(X)$. $\rhd$

4.1.13. Consider a set X with disjointness Δ. Let $\jmath$ be an isomorphism of $\mathfrak{K}_\Delta(X)$ onto a complete Boolean algebra B. Introduce a mapping $s : X \to B$ by the formula $s(x) := \jmath([x])$ $(x \in X)$. Assume that the least band is a singleton; i.e., $\Theta := \{\theta\} = [\theta]$ for some $\theta \in X$. Say that a B-metric d and the disjointness Δ on X *agree* provided that

$$d(x, \theta) = s(x) \quad (x \in X).$$

Consider another mapping

$$\delta : (x, y) \mapsto (s(x) \wedge s(y))^* \quad (x, y \in X).$$

Theorem. *Assume that X is a set equipped with disjointness and B-metric d that agree on X. Then the 3-tuple $\mathfrak{X} := (X, \delta, \theta)$ is an algebraic B-system satisfying the axioms of simple disjointness* (a)–(d) *of* 4.1.12 (2).

$\triangleleft$ First of all, note that

$$d(x, y)^* \wedge s(x) = d(x, y)^* \wedge d(x, \theta)$$
$$\leq d(x, y)^* \wedge (d(x, y) \vee d(y, \theta)) \leq d(y, \theta) = s(y).$$

Hence, s is a contractive mapping. Therefore, the mapping δ is contractive too, and so $\mathfrak{X}$ is an algebraic B-system with binary predicate δ and distinguished element θ. By definition, obtain

$$|x \delta y|^{\mathfrak{X}} = \delta(x, y), \quad |x \neq \theta|^{\mathfrak{X}} = s(x) \quad (x, y \in X).$$

Validate the axioms of disjointness for δ. Obviously, δ is symmetric. The set $\{\theta\}$ is the least δ-band as is immediate from the following:

$$|x \in \pi_\delta(X) \to x = \theta|^{\mathfrak{X}} = \Big(\bigwedge_{y \in X} \delta(x, y) \Big) \Rightarrow s(x^*)$$
$$= \bigvee_{y \in X} (s(x) \wedge s(y)) \vee s(x)^* = s(x)^* \vee \bigvee_{y \in X} s(y) = \mathbf{1}.$$

It is also obvious that

$$\delta(x, x) = |x \delta x|^{\mathfrak{X}} = s(x)^* = |x = \theta|^{\mathfrak{X}}$$

for all $x, y \in X$. Therefore, condition (b) of the definition of disjointness is fulfilled. Note further that

$$|u \in [x]|^{\mathfrak{X}} = s(u) \Rightarrow s(x) \quad (x, u \in X).$$

Using this, proceed with the calculation

$$
\begin{aligned}
|[x] \cap [y] = \{\theta\}|^{\mathfrak{X}} &= \Big(\bigwedge_{u \in X} (s(u) \Rightarrow s(x)) \wedge (s(u) \Rightarrow s(y)) \Big) \\
&\Rightarrow s(u)^* = \bigwedge_{u \in X} s(u)^* \vee (s(x) \bigwedge s(y))^* = \delta(x, y).
\end{aligned}
$$

Therefore, $|[x] \cap [y] = \{\theta\} \to x\delta y|^{\mathfrak{X}} = \mathbf{1}$. Hence, δ is a disjointness. Furthermore, δ is simple. Indeed, if $x, y \in X$ then

$$
|x\delta y \to x = \theta \vee y = \theta|^{\mathfrak{X}} = \mathbf{1}
$$

or, equivalently,

$$
\delta(x, y) \Rightarrow s(x)^* \vee s(y)^* = \mathbf{1},
$$

which ensues from the definition of δ. $\rhd$

Assume now that $\mathfrak{A} := (A, \nu)$ is an algebraic B-system and Δ is the same as in 4.1.13. Assume further that all operations of $\mathfrak{A}$ *preserve disjointness*, i.e., for each function symbol f and all $a \in A$, $x_0, \dots, x_{n-1} \in A$ $(n := \mathfrak{a}(f))$, from $x_k \perp a$ $(k := 0, 1, \dots, n-1)$ it follows that $f^\nu(x_0, \dots, x_{n-1}) \perp a$. If, moreover, the B-metric and disjointness Δ agree then the 3-tuple (A, ν, Δ) is called an *algebraic B-system with disjointness*.

4.1.14. Comments.

(1) While proving the Stone Theorem 1.2.4, we find that every Boolean algebra B is isomorphic to the algebra of continuous functions $C(\mathrm{St}(B), \mathbf{2})$, with $\mathrm{St}(B)$ a Boolean space. It seems reasonable to substitute an arbitrary universal algebra for the two-element field $\mathbf{2}$. This leads us to an important example of an algebraic B-system, the ***Boolean power*** of a universal algebra which was introduced by R. F. Arens and I. Kaplansky [5] (see also [51, 52, 202]).

(2) In the sequel we proceed along the lines of the present section, discussing only the problems pertinent to Boolean valued representation of algebraic B-systems and to relevant specification of ascending and descending. The logical-algebraic aspects of algebraic B-systems are expounded in full detail elsewhere [9, 54].

4.2. The Descent of an Algebraic System

In the present section we specify the technique of descent in the case of algebraic systems and give some illuminating examples.

4.2.1. Let $\sigma := (F, P, \mathfrak{a})$ be a signature. From the general properties of the canonical embedding of the von Neumann universe $\mathbf{V}$ into a Boolean valued universe $\mathbf{V}^{(B)}$ (cf. 3.1.6 and 3.1.9) it follows that $\mathbf{V}^{(B)} \models$ "$\mathfrak{a}^\wedge$ is a mapping from $F^\wedge \cup P^\wedge$ into the set of positive integers $\omega^\wedge$." Moreover, $\mathbf{V}^{(B)} \models \sigma^\wedge = (F^\wedge, P^\wedge, \mathfrak{a}^\wedge)$ and so

$$\mathbf{V}^{(B)} \models \text{"}\sigma^\wedge \text{ is a signature."}$$

If σ is a signature inside $\mathbf{V}^{(B)}$ then $\sigma{\downarrow}$ fails in general to be a signature in the conventional sense of the word. Indeed, assume that $\sigma = (F, P, \mathfrak{a})^B \in \mathbf{V}^{(B)}$ for some F, P, and $\mathfrak{a}$ in $\mathbf{V}^{(B)}$ satisfying $[\![\mathfrak{a} : F \cup P \to \omega^\wedge]\!] = \mathbf{1}$. Then, for every $u \in F{\downarrow} \cup P{\downarrow}$, we can find a countable partition of unity $(b_n)_{n \in \omega} \subset B$ such that $\mathfrak{a}{\downarrow}(u) = \operatorname{mix}(b_n n^\wedge)$.

Therefore, the descent of an algebraic system of arbitrary signature leads to function and predicate symbols of "mixed arity." It goes without saying that we can elaborate a theory that admits operations and predicates of mixed arity since this entails no principal difficulties. Another possible abstraction concerns algebraic systems with operations and predicates of infinite arity. The present exposition leaves these possibilities intact for better times.

4.2.2. Before giving general definitions, consider the descent of a very simple but important algebraic system, the two-element Boolean algebra. Choose two arbitrary elements, $0,\ 1 \in \mathbf{V}^{(B)}$, satisfying $[\![0 \neq 1]\!] = \mathbf{1}_B$. We can for instance assume that $0 := \mathbf{0}_B^\wedge$ and $1 := \mathbf{1}_B^\wedge$.

The descent C of the two-element Boolean algebra $\{0,1\}^B \in \mathbf{V}^{(B)}$ is a complete Boolean algebra isomorphic to B. The formulas

$$[\![\chi(b) = 1]\!] = b, \quad [\![\chi(b) = 0]\!] = b^* \quad (b \in B)$$

define an isomorphism $\chi : B \to C$.

$\lhd$ Since $0, 1 \in C$; for every $b \in B$, the mixing $c := \operatorname{mix}(b1, b^*0)$ belongs to C; moreover, $[\![c = 1]\!] \geq b$ and $[\![c = 0]\!] \geq b^*$. On the other hand,

$$[\![c = 1]\!] \wedge [\![c = 0]\!] = [\![c = 1 \wedge c = 0]\!] \leq [\![0 = 1]\!] = \mathbf{0}.$$

Hence, $[\![c = 1]\!] = b$ and $[\![c = 0]\!] = b^*$. Putting $\chi(b) := c$, obtain a mapping $\chi : B \to C$. Obviously, χ is injective. Check that χ is surjective. Indeed, if $c \in C$ then, letting $b := [\![c = 1]\!]$, note that

$$[\![\chi(b) = 0]\!] = b^* = [\![c = 0]\!], \quad [\![\chi(b) = 1]\!] = b,$$

and so

$$[\![\chi(b) = c]\!] \geq [\![\chi(b) = 1]\!] \wedge [\![c = 1]\!] = b.$$

By analogy, $[\![\chi(b) = c]\!] \geq b^*$ and so $\chi(b) = c$.

Descend the Boolean operations of $\{0,1\}^{(B)}$ to note that, for all x, y, $z \in C$, the following hold:

$$z = x \wedge y \leftrightarrow [\![z = 1 \leftrightarrow x = 1 \wedge y = 1]\!] = \mathbf{1},$$
$$z = x \vee y \leftrightarrow [\![z = 0 \leftrightarrow x = 0 \wedge y = 0]\!] = \mathbf{1},$$
$$x = y^* \leftrightarrow [\![x = 1 \leftrightarrow y = 0]\!] = \mathbf{1}.$$

These formulas make it easy to prove that C is a Boolean algebra while χ is a Boolean isomorphism. For instance, show that χ preserves joins. Assume that b_1, $b_2 \in B$, $b_0 := b_1 \vee b_2$, and $c_l := \chi(b_l)$ for $l = 0, 1, 2$. By definition,

$$[\![c_l = 1]\!] = b_l, \quad [\![c_l = 0]\!] = b_l^* \quad (l = 0, 1, 2),$$

and so

$$[\![c_0 = 0]\!] = b_0^* = b_1^* \wedge b_2^* = [\![c_1 = 0]\!] \wedge [\![c_2 = 0]\!]$$

or, which is the same, $[\![c_0 = 0 \leftrightarrow c_1 = 0 \wedge c_2 = 0]\!] = \mathbf{1}$. Therefore, $c_0 = c_1 \vee c_2$ or $\chi(b_0) = \chi(b_1) \vee \chi(b_2)$. By analogy, show that meets and complements are preserved too, so completing the proof. $\triangleright$

4.2.3. Consider now an algebraic system $\mathfrak{A}$ of signature $\sigma^\wedge$ inside $\mathbf{V}^{(B)}$, and let $[\![\mathfrak{A} = (A, \nu)^B]\!] = \mathbf{1}$ for some A, $\nu \in \mathbf{V}^{(B)}$. The *descent* of $\mathfrak{A}$ is the pair $\mathfrak{A}{\downarrow} := (A{\downarrow}, \mu)$, where μ is the function determined from the formulas:

$$\mu : f \mapsto (\nu{\Downarrow}(f)){\downarrow} \quad (f \in F),$$
$$\mu : p \mapsto \chi^{-1} \circ (\nu{\Downarrow}(p)){\downarrow} \quad (p \in P).$$

Here χ is the canonical isomorphism of the Boolean algebras B and the descent of $\{0,1\}^B$ (of 4.2.2).

In more detail, the modified descent $\nu{\Downarrow}$ is a mapping with domain $\mathrm{dom}(\nu{\Downarrow}) = F \cup P$. Given $p \in P$, observe $[\![\mathfrak{a}(p)^\wedge = \mathfrak{a}^\wedge(p^\wedge)]\!] = \mathbf{1}$, $[\![\nu{\Downarrow}(p) = \nu(p^\wedge)]\!] = \mathbf{1}$ and so

$$\mathbf{V}^{(B)} \models \nu{\Downarrow}(p) : A^{\mathfrak{a}(f)^\wedge} \to \{0,1\}^B.$$

It is now obvious that $(\nu{\Downarrow}(p)){\downarrow} : (A{\downarrow})^{\mathfrak{a}(f)} \to C := \{0,1\}^B{\downarrow}$ and we may put $\mu(p) := \chi^{-1} \circ (\nu{\Downarrow}(p)){\downarrow}$.

Let $\varphi(x_0, \dots, x_{n-1})$ be a fixed formula of signature σ in n free variables. Write down the formula $\Phi(x_0, \dots, x_{n-1}, \mathfrak{A})$ in the language of set theory which formalizes the proposition $\mathfrak{A} \models \varphi(x_0, \dots, x_{n-1})$. Recall that the formula $\mathfrak{A} \models \varphi(x_0, \dots, x_{n-1})$ determines an n-ary predicate on A or, which is the same, a mapping from A^n

to $\{0,1\}$. By the maximum and transfer principles, there is a unique element $|\varphi|^{\mathfrak{A}} \in \mathbf{V}^{(B)}$ such that

$$[\![|\varphi|^{\mathfrak{A}} : A^{n^{\wedge}} \to \{0,1\}^B]\!] = \mathbf{1},$$
$$[\![|\varphi|^{\mathfrak{A}}(a\uparrow) = 1]\!] = [\![\Phi(a(0), \dots, a(n-1), \mathfrak{A})]\!] = \mathbf{1}$$

for every $a : n \to A\downarrow$. Henceforth we write $|\varphi|^{\mathfrak{A}}\,(a_0, \dots, a_{n-1})$ instead of $|\varphi|^{\mathfrak{A}}(a\uparrow)$, where $a_l := a(l)$. Therefore, the formula

$$\mathbf{V}^{(B)} \models \text{"}\varphi(a_0, \dots, a_{n-1}) \text{ is satisfied in } \mathfrak{A}\text{"}$$

holds if and only if $[\![\Phi(a_0, \dots, a_{n-1}, \mathfrak{A})]\!] = \mathbf{1}$.

4.2.4. Theorem. *Let $\mathfrak{A}$ be an algebraic system of signature $\sigma^{\wedge}$ inside $\mathbf{V}^{(B)}$. Then $\mathfrak{A}\downarrow$ is a universally complete algebraic B-system of signature σ. In this event,*

$$\chi \circ |\varphi|^{\mathfrak{A}\downarrow} = |\varphi|^{\mathfrak{A}}\downarrow$$

for each formula φ of signature σ.

$\triangleleft$ As we already know, $A\downarrow$ is a universally complete B-set. Further, the modified descent ν' of $\nu \in \mathbf{V}^{(B)}$ is a mapping with $\operatorname{dom}(\nu') = F \cup P$ (see 3.5.5 (3)). Furthermore,

$$[\![\nu'(f) : A^{\mathfrak{a}(f)^{\wedge}} \to A]\!] = \mathbf{1} \quad (f \in F),$$
$$[\![\nu'(p) : A^{\mathfrak{a}(p)^{\wedge}} \to \{0,1\}]\!] = \mathbf{1} \quad (p \in P).$$

By 3.2.6 (10) and 3.2.12, the above formulas show that $\nu'(f)\downarrow$ and $\nu'(p)\downarrow$ are contractive mappings from $(A\downarrow)^{\mathfrak{a}(f)}$ to $A\downarrow$ and from $(A\downarrow)^{\mathfrak{a}(p)}$ to $C := \{0,1\}^B\downarrow$, respectively. Hence, $(A\downarrow, \mu)$ is a universally complete algebraic B-system.

Assume now that φ is a formula of signature σ and show that

$$[\![|\varphi|^{\mathfrak{A}}\,(a_0, \dots, a_{n-1}) = 1]\!] = |\varphi|^{\mathfrak{A}\downarrow}(a_0, \dots, a_{n-1})$$

for all $a_0, \dots, a_{n-1} \in A\downarrow$. Using 3.2.12 and the definition of χ in 4.2.2, obtain

$$|\varphi|^{\mathfrak{A}\downarrow}(a_0, \dots, a_{n-1}) = [\![|\varphi|^{\mathfrak{A}}\downarrow(a_0, \dots, a_{n-1}) = 1]\!]$$
$$= \chi^{-1}(|\varphi|^{\mathfrak{A}}\downarrow(a_0, \dots, a_{n-1})),$$

which implies the claim.

Induct on the length of φ. At first, assume that φ is atomic. If $q \in P$ and $\mathfrak{a}(q) = 0$ then $[\![\nu(q^\wedge) = 0 \vee \nu(q^\wedge) = 1]\!] = \mathbf{1}$, so that $\nu'(q) \in C$ and $\mu(q) = \chi^{-1}(\nu'(q)) \in B$. By 4.2.2, $\mu(q) = [\![\chi \circ \mu(q) = 1]\!] = [\![1 = \nu(q^\wedge)]\!]$. Now consider the terms $t_0, \dots, t_{m-1}$ of signature σ which assume the values $b_0, \dots, b_{m-1}$ when the variables $x_0, \dots, x_{n-1}$ take the values $a_0, \dots, a_{n-1}$. Assume that $p \in P$ and $\mathfrak{a}(p) = m$. If $\varphi(x_0, \dots, x_{n-1}) := p(t_0, \dots, t_{m-1})$ then

$$[\![|\varphi|^{\mathfrak{A}}(a_0, \dots, a_{n-1}) = 1]\!] = [\![\nu{\downarrow}(p)(b_0, \dots, b_{m-1}) = 1]\!]$$
$$= [\![\chi \circ p^{\mu}(b_0, \dots, b_{m-1}) = 1]\!] = p^{\mu}(b_0, \dots, b_{m-1}).$$

Whereas if $\varphi(x_0, \dots, x_{n-1}) := (t_0(x_0, \dots, x_{n-1}) = t_1(x_0, \dots, x_{n-1}))$ then

$$[\![|\varphi|^{\mathfrak{A}}(a_0, \dots, a_{n-1}) = 1]\!] = [\![b_0 = b_1]\!] = d(b_0, b_1)^*.$$

Suppose now that φ_1 and φ_2 are $\varphi \wedge \psi$ and $(\forall x_0)\varphi$ while the claim is already demonstrated for φ and ψ. In this event,

$$[\![|\varphi_1|^{\mathfrak{A}}(a_0, \dots, a_{n-1}) = 1]\!]$$
$$= [\![|\varphi|^{\mathfrak{A}}(a_0, \dots, a_{n-1}) = 1 \wedge |\psi|^{\mathfrak{A}}(a_0, \dots, a_{n-1}) = 1]\!]$$
$$= [\![|\varphi|^{\mathfrak{A}}(a_0, \dots, a_{n-1}) = 1]\!] \wedge [\![|\psi|^{\mathfrak{A}}(a_0, \dots, a_{n-1}) = 1]\!]$$
$$= |\varphi_1|^{\mathfrak{A}\downarrow}(a_0, \dots, a_{n-1});$$
$$[\![|\varphi_2|^{\mathfrak{A}}(a_0, \dots, a_{n-1}) = 1]\!] = [\![(\forall x_0 \in A)|\varphi|^{\mathfrak{A}}(a_0, \dots, a_{n-1}) = 1]\!]$$
$$= \bigwedge_{a_0 \in A\downarrow} [\![|\varphi|^{\mathfrak{A}}(a_0, \dots, a_{n-1}) = 1]\!] = |\varphi_2|^{\mathfrak{A}\downarrow}(a_0, \dots, a_{n-1}).$$

The cases of the remaining propositional connectives are settled in much the same way. $\rhd$

4.2.5. Theorem. *Let $\mathfrak{A}$ and $\mathfrak{B}$ be algebraic systems of the same signature $\sigma^\wedge$ inside $\mathbf{V}^{(B)}$. Put $\mathfrak{A}' := \mathfrak{A}{\downarrow}$ and $\mathfrak{B}' := \mathfrak{B}{\downarrow}$. If h is a homomorphism (strong homomorphism) from $\mathfrak{A}$ to $\mathfrak{B}$ inside $\mathbf{V}^{(B)}$ then $h' := h{\downarrow}$ is a homomorphism (strong homomorphism) of the B-systems $\mathfrak{A}'$ and $\mathfrak{B}'$.*

Conversely, if $h' : \mathfrak{A}' \to \mathfrak{B}'$ is a homomorphism (strong homomorphism) of algebraic B-systems then $h := h'{\uparrow}$ is a homomorphism (strong homomorphism) from $\mathfrak{A}$ to $\mathfrak{B}$ inside $\mathbf{V}^{(B)}$.

$\lhd$ We confine exposition to substantiating 4.1.10 (3) of the definition of homomorphism; i.e., we will consider only the case of a nonnullary function symbol, since reasoning for the other symbols of signature σ proceeds by analogy.

Let $\mathfrak{A} := (A, \nu)^B$ for some A, $\nu \in \mathbf{V}^{(B)}$, and $\mathfrak{A}' = (A', \nu')$. Assume that $\mu \in \mathbf{V}^{(B)}$ and $\mu' \in \mathbf{V}$ are the interpretations of $\mathfrak{B}$ and $\mathfrak{B}'$, respectively. Consider a function symbol f of arity $n = \mathfrak{a}(f)$ and elements $a_0, \dots, a_{n-1} \in A'$.

As before, the record $t = g(a_0, \dots, a_{n-1})$ for $g \in \mathbf{V}^{(B)}$ denotes the formula $t = g(a)$ where $a \in \mathbf{V}^{(B)}$ is a member of $\mathbf{V}^{(B)}$ such that $[\![a : n^\wedge \to A]\!] = \mathbf{1}$ and $a{\downarrow}(l) = a_l$ $(l < n)$. If $h \in \mathbf{V}^{(B)}$ is a homomorphism from $\mathfrak{A}$ to $\mathfrak{B}$ inside $\mathbf{V}^{(B)}$ then

$$[\![h(\nu(f^\wedge)(a_0, \dots, a_{n-1})) = \mu(f^\wedge)(h(a_0), \dots, h(a_{n-1}))]\!] = \mathbf{1}.$$

Moreover, by the definition of descents (see 3.5.5 (3))

$$\begin{gathered}
[\![\nu(f^\wedge) = \nu{\downarrow}(f)]\!] = [\![\mu(f^\wedge) = \mu{\downarrow}(f)]\!] = \mathbf{1}; \\
[\![\nu{\downarrow}(f)(a_0, \dots, a_{n-1}) = \nu'(f)(a_0, \dots, a_{n-1})]\!] = \mathbf{1}; \\
[\![\mu{\downarrow}(f)(b_0, \dots, b_{n-1}) = \mu'(f)(b_0, \dots, b_{n-1})]\!] = \mathbf{1}; \\
[\![h(t) = h'(t)]\!] = \mathbf{1} \quad (t \in A').
\end{gathered}$$

Combining the above formulas and recalling that $\mathbf{V}^{(B)}$ is a separated universe, we obtain

$$h'(\nu'(f)(a_0, \dots, a_{n-1})) = \mu'(f)(h(a_0), \dots, h(a_{n-1})).$$

Assume conversely that the last equality holds. By replacing h' in it with $h := h'{\uparrow}$, we arrive at a true formula inside $\mathbf{V}^{(B)}$. Substituting in the latter consecutively $\nu'(f)$ for $\nu{\downarrow}(f)$, $\nu{\downarrow}(f)$ for $\nu(f^\wedge)$, $\mu'(f)$ for $\mu{\downarrow}(f)$, and $\mu{\downarrow}(f)$ for $\mu(f^\wedge)$, come to another true formula inside $\mathbf{V}^{(B)}$. It is this new formula that has the required property inside $\mathbf{V}^{(B)}$. $\triangleright$

Corollary. *In the notation of Theorem 4.2.5* $[\![h$ *is an isomorphism between the algebraic systems* $\mathfrak{A}$ *and* $\mathfrak{B}]\!] = \mathbf{1}$ *if and only if* h' *is an isomorphism between the algebraic* B*-systems* $\mathfrak{A}'$ *and* $\mathfrak{B}'$.

4.2.6. As noted in 4.1.3, a universally complete algebraic B-system $\mathfrak{A} := (A, \nu)$ can be viewed as a conventional (i.e., $\{0, 1\}$-valued) algebraic system $\mathfrak{A}' := (A, \nu')$ of the same signature provided that the B-valued predicates p^ν are replaced with the sets $\nu'(p) := \{(x_0, \dots, x_{n-1}) \in A^n : p^\nu(x) = \mathbf{1}\}$. This does not mean however that if $\mathfrak{A}$ is a B-model of an arbitrary formula φ of signature $\sigma(\mathfrak{A})$ then $\mathfrak{A}'$ is a $\{\mathbf{0}, \mathbf{1}\}$-valued model; i.e., a model in the conventional sense for the same formula φ. On the other hand, this phenomenon may take place for some formulas.

We elaborate the details in the section to follow. Now, we confine exposition to some concrete examples of algebraic B-systems obtainable by descent.

If a formula φ is the conjunction of the axioms of a group (a ring, a module, etc.) and the algebraic system $\mathfrak{A}$ is a two-valued model for φ then we adopt the usual practice of calling $\mathfrak{A}$ a group (a ring, a module, etc.). Whereas if $\mathfrak{A}$ is a B-model for φ then we say that $\mathfrak{A}$ is a B-group (a B-ring, a B-module, etc.).

Consider an arbitrary group G. An endomorphism $\pi : G \to G$ is a *projection* or *idempotent* whenever $\pi \circ \pi = \pi$. Say that $\mathscr{B}$ is a Boolean algebra of projections in G if $\mathscr{B}$ consists of mutually commuting projections in G and presents a Boolean algebra with zero $\mathbf{0}_{\mathscr{B}} := 0$ and unity $\mathbf{1}_{\mathscr{B}} := I_G$ under the operations:

$$\pi_1 \vee \pi_2 := \pi_1 + \pi_2 - \pi_1 \circ \pi_2,$$
$$\pi_1 \wedge \pi_2 := \pi_1 \circ \pi_2, \ \pi^* := 1 - \pi \quad (\pi_1, \pi_2, \pi \in \mathscr{B}).$$

The order on $\mathscr{B}$ is defined as follows: $\pi_1 \leq \pi_2$ if and only if $\pi_1(G) \subset \pi_2(G)$. We call the algebraic system $(G, \mathscr{B})$, as well as the underlying group G, a *group with projections* or a BAP-*group*. Given a BAP-group we refer to $\mathscr{B}$ as the *distinguished Boolean algebra* of projections of $(G, \mathscr{B})$ or G. A BAP-group $(G, \mathscr{B})$ is *universally complete* if $\mathscr{B}$ is a complete Boolean algebra and, to each family $(x_\xi) \subset G$ and each partition of unity $(\pi_\xi) \subset \mathscr{B}$, there is a unique element $x \in G$ such that $\pi_\xi x_\xi = \pi_\xi x$ for all ξ.

Let $(G, \mathscr{B})$ and $(G', \mathscr{B}')$ be BAP-groups. A group homomorphism $h : G \to G'$ is a BAP-*homomorphism* if there is a Boolean isomorphism $\jmath : \mathscr{B} \to \mathscr{B}'$ such that $h \circ \pi = \jmath(\pi) \circ h$ for all $\pi \in \mathscr{B}$.

Assume that R is a ring whose additive group has a distinguished Boolean algebra of projections $\mathscr{B}$. If, moreover, each projection $\pi \in \mathscr{B}$ is a ring homomorphism then $(R, \mathscr{B})$ is a BAP-*ring* or a *ring with projections*.

Given $x \in R$, call the projection $[x] := \bigwedge\{\pi \in \mathscr{B} : \pi x = x\}$ the *carrier* of x. It is obvious that if the carriers of $[x]$ and $[y]$ are disjoint (as elements of the Boolean algebra $\mathscr{B}$) then $x \cdot y = 0$, but the converse proposition fails in general. If $x \cdot y = 0$ then x and y are *orthogonal*. An element x of R is *regular* if x is orthogonal only to the zero of R. A *zero divisor* is each element orthogonal to some nonzero element.

A ring is *semiprime*, if it has no nonzero nilpotent ideals. Recall that an ideal $J \subset K$ is *nilpotent* if $J^n := \underbrace{J \cdot \ldots \cdot J}_{n \text{ times}} = \{0\}$ for some natural n.

Let S be a *multiplicative subset* of a unital ring K; i.e., $1 \in S$ and $xy \in S$ for all $x, y \in S$. Furnish the set $K \times S$ with an equivalence, by letting

$$(x, s) \sim (x', s') \leftrightarrow (\exists\, t \in S)(t(sx' - s'x) = 0).$$

Let $S^{-1}K := (K \times S)/\!\sim$, and $(x, s) \mapsto x/s$ be the canonical mapping. The set $S^{-1}K$ becomes a ring under the operations

$$(x/s) + (y/t) := (tx + sy)/st, \quad (x/s)(y/t) := (xy)/(st).$$

The mapping $x \mapsto x/1$ $(x \in K)$ is a homomorphism from K to $S^{-1}K$ called *canonical*. The ring $S^{-1}K$ is the *ring of fractions* or *ring of quotients* of K by S.

4.2.7. Theorem. *Let $\mathscr{G}$ be a group inside $\mathbf{V}^{(B)}$ and $G := \mathscr{G}{\downarrow}$. Then G is a BAP-group with distinguished complete Boolean algebra of projections $\mathscr{B}$ and there is an isomorphism $\jmath : B \xrightarrow{\text{onto}} \mathscr{B}$ such that*

$$b \leq [\![x = 0]\!] \leftrightarrow \jmath(b)x = 0 \quad (x \in G,\ b \in B).$$

Moreover, $(G, \mathscr{B})$ is a universally complete BAP*-group and the following hold:*

(1) $\mathbf{V}^{(B)} \models$ "*$\mathscr{G}$ is commutative*" $\leftrightarrow$ "*G is commutative*";

(2) $\mathbf{V}^{(B)} \models$ "*$\mathscr{G}$ is torsion-free*" $\leftrightarrow$ "*G is torsion-free.*"

$\vartriangleleft$ By Theorem 4.2.4, $\mathscr{G}{\downarrow}$ is a universally complete algebraic B-system; namely, a B-group. Denote the descent of $+$ by the same symbol. Show that G is a group. We confine demonstration to the existence of inverses.

Put $\varphi := (\forall\, x)(\exists!\, y)(x + y = 0)$. Then, by 4.1.8,

$$|\varphi|^G := \bigwedge_{x \in G} \bigvee_{y \in G} |x + y = 0|^G = \mathbf{1}.$$

Since G is a universally complete B-set, to every $x \in G$ there is some y in G such that

$$\mathbf{1} = |x + y = 0|^G = d(x + y, 0)^* = [\![x + y = 0]\!],$$

and so $x + y = 0$. If $x + z = 0$ for some $z \in G$ then $|x + z = 0|^G = \mathbf{1}$. Recalling that G is a B-group, note

$$\mathbf{1} = |x + y = 0 \wedge x + z = 0|^G \Rightarrow |y = z|^G.$$

Hence, $|y = z|^G = [\![z = y]\!] = \mathbf{1}$ and $z = y$.

The congruences of G are exactly the equivalences determined by its various normal subgroups. Therefore, by Theorem 4.1.6, there is an isomorphism $\jmath$ from B onto some complete Boolean algebra $\mathscr{B}'$ of normal subgroups of G such that

$$b \leq [\![x = 0]\!] \leftrightarrow x \in \jmath(b^*) \quad (b \in B,\ x \in G).$$

If $b \in B$ then $f(b) \cap f(b^*) = 0$. On the other hand, given $x \in G$, we may arrange $x_1 := \operatorname{mix}\{bx, b^*0\}$ and $x_2 := \operatorname{mix}\{b^*x, b0\}$. Since $b^* \leq [\![x_1 = 0]\!]$ and $b \leq [\![x_2 = 0]\!]$; therefore, $x_1 \in \jmath(b)$, $x_2 \in \jmath(b^*)$. Moreover, $[\![x = x_1 + x_2]\!] \geq [\![x_1 = x]\!] \wedge [\![x_2 = 0]\!] \geq b$ and $[\![x = x_1 + x_2]\!] \geq [\![x_1 = 0]\!] \wedge [\![x_2 = x]\!] \geq b^*$, which gives $x = x_1 + x_2$.

Therefore, each subgroup of the type $\jmath(b)$ is a summand of G to which there corresponds the projection π_b on $\jmath(b)$ along the complementary subgroup $\jmath(b^*)$. To be more exact, π_b is determined from the conditions: $\pi_b x = x$ for all $x \in \jmath(b)$ and $\pi_b x = 0$ for all $x \in \jmath(b^*)$. Let the same letter $\jmath$ stand for the isomorphism

$b \mapsto \pi_b\, (b \in B)$, and put $\mathscr{B} := \jmath(B)$. Obviously, $\mathscr{B}$ and $\jmath$ obey the required conditions. The universal completeness of G amounts to the same property of the underlying B-set. Indeed, $x = \operatorname{mix}(b_\xi x_\xi)$ if and only if $\jmath(b_\xi)x = \jmath(b_\xi)x_\xi$ for all ξ.

Assume now that $\mathscr{G}$ is torsion-free. Then

$$[\![(\exists\, x \in \mathscr{G})(\exists\, n \in \omega^\wedge)(nx = 0) \wedge (0 \neq x) \wedge (0 < n)]\!] = \mathbf{1}.$$

Hence, there are an element $0 \neq x \in G$ and a partition of unity $(b_n)_{n \in \omega}$ in B such that $b_n \leq [\![n^\wedge x = 0]\!]$ for all $n \in \omega$. Note that $[\![n^\wedge x = nx]\!] = \mathbf{1}$ and so $b_n \leq [\![x \neq 0]\!]$, $b_n \leq [\![nx = 0]\!]$, and $\jmath(b_n)(nx) = n\jmath(b_n)x = 0$.

The projection $\jmath(b_n)$ is nonzero for at least one $0 \neq n \in \omega$, which implies that G is not torsion-free. Conversely, if $nx = 0$ for some $0 \neq x \in G$ and $n \in \omega$ then $[\![n^\wedge x = 0]\!] = [\![nx = 0]\!] = \mathbf{1}$ and $[\![(\exists\, n \in \omega^\wedge)(nx = 0) \wedge (n > 0)]\!] = \mathbf{1}$; i.e., $[\![\mathscr{G}$ is not torsion-free$\,]\!] = \mathbf{1}$.

The claim about commutativity is obvious. $\triangleright$

4.2.8. Theorem. *Let $\mathscr{K}$ be a ring inside $\mathbf{V}^{(B)}$ and $K := \mathscr{K}{\downarrow}$. Then K is a universally complete* BAP*-ring with distinguished Boolean algebra of projections $\mathscr{B}$ and there is an isomorphism $\jmath : B \xrightarrow{\text{onto}} \mathscr{B}$ such that*

$$b \leq [\![x = 0]\!] \leftrightarrow \jmath(b)x = 0 \quad (x \in K,\ b \in B).$$

Moreover, the following hold:

(1) $\mathbf{V}^{(B)} \models$ *"$\mathscr{K}$ is commutative (semiprime)" $\leftrightarrow$ "K is commutative (semiprime)";*

(2) $\mathbf{V}^{(B)} \models$ *"$\mathscr{K}$ has no zero divisors" $\leftrightarrow$ "every two elements of K are orthogonal only if their carriers are disjoint";*

(3) $\mathbf{V}^{(B)} \models$ *"$\mathscr{S}$ is a multiplicative subset of $\mathscr{K}$" $\leftrightarrow$ "$S := \mathscr{S}{\downarrow}$ is a multiplicative subset in K"; moreover, $(\mathscr{S}^{-1}\mathscr{K}){\downarrow} \simeq S^{-1}K$ (with $\simeq$ standing for a ring isomorphism);*

(4) $\mathbf{V}^{(B)} \models$ *"$\mathscr{K}$ is a field" $\leftrightarrow$ "K is semiprime, the orthogonality of the elements of K is equivalent to the disjointness of their carriers, and every regular element in K is invertible";*

(5) $\mathbf{V}^{(B)} \models$ *"$\mathfrak{R}$ is the radical of the unital ring $\mathscr{K}$" $\leftrightarrow$ "$\mathfrak{R}{\downarrow}$ is the radical of the unital ring K"; in other words, if $\mathscr{K}$ has unity then $\mathfrak{R}(\mathscr{K}){\downarrow} = \mathfrak{R}(K)$;*

(6) $\mathbf{V}^{(B)} \models$ *"$(\mathscr{K}, \mathscr{D})$ is a* BAP*-ring" $\leftrightarrow$ "the mapping $\pi \mapsto \pi{\downarrow}$ $(\pi \in \mathscr{D}{\downarrow})$ is an isomorphism of $\mathscr{D}{\downarrow}$ onto some Boolean algebra of projections D of K, in which case $\mathscr{B}$ is a regular subalgebra in D and (K, D) is a* BAP*-ring."*

$\triangleleft$ By Theorem 4.2.7, K is a universally complete BAP-group, and there is an isomorphism $\jmath$ from B onto a complete Boolean algebra $\mathscr{B}$ of additive projections obeying the necessary condition. Furnish K with multiplication by appealing to the general definition of 4.2.3: Given x, $y \in K$, note that $[\![x, y \in \mathscr{K}]\!] = \mathbf{1}$; and so to x and y in K there corresponds their product z in $\mathbf{V}^{(B)}$ which satisfies $[\![z \in \mathscr{K}]\!] = [\![z = x \cdot y]\!] = \mathbf{1}$. We let z be the product of x and y in K. Therefore,

$$z = x \cdot y \leftrightarrow [\![z = x \cdot y]\!] = \mathbf{1} \quad (x, y, z \in K).$$

Using Theorem 4.2.4, we easily see that K becomes a ring. Take an arbitrary element b in B and show that the projection $\jmath(b)$ is a ring homomorphism. Indeed, the multiplication of K, as the descent of an operation on $\mathscr{K}$, is extensional and so it preserves mixing. Therefore, by the definition of $\jmath(b)$ (see 4.2.7), given x, $y \in K$, we find

$$\begin{aligned}\jmath(b)xy &= \operatorname{mix}\{bxy, b^*0\}\\ &= \operatorname{mix}\{bx, b^*0\} \cdot \operatorname{mix}\{by, b^*0\} = \jmath(b)x \cdot \jmath(b)y.\end{aligned}$$

We now turn to demonstrating (1)–(6).

(1) The proof proceeds by analogy with 4.2.7 (1).

(2) The proposition $\mathbf{V}^{(B)} \models$ "$\mathscr{K}$ has no zero divisors" is equivalent to the fact that $b := [\![xy = 0]\!] = [\![x = 0]\!] \vee [\![y = 0]\!]$ for all x and y in $\mathscr{K}{\downarrow}$. If the last formula is fulfilled and $xy = 0$ then $b = \mathbf{1}$. Hence, letting $e := [\![x = 0]\!]$ and $c := [\![y = 0]\!]$, note that $e^* \wedge c^* = 0$. Moreover, $\jmath(e^*)x = x$ and $\jmath(c^*)y = y$. Therefore, $[x] \leq \jmath(e^*)$ and $[y] \leq \jmath(c^*)$. Clearly, the carriers $[x]$ and $[y]$ are disjoint. If, however, $[x] \circ [y] = 0$ then, as was mentioned in 4.2.6, $x \cdot y = 0$. Conversely, assume that the equality $xy = 0$ is equivalent to the fact that the carriers $[x]$ and $[y]$ are disjoint. Then for $b := [\![xy = 0]\!]$ the equalities $0 = \jmath(b)xy = (\jmath(b)x) \cdot (\jmath(b)y)$ yield that the projections $\pi := [\jmath(b)x]$ and $\rho := [f(b)y]$ are disjoint. Note that $\jmath(b) \circ \pi^* x = 0$ and $\jmath(b) \circ \rho^* y = 0$ and so

$$[\![x = 0]\!] \vee [\![y = 0]\!] \geq (b \wedge f^{-1}(\pi^*)) \vee (b \wedge \jmath^{-1}(\rho^*)) = b.$$

(3) The claim about multiplicativity is evident. Prove that the descent of a ring of fractions is a ring of fractions. Note first that $(\mathscr{S} \times \mathscr{K}){\downarrow} = S \times K$. Consider an equivalence relation $\mathscr{P} \in \mathbf{V}^{(B)}$ such that, for x, $x' \in K$ and s, $s' \in S$, we have

$$\mathbf{V}^{(B)} \models (x, s)\mathscr{P}(x', s') \leftrightarrow (\exists t \in \mathscr{S})(t(sx' - s'x) = 0).$$

If $P := \mathscr{P}{\downarrow}$ then P is an equivalence relation in $K \times S$, in which case

$$(x, s)P(x', s') \leftrightarrow (\exists t \in S)\,(t(sx' - s'x) = 0).$$

Then the descent of the factor set $(\mathscr{S}\times\mathscr{K})/\mathscr{P}$ is bijective with the set $(KS\times K)/P$. Finally, for $x, y \in K$ and $s, t \in S$, the formulas

$$(x/s) + (y/t) = (tx + sy)/st, \quad (x/s)(y/t) = (xy/st)$$

hold if and only if they are satisfied inside $\mathbf{V}^{(B)}$. All we have to do now is to compare this with the definition of ring of fractions.

(4) Assume that $[\![\mathscr{K}$ is a field $]\!] = \mathbf{1}$. In this case K is semiprime and $xy = 0$ yields that $[x] \circ [y] = 0$ for all x and y in K (see (1) and (2)). Given a regular element $x \in K$, note that $\jmath(b)xy = 0 \to \jmath(b)y = 0$ for all $b \in B$ and $y \in K$. However, $[\![xy = 0]\!] \leq [\![y = 0]\!]$; i.e., $[\![x \neq 0]\!] = \mathbf{1}$. Consequently, there is an element $u \in K$ such that $[\![xu = ux = 1]\!] = \mathbf{1}$. Hence, $xu = ux = \mathbf{1}$; i.e., x is invertible in K.

Conversely, assume that K is semiprime, every regular element in K is invertible, and the orthogonality of two elements in K is equivalent to the disjointness of their carriers. Then $\mathbf{V}^{(B)} \models$ "$\mathscr{K}$ is a commutative ring." Hence, $[\![\mathscr{K}$ is a field$]\!] = [\![(\forall x)(x \in \mathscr{K} \wedge x \neq 0 \to (x \text{ is invertible}))]\!] = \bigwedge\{[\![(\exists z)(z = x^{-1})]\!] : x \in K \wedge [\![x \neq 0]\!] = \mathbf{1}\}$. Therefore, it suffices to show that if $[\![x \neq 0]\!] = \mathbf{1}$, then $[\![x$ is invertible$]\!] = \mathbf{1}$ for all $x \in K$. Assume that $[\![x \neq 0]\!] = \mathbf{1}$ and $xy = 0$ for some $y \in K$. Then, putting $\pi := [x]$ and $\rho := [y]$, note $\pi \circ \rho = 0$. On the other hand, $\jmath(b)x = 0$ implies $b \leq [\![x = 0]\!] = [\![x \neq 0]\!]^* = \mathbf{1}^* = \mathbf{0}$, and so $\rho := \jmath(1) = I_K$. Therefore, $\pi \leq \rho^* = 0$ or $y = 0$. Hence, x is an invertible element of K. Whence it is immediate that $[\![x$ is invertible in $\mathscr{K}]\!] = \mathbf{1}$.

(5) An element x belongs to the radical of a ring if and only if for each y the element $1 - yx$ is left-invertible. Now, we have to note that $1 - yx$ is left-invertible in K if and only if $[\![1 - yx$ is left-invertible in $\mathscr{K}]\!] = \mathbf{1}$.

(6) If $[\![(\mathscr{K}, \mathscr{D})$ is a BAP-ring$]\!] = \mathbf{1}$ and $\pi \in \mathscr{D}{\downarrow}$ then, by 4.2.7, $\pi{\downarrow} : K \to K$ is a homomorphism. On the other hand, $[\![\pi \circ \pi = \pi]\!] = \mathbf{1}$. Hence, $(\pi{\downarrow}) \circ (\pi{\downarrow}) = (\pi \circ \pi){\downarrow} = \pi{\downarrow}$; i.e., $\pi{\downarrow}$ is a projection.

The fact that D is a Boolean algebra will be established in 4.2.9. Therefore, (K, D) is a BAP-ring. By definition, $\mathscr{B} = \{\pi{\downarrow} : \pi \in \{\mathbf{0}_{\mathscr{D}}, \mathbf{1}_{\mathscr{D}}\}^B{\downarrow}\}$ (see 4.2.7). Hence, $\mathscr{B} \subset D$. The converse implication is established by analogy. ▷

4.2.9. Theorem. *Let $\mathscr{D}$ be a complete Boolean algebra inside $\mathbf{V}^{(B)}$ and $D := \mathscr{D}{\downarrow}$. Then D is a complete Boolean algebra and there exists a complete monomorphism $\imath : B \to D$ such that*

$$b \leq [\![x \leq y]\!] \leftrightarrow \imath(b)x \leq \imath(b)y$$

for all $x, y \in D$ and $b \in B$.

◁ By virtue of 4.2.4, D is a universally complete algebraic B-system of signature $(\vee, \wedge, *, 0, 1)$. The fact that D is a Boolean algebra also follows from 4.2.4.

We temporarily denote the Boolean operations in D by $\widetilde{\vee}$, $\widetilde{\wedge}$ and check only distributivity to give an example of reasoning.

Take the terms $t_1(x,y,z) := (x \wedge y) \vee z$ and $t_2(x,y,z) := (x \vee z) \wedge (x \vee y)$. Consider the formula $\psi := (\forall\, x)(\forall\, y)(\forall\, z)\varphi(x,y,z)$ where $\varphi(x,y,z) := (t_1(x,y,z) = t_2(x,y,z))$. Using 4.2.4, note that

$$\mathbf{1} = [\![|\psi|^{\mathscr{D}} = 1]\!] = |\psi|^D = \bigwedge_{a,b,c \in D} |\varphi|^D(a,b,c),$$

and so $|\varphi|^D(a,b,c) = \mathbf{1}$ for all $a,b,c \in D$. Furthermore,

$$\begin{aligned}\mathbf{1} &= |\varphi|^D(a,b,c) = d(t_1(a,b,c), t_2(a,b,c))^* \\ &= [\![t_1(a,b,c) = t_2(a,b,c)]\!] = [\![(a\widetilde{\wedge}b)\widetilde{\vee}c = (a\widetilde{\vee}c)\widetilde{\wedge}(b\widetilde{\vee}c)]\!].\end{aligned}$$

Since $\mathbf{V}^{(B)}$ is separated, we thus obtain $(a\widetilde{\wedge}b)\widetilde{\vee}c = (a\widetilde{\vee}c)\widetilde{\wedge}(b\widetilde{\vee}c)$. In much the same way we demonstrate the remaining axioms of Boolean algebras. Therefore, D is a Boolean algebra.

The completeness of D is not expressible by a bounded formula. Consequently, the above approach is inapplicable, and so we proceed otherwise.

Let $\leq \in \mathbf{V}^{(B)}$ stand for the conventional order relation on $\mathscr{D}$; i.e.,

$$\mathbf{V}^{(B)} \models (\forall\, x \in \mathscr{D})(\forall\, y \in \mathscr{D})(x \leq y \leftrightarrow x \wedge y = x).$$

Put $\widetilde{\leq} := (\leq){\downarrow}$. Given $x,y \in D$, observe then that $x \widetilde{\leq} y$ if and only if $x\widetilde{\wedge}y = x$. Consider the correspondence $\Phi := (\widetilde{\leq}, D, D)$. It is obvious that Φ is fully contractive. Recall that if $A \subset D$ then $\pi_\Phi(A)$ $(\pi_\Phi^{-1}(A))$ is the set of all upper (lower) bounds of A (with respect to the order $\widetilde{\leq}$). Therefore,

$$\sup(A) = \pi_\Phi(A) \cap \pi_\Phi^{-1}(\pi_\Phi(A))$$

provided that $\sup(A)$ exists. If $\Psi := (\leq, \mathscr{D}, \mathscr{D})^B$ then Ψ is a correspondence inside $\mathbf{V}^{(B)}$ and $\Phi = \Psi{\downarrow}$. Since $\mathscr{D}$ is complete, there is an element $a \in D$ such that $[\![a = \sup(A)]\!] = \mathbf{1}$ or $[\![\pi_\Psi(A) \cap \pi_\Psi^{-1}(\pi_\Psi(A)) = a]\!] = \mathbf{1}$. Employing the rule for descending polars (cf. 3.2.13 (2)), carry out the simple calculations

$$\begin{aligned}a &= (\pi_\Psi^{-1}(\pi_\Psi(A{\uparrow})) \cap \pi_\Psi(A{\uparrow})){\downarrow} \\ &= \pi_\Psi^{-1}(\pi_\Psi(A{\uparrow}{\downarrow})) \cap \pi_\Phi(A{\uparrow}{\downarrow}) = \sup(\mathrm{mix}(A)) = \sup(A).\end{aligned}$$

Therefore, $a = \sup(A)$, and so D is complete.

Let $\lambda \in \mathbf{V}^{(B)}$ be the identical embedding of the algebra $\{\mathbf{0}_{\mathscr{D}}, \mathbf{1}_{\mathscr{D}}\}^B$ in $\mathscr{D}$ inside $\mathbf{V}^{(B)}$. Put $\imath_1 = \lambda{\downarrow}$ and $\imath := \imath_1 \circ \imath_2$, where $\imath_2$ is an isomorphism B on $\{\mathbf{0}_{\mathscr{D}}, \mathbf{1}_{\mathscr{D}}\}^B{\downarrow}$.

In this case $\imath$ is a monomorphism. The monomorphism $\imath$ is complete. Indeed, if $A \subset B$ then $\imath(\pi_{\Phi'}(A)) \subset \pi_{\Phi}(\imath(A))$, where $\Phi' := \imath^{-1} \circ \Phi \circ \imath$. Furthermore, using the obvious relation

$$\mathbf{V}^{(B)} \models (\forall\, x, y \in \mathscr{D})(\forall\, c \in \{\mathbf{0}_{\mathscr{D}}, \mathbf{1}_{\mathscr{D}}\})$$
$$(\lambda(c)x = \lambda(c)y \leftrightarrow (c = \mathbf{0}_{\mathscr{D}}) \vee (c = \mathbf{1}_{\mathscr{D}} \wedge x = y))$$

and given $x, y \in D$ and $b \in B$, note that

$$[\![\imath(b)x = \imath(b)y]\!] = b^* \vee (b \wedge [\![x = y]\!]).$$

Hence,

$$\imath(b)x = \imath(b)y \leftrightarrow b \leq [\![x = y]\!],$$

and so

$$d(x, y)^* = [\![x = y]\!] = \bigvee \{b \in B : \imath(b)x = \imath(b)y\}.$$

It is now evident that if $\varphi(x, y) := x \leq y$ then

$$|\varphi|^D(x, y) = \bigvee \{b \in B : \imath(b)x \leq \imath(b)y\},$$
$$[\![|\varphi|^{\mathscr{D}}(x, y) = \mathbf{1}]\!] = [\![x \leq y]\!],$$

which yields the equivalence in question. ▷

4.2.10. We now list a few corollaries for BAP-rings and Boolean algebras whose proofs are in fact implicit in 4.2.5, 4.2.7, 4.2.8, and 4.3.2.

Given BAP-rings K_1 and K_2, assume that $\jmath_1$ and $\jmath_2$ are isomorphisms of B to the distinguished Boolean algebras of K_1 and K_2 respectively.

A homomorphism $h : K_1 \to K_2$ is *B-homogeneous* if $h \circ \jmath_1(b) = \jmath_2(b) \circ h$ $(b \in B)$. We also say in this event that K_1 is a BAP*-ring with distinguished algebra* B and h commutes with the members of B.

(1) Theorem. *Let $\mathscr{K}_1$ and $\mathscr{K}_2$ be* BAP*-rings with distinguished algebra $\mathscr{D}$ inside $\mathbf{V}^{(B)}$. Put $D := \mathscr{D}{\downarrow}$ and $K_l := \mathscr{K}_l{\downarrow}$ for $l := 1, 2$. Then K_1 and K_2 are* BAP*-rings with distinguished algebra D.*

Moreover, if h is a homomorphism from $\mathscr{K}_1$ to $\mathscr{K}_2$ commuting with the members of $\mathscr{D}$ inside $\mathbf{V}^{(B)}$, then $h{\downarrow}$ is a homomorphism from K_1 to K_2 commuting with the members of D. If h is an isomorphism between $\mathscr{K}_1$ and $\mathscr{K}_2$ then $h{\downarrow}$ is an isomorphism between K_1 and K_2.

(2) Theorem. *Let $\mathscr{D}_1$ and $\mathscr{D}_2$ be complete Boolean algebras inside $\mathbf{V}^{(B)}$. Put $D_k := \mathscr{D}_k\downarrow$ and let $\imath_k : B \to D_k$ symbolize the canonical monomorphism for $k = 1, 2$ (cf. 4.2.9). If $h \in \mathbf{V}^{(B)}$ is an isomorphism of $\mathscr{D}_1$ onto $\mathscr{D}_2$ inside $\mathbf{V}^{(B)}$ then there is an isomorphism H of D_1 onto D_2 such that the following diagram commutes:*

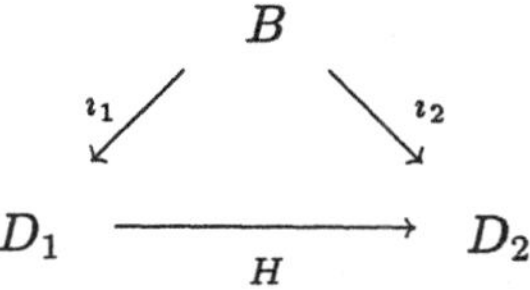

Conversely, if $H : D_1 \to D_2$ is an isomorphism of D_1 onto D_2 making the above diagram commutative then $\mathscr{D}_1$ and $\mathscr{D}_2$ are isomorphic Boolean algebras inside $\mathbf{V}^{(B)}$.

4.3. Immersion of Algebraic B-Systems

In the present section the immersion functor of Section 3.4 is extended to the category of algebraic B-systems.

4.3.1. Let $\mathfrak{A} := (A, \nu)$ be an algebraic B-system of signature $\sigma := (F, P, \mathfrak{a})$. Consider a mapping $\nu' : F \cup P \to \mathbf{V}^{(B)}$ acting by the rule

$$\nu' : s \mapsto \nu(s)^{\sim} := \mathscr{F}^{\sim}(\nu(s)) \quad (s \in F \cup P),$$

where $\mathscr{F}^{\sim}$ is the descent functor (cf. 3.4.12–3.4.16). In accordance with the general definition of immersion for correspondences (cf. 3.4.13), to each $f \in F$, $\mathfrak{a}(f) = n$, there corresponds the mapping $\lambda'(f) : (A^{\sim})^{n^{\wedge}} \to A^{\sim}$ inside $\mathbf{V}^{(B)}$ defined by the formula

$$[\![\nu'(f)(\imath_A(x_0), \dots, \imath_A(x_{n-1})) = \imath_A \circ \nu(f)(x_0, \dots, x_{n-1})]\!] = \mathbf{1},$$

where $\imath_A$ is the canonical embedding of A to $A' := A^{\sim}\downarrow$ (see 3.5.4). Analogously, for $p \in P$, $\mathfrak{a}(p) = m$, the element $\nu'(p) \in \mathbf{V}^{(B)}$ is the mapping from $(A^{\sim})^{m^{\wedge}}$ to $\{0, 1\}^B \in \mathbf{V}^{(B)}$ acting by the rule

$$[\![\nu'(p)(\imath_A(x_0), \dots, \imath_A(x_{m-1})) = \imath_B \circ \nu(p)(x_0, \dots, x_{m-1})]\!] = \mathbf{1}.$$

The modified ascent $\mu := (\nu')\Uparrow$ of $\nu' : F \cup P \to \operatorname{im}(\nu')$ is clearly an interpretation inside $\mathbf{V}^{(B)}$.

Given an algebraic B-system $\mathfrak{A}$, call the pair $(A^{\sim}, \mu)$ or the element $(A^{\sim}, \mu)^B \in \mathbf{V}^{(B)}$ the *Boolean valued representation of* $\mathfrak{A}$ and denote it by $\mathfrak{A}^{\sim}$.

4.3.2. Theorem. *For each algebraic B-system $\mathfrak{A}$ of signature σ the Boolean valued representation $\mathfrak{A}^{\sim}$ of $\mathfrak{A}$ is an algebraic system of signature $\sigma^{\wedge}$ inside $\mathbf{V}^{(B)}$.*

Moreover,

$$|\varphi|^{\mathfrak{A}}(a_0,\dots,a_{n-1}) = [\![|\varphi|^{\mathfrak{A}^{\sim}}(\imath_A(a_0),\dots,\imath_A(a_{n-1})) = 1]\!]$$

for every formula φ of signature σ with n free variables and all $a_0,\dots,a_{n-1} \in |\mathfrak{A}|$.

$\lhd$ Recall that, considering an arbitrary set U as a B-set, we imply the discrete B-metric on U. Therefore, $\sigma^{\sim} = \sigma^{\wedge}$ (see 3.4.12). By 3.5.5,

$$\mathbf{V}^{(B)} \models \text{“}\mu \text{ is a function with } \operatorname{dom}(\mu) = F^{\wedge} \cup P^{\wedge}.\text{”}$$

Let A stand for $|\mathfrak{A}|$. By Theorem 3.4.14, $\mathbf{V}^{(B)} \models$ “$\mu(f^{\wedge})$ is a mapping from $(A^{\sim})^{\mathfrak{a}(f)^{\wedge}}$ to $A^{\sim}$” for all $f \in F$, and $\mathbf{V}^{(B)} \models$ “$\mu(p)$ is a mapping from $(A^{\sim})^{\mathfrak{a}(p)^{\wedge}}$ to $\{0,1\}$” for every $p \in P$. Hence, $\mathbf{V}^{(B)} \models$ “$\mathfrak{A}^{\sim}$ is an algebraic system of signature $\sigma^{\wedge}$.”

Consider a formula φ of signature σ. By Theorem 3.5.5 (3), granted $f \in F$ and $p \in P$, observe

$$\imath_A \circ f^{\nu}(a_0,\dots,a_{n-1}) = \mu(f^{\wedge})\!\downarrow\!(\imath_A(a_0),\dots,\imath_A(a_{n-1})) \quad (a_l \in A),$$
$$\imath_B \circ p^{\nu}(a_0,\dots,a_{n-1}) = \mu(p^{\wedge})\!\downarrow\!(\imath_A(a_0),\dots,\imath_A(a_{n-1})) \quad (a_l \in A).$$

Using the above equalities and inducting on the length of φ, deduce

$$|\varphi|^{\mathfrak{A}}(a_0,\dots,a_{n-1}) = |\varphi|^{\mathfrak{A}'}(\imath_A(a_0),\dots,\imath_A(a_{n-1})) \quad (a_0,\dots,a_{n-1} \in A),$$

with $\mathfrak{A}' := \mathfrak{A}^{\sim}\!\downarrow$. To complete the proof, appeal to Theorem 4.2.4. $\rhd$

4.3.3. Theorem. *Let $\mathfrak{A} := (A,\nu)$ be an algebraic B-system of signature σ. Then there are $\mathscr{A}$ and $\mu \in \mathbf{V}^{(B)}$ such that the following are fulfilled:*

(1) $\mathbf{V}^{(B)} \models$ *“$(\mathscr{A},\mu)$ is an algebraic system of signature $\sigma^{\wedge}$”;*

(2) *If $\mathfrak{A}' := (A',\nu')$ is the descent of $(\mathscr{A},\mu)$ then $\mathfrak{A}'$ is a universally complete algebraic B-system of signature σ;*

(3) *There is an isomorphism $\imath$ from $\mathfrak{A}$ to $\mathfrak{A}'$ such that $A' = \operatorname{mix}(\imath(A))$;*

(4) *For every formula φ of signature σ in n free variables, the equalities hold*

$$|\varphi|^{\mathfrak{A}}(a_0,\dots,a_{n-1}) = |\varphi|^{\mathfrak{A}'}(\imath(a_0),\dots,\imath(a_{n-1}))$$
$$= \chi^{-1} \circ (|\varphi|^{\mathfrak{A}^{\sim}})\!\downarrow\!(\imath(a_0),\dots,\imath(a_{n-1}))$$

for all $a_0,\dots,a_{n-1} \in A$, where χ is the same as in 4.2.2.

$\lhd$ Put $\mathscr{A} := A^{\sim}$ and $\imath := \imath_A$. Define μ as in 4.3.1. Now, all claims ensue from 3.5.5 (3), 4.2.4, and 4.3.2. $\rhd$

4.3.4. Theorem. *Consider algebraic B-systems $\mathfrak{A}$ and $\mathfrak{B}$ of the same signature.*

(1) *Let h be a contractive mapping from $|\mathfrak{A}|$ to $|\mathfrak{B}|$. Then h is a homomorphism (strong homomorphism, or isomorphism) if and only if $\mathbf{V}^{(B)} \models$ "$h^\sim$ is a homomorphism (strong homomorphism, or isomorphism) from $\mathfrak{A}^\sim$ to $\mathfrak{B}^\sim$." A homomorphism $h^\sim$ is surjective inside $\mathbf{V}^{(B)}$ if and only if $|\mathfrak{B}| = \operatorname{mix}(h(|\mathfrak{A}|))$.*

(2) *Assume that $g \in \mathbf{V}^{(B)}$ and $\mathbf{V}^{(B)} \models$ "$g : \mathfrak{A}^\sim \to \mathfrak{B}^\sim$ is a homomorphism of algebraic B-systems." If $\mathfrak{B}$ is a universally complete algebraic B-system then there is a unique homomorphism $h : \mathfrak{A} \to \mathfrak{B}$ such that $g = h^\sim$.*

$\triangleleft$ (1) If $h' := h^\sim{\downarrow}$, $\mathfrak{A}' := \mathfrak{A}^\sim{\downarrow}$, $\mathfrak{B}' := \mathfrak{B}^\sim{\downarrow}$, $\imath := \imath_{|\mathfrak{A}|}$, and $\jmath := \imath_{|\mathfrak{B}|}$; then $h' \circ \imath = \jmath \circ h$ (cf. 3.5.4 (3)).

Show now that h is a homomorphism if and only if h' is a homomorphism. We agree to confine exposition to demonstrating 4.1.10 (3) with $n = 1$. In other words, we will demonstrate that h and h' preserve or fail to preserve unary operations simultaneously.

To this end, let ν, λ, $\mu(\nu)$, and $\mu(\lambda)$ be the interpretations of the systems $\mathfrak{A}$, $\mathfrak{B}$, $\mathfrak{A}^\sim$, and $\mathfrak{B}^\sim$. If h is a homomorphism then $h \circ f^\nu = f^\lambda \circ h$. Moreover, $\imath \circ f^\nu = (f^{\mu(\nu)}{\downarrow}) \circ \imath$ and $\jmath \circ f^\lambda = (f^{\mu(\lambda)}{\downarrow}) \circ \jmath$. Hence,

$$h' \circ (f^{\mu(\nu)}{\downarrow}) \circ \imath = \jmath \circ h \circ f^\nu = \jmath \circ f^\lambda \circ h = (f^{\mu(\lambda)}{\downarrow}) \circ h' \circ \imath.$$

Using the equality $|\mathfrak{A}^\sim{\downarrow}| = \operatorname{mix}(\imath(|\mathfrak{A}|))$, obtain $h' \circ (f^{\mu(\nu)}{\downarrow}) = (f^{\mu(\lambda)}{\downarrow}) \circ h'$. Conversely, if the last equality holds then, reasoning in the opposite direction, we find $h \circ f^\nu = f^\lambda \circ h$. The case of an arbitrary operation, as well as that of an arbitrary predicate, is more cumbersome but causes no principal difficulties. Consequently, h is a homomorphism, a strong homomorphism, or an isomorphism between $\mathfrak{A}$ and $\mathfrak{B}$ if and only if the mapping h' from $\mathfrak{A}'$ to $\mathfrak{B}'$ has the corresponding property. Therefore, all claims follow from 4.2.5 and 4.3.3.

(2) If $\mathfrak{A}$ is a universally complete algebraic system then the claims ensue from 3.5.8 (4). The general case is settled on appealing to 3.5.8 (2) at the beginning of the proof. The sought homomorphism has the shape $h := \jmath^{-1} \circ (g{\downarrow}) \circ \imath$. $\triangleright$

4.3.5. Note some corollaries to Theorems 4.3.3 and 4.3.4.

(1) Theorem. *If $\mathfrak{A}$ is an algebraic system of finite signature σ then $\mathbf{V}^{(B)} \models$ "$\mathfrak{A}^\wedge$ is an algebraic system of signature $\sigma^\wedge$." Moreover,*

$$\mathfrak{A} \models \varphi(a_0, \dots, a_{n-1}) \leftrightarrow [\![\mathfrak{A}^\wedge \models \varphi(a_0^\wedge, \dots, a_{n-1}^\wedge)]\!] = \mathbf{1}$$

for a formula φ of signature σ in n free variables and all $a_0, \dots, a_{n-1} \in A$.

◁ To demonstrate, it suffices to note that in case $\mathfrak{A} := (A, f_0, \dots, f_{k-1}, p_0, \dots, p_{m-1})$, the proposition $\mathfrak{A} \models \varphi(a_0, \dots, a_{n-1})$ is as a bounded set-theoretic formula $\psi(A^\wedge, f_0^\wedge, \dots, f_n^\wedge, p_0^\wedge, \dots, p_{m-1}^\wedge, a_0^\wedge, \dots, a_{n-1}^\wedge)$. Reference to 2.2.9 completes the proof. ▷

(2) Theorem. *To each algebraic B-system $\mathfrak{A}$ there are a universally complete algebraic B-system $\mathfrak{A}'$ of signature $\sigma(\mathfrak{A})$ and an isomorphism $\imath$ from $\mathfrak{A}$ to $\mathfrak{A}'$ such that*

(a) $|\mathfrak{A}'| = \mathrm{mix}(\imath(|\mathfrak{A}|))$;

(b) *if h is a homomorphism from $\mathfrak{A}$ to a universally complete algebraic B-system $\mathfrak{B}$ then there is a unique homomorphism $h' : \mathfrak{A}' \to \mathfrak{B}$ such that $h' \circ \imath = h$;*

(c) *if $\mathfrak{A}''$ is a universally complete algebraic B-system, and a homomorphism $\imath' : \mathfrak{A} \to \mathfrak{A}''$ obeys* (a) *with $\mathfrak{A}'$ substituted for $\mathfrak{A}''$; then there is a unique isomorphism h from $\mathfrak{A}'$ onto $\mathfrak{A}''$ such that $h \circ \imath = \imath'$.*

◁ Let $(\mathscr{A}, \mu)$ be the Boolean valued representation of $\mathfrak{A}$. Then the descent $\mathfrak{A}' := (\mathscr{A}, \mu){\downarrow}$ of $\mathfrak{A}$ obeys all requirements. Indeed, by 4.3.3 (3, 4) the canonical embedding $\imath := \imath_{|\mathfrak{A}|}$ is an isomorphism satisfying (a). If h and $\mathfrak{B}$ are the same as in (b) then, by Theorem 4.3.4, $g := h^{\sim}{\downarrow}$ is a homomorphism from $\mathfrak{A}'$ to $\mathfrak{B}' := \mathfrak{B}^{\sim}{\downarrow}$. Since $\mathfrak{B}$ is universally complete, the canonical mapping $\jmath := \imath_{|\mathfrak{B}|}$ is an isomorphism "onto." Obviously, $h' := \jmath^{-1} \circ g$ is a sought homomorphism. It stands to reason to remark that if $a \in |\mathfrak{A}'|$ and $a = \mathrm{mix}(b_\xi \imath(a_\xi))$ then $h'(a) = \mathrm{mix}(b_\xi h \circ \imath(a_\xi))$. The claim (c) results now from (a) and Theorem 4.3.4. ▷

Each pair $(\mathfrak{A}', \imath)$, where $\mathfrak{A}'$ is a universally complete algebraic B-system and $\imath$ is an isomorphism from $\mathfrak{A}$ to $\mathfrak{A}'$ obeying (a) of Theorem (2), is naturally called a *universal completion* of $\mathfrak{A}$. Consequently, Theorem (2) yields the following:

(3) *Each algebraic B-system has a universal completion unique up to isomorphism.*

Take a complete homomorphism π from B to a complete Boolean algebra C. Let $\mathfrak{A} := (A, f_0, \dots, f_{k-1}, p_0, \dots, p_{m-1})$ be an algebraic system of finite signature inside $\mathbf{V}^{(B)}$. Assign

$$\pi^*(\mathfrak{A}) := (\pi^*(A), \pi^*(f_0), \dots, \pi^*(p_{m-1}))^C, \quad \pi^*(\mathfrak{A}) \in \mathbf{V}^{(C)},$$

where $\pi^* : \mathbf{V}^{(B)} \to \mathbf{V}^{(C)}$ is the mapping associated with π (cf. Section 2.2).

As usual, these facts enable us to speak about *the* universal completion of an algebraic B-system (cf. 1.1.6 (7)).

(4) Theorem. *The element $\pi^*(\mathfrak{A})$ is an algebraic system of finite signature $\sigma(\mathfrak{A})$ inside $\mathbf{V}^{(C)}$. The mapping $a \mapsto \pi^*(a)$ $(a \in A{\downarrow})$ is a homomorphism from $\mathfrak{A}{\downarrow}$*

to $\pi^*(\mathfrak{A})\downarrow$. For each formula φ of signature $\sigma(\mathfrak{A})$ with n free variables and for all $a_0, \dots, a_{n-1} \in |\mathfrak{A}\downarrow|$, the following holds

$$\mathfrak{A}\downarrow \models \varphi(a_0, \dots, a_{n-1}) \to \pi^*(\mathfrak{A})\downarrow \models \varphi(\pi^*(a_0), \dots, \pi^*(a_{n-1})).$$

In particular, if $\mathfrak{B}$ is an algebraic B-system of finite signature and $\mathfrak{A} = \mathfrak{B}^\sim$ then, for $a_0, \dots, a_{n-1} \in |\mathfrak{B}|$,

$$\mathfrak{B} \models \varphi(a_0, \dots, a_{n-1}) \to \pi^*(\mathfrak{A})\downarrow \models \varphi(\pi^* \circ \imath(a_0), \dots, \pi^* \circ \imath(a_{n-1})),$$

with $\imath := \imath_{|\mathfrak{B}|}$. If π is a monomorphism then π^* is a monomorphism from $\mathfrak{A}\downarrow$ to $\pi^*(\mathfrak{A})\downarrow$ and the converse implication is also true in the above formulas. If π is an isomorphism of algebraic systems then π^* is an isomorphism of algebraic B-systems.

◁ To prove, combine 2.2.4, 2.2.5, 4.1.10, and 4.2.5, on using the reasoning of (1). ▷

(5) *If $\mathfrak{A}$ is an algebraic system inside $\mathbf{V}^{(B)}$ then $[\![\mathfrak{A}\downarrow^\sim \simeq \mathfrak{A}]\!] = \mathbf{1}$.*

(6) Theorem. *The Boolean valued representation $(\mathscr{A}, \nu, \delta)$ of an algebraic B-system with disjointness (A, ν, Δ) is an algebraic system with simple disjointness inside $\mathbf{V}^{(B)}$. If $(A', \nu') := (\mathscr{A}, \mu)\downarrow$ and $\Delta' := \{(x, y) \in A' \times A' : \delta\downarrow(x, y) = \mathbf{1}\}$ then (A', ν', Δ') is a universally complete algebraic B-system with disjointness and for all $x, y \in A$ the following hold:*

$$x \perp y \leftrightarrow \imath x \perp \imath y \leftrightarrow [\![\imath x = \theta \vee \imath y = \theta]\!] = \mathbf{1},$$

with $\imath = \imath_A : A \to A'$ the canonical embedding.

◁ It suffices to use 4.1.13 and 4.3.3. ▷

4.3.6. We now address the important problem that was mentioned in 4.2.6.

Take an algebraic B-system $\mathfrak{A}$ of signature σ. Given a formula φ of signature σ and elements $a_0, \dots, a_{n-1} \in |\mathfrak{A}|$, we will temporarily employ the record $\mathfrak{A} \models_B \varphi(a_0, \dots, a_{n-1})$ in place of $\mathfrak{A} \models \varphi(a_0, \dots, a_{n-1})$ since the latter is less informative.

Starting with the B-system $\mathfrak{A}$, arrange the two-valued algebraic system $\bar{\mathfrak{A}}$ by purification (cf. 4.1.3). We may speak about satisfaction of $\varphi(a_0, \dots, a_{n-1})$ both in $\mathfrak{A}$ and $\bar{\mathfrak{A}}$ since $|\mathfrak{A}| = |\bar{\mathfrak{A}}|$ and $\sigma(\bar{\mathfrak{A}}) = \sigma$. This gives rise to a natural question of interrelation between the statements $\mathfrak{A} \models_B \varphi(a_0, \dots, a_{n-1})$ and $\bar{\mathfrak{A}} \models \varphi(a_0, \dots, a_{n-1})$.

Theorems 4.2.7 and 4.2.8 provide examples of the formulas φ for which $\bar{\mathfrak{A}} \models \varphi$ results from $\mathfrak{A} \models_B \varphi$. On the other hand, we can easily exhibit an example that violates this implication. Indeed, assume that $B := \mathscr{P}([0, 1])$ and $A := \mathbb{R}^{[0,1]}$ is the set of all real functions on the interval $[0, 1]$ with the B-metric

$$d(f, g) := \{t \in [0, 1] : f(t) \neq g(t)\} \quad (f, g \in A).$$

Equip A with the B-valued binary predicate $[\![\cdot \leq \cdot]\!]$ as follows:

$$[\![f \leq g]\!] := \{t \in [0,1] : f(t) \leq g(t)\} \quad (f, g \in A).$$

Then $\mathfrak{A} := (A, [\![\cdot \leq \cdot]\!])$ is an algebraic B-system and $\mathfrak{A} \models_B \varphi$, where $\varphi := (\forall x)(\forall y)(x \leq y \vee y \leq x)$. Moreover, $\bar{\mathfrak{A}} := (A, \leq)$ is obviously the purification of $\mathfrak{A}$ if we assign

$$f \leq g \leftrightarrow (\forall t \in [0,1]) f(t) \leq g(t).$$

Evidently, $\bar{\mathfrak{A}} \models \neg\varphi$. Denote by $\mathscr{T}^B(\mathfrak{A})$ and $\mathscr{T}(\bar{\mathfrak{A}})$ the sets of all true formulas in the systems $\mathfrak{A}$ and $\bar{\mathfrak{A}}$, with the constants ranging over $|\mathfrak{A}|$. Clearly, none of these two sets is in general a subset of the other.

We may expect therefore that for a certain class Φ of formulas of signature σ there exist only relations of the type $\mathscr{T}^B(\mathfrak{A}) \cap \Phi(?)\mathscr{T}(\bar{\mathfrak{A}}) \cap \Phi$. Exact formulations require some syntactic analysis of the texts under study.

4.3.7. Here we select particular classes of formulas.

(1) Consider the classes of *generic* and *strictly generic* formulas. These are defined by recursion on the length of a formula. The rules are as follows:

(a) Every atomic formula is strictly generic.

(b) If φ and ψ are strictly generic formulas then $\varphi \wedge \psi$, $(\exists x)\varphi$, and $(\forall x)\varphi$ are also strictly generic.

(c) A strictly generic formula is generic.

(d) If φ and ψ are generic formulas then $\varphi \wedge \psi$, $(\exists x)\varphi$, and $(\forall x)\varphi$ are also generic.

(e) If φ is a strictly generic formula then $\neg\varphi$ is a generic formula.

(f) If φ is a strictly generic formula and ψ is a generic formula then $\varphi \to \psi$ is a generic formula.

(2) A *basis Horn formula* is a disjunction $\theta_1 \wedge \ldots \wedge \theta_n$ where at most one of the formulas θ_k is a basis formula and the remaining formulas are negations of atomic formulas. A formula is a *Horn* formula if it is built from basis Horn formulas with the connectives $\wedge$, $\exists$, and $\forall$.

(3) *Each generic formula is equivalent in predicate calculus to a Horn formula and conversely.*

4.3.8. Examples.

(1) Let φ be a formula of signature $\{\leq\}$ with a sole predicate symbol. If φ is the conjunction of the axioms of a lattice (cf. 1.1.1) then φ is a generic formula. It is impossible to express distributivity as a generic formula in the signature $\{\leq\}$. If, however, we take the signature $\sigma := \{\wedge, \vee\}$, where $\wedge$ and $\vee$ are binary function symbols; then the formula $x \wedge (y \vee z) = (x \wedge y) \vee (x \wedge z)$ is atomic and, hence, strictly generic. Moreover, the property of being a distributive lattice is a strictly generic formula of signature $\{\wedge, \vee\}$.

(2) Consider formulas φ and ψ of signature $\{\wedge, \vee, *, 0, 1\}$. Let φ be the conjunction of the axioms of a Boolean algebra (see 1.1.2), while $\psi :=$ "there exists at least one atom," i.e.,

$$\psi := (\exists\, x)(\forall\, y)(x \neq 0 \wedge y = y \to x = y \vee y = 0).$$

Then φ is a strictly generic formula, whereas ψ is not generic.

(3) Let $\sigma := \{+, 0\}$, where $+$ is a binary function symbol and 0 is a constant. If φ is the conjunction of the axioms of a group (associativity of the group operation, the axiom of zero, and existence of an inverse); then φ is a strictly generic formula of signature σ.

(4) Let $\sigma := \{+, \cdot, 0, 1\}$, where $+$ and $\cdot$ are binary function symbols, and 0 and 1 are constants. Let φ be the conjunction of the axioms of a ring and ψ, the conjunction of the axioms of an integral domain; i.e., $\psi := \varphi \wedge \theta$, with

$$\theta := (\forall\, x)(\forall\, y)(x \cdot y = 0 \to x = 0 \vee y = 0).$$

Then φ is a strictly generic formula, while ψ is a generic formula.

4.3.9. We continue our syntactic analysis with the following

(1) ***Jech Theorem.*** *Let $\mathfrak{A}$ be a universally complete algebraic B-system. Assume further that φ is a formula of signature $\sigma(\mathfrak{A})$ and $a_0, \dots, a_{n-1} \in |\mathfrak{A}|$. Then the following hold:*

$$\textbf{(a)}\ \mathfrak{A} \models_B \varphi(a_0, \dots, a_{n-1}) \leftrightarrow \bar{\mathfrak{A}} \models \varphi(a_0, \dots, a_{n-1})$$

in case φ is strictly generic;

$$\textbf{(b)}\ \mathfrak{A} \models_B \varphi(a_0, \dots, a_{n-1}) \to \bar{\mathfrak{A}} \models \varphi(a_0, \dots, a_{n-1})$$

in case φ is generic.

$\lhd$ The proof proceeds by induction on the length of φ. By Theorem 4.3.3, we may assume that $\mathfrak{A} = \mathscr{A}\!\downarrow$ where $\mathscr{A}$ is an algebraic system of signature $\sigma^\wedge$ inside $\mathbf{V}^{(B)}$.

We start with the case of a strictly generic φ. If φ is an atomic formula then (a) is immediate from the definition of purification. Indeed, given a predicate symbol $p \in \sigma(\mathfrak{A})$, $\mathfrak{a}(p) = n$, note that

$$p^\nu(a_0, \dots, a_{n-1}) = 1 \leftrightarrow (a_0, \dots, a_{n-1}) \in \bar{\nu}(p)$$

for all $a_0, \dots, a_{n-1} \in |\mathfrak{A}|$.

Regarding the conjunction $\varphi := \psi \wedge \theta$, use the induction hypothesis to obtain

$$[\![\psi \wedge \theta]\!]^{\mathfrak{A}} = \mathbf{1} \leftrightarrow |\psi|^{\mathfrak{A}} = \mathbf{1} \wedge |\theta|^{\mathfrak{A}} = \mathbf{1} \leftrightarrow \bar{\mathfrak{A}} \models \psi \wedge \bar{\mathfrak{A}} \models \theta \leftrightarrow \bar{\mathfrak{A}} \models \psi \wedge \theta.$$

The case of a universal quantifier $\varphi := (\forall\, x)\psi$ is settled by analogy:

$$|(\forall\, x)\varphi|^{\mathfrak{A}} = \mathbf{1} \leftrightarrow (\forall\, a \in |\mathfrak{A}|)\psi(a)|^{\mathfrak{A}} = \mathbf{1}$$
$$\leftrightarrow (\forall\, a \in |\mathfrak{A}|)\bar{\mathfrak{A}} \models \psi(a) \leftrightarrow \bar{\mathfrak{A}} \models (\forall\, x)\psi.$$

Consider the case of an existential quantifier $\varphi := (\exists\, x)\psi$. By the maximum principle, there is an element $z \in \mathbf{V}^{(B)}$ such that

$$[\![\mathscr{A} \models (\exists\, x)\psi]\!] = [\![z \in |\mathscr{A}| \wedge \mathscr{A} \models \psi(z)]\!].$$

By Theorem 4.3.3, the above formula may be rewritten as

$$[\![z \in |\mathscr{A}|]\!] \wedge |\psi(z)|^{\mathfrak{A}} = |(\exists\, x)\psi|^{\mathfrak{A}}.$$

This, together with the induction hypothesis, implies that the following equivalences hold:

$$|(\exists\, x)\psi|^{\mathfrak{A}} = \mathbf{1} \leftrightarrow (\exists\, z \in |\mathfrak{A}|)|\psi(z)|^{\mathfrak{A}} = \mathbf{1}$$
$$\leftrightarrow (\exists\, z \in |\bar{\mathfrak{A}}|)(\bar{\mathfrak{A}} \models \psi(z) \leftrightarrow \bar{\mathfrak{A}} \models (\exists\, x)\psi),$$

since $|\mathfrak{A}| = |\mathscr{A}|{\downarrow}$ by the definition of descent in 4.2.3. Therefore, the induction step is legitimate for a strictly generic φ, which settles (a).

Turning to (b), note that the cases of $\wedge$, $\exists$, and $\forall$ are settled in much the same way as above. We are left with considering negation and implication, cf. 4.1.7 (e, f).

Let $\varphi := \neg\psi$, where ψ is a strictly generic formula. If $|\varphi|^{\mathfrak{A}} = \mathbf{1}$ then $|\psi|^{\mathfrak{A}} = \mathbf{0}$ and from (a) it follows that ψ cannot be true in $\bar{\mathfrak{A}}$. However, $\bar{\mathfrak{A}} \models \varphi$.

Finally, consider a formula of the type $\varphi := \theta \to \psi$, where θ is a strictly generic formula and ψ is a generic formula. Assume that $|\theta \to \psi|^{\mathfrak{A}} = \mathbf{1}$. If $\bar{\mathfrak{A}} \models \theta$ then from (a) it follows that $|\theta|^{\mathfrak{A}} = \mathbf{1}$ and so $|\psi|^{\mathfrak{A}} = \mathbf{1}$. By the induction hypothesis, $\bar{\mathfrak{A}} \models \psi$. Therefore, $\bar{\mathfrak{A}} \models \theta \to \psi$. $\triangleright$

Note that the Jech Theorem makes it possible to replace the proofs of some fragments of Theorems 4.2.7–4.2.9 with syntactic analysis of the corresponding sentences. It goes without saying that we may proceed further in the abstract.

(2) Corollary. *Assume that $\mathscr{A}$ and $\bar{\mathfrak{A}}$ stand for the Boolean valued representation and the purification of a universally complete algebraic B-system. For every Horn sentence φ the following holds:*

$$[\![\mathscr{A} \models \varphi]\!] = \mathbf{1} \to \bar{\mathfrak{A}} \models \varphi.$$

4.3.10. Let Φ be some set of formulas of signature σ. Introduce the category $\mathrm{AS}^{(B)}(\Phi)$ as follows:

$$\mathrm{Ob}\,\mathrm{AS}^{(B)}(\Phi)$$
$$:= \{\mathfrak{A} \in \mathbf{V}^{(B)} : [\![\,\mathfrak{A} \text{ is an algebraic system of signature } \sigma^\wedge \text{ and } \mathfrak{A} \models \Phi\,]\!] = \mathbf{1}\};$$
$$\mathrm{AS}^{(B)}(\mathfrak{A}, \mathfrak{B})$$
$$:= \{h \in \mathbf{V}^{(B)} : [\![\, h \text{ is a homomorphism from } \mathfrak{A} \text{ to } \mathfrak{B}\,]\!] = \mathbf{1}\};$$
$$\mathrm{Com}(f, g) = h \leftrightarrow [\![h = g \circ f]\!] = \mathbf{1}.$$

The above assignments determine a category in view of the transfer and maximum principles, Theorem 4.3.2, and other properties of the embedding functor. As before, by $\mathscr{F}^\sim$ and $\mathscr{F}{\downarrow}$ we denote the mappings of immersion and descent which act in the categories of algebraic systems: $\mathscr{F}^\sim$: B-$\mathrm{AS}(\Phi) \to \mathrm{AS}^{(B)}(\Phi)$, $\mathscr{F}{\downarrow} : \mathrm{AS}^{(B)}(\Phi) \to B$-$\mathrm{AS}(\Phi)$.

Theorem. *The following hold:*

(1) *The mapping $\mathscr{F}{\downarrow}$ is a covariant functor from the category $\mathrm{AS}^{(B)}(\Phi)$ to the category B-$\mathrm{CAS}^{(B)}(\Phi)$;*

(2) *The mapping $\mathscr{F}^\sim$ is a covariant functor from the category B-$\mathrm{AS}(\Phi)$ (as well as from B-$\mathrm{CAS}(\Phi)$) to the category $\mathrm{AS}^{(B)}(\Phi)$;*

(3) *The functors $\mathscr{F}{\downarrow}$ and $\mathscr{F}^\sim$ carry out equivalence between the categories $\mathrm{AS}^{(B)}(\Phi)$ and B-$\mathrm{CAS}(\Phi)$.*

4.3.11. We now state two important theorems by R. Solovay and S. Tennenbaum.

(1) Theorem. *Assume that D is a complete Boolean algebra and $\jmath : B \to D$ is a complete monomorphism. Then there are a complete Boolean algebra $\mathscr{D}$ inside $\mathbf{V}^{(B)}$ and an isomorphism H from D onto $D' := \mathscr{D}{\downarrow}$ such that the following diagram commutes:*

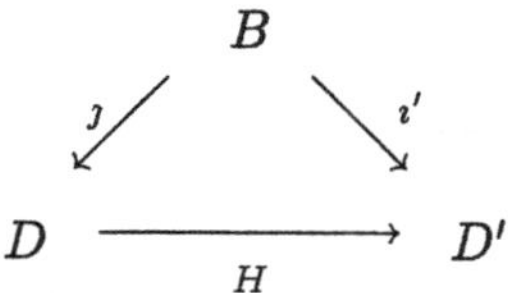

where $\imath'$ is the canonical monomorphism from B to D'.

(2) Theorem. *Let (K, D) be a BAP-ring and let $\jmath : B \to D$ be a complete homomorphism. Then there are a BAP-ring $(\mathscr{K}, \mathscr{D})$ inside $\mathbf{V}^{(B)}$ and an isomorphism h of K to $K' := \mathscr{K}{\downarrow}$ such that for each $b \in B$ the following diagram commutes:*

$$\begin{array}{ccc} K & \xrightarrow{h} & K' \\ {\scriptstyle \jmath(b)}\big\downarrow & & \big\downarrow{\scriptstyle \imath'(b)} \\ K & \xrightarrow[h]{} & K' \end{array}$$

where $\imath'$ is the canonical monomorphism from B to D'.

Analogous results hold also for BAP-groups.

4.3.12. COMMENTS.

(1) Assume that C and D are Boolean algebras. Assume further that P and Q are the Stone spaces of C and D. Define the *tensor product* $C \otimes D$ of C and D as the Boolean algebra of clopen subsets of the product $P \times Q$ (cf. 1.1.6 (6) and 1.2.6 (8)). Agree to denote by $C \widehat{\otimes} D$ the completion of $C \otimes D$ (cf. 1.1.6 (7) and 1.2.6 (9)).

If D is a Boolean algebra and $\mathscr{D} \in \mathbf{V}^{(B)}$ is such that $\mathbf{V}^{(B)} \models$ "$\mathscr{D}$ is the completion of $D^{\wedge}$," then $\mathscr{D}{\downarrow}$ and $B \widehat{\otimes} D$ are isomorphic Boolean algebras (cf. [227]).

(2) The Solovay–Tennenbaum Theorems 4.3.11 (1, 2) give grounds to iterating the construction of a Boolean valued model.

Assume that $\mathscr{D} \in \mathbf{V}^{(B)}$ and $\mathbf{V}^{(B)} \models$ "$\mathscr{D}$ is a complete Boolean algebra." Proceeding along the lines of Section 2.1 inside $\mathbf{V}^{(B)}$, we may construct the following $\mathbf{V}^{(B)}$-classes: the Boolean valued universe $(\mathbf{V}^{(B)})^{(\mathscr{D})}$, the corresponding Boolean truth values $[\![\cdot = \cdot]\!]^{\mathscr{D}}$ and $[\![\cdot \in \cdot]\!]^{\mathscr{D}}$, and the canonical embedding $(\,\cdot\,)^{\wedge}$ of the universal class $\mathbf{U}_B$ in $(\mathbf{V}^{(B)})^{\mathscr{D}}$. Put $D := \mathscr{D}{\downarrow}$, $\mathbf{W}^{(D)} := (\mathbf{V}^{(B)})^{(\mathscr{D})}{\downarrow}$, $[\![\cdot = \cdot]\!]^{D} := ([\![\cdot = \cdot]\!]^{\mathscr{D}}){\downarrow}$, $[\![\cdot \in \cdot]\!]^{D} := ([\![\cdot \in \cdot]\!]^{\mathscr{D}}){\downarrow}$, $\jmath := (\,\cdot\,)^{\wedge}{\downarrow}$. Let $\imath : B \to D$ be the canonical monomorphism, with $\imath^{*} : \mathbf{V}^{(B)} \to \mathbf{V}^{(D)}$ standing for the corresponding injection (cf. Section 2.2). Then there is a unique bijection $h : \mathbf{V}^{(D)} \to \mathbf{W}^{(D)}$ such that $[\![x = y]\!]^{D} = [\![h(x) = h(y)]\!]^{\mathscr{D}}$ and $[\![x \in y]\!]^{D} = [\![h(x) \in h(y)]\!]^{\mathscr{D}}$ for all $x, y \in \mathbf{V}^{(B)}$.

In this event the following diagram commutes:

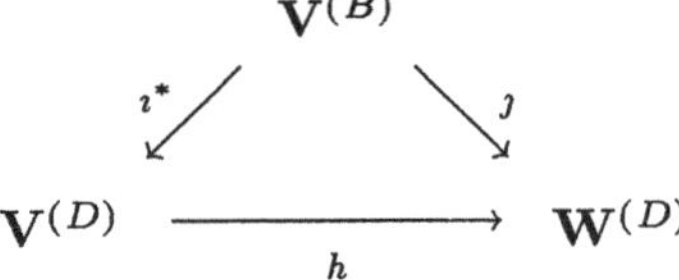

For more details, see [227].

As regards some related Boolean topics in the theory of universal algebras, cf. [202].

(3) Further iterations of the above construction lead to a transfinite sequence of Boolean valued extensions. In this way there appears an efficient method, the *iterated forcing*, which has been used to establish the relative consistency of the Suslin hypothesis with ZFC (cf. [227]).

4.4. Ordered Algebraic Systems

A complete Boolean algebra of congruences necessary for Boolean valued representation of an algebraic system is often generated by an order relation. This peculiarity brings about the possibility of Boolean valued representation for ordered algebraic systems. Supplementary information may be found in [13, 14, 56, 111].

4.4.1. An *ordered group* is an algebraic system $(G, +, 0, \leq)$ satisfying the following conditions:

(1) $(G, +, 0)$ is a group;

(2) $(G, \leq)$ is a poset;

(3) The group order structures of G are compatible, which means that group translations are isotonic mappings; i.e., G is a model for

$$(\forall\, x)(\forall\, y)(\forall\, a)(\forall\, b)(x \leq y \leftrightarrow a + x + b \leq a + y + b).$$

(Notice that the plus sign for the group operation does not imply commutativity.)

Say that G is a *totally ordered group* in the case when in addition to (1)–(3) the following condition is also fulfilled:

(4) $(G, \leq)$ is a totally ordered set; i.e., the formula $(\forall\, x)(\forall\, y)$ $(x \leq y \vee y \leq x)$ holds in G.

An element x in G is *positive* if $x \geq 0$. The set of all positive elements is called the *positive cone* of G and denoted by G^+. A subset K of G is the positive cone of some order on G provided that the following conditions are met:

(a) $K \cap (-K) = \{0\}$;

(b) $K + K = K$;

(c) $x + K = K + x$ $(x \in G)$.

In this case K and the order that K induces on G are related as follows:

$$x \leq y \leftrightarrow y - x \in K \leftrightarrow -x + y \in K.$$

A group G is totally ordered if and only if

(d) $G = G^+ \cup (-G^+)$.

The positive cone G^+ of G reproduces G or is *reproducing* provided that $G = G^+ - G^+$. In this event G is sometimes called a *directed group*. A homomorphism $h : G \to G'$, acting from an ordered group G to another ordered group G', is *positive* if $h(x) \geq 0$ for every $0 \leq x \in G$.

An ordered group G is *integrably-closed* if, for all x, $y \in G$, the inequalities $nx \leq y$, $n \in \omega$, imply that $x \leq 0$. An ordered group G is *Archimedean* if, for all x, $y \in G$, the inequalities $nx \leq y$, $\pm n \in \omega$, imply that $x = 0$.

4.4.2. A *lattice ordered group* is an ordered group G in which every nonempty finite set $\{x_0, \dots, x_{n-1}\} \subset G$ has the *join* $x_0 \vee \dots \vee x_{n-1} := \sup\{x_0, \dots, x_{n-1}\}$ and *meet* $x_0 \wedge \dots \wedge x_{n-1} := \inf\{x_0, \dots, x_{n-1}\}$. Given an element x of a lattice ordered group G, define the elements $|x| := x \vee (-x)$, $x^+ := x \vee 0$, and $x^- := (-x)^+ = -x \wedge 0$ which are called the *absolute value* or *module* of x, the *positive part* of x, and the *negative part* of x.

In every lattice ordered group the following hold:

(1) $x = x^+ - x^-$, $|x| = x^+ + x^-$, $x^+ \wedge x^- = 0$;

(2) $(x+y)^+ \leq x^+ + y^+$, $(x+y)^- \leq x^- + y^-$;

(3) $(nx)^+ = nx^+$, $(nx)^- = nx^-$, $|nx| = n|x|$ $(n \in \omega)$;

(4) $|x+y| \leq |x| + |y| + |x|$;

(5) $|x+y-x| = x + |y| - x$; $(x+y-x)^- = x + y^- - x$;

(6) $u \wedge x = 0, u \wedge y = 0 \to u \wedge (x+y) = 0$.

A lattice ordered group G is commutative if and only if (4) becomes $|x+y| \leq |x|+|y|$ for all x, $y \in G$. Recall that a commutative group is also referred to as *Abelian* or *abelian*.

Listing the properties of a lattice ordered group G, note that G is a torsion-free group and a distributive lattice. Moreover, the following identities hold:

$$a + \left(\bigvee x_\alpha\right) + b = \bigvee (a + x_\alpha + b),$$

$$a + \left(\bigwedge x_\alpha\right) + b = \bigwedge (a + x_\alpha + b).$$

A subgroup G_0 of a lattice ordered group G is an *o-ideal*, or an *order ideal* or a *convex subgroup* if, for all x and y in G, it follows from $|x| \leq |y|$ and $y \in G_0$ that $x \in G_0$. If, moreover, G_0 is a normal subgroup then G_0 is called an *l-ideal.*

4.4.3. From now on we assume G to be a lattice ordered group and equip G with the *disjointness* $\perp$ by the rule:

$$\perp := \{(x, y) \in G \times G : |x| \wedge |y| = 0\}.$$

There is no doubt that $\perp$ obeys all axioms of disjointness of 4.1.12 (2). The complete Boolean algebra $\mathfrak{K}_{\perp}(G)$ of $\perp$-bands of G is called the *base* of G and denoted by $\mathscr{B}(G)$. Assume that a band $K \in \mathscr{B}(G)$ is a summand of G. The corresponding band projection π_K is a positive endomorphism in G satisfying $\pi_K x \leq x$ for all $0 \leq x \in G$. If each band of K is a summand then the set $\mathfrak{Pr}(G)$ of all band projections π_K $(K \in \mathscr{B}(G))$ is a complete Boolean algebra isomorphic to $\mathscr{B}(G)$. In this event, say that G has the *band projection property*.

A lattice ordered group G with the band projection property is *universally complete* or *orthogonally complete* provided that G is universally complete with respect to the Boolean algebra $\mathfrak{Pr}(G)$. The *universal completion* of a lattice ordered group G is a universally complete lattice ordered group G' together with an order isomorphism $\imath : G \to G'$ such that (a) $G' = \mathrm{mix}(\imath(G))$, where mix is calculated with respect to the Boolean algebra $\mathfrak{Pr}(G)$, and (b) to each $0 < x' \in G'$ there is $0 < x \in G$ satisfying $\imath(x) \leq x'$

Recall that $[x]$ stands for the least band containing x. The properties, listed in 4.4.2, allow us to deduce that

(1) *The following hold:*

$$
\begin{aligned}
[x+y] = [x \vee y] = [x] \vee [y] \quad & (x, y \in G^+);\\
[x] = [|x|] = [x^+] \vee [x^-] \quad & (x \in G);\\
[x + y - x] = x + [y] - x \quad & (x, y \in G);\\
x \perp y \to x + y = y + x \quad & (x, y \in G).
\end{aligned}
$$

(2) *Each band, a member of $\mathscr{B}(G)$, is an order ideal of G.*

$\lhd$ Indeed, if x and y belong to $A^{\perp}$ for some $A \subset G$ then, using the second identity of (1) and 4.4.2, we may write

$$\{x+y\}^{\perp} \supset \{x\}^{\perp} \wedge \{y\}^{\perp} \wedge \{x\}^{\perp} \supset A.$$

Hence, $x + y \in \{x+y\}^{\perp\perp} \subset A^{\perp}$. Therefore, $A^{\perp}$ is a subgroup of G. On the other hand, if $y \in A^{\perp}$ and $|x| \leq |y|$ then $\{x\}^{\perp} \supset \{y\}^{\perp} \supset A$ and so $x \in \{x\}^{\perp\perp} \subset A^{\perp}$, which completes the proof. $\rhd$

4.4.4. If G is not commutative then the bands of G are not necessarily normal subgroups; i.e., they are not l-ideals in general. Therefore, the following definition is timely: A band $K \in \mathscr{B}(G)$ is *invariant* if $x + K - x \subset K$ for all $x \in G$. By 4.4.3 (2), this amounts to the property that K is an l-ideal. Let $\mathscr{B}_{\imath}(G)$ stand for the set of all invariant bands of G.

(1) *$\mathscr{B}_{\imath}(G)$ is a regular subalgebra of $\mathscr{B}(G)$.*

$\lhd$ It is obvious that the intersection of invariant bands is an invariant band too. Therefore, it suffices to show that the disjoint complement of an invariant band is invariant too. To this end, take $K \in \mathscr{B}_\imath(G)$ and $x \in K^\perp$. Granted $y \in K$ and $a \in G$, observe $0 = (a+|y|-a)\wedge|x| = -a+(a+|y|-a)\wedge|x|+a = |y|\wedge(-a+|x|+a)$. Hence, $-a+|x|+a \in K^\perp$, which means that $K^\perp$ is an invariant band. $\rhd$

(2) *The following are equivalent for a lattice ordered group G:*

(a) *Every band is invariant; i.e., $\mathscr{B}(G) = \mathscr{B}_\imath(G)$;*

(b) *For all $x, y \in G$,*

$$\{x\}^\perp = y + \{x\}^\perp - y;$$

(c) *If x in G is disjoint from all its conjugates $y+x-y$ then $x = 0$.*

$\lhd$ The condition (b) is an obvious consequence of (a). Assume that (b) holds and $x \perp (y+x-y)$ for some x and y in G. Then

$$x \in \{y+x-y\}^\perp = y + \{x\}^\perp - y = \{x\}^\perp,$$

which immediately yields $x = 0$.

Assume finally that (c) is fulfilled and a band K has the form $A^\perp$ for some $A \subset G$. Take $x \in K$, $y \in G$, and $a \in A$ and put $z := (y+|x|-y)\wedge|a|$. Obviously, $0 \le z \wedge (-y+z+y) \le |x| \wedge |a| = 0$, so that $z = 0$. This means that $|y+x-y| = y+|x|-y \in A^\perp = K$; i.e., $y + K - y \subset K$. $\rhd$

Furnish G with the symmetric relation defined as follows:

$$\Delta := \{(x,y) \in G\times G : (\forall\, a)(\forall\, b)(a+|x|-a)\wedge(b+|y|-b) = 0\}.$$

If, for some x and y in G, it is false that $x \,\Delta\, y$; then there are a_0 and b_0 in G such that $u_0 := (a_0+|x|-a_0)\wedge(b_0+|y|-b_0) \neq 0$. Obviously, $u_0 \in \{a_0+|x|-a_0\}^{\Delta\Delta}$. On the other hand, $\{a_0+|x|-a_0\}^{\Delta\Delta} = \{x\}^{\Delta\Delta}$. Therefore, $u_0 \in \{x\}^{\Delta\Delta}$. Analogously, $u_0 \in \{y\}^{\Delta\Delta}$. Note also that the least Δ-band is $\{0\}$, and $\Delta \cap I_G \subset \perp \cap I_G = \{(0,0)\}$. Hence, Δ is a disjointness on G (cf. 4.1.12 (2)).

(3) *The set of all Δ-bands coincides with the complete Boolean algebra of invariant $\perp$-components: $\mathfrak{K}_\Delta(G) = \mathscr{B}_\imath(G)$.*

4.4.5. Assume given a group G with invariant base; i.e., all bands of G are invariant. This means exactly that $\Delta{=}\perp$. Clearly, each commutative lattice ordered group has invariant base. In this event, G may be transformed into an algebraic B-system.

Let $\jmath$ be an isomorphism of a complete Boolean algebra B on the (invariant) base $\mathscr{B}(G)$ of G. Assign

$$p(x) := \jmath^{-1}(\{x^-\}^\Delta) \quad (x \in G).$$

The mapping $p : G \to B$ has a few important properties.

(1) *For all x and y in G the following hold:*

(a) $0 \le x \to p(x) = \mathbf{1}$;

(b) $p(x) \wedge p(-x) = \jmath^{-1}(\{x\}^{\perp})$;

(c) $p(x) \wedge p(y) \le p(x+y)$;

(d) $p(x) = p(y + x - y)$;

(e) $p(x) \vee p(-x) = \mathbf{1}$.

$\lhd$ Claim (a) is evident. To prove (b), note that $\{x\}^{\perp} = \{x^+\}^{\perp} \wedge \{x^-\}^{\perp} = \{x^-\}^{\perp} \wedge \{(-x)^-\}^{\perp}$ since x^+ and x^- are disjoint. It is now clear that $\jmath^{-1}(\{x^-\}^{\perp}) = \jmath^{-1}(\{x^-\}^{\perp}) \wedge \jmath^{-1}(\{(-x)^-\}^{\perp}) = p(x) \wedge p(-x)$. Analogous arguments will lead to (c) if we begin with 4.4.2 (2, 6). Claim (d) ensues from 4.4.2 (5) since every band is invariant.

Considering again that x^+ and x^- are disjoint, we may write

$$(\{x^+\}^{\perp} \vee \{x^-\}^{\perp})^{\perp} = \{x^+\}^{\perp\perp} \wedge \{x^-\}^{\perp\perp} = \{0\}.$$

Whence we infer that $\{x^+\}^{\perp} \vee \{x^-\}^{\perp} = G$, which amounts to (e). $\rhd$

Introduce the two mappings $\sigma, d : G \times G \to B$ by the rules:

$$\sigma(x,y) := p(y - x), \quad d(x,y) := \jmath^{-1}(\{x - y\}^{\Delta}) \quad (x, y \in G).$$

From 4.4.5 (1) (a–e) we immediate derive

(2) *The mapping σ possesses the following properties:*

(a) $\sigma(x,x) = 0$ (*reflexivity*);

(b) $\sigma(x,y) \wedge \sigma(y,z) \le \sigma(x,z)$ (*transitivity*);

(c) $\sigma(x,y) = \sigma(a + x - b, a + y - b)$ (*invariance*);

(d) $\sigma(x,y) \wedge \sigma(y,x) = d(x,y)^*$ (*antisymmetry*).

By virtue of (d), $d(x,y) = \sigma(x,y)^* \vee \sigma(y,x)^*$. Hence, d is a B-metric on G invariant under left and right translations, while σ is a B-predicate. Finally, it is obvious that $d(x,0) = \jmath^{-1}(\{x\}^{\perp\perp})$, i.e., the B-metric d agrees with the disjointness $\perp$ (cf. 4.1.13).

4.4.6. Theorem. *Let G be a lattice ordered group with invariant base. Denote by $\mathfrak{G}$ the algebraic system that results from furnishing G with the B-predicate σ and the corresponding B-metric d. Then $\mathfrak{G}$ is an algebraic B-system of signature $(+, 0, \le)$ which satisfies the axioms of a totally ordered group.*

$\lhd$ As was mentioned above, the B-metric d is translation-invariant. Using this, deduce

$$d(x + y, u + v) = d(x, -y + u + v) \le d(x,u) \vee d(u, -y + u + v),$$
$$d(u, -y + u + v) = d(u + y - u, v) \le d(y, v) \vee d(u + y - u, y),$$
$$d(u + y - u, y) = d(u + y, u + y) = 0.$$

These formulas show that $d(x+y,u+v) \leq d(x,u) \vee d(y,v)$; i.e., addition is a contraction. Using 4.4.5 (1) (c) and the definition of d, obtain

$$d(x,y)^* \wedge p(x) = p(x) \wedge p(x-y) \wedge p(y-x) \leq p(y)$$

for all $x, y \in G$. Whence it follows easily that $\sigma(x,y) \wedge d(x,u)^* \wedge d(y,v)^* \leq \sigma(u,v)$, which implies that σ is a contraction too. Hence, $(G, +, 0, \sigma)$ is an algebraic B-system of signature $(+, 0, \leq)$. By implication, we interpret the symbol $\leq$ as follows: given $x, y \in G$, we let $|x \leq y|^G := \sigma(x,y)$.

The unary B-predicate p on G is obviously the interpretation of the positivity property; i.e., $|0 \leq x|^G = p(x)$ for all $x \in G$. The fact that G is a B-model for the axioms of a totally ordered group is just a paraphrase of the properties 4.4.5 (1) (a–e). For instance, we demonstrate that the order σ is total and compatible with the group structure.

Let φ be the axiom of total order 4.4.1 (4) (d). Using 4.1.8, write

$$|\varphi|^G = \bigwedge_{x,y \in G} |x \leq y \vee y \leq x|^G = \bigwedge_{x,y \in G} \sigma(x,y) \vee \sigma(y,x).$$

Applying 4.4.5 (1) (e), note that

$$\sigma(x,y) \vee \sigma(y,x) = p(y-x) \vee p(x-y) = \mathbf{1},$$

and so $|\varphi|^G = \mathbf{1}$.

If φ is now the closed formula 4.4.1 (4) (c); then, developing the Boolean truth values for quantifiers according to 4.1.8, obtain

$$|\varphi|^G = \bigwedge_{x,y,a,b \in G} |x \leq y \to a+x+b \leq a+y+b|^G.$$

Since σ interprets $\leq$, infer

$$|x \leq y \to a+x+b \leq a+y+b|^G = \sigma(x,y) \Rightarrow \sigma(a+x+b,\ a+y+b).$$

On the other hand, by 4.4.5 (1) (d),

$$\begin{aligned} \sigma(a+x+b,\ a+y+b) &= p(a+y+b-(a+x+b)) \\ &= p(a+(y-x)-a) = p(y-x) = \sigma(x,y). \end{aligned}$$

Therefore, $\mathbf{1} = \sigma(x,y) \Rightarrow \sigma(a+x+b,\ a+y+b)$ and so $|\varphi|^G = \mathbf{1}$. $\triangleright$

4.4.7. We now turn to lattice ordered rings.

An algebraic system $(A, +, \cdot, 0, \leq)$ is an *ordered ring* if the following conditions are satisfied:

(1) $(K, +, 0, \leq)$ is a commutative ordered group;

(2) $(K, +, \cdot, 0)$ is a ring (not necessarily commutative or associative);

(3) multiplication and order are compatible on K so that $0 \leq x,\ y \in K$ yields $0 \leq xy$; i.e., K is a model for the formula

$$(\forall\, x)(\forall\, y)(x \geq 0 \wedge y \geq 0 \to x \cdot y \geq 0).$$

In other words, an ordered ring is a ring whose additive group is ordered and, moreover, the ring homomorphisms corresponding to positive elements are positive endomorphisms of the ordered additive group.

We often ascribe to a ring the properties of the underlying ordered additive group. For instance, speaking about a lattice ordered ring, a totally ordered ring, the positive cone of a ring, etc., we bear in mind the properties of the ordered group of the ring under study, avoiding further specification. The order on a ring is called a *ring order* provided that it obeys all conditions from (1) to (3).

An ordered ring K is *commutative* if, alongside with (1)–(3), the following axiom holds:

(4) $(\forall\, x)(\forall\, y)(xy = yx)$.

A subset P of a ring K is the positive cone of some ring order if and only if

$$P \cap (-P) = \{0\}; \quad P + P \subset P; \quad P \cdot P \subset P.$$

Every lattice ordered ring K, having the properties indicated in 4.4.2, obeys the conditions: $(xy)^+ \leq x^+y^+ + x^-y^-$; $(xy)^- \leq x^+y^- + x^-y^+$; $|xy| \leq |x| \cdot |y|$.

4.4.8. Each lattice ordered ring K may be transformed into an ordered B-group, in which case K may fail to become a B-ring in general. The point is that the ring multiplication on K is not necessarily a contraction with respect to the relevant B-metric. In order to exclude this undesirable phenomenon, we need a closer compatibility between multiplication and order.

A lattice ordered ring K is an *f-ring* provided that K satisfies the following condition: if $x,\ y \in K$ and $x \wedge y = 0$ then $(ax) \wedge y = 0$ and $(xa) \wedge y = 0$ for all $0 \leq a \in K$. Note that in every f-ring the following condition is fulfilled: $|x| \wedge |y| = 0 \to xy = 0$.

If an f-ring K has no nilpotent elements then the converse statement is also true, which is expressed customarily as K is a *faithful* f-ring. In particular, an f-ring without zero divisors is totally ordered. Also, every totally ordered ring without nilpotent elements contains no zero divisors. Among the other properties of an f-ring, we mention a few:

$$(x \vee y)z = (xz) \vee (yz); \quad z(x \vee y) = (zx) \vee (zy);$$
$$(x \wedge y)z = (xz) \wedge (yz); \quad z(x \wedge y) = (zx) \wedge (zy);$$
$$|xy| = |x| \cdot |y|.$$

For every lattice ordered ring K the following are equivalent:

(1) *K is an f-ring;*

(2) $\{xy\}^{\perp\perp} \le \{x\}^{\perp\perp} \wedge \{y\}^{\perp\perp}$;

(3) $d(xy, uv) \le d(x, u) \vee d(y, v)$.

$\lhd$ Assume that K is an f-ring. If $|x| \wedge |u| = 0$ or $|y| \wedge |u| = 0$ then $|xy| \wedge |u| = (|x| \cdot |y|) \wedge |u| = 0$. Therefore, $u \in \{x\}^{\perp}$ or $u \in \{y\}^{\perp}$ yields $u \in \{x \cdot y\}^{\perp}$, i.e., $\{x\}^{\perp} \cup \{y\}^{\perp} \subset \{xy\}^{\perp}$. Hence, $\{xy\}^{\perp\perp} \le (\{x\}^{\perp} \cup \{y\}^{\perp})^{\perp} = \{x\}^{\perp\perp} \wedge \{y\}^{\perp\perp}$.

Assuming (2), note that $|xy - uv| = |x(y-v) + (x-u)v| \le |x| \cdot |y-v| + |x-u| \cdot |v|$. Hence,

$$\{xy - uv\}^{\perp\perp} \le \{y - v\}^{\perp\perp} \vee \{x - u\}^{\perp\perp}.$$

This amounts to (3) by the definition of the B-metric d in 4.4.5.

Assume finally that $(x, y) \mapsto xy$ is a contraction. Put $u := 0$, and $v := y := a$ in (3) and rewrite the result as $\{x \cdot a\}^{\perp\perp} \subset \{x\}^{\perp\perp} \vee \{0\}^{\perp\perp} = \{x\}^{\perp\perp}$ or $\{xa\}^{\perp} \supset \{x\}^{\perp}$. By analogy, show that $(ax) \wedge y = 0$, implying that K is an f-ring. $\rhd$

4.4.9. Theorem. *Each (associative, commutative) f-ring K with B-predicate σ and B-metric d is a B-ring, i.e., an algebraic B-system that is a B-model for the axioms of an (associative, commutative) totally ordered ring. Moreover, an element $0 \ne e \in K$ is a ring unity of this B-ring if and only if e is an order and ring unity of K.*

$\lhd$ As shown in 4.4.6, K, furnished with σ and d, is a totally ordered B-group. Enrich this group with the contractive mapping $(x, y) \mapsto xy$ and prove that the so-obtained algebraic B-system is an f-ring. Associativity, commutativity, and distributivity in the B-system K follow trivially from the corresponding properties of the ring K. Check the compatibility condition 4.4.7 (3). To this end, note that, by 4.4.7 and 4.4.8 (2),

$$\{(xy)^-\}^{\perp} \ge \{x^+y^-\}^{\perp} \wedge \{x^-y^+\}^{\perp} \ge \{x^-\}^{\perp} \wedge \{y^-\}^{\perp}.$$

Recalling the definition of p, conclude that $p(x) \wedge p(y) \le p(xy)$. We are left with calculating the Boolean truth values by using 4.1.8:

$$
\begin{aligned}
&|(\forall\, x)(\forall\, y)(x \geq 0 \wedge y \geq 0 \rightarrow xy \geq 0)|^K \\
&= \bigwedge_{x,y\in K} |x \geq 0|^K \wedge |y \geq 0|^K \Rightarrow |xy \geq 0|^K \\
&= \bigwedge_{x,y\in K} p(x) \wedge p(y) \Rightarrow p(x \cdot y) = \mathbf{1}.
\end{aligned}
$$

Given $e \in K$, note further that the equality $\mathbf{1} = |\theta < e|^K = |e \geq 0 \wedge e \neq 0|^K$ implies $p(e) \wedge d(e, 0) = \mathbf{1}$; i.e., $e \geq 0$, and e is an order unit. On the other hand,

$$
|(\forall\, x)(xe = ex = x)|^K = \bigwedge_{x\in K} d(x, ex)^* \wedge d(x, xe)^*.
$$

Hence, e is the unity of the B-ring K if and only if e is an order unit in K. In other words, for all $x \in K$, the equalities $d(xe, x) = d(ex, x) = 0$ hold, meaning that $x = ex = xe$. This completes the proof. ▷

4.4.10. Theorem. *Let $\mathscr{G}$ be an ordered group inside $\mathbf{V}^{(B)}$, and put $G := \mathscr{G}\!\downarrow$. Then G is a universally complete ordered group with respect to the Boolean algebra of projections $\mathscr{B}$, and there is an isomorphism $\jmath$ from B to $\mathscr{B}$ such that*

$$
b \leq [\![0 \leq x]\!] \leftrightarrow 0 \leq \jmath(b)x \quad (x \in G,\ b \in B).
$$

In this event the following are equivalent:

(1) $\mathbf{V}^{(B)} \models$ *"$\mathscr{G}$ is directed (integrally-closed, or Archimedean)" $\leftrightarrow$ "G is directed (integrally-closed, or Archimedean)";*

(2) $\mathbf{V}^{(B)} \models$ *"$\mathscr{G}$ is lattice ordered (Dedekind complete)" $\leftrightarrow$ "G is lattice ordered (Dedekind complete)";*

(3) $\mathbf{V}^{(B)} \models$ *"$\mathscr{G}$ is an ordered ring" $\leftrightarrow$ "G is a universally complete ordered BAP-ring with distinguished Boolean algebra $\mathscr{B}$";*

(4) $\mathbf{V}^{(B)} \models$ *"$\mathscr{G}$ is a totally ordered skew field" $\leftrightarrow$ "G is a universally complete f-ring without nilpotent elements, $\mathscr{B}$ is the algebra of band projections of G, and every regular element in G is invertible."*

◁ The fact is established in 4.2.7 that G is a universally complete BAP-group with distinguished Boolean algebra $\mathscr{B}$. Denote by $\mathscr{G}^+$ the positive cone of $\mathscr{G}$ inside $\mathbf{V}^{(B)}$. Then

$$
\begin{aligned}
&[\![\mathscr{G}^+ + \mathscr{G}^+ \subset \mathscr{G}^+]\!] = [\![\mathscr{G}^+ \cap -\mathscr{G}^+ = \{0\}]\!] \\
&= [\![(\forall\, x \in \mathscr{G})(x + \mathscr{G}^+ = \mathscr{G}^+ + x)]\!] = \mathbf{1}.
\end{aligned}
$$

Assign $G^+ := \mathscr{G}^+\!\downarrow$ and note that, by the rules for the descending intersection and image, $G^+ + G^+ \subset G^+$, $G^+ \cap -G^+ = \{0\}$. So, given $x \in G$, note $[\![x + \mathscr{G}^+ = \mathscr{G}^+ + x]\!] = \mathbf{1}$; i.e., $x + \mathscr{G}^+ = \mathscr{G}^+ + x$. But then

$$(x + G^+) = (x + \mathscr{G}^+)\!\downarrow = (\mathscr{G}^+ + x)\!\downarrow = G^+ + x.$$

Therefore, G is an ordered group with positive cone G^+. The existence of the isomorphism $\jmath : B \to \mathscr{B}$ is proven in 4.2.7. Moreover, the equalities $b \leq [\![x = y]\!]$ and $\jmath(b)x = \jmath(b)y$ are equivalent. Take $x \in G$ and note that $[\![0 \leq x \leftrightarrow (\exists\, y \in \mathscr{G}^+)(x = y)]\!] = \mathbf{1}$. This implies that $b \leq [\![0 \leq x]\!]$ if and only if $b \leq [\![(\exists\, y \in \mathscr{G}^+)(x = y)]\!]$. The last statement is equivalent to the existence of some $y \in \mathscr{G}^+\!\downarrow =: G^+$ such that either $b \leq [\![x = y]\!]$ or $\jmath(b)x = \jmath(b)y \geq 0$.

We now prove the equivalence of the propositions from (1) to (4).

(1) If $\mathscr{G}$ is directed then $[\![\mathscr{G}^+ - \mathscr{G}^+ = \mathscr{G}]\!] = \mathbf{1}$, which is equivalent to the fact that G is directed since $(\mathscr{G}^+ - \mathscr{G}^+)\!\downarrow = \mathscr{G}^+\!\downarrow - \mathscr{G}^+\!\downarrow = G^+ - G^+$. The integral closure of $\mathscr{G}$ is nothing else but

$$\bigwedge\{[\![x \leq 0]\!] : [\![(\exists\, y \in \mathscr{G})(\forall\, n \in \omega^\wedge)(nx \leq y)]\!] = \mathbf{1}\} = \mathbf{1}.$$

Hence, $\mathscr{G}$ is integrally-closed if and only if, for every $x \in G$, the following implication holds:

$$(\exists\, y \in G)([\![(\forall\, n \in \omega^\wedge)(nx \leq y)]\!] = \mathbf{1} \to [\![x \leq 0]\!] = \mathbf{1}),$$

or

$$((\exists\, y \in G)(\forall\, n \in \omega)[\![n^\wedge x \leq y]\!] = \mathbf{1}) \to [\![x \leq 0]\!] = \mathbf{1}.$$

The last line is an equivalent paraphrase of the property that G is integrally-closed.

The claim about the Archimedean property of G is proven analogously.

(2) Let $\mathscr{G}$ be a lattice ordered group. Prove that the closed formula $(\forall\, x)(\forall\, y)(\exists\, z)(z = \sup\{x, y\})$ holds on G; i.e., every two elements of G has a least upper bound. If x and y belong to G then $[\![\{x, y\} \subset \mathscr{G}]\!] = \mathbf{1}$. Therefore, $[\![(\exists\, u \in \mathscr{G})(u = \sup\{x, y\})]\!] = \mathbf{1}$. By the maximum principle, there is some $z \in \mathbf{V}^{(B)}$ such that

$$[\![z \in \mathscr{G}]\!] \wedge [\![z = \sup\{x, y\}]\!] = \mathbf{1}.$$

This implies that, on the one hand, $z \in G$; whereas, on the other hand,

$$|z = \sup\{x, y\}|^{\mathscr{G}\downarrow} = \mathbf{1}.$$

By definition, $z = x \vee y$. Analogous reasoning enables us to proclaim the existence of $x \wedge y$.

Assume now that $[\![\mathscr{G}$ is a Dedekind complete group $]\!] = \mathbf{1}$. Show that in this case G is also Dedekind complete. We first recall the following equivalent definition of the least upper bound $\sup(A)$ of a set A in an arbitrary ordered set:

$$\sup(A) = \pi_{\leq}(A) \cap \pi_{\leq}^{-1}(\pi_{\leq}(A)).$$

Choose an arbitrary upper bounded subset A of $\mathscr{G}{\downarrow}$, this means that $\pi_{\leq}(A) \neq \varnothing$. In this case, however, by the rules for ascending and descending polars infer that $[\![\pi_{\leq}(A{\uparrow}) \neq \varnothing]\!] = \mathbf{1}$ or, which is equivalent, $[\![A{\uparrow}$ is an upper bounded subset of $\mathscr{G}]\!] = \mathbf{1}$. Using the maximum principle, find $a \in \mathscr{G}{\downarrow}$ so that

$$[\![a = \sup(A{\uparrow}) = \pi_{\leq}(A{\uparrow}) \cap \pi_{\leq}^{-1}(\pi_{\leq}(A{\uparrow}))]\!] = \mathbf{1}.$$

Applying now the rules for ascending and descending, deduce $a = \sup(\mathrm{mix}(A))$. Since the relation $\leq$ is fully extensional, conclude that $\sup(\mathrm{mix}(A)) = \sup(A)$. Therefore, A has a least upper bound, and so G is a Dedekind complete group.

(3) This follows from 4.2.8 and the properties of G we have established earlier.

(4) Assume that $\mathbf{V}^{(B)} \models$ "$\mathscr{G}$ is a totally ordered skew field." By (3) and 4.2.8, conclude that G is a universally complete associative ordered BAP-ring with distinguished Boolean algebra of positive projections $\mathscr{B}$, and G has no nilpotent elements.

Since $\mathscr{G}$ is a model for $(\forall\, x)(\forall\, y)(x \wedge y = 0 \to x = 0 \vee y = 0)$; therefore, $[\![x \wedge y = 0]\!] \leq (x = 0) \vee (y = 0)$ for all $x, y \in G$. If $x \wedge y = 0$ then $b^* \leq [\![x = 0]\!]$ and $b \leq [\![y = 0]\!]$, or $\jmath(b)x = x$ and $\jmath(b)y = 0$ for a suitable $b \in B$.

Hence, we easily deduce that $\mathscr{B}$ is a Boolean algebra of band projections. But then the orthogonal completeness of G amounts to the universal completeness of G with respect to $\mathscr{B}$. Since the projections $\jmath(b)$ $(b \in B)$ are multiplicative (see 4.2.8), the kernel of each of them is a ring ideal. From this it is immediate that the defining property of an f-ring holds for G (cf. 4.4.8 (2)).

Conversely, if G obeys (4) then, by virtue of (2), $[\![\mathscr{G}$ is a lattice ordered ring $]\!] = \mathbf{1}$. As is readily seen, $\mathscr{G}$ is also an f-ring without nilpotent elements inside $\mathbf{V}^{(B)}$. In this case, however, for $x, y \in G$ it follows from $[\![xy = 1]\!] = \mathbf{1}$ that $[\![|x| \wedge |y| = 0]\!] = \mathbf{1}$, or $|x| \wedge |y| = 0$. Hence, there is an element b in B such that $\jmath(b)x = 0$ and $\jmath(b^*)y = 0$. Therefore, $b \leq [\![x = 0]\!]$ and $b^* \leq [\![y = 0]\!]$, yielding $[\![x = 0 \vee y = 0]\!] \geq b \vee b^* = \mathbf{1}$. We have thus established that $\mathbf{V}^{(B)} \models$ "$\mathscr{G}$ has no zero divisors." An f-ring with no zero divisors is, however, known to be totally ordered. Hence, $\mathbf{V}^{(B)} \models$ "$\mathscr{G}$ is totally ordered."

Finally, by 4.2.8, the nonzero elements of $\mathscr{G}$ are invertible. Hence, $\mathbf{V}^{(B)} \models$ "$\mathscr{G}$ is a totally ordered skew field." $\triangleright$

4.4.11. The above shows that totally ordered groups and f-rings both transform somehow into B-groups and B-rings. By the results of Section 4.3, this implies that the Boolean valued representations of these groups are totally ordered groups and rings, respectively. We may thus translate the available information on the structure of totally ordered groups and rings to more general classes of groups and rings. We will illustrate the last statement with the examples of the well known facts (see [14, 56]):

(1) Hölder Theorem. *Each Archimedean totally ordered group is isomorphic to a subgroup of the additive group of the reals.*

(2) *Every Archimedean directed group is commutative.*

(3) Theorem. *An Archimedean totally ordered ring R is either a zero field; i.e., the product of every two elements of R is zero; or R is order and algebraically isomorphic to a uniquely determined subring of the reals.*

4.4.12. Theorem. *Let G be an Archimedean lattice ordered group with base isomorphic to a Boolean algebra B. Then there is a subgroup $\mathscr{G}$ of the additive group of the reals inside $\mathbf{V}^{(B)}$ such that the lattice ordered group $G' := \mathscr{G}{\downarrow}$ is the universal completion of G.*

$\lhd$ By 4.4.6, the group G can be transformed into an ordered B-group. Let $\mathscr{G}$ be the Boolean valued representation of this algebraic B-system. Then, by 4.3.3, $\mathscr{G}$ is a totally ordered group inside $\mathbf{V}^{(B)}$. In line with Theorem 4.4.10, $G' := \mathscr{G}{\downarrow}$ is a lattice ordered group, in which case $G' = \operatorname{mix}(\imath(G))$, with $\imath$ the canonical isomorphism from G to G'. If $b \in B$ and $L_b \in \mathscr{B}(G)$ and $\pi_b \in \mathfrak{Pr}(G')$ are the corresponding band and band projection, respectively; then the conditions $x \in L_b$ and $(I - \pi_b)(\imath(x)) = 0$ are equivalent for all $x \in G$.

Indeed, by the definition of a B-metric on G (see 4.4.5), the containment $x \in L_b$ is fulfilled if and only if $d(x, 0) \leq b$. However, Theorem 4.4.10 implies that the equality $\pi_b \imath(x) = \imath(x)$ holds if and only if $b^* \leq [\![\imath(x) = 0]\!]$. We know in this event that

$$[\![\imath(x) = 0]\!] = [\![\imath(x) \neq 0]\!]^* = d(x, 0)^*.$$

We have thus established that the correspondence $L' \mapsto \imath^{-1}(L')\,(L' \in \mathfrak{B}(G'))$ is an isomorphism between the bases $\mathscr{B}(G')$ and $\mathscr{B}(G)$. Choose $0 < x \in G'$. If $x = \operatorname{mix}(\pi_\xi \imath(x_\xi))$ then $0 < \pi_\xi \circ \imath(x_\xi) \leq \imath(x_\xi)$ for some ξ. In view of the isomorphism between the bases, there is $0 < z \in G$ for which $z \in \{\pi_\xi \circ \imath(x_\xi)\}^{\perp\perp}$. Putting $x_0 := x_\xi \wedge z$, note that

$$0 < \imath(x_0) \leq \imath(z) \wedge \pi_\xi \circ \imath(x_\xi) \leq \pi_\xi \circ \imath(x_\xi) \leq x.$$

Therefore, $\imath(G)$ minorizes G'. Assume that $n|x| \leq y$ $(n \in \omega)$ for some x, $y \in G'$. Let $y = \operatorname{mix}(\pi_\xi \imath(y_\xi))$ and $x = \operatorname{mix}(\pi_\xi \imath(x_\xi))$ for some families (x_ξ) and (y_ξ) in G and a partition of unity (π_ξ) in $\mathfrak{Pr}(G')$.

Put $\Xi_0 := \{\xi \in \Xi : \pi_\xi \circ \imath(|x_\xi|) = 0\}$. Since $\imath(G)$ is minorant, for all $\xi \in \Xi \setminus \Xi_0$ there is some $u_\xi \in G$, $u_\xi > 0$, satisfying $\imath(u_\xi) \le \pi_\xi(\imath|x_\xi|)$. Then, for the same ξ and for all $n \in \omega$, obtain

$$\imath(nu_\xi) \le \pi_\xi \circ \imath(n|x_\xi|) = \pi_\xi(n|x|) \le \pi_\xi y = \pi_\xi \circ \imath(y_\xi) \le \imath(y_\xi)$$

or $nu_\xi \le y_\xi$.

Since G is Archimedean, conclude that $u_\xi = 0$, which implies that $\Xi_0 = \Xi$ and so $x = 0$. Therefore, the group G' is Archimedean and, by 4.4.10, $[\![\mathscr{G}$ is Archimedean$]\!] = \mathbb{1}$. In line with the Hölder Theorem 4.4.11 (1), $\mathscr{G}$ is isomorphic to an additive subgroup of the reals $\mathscr{R}$. By Theorem 4.3.4, we may assume $\mathscr{G}$ to be a totally ordered subgroup of $\mathscr{R}$. $\triangleright$

4.4.13. Theorem. *Let K be an Archimedean f-ring. Then K splits into the direct sum of two bands K_0 and K_1 complementary to each other such that if the bases $\mathscr{B}(K_0)$ and $\mathscr{B}(K_1)$ are isomorphic to the Boolean algebras B_0 and B_1, then the following hold:*

(1) *There is a subgroup $\mathscr{K}_0$ of the reals inside $\mathbf{V}^{(B_0)}$ such that the lattice ordered group $K_0' := \mathscr{K}_0\!\downarrow$ with zero multiplication is the universal completion of f-ring K_0;*

(2) *There is a subring $\mathscr{K}_1$ of the reals inside $\mathbf{V}^{(B)}$ such that the f-ring $K_1' := \mathscr{K}_1'\!\downarrow$ is the universal completion of K.*

In this event the f-ring $K_0' \oplus K_1'$ is the universal completion of K.

$\triangleleft$ As we have seen in 4.4.12, the representation of the additive group of the f-ring K in $\mathbf{V}^{(B)}$, with $B = \mathscr{B}(K)$, is a subgroup of the additive group of the reals. According to 4.4.9, K is a B-ring; while, by Theorem 4.3.3, $[\![\mathscr{K}$ is a ring$]\!] = \mathbb{1}$. Put $b_0 := [\![\mathscr{K}$ is a zero ring$]\!]$ and $b_1 := [\![\mathscr{K}$ is a subring of the reals$]\!]$. By the transfer principle and Theorem 4.4.11 (3), $b_0 \vee b_1 = \mathbb{1}$. On the other hand, $b_0 \wedge b_1 = \mathbb{0}$, since a ring cannot be simultaneously a zero ring and a subring of the reals. Let K_0 and K_1 be the bands of K corresponding to b_0 and b_1; i.e., K_0 and K_1 are determined from the conditions

$$x \in K_\imath \leftrightarrow d(x,0) \le b_\imath \quad (\imath = 0,1),$$

where d is the B-metric of the B-system K. Assign $B_\imath := [0, b_\imath]$ and observe that the base $\mathscr{B}(K_\imath)$ is isomorphic to $B_\imath$, in which case $b_\imath$ is the unity of the algebra $B_\imath$. Put $\mathscr{K}_\imath := \pi_\imath^*(\mathscr{K}) \in \mathbf{V}^{(B_\imath)}$, where $\pi_\imath : b \mapsto b \wedge b_\imath$, $b \in B$. Since $\pi_\imath$ is an epimorphism of B onto $B_\imath$; therefore, $\mathbf{V}^{(B_0)} \models$ "$\pi_0^*(\mathscr{K})$ is a subgroup of the additive group of the reals" and $\mathbf{V}^{(B_1)} \models$ "$\pi_1^*(\mathscr{K})$ is a subring of the reals." By Theorem 4.4.12, $K' := K\!\downarrow$ is the universal completion of the ordered group K. As far as $b_\imath = [\![\pi_\imath^*(\mathscr{K}) \simeq \mathscr{K}]\!]$, where $K_l' := \mathscr{K}_\imath\!\downarrow \simeq \jmath(b_\imath)(K_\imath)$, and so $K' \simeq K_0' \oplus K_1'$. Therefore, K' is the universal completion of K. $\triangleright$

4.5. The Descent of a Field

Here we prove that rationally complete semiprime commutative rings are in one-to-one correspondence with fields in Boolean valued universes. This implies in particular the possibility of transferring the Horn properties of fields to these rings. All preliminaries to ring theory we need are explicit, for instance, in [50, 150].

4.5.1. Throughout this section, we let K stand for a commutative ring with unity 1, presuming that $0 \neq 1$. In this event K is a semiprime ring if K is free of nilpotent elements other than zero. Recall that x is *nilpotent* provided that $x^n = 0$ for some $n \in \mathbb{N}$. Recall also that a commutative ring K is an *integral domain* if $0 \neq 1$ and 0 is the only zero divisor of K.

(1) *Given a semiprime ring K, define $\perp$ as follows*

$$\perp := \{(x, y) \in K \times K : xy = 0\}.$$

Then $\perp$ is a disjointness relation on K and the least $\perp$-band is the singleton $\{0\}$. The disjointness $\perp$ is simple if and only if K is an integral domain.

$\lhd$ The relation $\perp$ is symmetric since K is commutative. Considering $x \in \pi_\perp(K)$, note that $x^2 = 0$, and so $x = 0$. Hence, the second defining property of disjointness (cf. 4.1.12 (2)) follows on recalling that K is semiprime. If $z = xy \neq 0$ then $uz = (ux)y = 0$ and $zv = x(yv) = 0$ for all $u \in \pi_\perp(x)$ and $v \in \pi_\perp(y)$. Therefore,

$$z \in \pi_\perp\big(\pi_\perp(x) \cup \pi_\perp(y)\big) = [x] \cap [y].$$

Alternatively, the third defining property of disjointness is available too. So, $\perp$ is a disjointness on K. From 4.1.12 (2) it follows that $\perp$ is a simple disjointness only if the equality $xy = 0$ implies either $x = 0$ or $y = 0$. $\rhd$

Evidently, the *annihilator* $L^\perp$ of a nonempty $L \subset K$, defined as

$$L^\perp := \pi_\perp(L) := \big\{k \in K : kL = \{0\}\big\},$$

is an ideal of K. An ideal of this provenance is called an *annihilator ideal.* It is an easy matter to show that a subset J of K is an annihilator ideal of K if and only if $J = J^{\perp\perp}$, where $J^{\perp\perp} := (J^\perp)^\perp$. From 4.1.12 (3) we infer the following:

(2) *The annihilator ideals of each semiprime ring K comprise the complete Boolean algebra $\mathscr{B}(K)$ with the following meet and join:*

$$L \wedge M := L \cap M, \quad L \vee M := (L \cup M)^{\perp\perp} \quad (L, M \in \mathscr{B}(K)),$$

while the Boolean complement L^* of an ideal $L \in \mathscr{B}(K)$ is the annihilator $L^\perp$ of L.

4.5.2. Let B stand for the complete Boolean algebra $\mathscr{B}(K)$ of the annihilator ideals of a ring K. Equip K with a B-metric by putting

$$d(k_1,k_2):=\{k_1-k_2\}^{\perp\perp}\quad (k_1,k_2\in K).$$

(1) *A semiprime commutative ring K with B-metric d and disjointness $\perp$ is a B-ring with disjointness.*

$\lhd$ Show first that D satisfies the properties of a Boolean metric in 3.4.1. The properties (1) and (2) are immediate from the definition of d. To show (3), take $k\in\{k_1-k_2\}^{\perp}\cap\{k_2-k_3\}^{\perp}$ and note that $k(k_1-k_2)=0$ and $k(k_2-k_3)=0$; i.e., $k(k_1-k_3)=0$, which amounts to $k\in\{k_1-k_3\}^{\perp}$. Whence,

$$\begin{aligned} d(k_1,k_3)&=\{k_1-k_3\}^{\perp\perp}\subset(\{k_1-k_2\}^{\perp}\cap\{k_2-k_3\}^{\perp})^{\perp}\\ &=\{k_1-k_2\}^{\perp\perp}\vee\{k_2-k_3\}^{\perp\perp}=d(k_1,k_2)\vee d(k_2,k_3). \end{aligned}$$

If $d(k_1,k_2)=\mathbf{0}$ then $\{k_1-k_2\}^{\perp}=K$, and so $(k_1-k_2)^2=0$. Since K has no nonzero nilpotents, infer that $k_1=k_2$.

Show now that the ring operations of K are contractive. To this end, demonstrate that

$$\begin{aligned} \{k_1-k_1'\}^{\perp}\cap\{k_2-k_2'\}^{\perp}&\subseteq\{(k_1+k_2)-(k_1'+k_2')\}^{\perp};\\ \{k_1-k_1'\}^{\perp}\cap\{k_2-k_2'\}^{\perp}&\subseteq\{k_1k_2-k_1'k_2'\}^{\perp}. \end{aligned}$$

The first inclusion is obvious. Further, note the evident equalities $k_1k_2-k_1'k_2'=k_1k_2-k_1k_2'+k_1k_2'-k_1'k_2'=k_1(k_2-k_2')+k_2'(k_1-k_1')$ which imply the second inclusion.

Obviously, the ring operations are disjointness-preserving; i.e., from $x,y\in a^{\perp}$ it follows that $xy,x+y\in a^{\perp}$. The fact that the disjointness and B-metric d agree is easy from the definitions, since $d(x,0)=x^{\perp\perp}$ (cf. 4.1.13). $\rhd$

(2) *For all $x,y\in K$, the equality holds: $d(xy,0)=d(x,0)\wedge d(y,0)$.*

$\lhd$ It suffices to show the equality $\{xy\}^{\perp\perp}=\{x\}^{\perp\perp}\wedge\{y\}^{\perp\perp}$ in which the inclusion $\subset$ is evident. Take $u\in\{x\}^{\perp\perp}\wedge\{y\}^{\perp\perp}=(\{x\}^{\perp}\cup\{y\}^{\perp})^{\perp}$. This means that, for all $a,b\in K$, from $ax=0$ it follows that $au=0$; and $by=0$ implies that $bu=0$. Using this with $b:=v^2x$ and $a:=v^2u$, consider an arbitrary $v\in K$ and deduce

$$\begin{aligned} &v\perp xy\to(v^2x)y=0\to(v^2u)y=0\\ &\to v^2u^2=0\to(vu)^2=0\to vu=0. \end{aligned}$$

Thus, $v\perp u$ holds for all $v\in\{xy\}^{\perp}$, and so $u\in\{xy\}^{\perp\perp}$. $\rhd$

4.5.3. An element e in K is an *idempotent* of K provided that $e^2 = e$. The idempotents of K comprise the Boolean algebra $\mathfrak{P}(K)$ with the Boolean operations as follows

$$e \wedge d = e \cdot d, \quad e \vee d = e + d - e \cdot d, \quad e^{\perp} = 1 - e \quad (e, d \in \mathfrak{P}(K)).$$

A ring K is *regular* (in the sense of von Neumann), if each principal ideal of K is generated by an idempotent or equivalently, each finitely generated ideal is a summand of K. The regularity of K amounts to solvability of the equation $a^2x = a$ for each $a \in K$ (the equation $aa'a = a$ in the case of a noncommutative K).

If a semiprime commutative ring K is finitely complete with respect to the B-metric d then each annihilator ideal of K is generated by an idempotent, implying that K is regular. In this event the mapping $\jmath : e \mapsto e \cdot K$ is a Boolean isomorphism of $\mathfrak{P}(K)$ to $\mathscr{B}(K)$.

$\lhd$ Take an annihilator ideal $b \in \mathscr{B}(K)$. Since the B-ring K is finitely complete, there is an element e in K such that $b \wedge d(1, e) = 0$ and $b^* \wedge d(0, e) = 0$; i.e., $e := \operatorname{mix}\{b1, b^*0\}$. This element is an idempotent, since from 4.5.2 (2) it follows that $d(e^2, e) = d(e, 0) \wedge d(1, e) \leq b \wedge b^{\perp} = 0$. In particular, $e \perp (1 - e)$. Hence, the annihilator ideals $d(e, 0) = \{e\}^{\perp\perp}$ and $d(1, e) = \{1 - e\}^{\perp\perp}$ are disjoint, yielding $d(e, 0) = b$ and $d(1, e) = b^{\perp}$. Now, using the equality $d(ex, x) = d(1, e) \wedge d(x, 0)$ (cf. 4.5.2 (2)) and given $x \in K$, infer

$$x \in b \leftrightarrow d(x, 0) \leq b \leftrightarrow d(ex, x) = 0 \leftrightarrow ex = x.$$

Consequently, $b = eK$. The remaining details are all evident. $\rhd$

4.5.4. A subset S of K is *dense* if $S^{\perp} = \{0\}$; i.e., the equality $k \cdot S = \{0\}$ implies $k = 0$ for all $k \in K$. A ring K is *rationally complete* if, to each dense ideal $J \subset K$ and each group homomorphism $h : J \to K$ satisfying $h(kx) = kh(x)$ for all $k \in K$ and $x \in J$, there is an element r in K such that $h(x) = rx$ for all $x \in J$.

Theorem. *Each rationally complete ring K is a universally complete B-ring, with $B = \mathscr{B}(K)$. If K is regular then the converse holds: Every universally complete B-ring is rationally complete.*

$\lhd$ Let (b_ξ) be a partition of unity in the Boolean algebra B of the annihilator ideals of K. Assume also that (k_ξ) is a family in K. Denote by J the set of all sums like $\sum_\xi x_\xi$, with $x_\xi \in b_\xi$ and at most finitely many of x_ξ are nonzero. Then J is a dense ideal. Define the mapping $h : J \to K$ by the formula $h(x) := k_\xi x$ for $x \in b_\xi$. Clearly, h obeys the needed conditions in the definition of rational completeness. Therefore, we may find $r \in K$ satisfying $h(x) = rx$ for all $x \in J$. If $x \in b_\xi$ then $h(x) = rx = k_\xi x$ and $x(r - k_\xi) = 0$. Hence, $b_\xi \subset \{r - k_\xi\}^{\perp} = d(r, k_\xi)$, implying that $b_\xi \wedge d(r, k_\xi) = 0$ and $r = \operatorname{mix}(b_\xi k_\xi)$.

Assume now that K is a regular ring. Take an ideal $J \subset K$ and a ring homomorphism $h : J \to K$. Using the Kuratowski–Zorn Lemma, choose an inclusion maximal disjoint family (e_ξ) in $J \cap \mathfrak{P}(K)$. Since our B-ring K is universally complete, there is an element k in K satisfying $e_\xi k = e_\xi h(e_\xi) = h(e_\xi)$. Note that $e_\xi kx = xh(e_\xi) = e_\xi h(x)$; i.e., $e_\xi(h(x) - kx) = 0$ for all ξ and $x \in J$. Now, if $h(x) \neq kx$ then $e_0(h(x) - kx) \neq 0$ for some nonzero idempotent $e_0 \in \mathfrak{P}(K)$. But then we would have $e_0 \perp e_\xi$ for all ξ, which contradicts the maximality of (e_ξ). $\rhd$

4.5.5. Notice the three corollaries to the just-established fact:

(1) *Every rationally complete semiprime ring is regular.*

(2) *Each annihilator ideal of a rationally complete semiprime commutative ring is a rationally complete ring.*

Say that a ring K *selfinjective* if K is an injective K-module. Recall that a K-module M is *injective* if, to whatever K-module N, a K-submodule N_0 of N, and a K-homomorphism $h_0 : N_0 \to M$, there is an extension of h_0 to a K-homomorphism $h : N \to M$. Baer's Criterion asserts that a K-module M is injective if and only if, to $J \subset K$ and a K-homomorphism $h : J \to M$, there is an element m in M such that $h(x) = mx$ for all $x \in J$ (see, for instance, [50] or [150]).

(3) *A ring K is rationally complete if and only if K is selfinjective.*

$\lhd$ Consider a homomorphism $h : J \to K$, with J an ideal of a rationally complete ring K. By 4.5.4, $K_0 := J^{\perp\perp} = eK$ for some idempotent $e \in K$. Since K_0 is a rationally complete ring and $eh : J \to K_0$ is a homomorphism, there is an element k in K such that $eh(x) = kx$ for all $x \in J$. It suffices to note that $eh(x) = h(ex) = h(x)$ for all $(x \in J)$ and complete proving $\to$. The implication $\leftarrow$ follows from Baer's Criterion. $\rhd$

4.5.6. Theorem. *Let $\mathscr{K} \in \mathbf{V}^{(B)}$ satisfy $[\![\mathscr{K}$ is a field$]\!] = \mathbf{1}$. Then $\mathscr{K}{\downarrow}$ is a rationally complete semiprime commutative ring and there is an isomorphism $\jmath$ of the Boolean algebra B to the Boolean algebra $\mathscr{B}(\mathscr{K}{\downarrow})$ of the annihilator ideals of $\mathscr{K}$ such that*

$$b \leq [\![x = 0]\!] \leftrightarrow x \in \jmath(b^*) \quad (x \in K,\ b \in B).$$

$\lhd$ Everything follows from 4.2.8, 4.5.3, and 4.5.4. It suffices to note that, by 4.2.8 (4), the projection $\jmath(b)$ corresponds in a one-to-one manner to the annihilator ideal $\jmath(b)$. $\rhd$

4.5.7. We proceed with Boolean valued analysis "in the field."

(1) Theorem. *Let K be a rationally complete semiprime commutative ring. Also, let B stand for the complete Boolean algebra $\mathscr{B}(\mathscr{K}{\downarrow})$ of the annihilator ideals of K. Then there is a field $\mathscr{K}$ inside $\mathbf{V}^{(B)}$ such that the rings K and $\mathscr{K}{\downarrow}$ are isomorphic.*

◁ Appeal to Theorem 4.3.3. The ring K is a universally complete algebraic B-system by 4.5.4. Consequently, the isomorphism $\imath$ of 4.3.3 (3) (in the sense of algebraic B-systems) is a bijection. Since K is a commutative B-ring, from 4.3.3 (4) it follows that $[\![\mathscr{K}$ is a commutative ring$]\!] = \mathbf{1}$. We are left with showing that every nonzero member of $\mathscr{K}$ is invertible; i.e., $[\![\mathscr{K} \models \varphi]\!] = \mathbf{1}$, with φ standing for $(\forall y)(\exists x)(y \neq 0 \to xy = 1)$. By 4.3.3 (4) it suffices to check that $|\varphi|^K = \mathbf{1}$; i.e., $K \models_B \varphi$.

Since K is a regular ring (cf. 4.5.3 and 4.5.4), to each $y \in K$ there is some $x \in K$ satisfying $y^2x = y$. The following implications are evident:

$$
\begin{aligned}
y^2x = y \to y(yx-1) = 0 \to y \in \{yx-1\}^{\perp} \\
\to \{y\} \subset \{yx-1\}^{\perp} \to \{y\}^{\perp\perp} \subset \{yx-1\}^{\perp\perp\perp} \\
\to \{y\}^{\perp\perp} \subset \{yx-1\}^{\perp}.
\end{aligned}
$$

Recalling the definition of d, infer $d(y,0) \leq d(yx,1)^{\perp}$. Using the definition of the B-valued interpretation of atomic formulas in 4.1.8, conclude that, to each $y \in K$, there is some $X \in K$ satisfying $|y \neq 0 \to yx = 1|^K = \mathbf{1}$. Using the definitions of 4.1.8 again, arrive at $|\varphi|^K = \mathbf{1}$, as desired. ▷

(2) Corollary. *The Horn theories of rationally complete semiprime commutative rings and fields coincide.*

4.5.8. We now give the construction of the so-called complete ring of fractions on using the above results on Boolean valued representation. We start with recalling a few definitions.

A ring $\widehat{K}$ is a *classical ring of fractions* of a ring K provided that there is a ring monomorphism $\lambda : K \to \widehat{K}$ such that $\lambda(x)$ is invertible in $\widehat{K}$ for each regular $x \in K$ and, moreover,

$$
\widehat{K} = \{\lambda(x)\lambda(y)^{-1} : x, y \in K;\ y \text{ is regular in } K\}.
$$

Considering $\widehat{K}$ up to isomorphism, we speak about *the* classical ring of fractions. If K is an integral domain then $\widehat{K}$ is a field called the *field of fractions* of K. Denote the classical ring of fractions of K by $Q_{\mathrm{cl}}(K) := \widehat{K}$. Note that $Q_{\mathrm{cl}}(K) = S^{-1}h(K)$ if we take the set of regular elements of K as the multiplicative set S in the definition of 4.2.6.

Since K is an algebraic B-system; therefore, by 4.3.5 (2) K possesses a universal completion $(K', \imath)$, where $\imath : K \to K'$ is a ring monomorphism. The ring $Q_B(K) :=$ K' is also referred to as *orthogonal completion* of K.

The ring $Q(K) := Q_{\mathrm{cl}}(Q_B(K))$, together with the monomorphism $\kappa := \lambda \circ \imath$, is the *complete ring of fractions* of K.

Theorem. *Assume that K is a semiprime commutative ring and B stands for the Boolean algebra $\mathscr{B}(K)$ of the annihilator ideals of K. Denote by K the Boolean valued representation of K viewed as an algebraic B-system. Then $[\![\mathscr{K}$ is an integral domain$]\!] = \mathbf{1}$. Moreover, there are elements $\mathscr{F}, \lambda \in \mathbf{V}^{(B)}$ such that the following hold:*

(1) $\mathbf{V}^{(B)} \models$ *"$\mathscr{F}$ is the field of fractions of the integral domain $\mathscr{K}$, and $\lambda : \mathscr{K} \to \mathscr{F}$ is an embedding of $\mathscr{K}$ into the ring of fractions of $\mathscr{K}$";*

(2) *$(\mathscr{F}{\downarrow}, \lambda{\downarrow} \circ \imath)$ is the complete ring of fractions of K, where $\imath : K \to K' := K{\downarrow}$ is the canonical embedding of K to K'.*

$\triangleleft$ The Boolean valued representation $\mathscr{K} := K^{\sim}$ of the algebraic B-system (B-ring) K is a ring inside $\mathbf{V}^{(B)}$, cf. 4.3.1, 4.3.3, and 4.5.2 (1). In accord with 4.1.13, the B-valued disjointness Δ on K is defined by the formula $\Delta(x,y) := \big(d(x,0) \wedge d(y,0)\big)^*$. From 4.5.2 (2) it follows now that $\Delta(x,y) = \big(d(xy,0)\big)^* = [\![xy = 0]\!]$. Hence, the Boolean valued representation δ of Δ satisfies $[\![\delta(x,y) \leftrightarrow xy = 0]\!]$. Consequently, δ relates to the ring multiplication of $\mathscr{K}$ in much the same way as Δ relates to the ring multiplication of K'. By 4.3.5 (6), δ is a simple disjointness, which means that $[\![\mathscr{K}$ is an integral domain$]\!] = \mathbf{1}$ in view of 4.5.1 (1).

The existence of $\mathscr{F}, \lambda \in \mathbf{V}^{(B)}$ satisfying (1) follows from the maximum principle and fact that the ring of fractions of an integral domain is a field. Put $K' := \mathscr{K}{\downarrow}$, and let $\imath : K \to K'$ stand for the respective canonical monomorphism (cf. 4.3.3). Then K' is the orthogonal completion of K; i.e., $K' = Q_B(K)$. Moreover, from 4.2.8 (3) it follows that $\mathscr{F}{\downarrow} = Q_{\mathrm{cl}}(K')$. Thus, $\mathscr{F}{\downarrow} = Q(K)$. $\triangleright$

4.5.9. The above theorem provides various corollaries on the structure of a ring of fractions. A few of them follow.

(1) *The complete ring of fractions of a semiprime commutative ring is rationally complete (consequently, selfinjective and regular).*

$\triangleleft$ The claim is immediate from 4.5.5 (1, 3), 4.5.6, and 4.5.8. $\triangleright$

(2) *The Boolean algebra $B := \mathscr{B}(K)$ of the annihilator ideals of a semiprime commutative ring K is isomorphic with the Boolean algebra of the annihilator ideals of each of the rings K' and $Q(K)$. The isomorphisms are carried out as follows:*

$$g_\imath : L \mapsto \imath^{-1}(L) \; (L \in \mathscr{B}(K')), \quad g_\kappa : L \mapsto \kappa^{-1}(L) \quad (L \in \mathscr{B}(Q(K))).$$

$\triangleleft$ A consequence of 4.2.8 and 4.3.5 (6). $\triangleright$

(3) *The complete ring of fractions $Q(K)$ of a semiprime commutative ring K is an injective K-module.*

$\lhd$ By Baer's Criterion (cf. 4.5.5) it suffices to prove that if J is an ideal of K and $h : J \to Q(K)$ is a K-homomorphism then, for some $q \in Q(K)$, the following holds: $h(x) = qx$ for all $x \in J$. By Theorem 4.5.8, there is no loss of generality in assuming that $K \subset K' := \mathscr{K}{\downarrow} \subset Q(K) = \mathscr{F}{\downarrow}$. Given $x \in J$ and $k \in K$, note that $x \perp k$ implies $h(x) \perp k$. Thus, $x \in b \to h(x) \in g_\kappa^{-1}(b)$ for all $b \in B$, and so h is an extensional mapping. Assign $\mathscr{J} := J{\uparrow}$ and $\eta := h'{\uparrow}$. Then $\mathscr{J}$ is an ideal of $\mathscr{K}$, and $\eta : \mathscr{J} \to \mathscr{F}$ is a $\mathscr{K}$-homomorphism.

It suffices now to show that we may find $q \in \mathscr{F}$ so that $\eta(x) = qx$ for all $x \in \mathscr{J}$. The last claim is immediate from the evident formula $a\eta(x) = \eta(ax) = x\eta(a)$ holding for all $a, x \in \mathscr{J}$. Indeed, if $a \neq 0$ then we may put $q := \eta(a)a^{-1} \in \mathscr{F}$. $\rhd$

A submodule M of a K-module $\widehat{M}$ is *massive* or *essential* if to each $0 \neq x \in \widehat{M}$ there is some $k \in K$ such that $kx \neq 0$ and $kx \in M$. An *injective hull* of a ring K is a pair $(\widehat{K}, \tau)$ such that $\widehat{M}$ is an injective K-module, $\tau : M \to \widehat{M}$ is a monomorphism, and $\tau(M)$ is a massive submodule of $\widehat{M}$.

(4) *The $(Q(K), \kappa)$ is an injective hull of a semiprime commutative ring K viewed as a K-module.*

$\lhd$ By (3) it suffices to check only that $\kappa(K)$ is a massive submodule of the K-module $Q(K)$. We may moreover assume that $K \subset Q(K)$. Hence, we are left with demonstrating that to each $0 \neq q \in Q(K)$ there is some $k \in K$ such that $kq \neq 0$ and $kq \in K$.

By the definition of $Q(K)$, there are families $(x_\xi) \subset K$ and $(y_\xi) \subset K$ and a partition of unity $(b_\xi) \subset B$ satisfying $q = xy^{-1}$, $x = \operatorname{mix}(b_\xi x_\xi)$, and $y = \operatorname{mix}(b_\xi y_\xi)$. Since $q \neq 0$; therefore, for some index ξ we have $ex_\xi \neq 0$, where e is the idempotent of K' corresponding to the ideal $b = b\xi$. It is also clear that $ey_\xi \neq 0$ because y is a regular element. Let a be an arbitrary nonzero member of the ideal b, with $ax_\xi \neq 0$. Put $k := ay_\xi = aey_\xi$. Then $qk = a(ex)(y_\xi y^{-1}) = ax_\xi = aex_\xi \in b \subset K$. $\rhd$

A *fraction* we call a homomorphism of K-modules $J \to K$, where J is a dense ideal of K. Equip the set of fractions with the following equivalence: Two fractions are equivalent if they agree on the intersection of their domains. It is an easy matter to make the resultant factor set into a ring (for details, see [150]). Denote the new ring by $Q'(K)$.

(5) *The rings $Q(K)$ and $Q'(K)$ are isomorphic.*

$\lhd$ We again consider K as a subring of $Q(K)$. Using (4), to each fraction $h \in Q'(K)$ we may assign the element $\sigma(h)$ such that $h(x) = \sigma(h)x$ for all x in the domain of h. Clearly, $h \mapsto \sigma(h)$ is a ring monomorphism. We are left with demonstrating that h is a surjection. To this end, take $q \in Q'(K)$ and put $J := \{k \in K : qk \in K\}$. Then J is a dense ideal of K. If the fraction h_q is defined by the formula $h_q : x \mapsto qx$ then $\sigma(h_q) = q$, which completes the proof. $\rhd$

A *ring of fractions of K in the Utami sense* is a pair (R, ν), with R a ring and $\nu : K \to R$ a ring monomorphism provided that there is a monomorphism $\tau : R \to Q(K)$ satisfying $\kappa = \tau \circ \nu$.

(6) *Let K' stand for the universal completion of a semiprime commutative ring K considered as an algebraic B-system. Then K' is a ring of fractions of K in the Utami sense.*

$\lhd$ The claim is immediate from the definition of a ring of fractions on letting $\nu := \imath$ and $\tau := \lambda$. $\rhd$

(7) *There is a unique (up to isomorphism) rationally complete ring of fractions $Q(K)$ of a semiprime commutative ring K.*

$\lhd$ This follows for instance from the fact that the injective hull is unique up to isomorphism. $\rhd$

4.5.10. COMMENTS.

(1) It is not a new idea to study regular commutative rings by considering the properties of appropriate fields. For instance, these rings were studied by representing them as subproducts of fields or as the ring of global sections of a ring bundle over a Boolean topological space [203, 212]. The approach of the current section unifies this idea and is advantageous as regards technique and methodology.

(2) Theorem 4.5.8 shows that, from the standpoint of $\mathbf{V}^{(B)}$, the complete ring of fractions of a semiprime ring K is simply the field of fractions of the integral domain we obtain by embedding K in $\mathbf{V}^{(B)}$, with B the Boolean algebra of annihilator ideals of K.

(3) A more explicit exposition is available of all preliminaries to ring theory, see for instance [50, 61, 150]. The results of 4.5.6 and 4.5.7 belong to E. I. Gordon [65]. Similar results were published somewhat later by K. Smith [222] who factually established equivalence of the category of regular commutative rings and the category of Boolean valued fields. Using this fact, K. Smith demonstrated that a regular commutative ring has an algebraic closure.

(4) The above methods apply to more general classes of rings. For instance, the relation of 4.5.1 is a disjointness also in the case of a noncommutative ring without nonzero nilpotents. Consequently, the set of annihilator ideals of such a ring K provides a complete Boolean algebra, and K itself ascends to $\mathbf{V}^{(B)}$ becoming a ring without zero divisors.

(5) Starting with the results of this section and using the same technique, we may come to analogous results about modules, cf. [66].

A module M over a ring K is *separated* if the equality $J \cdot x = \{0\}$ implies that $x = 0$ for every $x \in M$ and every dense ideal $J \subset K$.

Theorem. *Let $\mathscr{M}$ be a vector space over a field $\mathscr{K}$ inside $\mathbf{V}^{(B)}$. Let also $\imath : B \to \mathscr{B}(\mathscr{K}\!\downarrow)$ stand for the Boolean isomorphism of 4.5.3 (2). Then $\mathscr{M}\!\downarrow$ is a unital separated injective module over $\mathscr{K}$ satisfying*

$$b \le [\![x = 0]\!] \leftrightarrow \imath(b)x = 0 \quad (x \in \mathscr{M}\!\downarrow,\ b \in B).$$

(6) If a K-module M is separated then the B-semimetric d acting by the rule

$$d(x, y) := \bigwedge \{b \in B : b^*x = b^*y\} \quad (x, y \in M)$$

is a B-metric. Consequently, a separated K-module may be treated as an algebraic B-system, which leads to the following result (cf. [66]).

Theorem. *Assume that K is a rationally complete commutative ring. Assume also that $B = \mathscr{B}(K)$ and $\mathscr{K}$ is the Boolean valued representation of K. Let M be a unital separated injective K-module. Then there is some $\mathscr{M} \in \mathbf{V}^{(B)}$ such that $[\![\, \mathscr{M}$ is a vector space over $\mathscr{K}\,]\!]$. In this event, there are isomorphisms of algebraic B-systems $\imath_K : K \to \mathscr{K}\!\downarrow$ and $\imath_M : M \to \mathscr{M}\!\downarrow$ such that*

$$\imath_M(ax) = \imath_K(a)\imath_M(x) \quad (a \in K,\ x \in M).$$

Chapter 5

Boolean Valued Analysis of Banach Spaces

The Boolean valued inverse $\mathbf{V}^{(B)}$ associated with a fixed Boolean algebra B is one of the arenas of mathematical events. Indeed, by virtue of the transfer and maximum principles, $\mathbf{V}^{(B)}$ contains numbers and groups as well as the Lebesgue and Riemann integrals, with the Radon–Nikodým theorem and the Jordan normal form of a matrix available.

The elementary technique of ascending and descending which we become acquainted with when considering algebraic systems shows each of mathematical objects in $\mathbf{V}^{(B)}$ to be a representation of an analogous classical object with an additional structure induced by the algebra B. This relates in particular to functional-analytical objects.

In this chapter we present the facts that are associated with Boolean valued representation of the latter objects. Our main topic is Banach spaces in Boolean valued universes. It turns out that these spaces are inseparable from ordered vector spaces and, above all, K-spaces which were introduced by L. V. Kantorovich at the beginning of the thirties.

The fundamental result of Boolean valued analysis in regard to this aspect is Gordon's Theorem 5.2.2 which we may read as follows: *Every universally complete K-space is an interpretation of the reals in an appropriate Boolean valued universe.* Moreover, each theorem about the reals within Zermelo–Fraenkel set theory has an analog in the original K-space. Translation of theorems is carried out by appropriate general operations of Boolean valued analysis.

Theorems 5.2.4, 5.4.2, and 5.5.11 also rank among the principal results of the current chapter. The first of them claims that every Archimedean vector lattice embeds in a suitable universe $\mathbf{V}^{(B)}$, becoming a vector sublattice of the reals viewed as a vector space over some dense subfield of the reals. The second declares that every lattice normed space may be represented as a dense subspace of a Banach space viewed a vector space over some field, e.g., the rationals, in an appropriate $\mathbf{V}^{(B)}$. Finally, the third theorem means essentially that a Banach space X appears in result of bounded descent from a Boolean valued model if and only if X includes a complete Boolean algebra of norm one projections which possesses the cyclicity property. In other words, X is a Dedekind complete lattice normed space and the norm of X is a mixed norm. This fact serves as a starting point for the approach to involutive algebras which we pursue in the next chapter.

5.1. Vector Lattices

In this section we give some preliminaries to the theory of vector lattices; a more explicit exposition may be found elsewhere [1, 4, 103, 104, 158, 214, 253, 258].

5.1.1. Let $\mathbb{F}$ be a totally ordered field. Consider an algebraic system E whose signature contains the symbols $+, 0, \leq$, and λ, with λ ranging over $\mathbb{F}$ and denoting a unary operation. Given $\lambda \in \mathbb{F}$, call this operation λ-*scaling* or (*scalar*) *multiplication by* λ. Assume that E obeys the conditions:

(1) $(E, +, 0, \leq)$ is an ordered commutative group;

(2) E is a vector space over $\mathbb{F}$;

(3) Each multiplication by a positive λ in $\mathbb{F}$ is a positive endomorphism of the ordered group $(E, +, 0, \leq)$.

Say in this event that E is an *ordered vector space*.

Therefore, an ordered vector space E may be defined as a pair $(E, \leq)$, with E a vector space over $\mathbb{F}$ and $\leq$ a *vector order* on E; i.e., an order relation on E compatible with vector structure. Informally speaking, we may "sum inequalities in E and multiply them by positive members of $\mathbb{F}$." Formally, a vector order on E must be a cone in E^2 as well as an order on E.

Equipping a vector space E over $\mathbb{F}$ with a vector order amounts to defining some *positive cone* E^+ of E, that is a subset of E satisfying the conditions: $E^+ + E^+ \subset E^+$; $\lambda E^+ \subset E^+$ $(0 \leq \lambda \in \mathbb{F})$; and $E^+ \cap (-E^+) = 0$. The order $\leq$ on E and the positive cone E^+ are connected as follows:

$$x \leq y \leftrightarrow y - x \in E^+ \quad (x, y \in E).$$

Clearly, all notions and results of the theory of ordered groups apply freely to ordered vector spaces. For instance, when we say that an ordered vector space E is Archimedean or speak about some ideal of E, we imply the underlying ordered group of E.

5.1.2. A *vector lattice* is an ordered vector space whose underlying group is lattice ordered. Hence, each finite set $\{x_1, \dots, x_n\}$ in a vector lattice E has the *join*, i.e. the least upper bound $x_1 \vee \ldots \vee x_n := \sup\{x_1, \ldots, x_n\}$, and the *meet*, i.e. the greatest lower bound $x_1 \wedge \ldots \wedge x_n := \inf\{x_1, \ldots, x_n\}$. In particular, each member x of a vector lattice has the *positive part* $x^+ := x \vee \mathbf{0}$, the *negative part* $x^- := (-x)^+ := -x \wedge \mathbf{0}$, and the *absolute value* or *modulus* $|x| := x \vee (-x)$.

Recall that we introduce the *disjointness* of E by the formula

$$\perp := \{(x, y) \in E \times E : |x| \wedge |y| = 0\}.$$

A set K is a *band* of E (or *component* in the Russian literature) provided that K coincides with the disjoint complement of a subset of E; i.e., K equals to

$$M^\perp := \{x \in E : (\forall\, y \in M)(x \perp y)\}$$

where M is some nonempty subset of E. If K has the shape $\{u\}^{\perp\perp}$ then K is a *principle band* and $|u|$ is an *order unity* or *order unit* of K.

The inclusion ordered set $\mathscr{B}(E)$ of all bands of E is a complete Boolean algebra. The Boolean operations of $\mathscr{B}(E)$ take the shape:

$$L \wedge K = L \cap K, \quad L \vee K = (L \cup K)^{\perp\perp}, \quad L^* = L^\perp \quad (L, K \in \mathscr{B}(E)).$$

The Boolean algebra $\mathscr{B}(E)$ is the *base* of E.

Let K be a band of a vector lattice E and $0 \leq x \in E$. Assume that the set $\{u \in K : 0 \leq u \leq x\}$ has a supremum in E. This supremum $\sup\{u \in K : 0 \leq u \leq x\}$ is unique. Call it the *projection* of x to K and denote it by $[K]x$ or $\mathrm{Pr}_K\, x$. Given an arbitrary $x \in E$, put $[K]x := [K]x^+ - [K]x^-$.

The projection of an element $x \in E$ to a band K exists if and only if we have the decomposition $x = y + z$ with $y \in K$ and $z \in K^\perp$, in which case $y = [K]x$ and $z = [K^\perp]x$. Assume that each element $x \in E$ has a projection to K. Then $x \mapsto [K]x$ $(x \in E)$ is a linear idempotent operator and $0 \leq [K]x \leq x$ for all $0 \leq x \in E$. We call $[K]$ the *band projection* to K and say that K is a *projection band.* Say that a vector lattice E possesses the *projection property* (*principal projection property*) if every band (principal band) of E is a projection band. If a vector lattice E has the projection property and each *disjoint positive subset* of E (i.e., a subset composed of disjoint positive elements) has a supremum in E then E is a *universally complete vector lattice* or an *extended vector lattice* in the Russian literature.

5.1.3. An element $\mathbf{1} \in E$ is called an *order unity* or *order unit* of E provided that $\{\mathbf{1}\}^{\perp\perp} = E$; i.e., if E has no nonzero elements disjoint from $\mathbf{1}$. In other words, an order unit $\mathbf{1}$ of E is an order unit of the band E of E. Assume that some $0 \leq e \in E$ satisfies $e \wedge (\mathbf{1} - e) = 0$. We then say that e is a *unit element* relative to $\mathbf{1}$. The set $\mathfrak{C}(\mathbf{1}) := \mathfrak{C}(E)$ of all unit elements is a Boolean algebra under the induced order from E. The lattice operations of $\mathfrak{C}(\mathbf{1})$ are inherited from E, while the Boolean complement has the form $e^* = 1 - e$ for $e \in \mathfrak{C}(\mathbf{1})$. A disjoint positive family in E is *total* or *complete* provided that E has no nonzero element disjoint from every member of the family. Clearly, a total family pretends to play the role of a "compound" order unit of E.

Henceforth, unless specifying $\mathbb{F}$ explicitly, we imply only vector lattices over the reals $\mathbb{R}$ equipped with the natural total order. Considering the ideal $I(u) := \bigcup_{n=1}^{\infty}[-nu, nu]$ generated by an element $0 \leq u \in E$, we may introduce the following seminorm:

$$\|x\|_u := \inf\{\lambda \in \mathbb{R} : |x| \leq \lambda u\} \quad (x \in I(u)).$$

If $I(u) = E$ then call u a *strong unity* or *strong order unit* and E, a *vector lattice of bounded elements*. The seminorm $\|\cdot\|_u$ is a norm if and only if E is Archimedean.

An element $x \geq 0$ of a lattice is *discrete*, if $[0, x] = [0, 1]x$; i.e., in the case when from $0 \leq y \leq x$ it follows that $y = \lambda x$ for some $0 \leq \lambda \leq 1$. A vector lattice E is *discrete* if to each $0 < y \in E$ there is a discrete element $x \in E$ satisfying $0 < x \leq y$. If E has no nonzero discrete elements then E is *continuous*.

5.1.4. A *Kantorovich space* or, briefly, a *K-space* is a Dedekind complete vector lattice; i.e., a vector lattice whose every nonempty order bounded subset has a supremum and an infimum. Sometimes, a K-space is also referred to as boundedly order complete vector lattice. A vector lattice E is a *K_σ-space* if each countable nonempty bounded subset of E has a supremum and an infimum in E. Every K_σ-space, as well as every K-space, is Archimedean.

Denote the set of all band projections of E by $\mathfrak{P}\mathfrak{r}(E)$. Given π and ρ in $\mathfrak{P}\mathfrak{r}(E)$, put $\pi \leq \rho$ if and only if $\pi x \leq \rho x$ for all $0 \leq x \in E$.

Theorem. *Let E be an arbitrary K-space. Then the mapping $K \mapsto [K]$, sending a band K to the band projection $[K]$, is an isomorphism between the Boolean algebras $\mathscr{B}(E)$ and $\mathfrak{P}\mathfrak{r}(E)$.*

If E has an order unit then the mappings $\pi \mapsto \pi\mathbf{1}$ from $\mathfrak{P}\mathfrak{r}(E)$ to $\mathfrak{C}(E)$ and $e \mapsto \{e\}^{\perp\perp}$ from $\mathfrak{C}(E)$ to $\mathscr{B}(E)$ are also isomorphisms of the respective Boolean algebras.

The band projection π_u to the principal band $\{u\}^{\perp\perp}$, with $0 \leq u \in E$, can be obtained by a simpler rule than that in 5.1.2; namely,

$$\pi_u x = \sup\{x \wedge (nu) : n \in \mathbb{N}\} \quad (0 \leq x \in E).$$

In particular, *every K_σ-space has the principal projection property.*

Let E be a K_σ-space with order unit $\mathbf{1}$. We call the projection of the unit to the band $\{x\}^{\perp\perp}$ the *trace* of x and denoted it by e_x. Therefore, $e_x := \sup\{\mathbf{1} \wedge (n|x|) : n \in \mathbb{N}\}$. The trace e_x serves both as an order unit of $\{x\}^{\perp\perp}$ and a unit element of E. Given a real λ, denote the trace of the positive part of $\lambda\mathbf{1} - x$ by e^x_λ; i.e., $e^x_\lambda := e_{(\lambda\mathbf{1}-x)^+}$. The function $\lambda \mapsto e^x_\lambda$, with $\lambda \in \mathbb{R}$, arising in this case is called the *spectral function* or *characteristic* of x.

5.1.5. We now turn to algebra endowed with compatible order.

(1) Assume that E is an algebra over a field $\mathbb{F}$. Assume further that E is furnished with some order so that E becomes an ordered vector space whose positive cone is closed under multiplication. In this event E is an *ordered algebra* or an *ordered* $\mathbb{F}$*-algebra*. We may say that an ordered algebra E is an algebraic system E whose signature contains the symbols $+, 0, \leq, \lambda, \cdot$, with λ ranging over $\mathbb{F}$ and standing for λ-scaling, provided that

(a) E is an ordered vector space;

(b) $(E, +, \mathbf{0}, \leq, \cdot\,)$ is an ordered ring.

Say that E is a *lattice ordered algebra* (*f-algebra*) if the underlying ring of E is a lattice ordered ring (an f-ring). An f-algebra is *faithful* if for whatever x and y the equality $x \cdot y = 0$ implies that $x \perp y$. Clearly, an f-algebra is faithful if and only if it has no nonzero nilpotents. Also, an f-algebra is faithful if and only if it has nonzero positive element whose square is zero (cf. 4.4.8).

(2) A *complex vector lattice* is the *complexification* $E \otimes iE$ of a real vector lattice E. As usual, we let i stand for the *imaginary unity* in any appropriate context here and in the sequel. Furthermore, it is a routine to require additionally that every member z of $E \otimes iE$ has the *absolute value* or *modulus*

$$|z| := \sup\{\mathrm{Re}(e^{i\theta}z) : 0 \leq \theta \leq \pi\}.$$

The conditions for E to ensure existence for the absolute value of each element in $E \otimes iE$ are easy to formulate. Any proviso is perfectly excessive for a K-space and even for a K_σ-space. So, a *complex K-space* is simply the complexification of a real K-space.

Speaking about the order properties of a complex vector lattice $E \otimes iE$, we always mean its real part E. The definitions of sublattice, ideal, projection band, etc. are all naturally abstracted to the case of a complex vector lattice by way of due complexification.

5.1.6. The order of a vector lattice provides various types of convergence.

Let $(\mathrm{A}, \leq)$ be an upward-directed set; i.e., $\leq \circ \leq^{-1} = \mathrm{A}^2$. Consider a *net* $(x_\alpha) := (x_\alpha)_{\alpha \in \mathrm{A}}$ in E. Call (x_α) an *increasing* (*decreasing*) net provided that $x_\alpha \leq x_\beta$ ($x_\beta \leq x_\alpha$) for all $\alpha \leq \beta$, $\alpha, \beta \in \mathrm{A}$.

A net (x_α) *converges in order* or is *order convergent* or *o-convergent* to $x \in E$ if there is a decreasing net $(e_\alpha)_{\alpha \in \mathrm{A}}$ in E satisfying $\inf_{\alpha \in \mathrm{A}} e_\alpha = 0$ and $|x - x_\alpha| \leq e_\alpha$ ($\alpha \in \mathrm{A}$). In this case x is the *order limit* or *o-limit* of (x_α); in symbols, $x = o\text{-}\lim x_\alpha$ or $x_\alpha \xrightarrow{(o)} x$.

Given an order bounded net (x_α) in a K-space for E, define the *upper o-limit* and *lower o-limit* (or *limit superior* and *limit interior*) of (x_α) by the formulas:

$$\limsup_{\alpha\in A} x_\alpha := \overline{\lim_{\alpha\in A}}\, x_\alpha := \inf_{\alpha\in A}\sup_{\beta\geq\alpha} x_\beta,$$

$$\liminf_{\alpha\in A} x_\alpha := \underline{\lim_{\alpha\in A}}\, x_\alpha := \sup_{\alpha\in A}\inf_{\beta\geq\alpha} x_\beta.$$

Clearly,

$$x = o\text{-}\lim x_\alpha \leftrightarrow \limsup x_\alpha = x = \liminf x_\alpha.$$

The net $(x_\alpha)_{\alpha\in A}$ *converges with regulator* or is *relatively uniform convergent* (*r-convergent*) to $x \in X$ if there are an element $0 \leq u \in E$, called the *regulator of convergence*, and a numerical net $(\lambda_\alpha)_{\alpha\in A} \subset \mathbb{R}$ satisfying $\lim \lambda_\alpha = 0$ and $|x - x_\alpha| \leq \lambda_\alpha u$ $(\alpha \in A)$. In this event, call x the *r-limit* of (x_α) and write $x = r\text{-}\lim x_\alpha$ or $x_\alpha \xrightarrow{(r)} x$. Clearly, relative uniform convergence with regulator u is convergence in norm in the normed space $(I(u), \|\cdot\|_u)$.

The presence of order convergence in a K-space allows us to determined the sum of an infinite family $(x_\xi)_{\xi\in\Xi}$. Indeed, given $\theta := \{\xi_1, \dots, \xi_n\} \in \mathscr{P}_{\text{fin}}(\Xi)$, put $y_\theta := x_{\xi_1} + \ldots + x_{\xi_n}$. So, we arrive at the net $(y_\theta)_{\theta\in\Theta}$, where $\Theta := \mathscr{P}_{\text{fin}}(\Xi)$ is naturally ordered by inclusion. Assuming that there is some x satisfying $x = o\text{-}\lim_{\theta\in\Theta} y_\theta$, we call the family (x_ξ) *summable in order*, or *order summable*, or *o-summable.* The element x is the *o-sum* of (x_ξ); in symbols, $x = o\text{-}\sum_{\xi\in\Xi} x_\xi$. Obviously, if $x_\xi \geq 0$ $(\xi \in \Xi)$ then for the o-sum of the family (x_ξ) to exist it is necessary and sufficient that the net $(y_\theta)_{\theta\in\Theta}$ be order bounded, in which case $o\text{-}\sum_{\xi\in\Xi} x_\xi = \sup_{\theta\in\Theta} y_\theta$. If (x_ξ) is a disjoint family then

$$o\text{-}\sum_{\xi\in\Xi} x_\xi = \sup_{\xi\in\Xi} x_\xi^+ - \sup_{\xi\in\Xi} x_\xi^-.$$

Every K-space E is *order complete* (*o-complete*) in the following sense: If $(x_\alpha)_{\alpha\in A}$ is a net in E satisfying the condition

$$\limsup |x_\alpha - x_\beta| = \inf_{\gamma\in A}\sup_{\alpha,\beta\geq\gamma} |x_\alpha - x_\beta| = 0,$$

then there is an $x \in E$ such that $x = o\text{-}\lim x_\alpha$.

5.1.7. EXAMPLES.

(1) Assume given a family $(E_\alpha)_{\alpha\in A}$ of vector lattices (f-algebras) over the same ordered field $\mathbb{F}$. Furnish the product $E := \Pi_{\alpha\in A} E_\alpha$ with the coordinatewise operations and order. Then E becomes a vector lattice (f-algebra) over $\mathbb{F}$.

In this case E is a Dedekind complete, universally complete, or discrete vector lattice if and only if all factors E_α have the same property. The base $\mathscr{B}(E)$ is isomorphic with the product of the family of the Boolean algebras $(\mathscr{B}(E_\alpha))_{\alpha \in \mathrm{A}}$. An element $e \in E$ is an order unit if and only if $e(\alpha)$ is an order unit in E_α for all $\alpha \in \mathrm{A}$.

In particular, the set $\mathbb{R}^{\mathrm{A}}$ ($\mathbb{C}^{\mathrm{A}}$) of all real (complex) functions on a nonempty set A is a universally complete discrete K-space (complex K-space).

(2) *Every ideal of a vector lattice is a vector lattice. Moreover, every ideal of a K-space is a K-space.*

In particular, $l_p(A)$, the space of p-summable families, is a K-space for $1 \leq p \leq \infty$ (cf. (1)). The same applies to the most important instance of an ideal in a vector lattice E which is an order dense ideal or a *foundation* of E in the Russian literature. The base of a vector lattice E is isomorphic with that of each order dense ideal of E.

(3) Let N be an ideal of a vector lattice E. Then the factor space $\widetilde{E} := E/N$ is also a vector lattice provided that the order on $\widetilde{E}$ is determined by the positive cone $\varphi(E^+)$, with $\varphi : E \to \widetilde{E}$ standing for the factor mapping.

The factor lattice E/N is Archimedean if and only if N is closed under relative uniform convergence. If E is an f-algebra and N is a ring and order ideal then E/N is an f-algebra. If E is a K_σ-space and N is sequentially order closed then E/N is a K_σ-space and φ is sequentially order continuous. The base of $\widetilde{E}$ is isomorphic to the complete Boolean algebra of Δ-bands $\mathfrak{R}_\Delta(E)$, where $\Delta := \{(x, y) \in E \times E : |x| \wedge |y| \in N\}$.

(4) Let $(\Omega, \mathscr{A})$ be a *measurable space*; i.e., Ω is a nonempty set and $\mathscr{A}$ is a σ-algebra of its subsets. Denote by $\mathscr{M}(\Omega, \mathscr{A})$ the set of all real (complex) measurable functions on Ω and equip $\mathscr{M}(\Omega, \mathscr{A})$ with the pointwise operations and order induced from $\mathbb{R}^\Omega$ (from $\mathbb{C}^\Omega$). Choose some σ-complete ideal $\mathscr{N}$ of the algebra $\mathscr{A}$. Let N comprise the functions $f \in \mathscr{M}(\Omega, \mathscr{A})$ such that $\{t \in \Omega : f(t) \neq 0\} \in \mathscr{N}$. Assign $M(\Omega, \mathscr{A}, N) := \mathscr{M}(\Omega, \mathscr{A})/N$. Then $\mathscr{M}(\Omega, \mathscr{A})$ and $M(\Omega, \mathscr{A}, \mathscr{N})$ are real (complex) K_σ-spaces and f-algebras at the same time.

Assume now that $\mu : \mathscr{A} \to \mathbb{R} \cup \{+\infty\}$ is a countably additive positive measure. The vector lattice $M(\Omega, \mathscr{A}, \mu) := M(\Omega, \mathscr{A}, \mu^{-1}(0))$ is a universally complete K-space provided that μ is a totally finite or σ-finite measure. In general, the Dedekind completeness property of $M(\Omega, \mathscr{A}, \mu)$ relates to the direct sum property for μ [82, 103]. However, we will confine exposition to the case of a σ-finite measure μ for the sake of simplicity.

The space $M(\Omega, \mathscr{A}, \mu)$ is continuous if and only if μ has no atoms. Recall that an *atom* of a measure μ is a set $A \in \mathscr{A}$ such that $0 < \mu(A)$ and if $A' \in \mathscr{A}$, $A' \subset A$, then $\mu(A') = 0$ or $\mu(A') = \mu(A)$.

If $M(\Omega, \mathscr{A}, \mu)$ is discrete then μ is a *purely atomic* measure; i.e., each set of nonzero measure contains an atom of μ. The coset of the identically one function is an order and ring unity in $M(\Omega, \mathscr{A}, \mu)$.

The base of the K-space $M(\Omega, \mathscr{A}, \mu)$ is isomorphic to the Boolean algebra $\mathscr{A}/\mu^{-1}(0)$ of measurable sets modulo zero measure sets.

By (2), the $L_p(\Omega, \mathscr{A}, \mu)$-space, $1 \leq p \leq \infty$, presenting an order dense ideal of $M(\Omega, \mathscr{A}, \mu)$, is a K-space.

(5) Assume that H is a complex Hilbert space and A is a strongly closed commutative algebra of bounded selfadjoint operators on H. Denote by $\mathfrak{P}(A)$ the set of all orthoprojections in H belonging to A. Then $\mathfrak{P}(A)$ is a complete Boolean algebra.

We now let A_∞ stand for the set of all densely defined selfadjoint operators a in H such that the spectral function $\lambda \mapsto e_\lambda^a$ of a takes values in $\mathfrak{P}(A)$. Denote by $\overline{A}_\infty$ the set of densely defined normal operators a in H such that if $a = u|a|$ is the *polar decomposition* of a then $|a| \in A_\infty$.

Furnish the sets A_∞ and $\overline{A}_\infty$ with the structure of an ordered vector space in a natural way. Indeed, given a and b in A_∞, define the sum $a + b$ and the product $a \cdot b$ as the unique selfadjoint extensions of the operators $h \mapsto ah + bh$ and $h \mapsto a \cdot bh$ with $h \in \operatorname{dom}(a) \cap \operatorname{dom}(b)$ and $\operatorname{dom}(c)$ standing for the domain of c. Moreover, granted $a \in A_\infty$, we say that $a \geq 0$ if and only if $\langle ah, h \rangle \geq 0$ for all $h \in \operatorname{dom}(a)$. The operations and order on $\overline{A}_\infty$ result from complexifying A_∞. The sets A_∞ and $\overline{A}_\infty$ with the above operations and order are a universally complete K-space and a universally complete complex K-space with base $\mathfrak{P}(A)$, respectively. In this case A is the K-space of bounded elements of A_∞.

(6) Take a topological space Q and denote by $\operatorname{Bor}(Q) := \operatorname{Bor}(Q, \mathbb{R})$ the set of all Borel functions from Q to $\mathbb{R}$ with addition, multiplication, and order introduced pointwise. Then $\operatorname{Bor}(Q, \mathbb{R})$ is a K_σ-space.

By N we denote the set of such Borel functions $f \in \operatorname{Bor}(Q)$ that $\{t \in Q : f(t) \neq 0\}$ is a *meager* set (i.e., a set of the first category). Let $B(Q)$ stand for the factor space $\operatorname{Bor}(Q)/N$ with the operations and order induced from $\operatorname{Bor}(Q)$. Then $B(Q)$ is a K-space whose base is isomorphic to the Boolean algebra of Borel subsets Q modulo meager sets.

If Q is a *Baire* space (i.e., every nonempty open subset of Q is not meager), then the base $\mathscr{B}(B(Q))$ is isomorphic to the Boolean algebra of all regular open (or regular closed) subsets of Q. Each of the spaces $\operatorname{Bor}(Q)$ and $B(Q)$ is a faithful f-algebra. The identically one function serves as an order and ring unity in these spaces. Replacing $\mathbb{R}$ with $\mathbb{C}$, we arrive at the complex K-space $B(Q)$.

(7) Let Q be a topological space again. Denote by $C(Q)$ the space of continuous real functions on Q. Then $C(Q)$ is a sublattice and subalgebra of $\operatorname{Bor}(Q)$. In particular, $C(Q)$ is a faithful Archimedean f-algebra. Generally speaking, $C(Q)$

is not a K-space. The Dedekind completeness property of $C(Q)$ amounts to the extremal disconnectedness property of Q (see 1.2.5). In the case of a uniformizable space Q the base of $C(Q)$ is isomorphic to the algebra of regular open sets.

We now let $\mathrm{LSC}(Q)$ stand for the set of (the cosets of) lower semicontinuous functions $f : Q \to \overline{\mathbb{R}} := \mathbb{R}\cup\{\pm\infty\}$ such that $f^{-1}(-\infty)$ is nowhere dense whereas the interior of the set $f^{-1}([-\infty, \infty))$ is dense in Q. As usual, two functions are equivalent if they agree on the complement of a meager set. The sum $f + g$ (the product $f \cdot g$) of $f, g \in \mathrm{LSC}(Q)$ is defined as the lower semicontinuous regularization of the pointwise sum $t \mapsto f(t) + g(t)$ $(t \in Q_0)$ (the pointwise product $t \mapsto f(t) \cdot g(t)$ $(t \in Q_0)$) where Q_0 is some dense subset of Q on which f and g are both finite. We this make $\mathrm{LSC}(Q)$ into a universally complete K-space and an f-algebra, with the base of $\mathrm{LSC}(Q)$ isomorphic to the algebra of regular open sets. Hence, if Q is Baire then $B(Q)$ and $\mathrm{LSC}(Q)$ are isomorphic K-spaces; if Q is uniformizable then $C(Q)$ is an (order) dense sublattice of $\mathrm{LSC}(Q)$.

5.1.8. A special role in the theory of vector lattices is played by the spaces of continuous functions assuming possibly infinite values on a nowhere dense set depending on a function. Before introducing these spaces, we need some preliminaries.

Given a function $f : Q \to \overline{\mathbb{R}}$ and $\lambda \in \overline{\mathbb{R}}$, put

$$\{f < \lambda\} := \{t \in Q : f(t) < \lambda\}, \quad \{f \leq \lambda\} := \{t \in Q : f(t) \leq \lambda\}.$$

(1) *Assume that Q is a topological space, Λ is a dense set in $\overline{\mathbb{R}}$, and $\lambda \mapsto U_\lambda$ $(\lambda \in \Lambda)$ is an increasing mapping from Λ to the inclusion ordered set $\mathscr{P}(Q)$. Then the following are equivalent:*

(a) *There is a unique continuous function $f : Q \to \overline{\mathbb{R}}$ satisfying*

$$\{f < \lambda\} \subset U_\lambda \subset \{f \leq \lambda\} \quad (\lambda \in \Lambda),$$

(b) *If λ, $\mu \in \Lambda$, and $\lambda < \mu$ then*

$$\mathrm{cl}(U_\lambda) \subset \mathrm{int}(U_\mu).$$

$\triangleleft$ The implication (a) $\to$ (b) is evident.

Prove (b) $\to$ (a). To this end, given $t \in Q$, put $f(t) := \inf\{\lambda \in \Lambda : t \in U_\lambda\}$. For the so-defined $f : Q \to \overline{\mathbb{R}}$, we easily see that $\{f < \lambda\} \subset U_\lambda \subset \{f \leq \lambda\}$. It is also clear that

$$\{f < \lambda\} = \bigcup\{U_\mu : \mu < \lambda \wedge \mu \in \Lambda\}, \{f \leq \lambda\} = \bigcap\{U_\nu : \lambda < \nu \wedge \nu \in \Lambda\}.$$

Note that by now we have used only the fact that $\lambda \mapsto U_\lambda$ is an increasing mapping.

Consider the mappings

$$\lambda \mapsto V_\lambda := \operatorname{int}(U_\lambda), \quad \lambda \mapsto W_\lambda := \operatorname{cl}(U_\lambda) \quad (\lambda \in \Lambda).$$

These are also increasing mappings. So, the above implies that there are functions g and $h : Q \to \overline{\mathbb{R}}$ such that

$$\{g < \lambda\} \subset V_\lambda \subset \{g \leq \lambda\}, \quad \{h < \lambda\} \subset W_\lambda \subset \{h \leq \lambda\} \quad (\lambda \in \Lambda).$$

From the definition of W_λ it follows that $U_\mu \subset W_\lambda$ for $\mu < \lambda$. Since Λ is dense in $\mathbb{R}$, to all $t \in Q$ and $\nu > f(t)$ there are λ, $\mu \in \Lambda$ such that $f(t) < \mu < \lambda < \nu$ and so $t \in U_\mu \subset W_\lambda$ and $h(t) < \lambda < \nu$. Letting ν tend to $f(t)$, obtain $h(t) \leq f(t)$. The same inequality is immediate for $f(t) = +\infty$. By analogy, $V_\mu \subset U_\lambda$ for $\mu < \lambda$. Hence, $f(t) \leq g(t)$ for all $t \in Q$.

Writing (b) as $W_\mu \subset V_\lambda$ ($\mu < \lambda$), and arguing as above, conclude that $g(t) \leq h(t)$ for all $t \in Q$. Therefore, $f = g = h$.

The fact that f is continuous follows from the equalities

$$\{f < \lambda\} = \{g < \lambda\} = \bigcup\{V_\mu : \mu < \lambda,\, \mu \in \Lambda\},$$
$$\{f \leq \lambda\} = \{h \leq \lambda\} = \bigcap\{W_\mu : \mu > \lambda,\, \mu \in \Lambda\},$$

since V_μ is open whereas W_μ is closed for all $\mu \in \Lambda$. ▷

(2) *Let Q be an extremally disconnected compact space; i.e., Q is a compact topological space wherein the closure of every open set is open. Assume that Q_0 is a dense open subset of Q and $f : Q_0 \to \mathbb{R}$ is a continuous function. Then there is a unique continuous function $\bar{f} : Q_0 \to \overline{\mathbb{R}}$ such that $f(t) = \bar{f}(t)$ $(t \in Q_0)$.*

◁ Indeed, if $U_\mu := \operatorname{cl}(\{f < \mu\})$ then the mapping $\mu \mapsto U_\mu$, with $\mu \in \mathbb{R}$, increases and meets the condition (b) of (1). Hence, there is a unique function $\bar{f} : Q \to \overline{\mathbb{R}}$ satisfying $\{\bar{f} < \mu\} \subset U_\mu \subset \{\bar{f} \leq \mu\}$ $(\mu \in \mathbb{R})$. Obviously, in this case $\bar{f} \restriction Q_0 = f$, i.e. the restriction of $\bar{f}$ to Q_0 coincides with f. ▷

(3) Denote by $C_\infty(Q)$ the set of all continuous functions $x : Q \to \overline{\mathbb{R}}$ assuming the values $\pm\infty$ possibly on a nowhere dense set. Order $C_\infty(Q)$ by assigning $x \leq y$ whenever $x(t) \leq y(t)$ for all $t \in Q$. Then, take $x, y \in C_\infty(Q)$ and put $Q_0 := \{|x| < +\infty\} \cap \{|y| < +\infty\}$. In this case Q_0 is open and dense in Q. According to (2), there is a unique continuous function $z : Q \to \overline{\mathbb{R}}$ such that $z(t) = x(t) + y(t)$ for $t \in Q_0$. It is this function z that we declare the sum of x and y.

In an analogous way we define the product of a pair of elements. Identifying the number λ with the identically λ function on Q, we obtain the product of $x \in C_\infty(Q)$ and $\lambda \in \mathbb{R}$.

Clearly, the space $C_\infty(Q)$ with the operations and order introduced above is a vector lattice and a faithful f-algebra. The identically one function is a ring and order unity.

We shall prove in the sequel that $C_\infty(Q)$ is a universally complete K-space. The base of $C_\infty(Q)$ is isomorphic with the Boolean algebra of all clopen subsets of the compact set Q.

5.1.9. Let E and F be vector lattices.

(1) A linear operator $U : E \to F$ is *positive* if $U(E_+) \subset F_+$; U is *regular* if it is representable as a difference of two positive operators; and, finally, U is *order bounded* or *o-bounded* if U sends every order bounded subset of E into an order bounded subset of F.

If F is a K-space then an operator is regular if and only if it is order bounded. The set of all regular (positive) operators from E into F is denoted by $L^\sim(E,F)$ $(L^\sim(E,F)_+)$.

Riesz–Kantorovich Theorem. *Assume that E is a vector lattice and F is a K-space. Then the space $L^\sim(E,F)$ of regular operators with positive cone $L^\sim(E,F)_+$ is a K-space.*

(2) Recall that an operator $U : E \to F$ is *order continuous* (or *o-continuous*) if, for every net $(x_\alpha)_{\alpha\in\mathrm{A}}$ in E, the equality $o\text{-}\lim_{\alpha\in\mathrm{A}} x_\alpha = 0$ yields $o\text{-}\lim_{\alpha\in\mathrm{A}} Ux_\alpha = 0$. Sequential o-continuity is understood likewise. The set of all order continuous regular operators equipped with the operations and order induced from $L^\sim(E,F)$ is denoted by $L_n^\sim(E,F)$. If $U \in L_n^\sim(E,F)$ then the band $\mathscr{N}(U)^\perp$, where $\mathscr{N}(U) := \{x \in E : U(|x|) = 0\}$, is the *carrier* or *band of essential positivity* of U. If $F = \mathbb{R}$ then we write $E_n^\sim$ rather than $L_n^\sim(E,\mathbb{R})$.

The space $L_n^\sim(E,F)$ is a band in $L^\sim(E,F)$ and so $L_n^\sim(E,F)$ is a K-space. If $f \in E_n^\sim$ and E_f is the carrier of f then the Boolean algebras $\mathscr{B}(f) := \mathscr{B}(\{f\}^{\perp\perp})$ and $\mathscr{B}(E_f)$ are isomorphic. A functional f is a unity in $E_n^\sim$ if and only if $\mathscr{N}(f)^\perp = E$.

(3) Consider a vector lattice E and a vector sublattice $D \subset E$. A linear operator U from D into E is said to be a *stabilizer* if $Ux \in \{x\}^{\perp\perp}$ for every $x \in D$. A stabilizer may fail to be regular. A regular stabilizer is called an *orthomorphism*.

Denote by $\mathrm{Orth}(E)$ the subspace of $L^\sim(E)$ comprising the orthomorphisms with domain E. We also let $\mathscr{Z}(E)$ stand for the order ideal generated by the *identity operator* I_E in $L^\sim(E)$. The space $\mathscr{Z}(E)$ is often called the *center* of E.

We now define the *orthomorphism algebra* $\mathrm{Orth}^\infty(E)$ of E as follows. First we denote by $\mathfrak{M}$ the collection of all pairs (D,π), where D is an order dense ideal in E and π is an orthomorphism from D into E. Elements (D,π) and (D',π') in $\mathfrak{M}$ are declared *equivalent* if the orthomorphisms π and π' agree on the intersection $D \cap D'$. The factor set of $\mathfrak{M}$ by the equivalence relation is exactly $\mathrm{Orth}^\infty(E)$. Identify every orthomorphism $\pi \in \mathrm{Orth}(E)$ with the corresponding coset in $\mathrm{Orth}^\infty(E)$. Then

$\mathscr{Z}(E) \subset \text{Orth}(E) \subset \text{Orth}^{\infty}(E)$. The set $\text{Orth}^{\infty}(E)$ can be naturally furnished with the structure of an ordered algebra justifying the term "orthomorphism algebra."

(4) *Theorem.* *If E is an Archimedean vector lattice then* $\text{Orth}^{\infty}(E)$ *is a faithful f-algebra with unity I_E. Moreover,* $\text{Orth}(E)$ *is an f-subalgebra in* $\text{Orth}^{\infty}(E)$ *and* $\mathscr{Z}(E)$ *is an f-subalgebra of bounded elements in* $\text{Orth}(E)$.

(5) *Theorem.* *Every Archimedean f-algebra E with unity* $\mathbf{1}$ *is algebraically and latticially isomorphic to the f-algebra of orthomorphisms. Moreover, the ideal $I(\mathbf{1})$ is mapped onto $\mathscr{Z}(E)$.*

If E is an Archimedean vector lattice then the base of each of the f-algebras $\text{Orth}^{\infty}(E)$, $\text{Orth}(E)$, and $\mathscr{Z}(E)$ is isomorphic to the base of E. If E is a K-space then $\text{Orth}^{\infty}(E)$ is a universally complete K-space and $\text{Orth}(E)$ is an order dense ideal of it.

5.1.10. COMMENTS.

(1) The rise of the theory of ordered vector spaces is commonly attributed to the contribution by G. Birkhoff, L. V. Kantorovich, M. G. Kreĭn, H. Nakano, F. Riesz, H. Freudenthal, et al. in the 1930s. At present, the theory of ordered vector spaces and its applications occupy a vast field of mathematics, serving as one of the main sections of contemporary functional analysis. The theory is well expounded in many monographs, cf. [1, 3, 4, 91, 103, 104, 114, 149, 154, 158, 214, 216, 253, 258]. Also, notice the surveys [22, 23].

(2) The contents of this subsection are the preliminaries to vector lattice theory whose exposition is given in each of the following sources [4, 103, 158, 214, 253]. Another title for a vector lattice is a *Riesz space*, see [158, 258].

(3) It was L. V. Kantorovich who initiated research into Dedekind complete vector lattices, alternatively, K-spaces. The notion of K-space appeared in Kantorovich's first article on this topic [96]. Therein he treated the members of a K-space as generalized numbers and propounded the *heuristic transfer principle*. He wrote: "In this note, I define a new type of space that I call a semiordered linear space. The introduction of such a space allows us to study linear operations of one abstract class (those with values in these spaces) in the same way as linear functionals."

(4) The heuristic transfer principle by L. V. Kantorovich was corroborated many times in the works of L. V. Kantorovich and his followers, cf. [97–102, 104]. Attempts at formalizing the heuristic ideas by L. V. Kantorovich have started at the initial stages of K-space theory, resulting in the so-called theorems of identity preservation (sometimes a less exact term "conservation" is also employed). They assert that if a proposition with finitely many function variables is proven for the reals then a similar fact holds for the members of an arbitrary K-space (see [104, 253]).

Unfortunately, no satisfactory explanation was suggested for the internal mechanism controlling the phenomenon of identity preservation. Insufficiently clear remained the limits on the heuristic transfer principle. The same applies to the general reasons for similarity and parallelism between the reals and their analogs in K-space. The omnipotence and omnipresence of Kantorovich's transfer principle found its full explanation within Boolean valued analysis (cf. 5.2.15 (1)).

5.2. Representation of Vector Lattices

In this section we prove that Archimedean vector lattices are represented as subgroups of the additive group of the reals in an appropriate Boolean valued universe. This enables us to deduce the basic structural properties of vector lattices: the functional calculus, spectral decomposition of elements, representation by function spaces, etc.

5.2.1. Denote by $\mathbb{R}$ the reals viewed as a totally ordered field and let $\mathbb{R}^\wedge$ be the standard name of $\mathbb{R}$; i.e., the value at $\mathbb{R}$ of the canonical embedding of the von Neumann universe into $\mathbf{V}^{(B)}$ (see 2.2.7).

Since $\mathbb{R}$ is an algebraic system of signature $\sigma := (+, \cdot, 0, 1, \leq)$; therefore, by virtue of Corollary 4.3.5 (1), $\mathbb{R}^\wedge$ is an algebraic system of signature $\sigma^\wedge$ inside $\mathbf{V}^{(B)}$. Moreover, given a formula $\varphi(u_0, \dots, u_{n-1})$ of signature σ and $x_0, \dots, x_{n-1} \in \mathbb{R}$, note that $\varphi(x_0, \dots, x_{n-1})$ holds if and only if $\varphi(x_0^\wedge, \dots, x_{n-1}^\wedge)$ holds inside $\mathbf{V}^{(B)}$.

Choosing as φ the axioms of an Archimedean totally ordered field, we note in particular that $\mathbf{V}^{(B)} \models$ "$\mathbb{R}^\wedge$ is an Archimedean totally ordered field."

However, we cannot claim that $\mathbb{R}^\wedge$ stands for the reals inside $\mathbf{V}^{(B)}$ (cf. [72]). The reason behind this is that the completeness postulate for the reals is not expressed by a bounded formula. In fact, one of the equivalent formulations of the completeness postulate reads:

$$(\forall A)(A \subset \mathbb{R} \wedge A \neq \varnothing \wedge \pi_{\leq}(A) \neq \varnothing \to (\exists x \in \mathbb{R})(x = \sup(A)));$$

i.e., each upper bounded nonempty set of reals has a least upper bound. This formula uses generalization over the powerset of $\mathbb{R}$.

Recall (cf. 3.1.1) that $B_0(\mathbb{R}) := \mathbb{R}^\wedge\!\downarrow$ consists of all mixings $\operatorname{mix}_{t\in\mathbb{R}}(b_t t^\wedge)$, where $(b_t)_{t\in\mathbb{R}}$ is a partition of unity in B. Theorem 4.4.10 shows that $B_0(\mathbb{R})$ is a universally complete faithful f-ring.

The f-ring $B_0(\mathbb{R})$ may be identified with the f-ring of all continuous functions x from the Stone space Q of the Boolean algebra B to the set $\overline{\mathbb{R}} := \mathbb{R} \cup \{\pm\infty\}$ with the discrete topology each of which takes the values $\pm\infty$ on a nowhere dense set. Obviously, $B_0(\mathbb{R})$ is indeed an f-algebra, since we may assume $\mathbb{R} \subset B_0(\mathbb{R})$ on identifying λ in $\mathbb{R}$ with the identically λ function on Q.

5.2.2. By the transfer and maximum principles, there is an element $\mathscr{R} \in \mathbf{V}^{(B)}$ such that $\mathbf{V}^{(B)} \models$ "$\mathscr{R}$ is an ordered field of the reals." It is obvious that inside $\mathbf{V}^{(B)}$ the field $\mathscr{R}$ is unique up to isomorphism; i.e., if $\mathscr{R}'$ is another field of the reals inside $\mathbf{V}^{(B)}$ then $\mathbf{V}^{(B)} \models$ "$\mathscr{R}$ and $\mathscr{R}'$ are isomorphic."

As was pointed out above, $\mathbb{R}^\wedge$ is an Archimedean ordered field inside $\mathbf{V}^{(B)}$ and so we may assume that $\mathbf{V}^{(B)} \models$ "$\mathbb{R}^\wedge \subset \mathscr{R}$ and $\mathscr{R}$ is the (metric) completion of $\mathbb{R}^\wedge$." Regarding the unity 1 of $\mathbb{R}$, notice that $\mathbf{V}^{(B)} \models$ "$1 := 1^\wedge$ is an order unit of $\mathscr{R}$."

Consider the descent $\mathscr{R}\!\downarrow$ of the algebraic system $\mathscr{R} := (|\mathscr{R}|, +, \cdot, 0, 1, \leq)$. By implication, we equip the descent of the underlying set of $\mathscr{R}$ with the descended operations and order of $\mathscr{R}$. In more detail, addition, multiplication, and order on $\mathscr{R}\!\downarrow$ appear in accord with the following rules (cf. 4.2.3):

$$
\begin{gathered}
x + y = z \leftrightarrow [\![x + y = z]\!] = \mathbf{1}, \\
xy = z \leftrightarrow [\![xy = z]\!] = \mathbf{1}, \\
x \leq y \leftrightarrow [\![x \leq y]\!] = \mathbf{1}, \\
\lambda x = y \leftrightarrow [\![\lambda^\wedge x = y]\!] = \mathbf{1} \\
(x, y, z \in \mathscr{R}\!\downarrow,\ \lambda \in \mathbb{R}).
\end{gathered}
$$

Gordon Theorem. *Let $\mathscr{R}$ be the reals in $\mathbf{V}^{(B)}$. Assume further that $\mathscr{R}\!\downarrow$ stands for the descent $|\mathscr{R}|\!\downarrow$ of the underlying set of $\mathscr{R}$ equipped with the descended operations and order. Then the algebraic system $\mathscr{R}$ is a universally complete K-space.*

Moreover, there is a (canonical) isomorphism χ from the Boolean algebra B onto the Boolean algebra of band projections $\mathfrak{Pr}(\mathscr{R}\!\downarrow)$ or onto the Boolean algebra of the unit elements $\mathfrak{C}(\mathscr{R}\!\downarrow)$ such that the following hold:

$$
\begin{gathered}
\chi(b)x = \chi(b)y \leftrightarrow b \leq [\![x = y]\!], \\
\chi(b)x \leq \chi(b)y \leftrightarrow b \leq [\![x \leq y]\!]
\end{gathered}
$$

for all $x, y \in \mathscr{R}\!\downarrow$ and $b \in B$.

$\lhd$ This is already proven in 4.4.10.

Indeed, by 4.4.10 (2, 4), $\mathscr{R}\!\downarrow$ is a universally complete and Dedekind complete f-ring with unity $\mathbf{1} := 1^\wedge$.

The mapping $\lambda \mapsto \lambda^\wedge \cdot 1$ is an isomorphism of $\mathbb{R}$ to $\mathscr{R}\!\downarrow$. Putting $\lambda x := \lambda^\wedge x$ $(x \in \mathscr{R}\!\downarrow,\ \lambda \in \mathbb{R})$, obtain a sought vector structure on $\mathscr{R}\!\downarrow$. Therefore, $\mathscr{R}\!\downarrow$ is a universally complete K-space. $\rhd$

5.2.3. With the notation of 5.2.2, we elaborate some general propositions about vector lattices in terms of the K-space $\mathscr{R}\!\downarrow$.

(1) *Assume that* $(b_\xi)_{\xi\in\Xi}$ *is a partition of unity in* B *and* $(x_\xi)_{\xi\in\Xi}$ *is a family in* $\mathscr{R}{\downarrow}$. *Then*

$$\operatorname*{mix}_{\xi\in\Xi}(b_\xi x_\xi) = o\text{-}\sum_{\xi\in\Xi}\chi(b_\xi)x_\xi.$$

$\triangleleft$ Indeed, if $x = \operatorname{mix}(b_\xi x_\xi)$ then the definition of mixing, together with Theorem 5.2.2, implies that $\chi(b_\xi)x = \chi(b_\xi)x_\xi$ for all ξ. Summing these formulas with respect to ξ, complete the proof. $\triangleright$

(2) *For a set* $A \subset \mathscr{R}{\downarrow}$ *and arbitrary* $a \in \mathscr{R}{\downarrow}$ *and* $b \in B$ *the following equivalence holds:*

$$\chi(b)a = \sup(\chi(b)(A)) \leftrightarrow b \leq [\![a = \sup(A{\uparrow})]\!].$$

$\triangleleft$ Indeed, by 5.2.2, the equality $\chi(b)a = \sup\{\chi(b)x : x \in A\}$ holds if and only if $b \leq [\![x \leq a]\!]$ for all $x \in A$ and for every $y \in \mathscr{R}{\downarrow}$ the formula $(\forall\, x \in A)(b \leq [\![x \leq y]\!])$ implies $b \leq [\![a \leq y]\!]$. The last statement is just another expression of the estimate $b \leq [\![\sup(A{\uparrow}) = a]\!]$. $\triangleright$

(3) *Consider a net* $s : \mathrm{A} \to \mathscr{R}{\downarrow}$, *with* A *a directed set. The modified ascent* $s{\uparrow} : \mathrm{A}^\wedge \to \mathscr{R}$ *is a net inside* $\mathbf{V}^{(B)}$. *Moreover,*

$$\chi(b)x = o\text{-}\lim(\chi(b) \circ s) \leftrightarrow b \leq [\![x = \lim(s{\uparrow})]\!]$$

for all $x \in \mathscr{R}{\downarrow}$ *and* $b \in B$.

$\triangleleft$ The equality $\chi(b)x = o\text{-}\lim(\chi(b) \circ s)$ amounts to the existence of a net $r : \mathrm{A} \to \mathscr{R}{\downarrow}$ such that $r(\alpha) \leq r(\beta)$ for $\alpha \leq \beta$, $\inf\{r(\alpha) : \alpha \in \mathrm{A}\} = 0$ and $|\chi(b)x - \chi(b)s(\alpha)| \leq \chi(b)r(\alpha)$ for all $\alpha \in \mathrm{A}$.

In view of 5.2.3 (2) and the equality $r(\mathrm{A}){\uparrow} = r{\uparrow}(\mathrm{A}^\wedge)$, the last three formulas imply the inequalities:

$$b \leq [\![(\forall\, \alpha \in \mathrm{A}^\wedge)(|x - s{\uparrow}(\alpha)| \leq r(\alpha))]\!],$$
$$b \leq [\![\inf(r{\uparrow}(\mathrm{A}^\wedge) = 0)]\!],$$
$$b \leq [\![(\forall\, \alpha, \beta \in \mathrm{A}^\wedge)(\alpha \leq \beta \to r{\uparrow}(\alpha) \leq r{\uparrow}(\beta)]\!].$$

These may be rewritten briefly as $b \leq [\![x = \lim(s{\uparrow})]\!]$, as claimed. $\triangleright$

The following proposition is proven along the same lines.

(4) *Assume given* s *and* A *in* $\mathbf{V}^{(B)}$ *such that* $[\![s : \mathrm{A} \to \mathscr{R}$ *is a net*$\,]\!] = \mathbf{1}$. *Then the descent* $s{\downarrow}: \mathrm{A}{\downarrow} \to \mathscr{R}{\downarrow}$ *is a net. Moreover,*

$$\chi(b)x = o\text{-}\lim(\chi(b) \circ (s{\downarrow})) \leftrightarrow b \leq [\![x = \lim(s)]\!]$$

for all $x \in \mathscr{R}{\downarrow}$ *and* $b \in B$.

(5) *For every element* $x \in \mathscr{R}{\downarrow}$ *the following equalities hold:*

$$e_x = \chi([\![x \neq 0]\!]), \quad e_\lambda^x = \chi([\![x < \lambda]\!]) \quad (\lambda \in \mathbb{R}).$$

$\lhd$ Note that a real t is other than zero if and only if the least upper bound of the set $\{1 \wedge (n|t|) : n \in \omega\}$ is 1. Given $x \in \mathscr{R}{\downarrow}$, by the transfer principle, we thus have $[\![x \neq 0]\!] = [\![\sup\{1^\wedge \wedge (n|x|) : n \in \omega^\wedge\} = 1]\!]$.

If $A := \{1 \wedge (n|x|) : n \in \omega\}$ then $[\![\sup(A{\uparrow}) = \sup\{1^\wedge \wedge (n|x|) : n \in \omega^\wedge\}]\!] = \mathbf{1}$ and $e_x = \sup(A)$. Therefore, $b := [\![x \neq 0]\!] = [\![e_x = \mathbf{1}]\!]$ and, analogously, $b^* = [\![e_x = 0]\!]$. Using the properties of χ, deduce $e_x = \chi(b)$.

Now, choose an arbitrary number $\lambda \in \mathbb{R}$ and note that $\lambda^\wedge = \lambda^\wedge \mathbf{1}$, and so $e_\lambda^x = e_{(\lambda^\wedge - x)^+}$. Whence,

$$\chi^{-1}(e_\lambda^x) = [\![(\lambda^\wedge - x) \vee 0 \neq 0]\!] = [\![\lambda^\wedge - x > 0]\!] = [\![x < \lambda^\wedge]\!]. \ \rhd$$

5.2.4. Theorem. *Assume that* X *is an Archimedean vector lattice with base* $B := \mathscr{B}(X)$ *and let* $\mathscr{R}$ *stand as before for the reals in* $\mathbf{V}^{(B)}$. *Then there is a linear and lattice isomorphism* $\imath$ *from* X *into the universally complete* K*-space* $\mathscr{R}{\downarrow}$ *such that the following conditions are met:*

(1) *The isomorphism* $\imath$ *preserves suprema and infima;*

(2) *The order ideal* $J(\imath(X))$ *generated by* $\imath(X)$ *is an order dense ideal of* $\mathscr{R}{\downarrow}$*;*

(3) $\inf\{\imath(x) : x \in X, \imath(x) \geq y\} = y = \sup\{\imath(x) : x \in X, \imath(x) \leq y\}$ *for all* $y \in J(\imath(X))$*;*

(4) *If* $x \in X$ *and* $b \in B$ *then* $b \leq [\![\imath(x) = 0]\!]$ *whenever* $x \in b^\perp$.

$\lhd$ By Theorem 4.4.12 there are a subgroup $\mathscr{X}$ of the additive group of the reals $\mathscr{R} \in \mathbf{V}^{(B)}$ and a group and lattice isomorphism $\imath := \imath_X$ from X to $\mathscr{X}$.

Let e be a nonzero positive element of $\mathscr{X}$. Replacing, if need be, $\mathscr{X}$ with the group $e^{-1}\mathscr{X}$ isomorphic to $\mathscr{X}$, assume that $e = \mathbf{1} \in \mathscr{X}$.

Note that $X^\wedge$ is a vector space over $\mathbb{R}^\wedge$. In these circumstances the factor mapping $\varphi := \varphi_X : X^\wedge \to \mathscr{X}$ is $\mathbb{R}^\wedge$-linear. In particular, $[\![\varphi((\lambda x)^\wedge) = \lambda^\wedge \varphi(x^\wedge)]\!] = \mathbf{1}$ for all $\lambda \in \mathbb{R}$ and $x \in X$. Therefore, $[\![\imath(\lambda x) = \lambda^\wedge \imath(x)]\!] = \mathbf{1}$, or $\imath(\lambda x) = \lambda \imath(x)$ (cf. 5.2.2).

Considering $\mathbf{1} = \operatorname{mix}(b_\xi \imath(e_\xi))$, $(e_\xi) \subset X$ and $\lambda \in \mathbb{R}$, we may write

$$b_\xi \leq [\![\lambda^\wedge = \lambda^\wedge \cdot \imath e_\xi]\!] \wedge [\![\lambda^\wedge \cdot \imath e_\xi = \imath(\lambda e_\xi)]\!] \wedge [\![\imath(\lambda e_\xi) \in \mathscr{X}]\!] \leq [\![\lambda^\wedge \in \mathscr{X}]\!].$$

Therefore, $\lambda^\wedge \in \mathscr{X}$ and so $[\![\mathbb{R}^\wedge \subset \mathscr{X} \subset \mathscr{R}]\!] = \mathbf{1}$.

Moreover, $\mathbf{V}^{(B)} \models$ "$\mathscr{X}$ is a vector sublattice of $\mathscr{R}$ viewed as a vector lattice over $\mathbb{R}^\wedge$." In this case, however, $\mathscr{X}{\downarrow}$ is a vector sublattice of the universally complete K-space $\mathscr{R}{\downarrow}$, while $\imath$ may be considered as an embedding of X in $\mathscr{R}{\downarrow}$.

The task we are left with now is to check that the claims of (1)–(4) hold.

(1) Take $A \subset X$ and $a \in X$ so that $a = \sup(A)$. Put $z = \sup(\imath(A))$ where the supremum is calculated in $\mathscr{R}{\downarrow}$. From the obvious equality $[\![\mathscr{X} \text{ minorizes } \mathscr{R}]\!] = \mathbf{1}$ it is easy that $\mathscr{X}{\downarrow}$ minorizes $\mathscr{R}{\downarrow}$. In this case, however, $\imath(X)$ also minorizes $\mathscr{R}{\downarrow}$ (see 4.4.12). If $\imath(a) \geq z$ then $\imath(x) \leq \imath(a) - z$ or $z \leq \imath(a - x)$ for some $0 < x \in X$, which implies that $a - x$ is an upper bound of A and the equality $a = \sup(A)$ implies $a - x \geq a$ or $x \leq 0$. This contradiction yields $z = \imath(a)$.

(2) Since $\imath(X)$ minorizes $\mathscr{R}{\downarrow}$; therefore, $\mathscr{R}{\downarrow} = \imath(X)^{\perp\perp}$. Hence, the equality $\mathscr{R}{\downarrow} = J(\imath(X))^{\perp\perp}$ holds, where $J(\imath(X))$ is the order ideal generated by $\imath(X)$.

(3) The formula $[\![\mathbb{R}^\wedge \subset \mathscr{X} \subset \mathscr{R}]\!] = \mathbf{1}$ allows us to conclude that $\mathbf{V}^{(B)} \models$ "$\mathscr{X}$ is a dense subgroup in $\mathscr{R}$." Hence, arguing inside $\mathbf{V}^{(B)}$, note that

$$\inf\{x' \in X : x' \geq x\} = x = \sup\{x' \in X : x' \leq x\}$$

for every $x \in \mathscr{R}{\downarrow}$. Applying 5.2.3 (2), immediately obtain

$$\inf\{x' \in \mathscr{X}{\downarrow} : x' \geq x\} = x = \sup\{x' \in \mathscr{X}{\downarrow} : x' \leq x\}.$$

To complete the proof, recall that $\imath(X)$ minorizes $\mathscr{X}{\downarrow}$.

(4) This is proven in 4.4.12. ▷

5.2.5. We now list a few corollaries to the above representation theorem.

(1) *Let X be an Archimedean vector lattice with base $\mathscr{B}(X)$ isomorphic to a Boolean algebra B. Then there is an element $\mathscr{X} \in \mathbf{V}^{(B)}$ obeying the conditions:*

(a) $\mathbf{V}^{(B)} \models$ *"$\mathscr{X}$ is a vector sublattice of the reals $\mathscr{R}$ viewed as a vector space over $\mathbb{R}^\wedge$";*

(b) *$X' := \mathscr{X}{\downarrow}$ is a universally complete vector lattice with the projection property which is an r-dense sublattice of the K-space $\mathscr{R}{\downarrow}$;*

(c) *There is a linear and lattice isomorphism $\imath : X \to X'$ preserving suprema and infima. Moreover, for each $x \in X'$ there are a partition of unity $(\pi_\xi)_{\xi\in\Xi}$ in $\mathfrak{Pr}(X')$ and a family $(x_\xi)_{\xi\in\Xi}$ in X such that*

$$x = o\text{-}\sum_{\xi\in\Xi} \pi_\xi \circ \imath(x_\xi).$$

◁ All claims are in fact immediate from 5.2.4. Prove for instance that X' is r-dense in $\mathscr{R}{\downarrow}$.

If $x \in \mathscr{R}{\downarrow}$ then $\mathbf{V}^{(B)} \models$ "x is a real and x may be approximated with any accuracy by the elements of $\mathscr{X}$." In other words, the following holds:

$$[\![(\forall\, \varepsilon \in \mathbb{R}^\wedge)(\varepsilon > 0 \to (\exists\, \lambda \in \mathscr{X})(|\lambda - x| < \varepsilon))]\!] = \mathbf{1}.$$

Writing out Boolean truth values for the quantifiers, observe that to every $\varepsilon > 0$ there is some $\lambda \in X'$ satisfying $|\lambda - x| \leq \varepsilon\mathbf{1}$. The proof is complete. ▷

(2) *If X is a K-space then $\mathscr{X} = \mathscr{R}$ and $\imath(X)$ is an order dense ideal of $\mathscr{R}{\downarrow}$. The image of X under the isomorphism $\imath$ is the whole of $\mathscr{R}{\downarrow}$ if and only if X is a universally complete K-space.*

$\lhd$ The proof results from 5.2.2 and 5.2.4 (2, 3). $\rhd$

(3) *Universally complete K-spaces are order isomorphic if and only if they have isomorphic bases.*

$\lhd$ Indeed, if X and Y are universally complete K-spaces, while h is an order isomorphism between X and Y; then the mapping $K \mapsto h(K)$ $(K \in \mathscr{B}(X))$ is an isomorphism between the respective bases.

Conversely, if $\mathscr{B}(X)$ and $\mathscr{B}(Y)$ are isomorphic to a Boolean algebra B then, by (2), each of the spaces X and Y is order isomorphic to the universally complete K-space $\mathscr{R}{\downarrow}$. $\rhd$

A *completion* of a K-space X is a pair $(Y, \imath)$, with Y another K-space and $\imath$ an isomorphism of X onto an order dense ideal of Y. Furnish the class $\mathrm{Ext}(X)$ of all completions of a K-space X with some order as follows: Given $(Y, \imath)$ and $(Z, \jmath) \in \mathrm{Ext}(X)$, put $(Y, \imath) \prec (Z, \jmath)$ provided that there is an isomorphism h of Y onto some order closed ideal of Z such that $h \circ \imath = \jmath$. A maximal element of the preordered class $\mathrm{Ext}(X)$ is a *universal completion* of X.

The following result ensues from (1) and (2).

(4) *Each K-space has a universal completion which is unique up to order isomorphism and presents a universally complete K-space.*

This proposition allows us to use the same symbol mX for every universal completion of X and speak about *the* universal completion of X (cf. 1.1.6 (7)). Note that the Russian literature uses the term "extension" for "completion" and "maximal extension" for "universal completion."

(5) *Assume that X is a universally complete K-space with order unit* $\mathbf{1}$. *There is a unique multiplication in X making X into a faithful f-ring with* $\mathbf{1}$ *the ring unity.*

$\lhd$ Identify $\lambda \in \mathbb{R}$ with $\lambda \cdot \mathbf{1}$. By virtue of (2), X is isomorphic to $\mathscr{R}{\downarrow}$ with 1 becoming $\mathbf{1} := 1^{\wedge} \in \mathscr{R}{\downarrow}$, since $[\![1^{\wedge}$ is the unity of $\mathscr{R}]\!] = \mathbf{1}$. The descent of multiplication in $\mathscr{R}$ brings about with the sought multiplicative structure. If $\times : X^2 \to X$ is another multiplication in X satisfying the above conditions then it is extensional and its ascent $(\times){\uparrow}$ is some multiplication in $\mathscr{R}$ whose unity is 1. We clearly see that $\times = \cdot$ in this event, since the multiplicative structure of the field $\mathscr{R}$ is unique when we have fixed a unity. $\rhd$

(6) *To each Archimedean vector lattice X there are a K-space oX, unique up to linear and lattice isomorphism, and a linear isomorphism $\imath : X \to oX$ preserving suprema and infima such that*

$$\sup\{\imath(x) : x \in X, \imath(x) \leq y\} = y = \inf\{\imath(x) : x \in X, \imath(x) \geq y\}$$

for all $y \in oX$.

$\lhd$ Let $\mathscr{R}$ and $J(\imath(X))$ be the same as in 5.2.4. Then the pair $(J(\imath(X)), \imath)$ meets all requirements.

If $(Y, \jmath)$ is another pair with the same properties then the bases $\mathscr{B}(Y)$ and $\mathscr{B}(\mathscr{R}{\downarrow})$ are isomorphic and so the K-spaces mY and $\mathscr{R}{\downarrow}$ are isomorphic by virtue of (2). We may thus assume that $\imath(X) \subset Y \subset \mathscr{R}{\downarrow}$ in which case Y is an order dense ideal of $\mathscr{R}{\downarrow}$. Then $J(\imath(X)) \subset Y$. On the other hand, to every $y \in Y$ there must exist x' and x'' in X satisfying $\imath(x') \leq y \leq \imath(x'')$; i.e., $Y \subset J(\imath(X))$. $\rhd$

Assume that F is a K-space and $A \subset F$. Denote by dA the set that consists of $c \in F$ presentable as $o\text{-}\sum_{\xi\in\Xi} \pi_\xi a_\xi$, with $(a_\xi)_{\xi\in\Xi} \subset A$ and $(\pi_\xi)_{\xi\in\Xi}$ a partition of unity in $\mathfrak{Pr}(F)$. Assume further that rA stands for the set comprising $x \in F$ of the form $r\text{-}\lim_{n\to\infty} a_n$, where (a_n) is an arbitrary r-convergent sequence in A.

(7) $oX = rdX$ *for every Archimedean vector lattice* X.

5.2.6. Theorem. *Let* X *be an arbitrary* K_σ*-space with order unit* $\mathbf{1}$. *The spectral function* $\lambda \mapsto e^x_\lambda$ $(\lambda \in \mathbb{R})$ *of* $x \in X$ *has the following properties:*

(1) $e^x_\lambda \leq e^x_\mu$ *for* $\lambda \leq \mu$;

(2) $e^x_{+\infty} := \bigvee_{\mu\in\mathbb{R}} e^x_\mu = \mathbf{1}$ *and* $e^x_{-\infty} := \bigwedge_{\mu\in\mathbb{R}} e^x_\mu = 0$;

(3) $\bigvee_{\mu<\lambda} e^x_\mu = e^x_\lambda$ $(\lambda \in \mathbb{R})$;

(4) $x \leq y \leftrightarrow (\forall\, \lambda \in \mathbb{R})\ (e^y_\lambda \leq e^x_\lambda)$;

(5) $e^{x+y}_\lambda = \bigvee\{e^x_\mu \cdot e^y_\nu : \mu, \nu \in \mathbb{R}, \mu + \nu = \lambda\}$;

(6) $e^{x\cdot y}_\lambda = \bigvee\{e^x_\mu \cdot e^y_\nu : 0 \leq \mu, \nu \in \mathbb{R}, \mu\nu = \lambda\}$ $(x \geq 0,\ y \geq 0)$;

(7) $e^{-x}_\lambda = \bigvee\{\mathbf{1} - e^x_{-\mu} : \mu \in \mathbb{R}, \mu < \lambda\} = (\mathbf{1} - e^x_{-\lambda}) \cdot e_{(x+\lambda\mathbf{1})}$;

(8) $x = \inf(A) \leftrightarrow (\forall\, \lambda \in \mathbb{R})(e^x_\lambda = \bigvee\{e^a_\lambda : a \in A\})$;

(9) $e^{x\vee y}_\lambda = e^x_\lambda \cdot e^y_\lambda$;

(10) $e^{cx}_\lambda = ce^x_\lambda + c^*$ *for* $\lambda > 0$, $e^{cx}_\lambda = ce^x_\lambda$ *for* $\lambda \leq 0$ $\quad (c \in \mathfrak{C}(X))$.

Moreover, the numbers μ *and* ν *in* (2), (3), *and* (5)–(7) *may range over some dense subfield* $\mathbb{P}$ *of* $\mathbb{R}$.

$\lhd$ Suppose first that X is a K-space. By Theorem 5.2.4, assume without loss of generality that $X = \mathscr{R}{\downarrow}$. In this case, the desired claims ensue easily from 5.2.3 (5) and the appropriate properties of the reals.

By way of example, prove (6) and (8).

(6) Assume that $x \geq 0$ and $y \geq 0$ with the product $x \cdot y$. Evidently, x and y are nonnegative reals inside $\mathbf{V}^{(B)}$. By 5.2.3 (5), $e^{x\cdot y}_\lambda = \chi([\![x \cdot y < \lambda^\wedge]\!])$, $e^x_\lambda = \chi([\![x < \lambda^\wedge]\!])$, and $e^y_\lambda = \chi([\![y < \lambda^\wedge]\!])$. Working inside $\mathbf{V}^{(B)}$, note that

$$(\forall\, x \in \mathscr{R})(\forall\, y \in \mathscr{R})\big(x \geq 0 \wedge y \geq 0 \to (x \cdot y < \lambda$$
$$\leftrightarrow (\exists\, 0 < \mu,\ \nu \in \mathbb{P}^\wedge)(x < \mu) \wedge (y < \nu) \wedge (\lambda = \mu\nu))\big),$$

and so

$$[\![x \cdot y < \lambda^\wedge]\!] = \bigvee_{\substack{0<\mu,\nu\in\mathbb{P}\\ \lambda=\mu\nu}} \{[\![x < \mu^\wedge]\!] \wedge [\![y < \nu^\wedge]\!]\},$$

whence the desired result follows.

(8) Now, given $A \subset X$ assume that $x = \inf(A)$. Then $e^x_\lambda = \chi([\![x < \lambda^\wedge]\!]) = \chi([\![\inf(A\!\uparrow) < \lambda^\wedge]\!])$ (see 5.2.3 (1, 5)). However, $A\!\uparrow$ is a certain nonempty subset of the reals inside $\mathbf{V}^{(B)}$. Hence,

$$\mathbf{V}^{(B)} \models \inf(A\!\uparrow) < \lambda^\wedge \leftrightarrow (\exists\, a \in A\!\uparrow)(a < \lambda^\wedge).$$

Calculating Boolean truth values, find

$$[\![x < \lambda^\wedge]\!] = \bigvee_{a\in A} [\![a < \lambda^\wedge]\!],$$

and so

$$e^x_\lambda = \bigvee\{\chi([\![a < \lambda^\wedge]\!]) : a \in A\} = \bigvee\{e^a_\lambda : a \in A\}.$$

Conversely, assume that e^x_λ is the supremum of the set $\{e^a_\lambda : a \in A\}$ for $\lambda \in \mathbb{R}$. Then

$$[\![x < \lambda^\wedge]\!] = [\![(\exists\, a \in A\!\uparrow)(a < \lambda^\wedge)]\!] = [\![\inf(A\!\uparrow) < \lambda^\wedge]\!]$$

for every $\lambda \in \mathbb{R}$ and so

$$[\![(\forall\, \lambda \in \mathbb{R}^\wedge)(x < \lambda \leftrightarrow \inf(A\!\uparrow) < \lambda)]\!] = \mathbf{1}.$$

Whence $[\![x = \sup(A\!\uparrow)]\!] = \mathbf{1}$. Applying 5.2.3 (2), note that $x = \inf(A)$. The last claim of the theorem results from the fact that if $\mathbb{P}$ is a dense subfield of $\mathbb{R}$ then $\mathbf{V}^{(B)} \models$ "$\mathbb{P}^\wedge$ is dense in $\mathscr{R}$."

In the case when X is a K_σ-space, we may assume that $X \subset \mathscr{R}\!\downarrow$. If we put the rationals $\mathbb{Q}$ in place of $\mathbb{P}$ then each of the suprema and infima above ranges over a countable set. Consequently, a supremum taken in $\mathscr{R}\!\downarrow$ belongs in fact to X, serving so as the supremum in X. $\triangleright$

5.2.7. Here we establish the following three useful properties of order convergence.

(1) *Assume again that X is a K-space with order unit $\mathbf{1}$. Consider an order bounded net $(x_\alpha)_{\alpha\in\mathrm{A}}$ of positive elements in X. Then (x_α) vanishes in order, i.e., converges in order to zero, if and only if for every $0 < \varepsilon \in \mathbb{R}$ the net of unit elements $(e^{x_\alpha}_\varepsilon)$ converges in order to $\mathbf{1}$.*

◁ Indeed, by Theorem 5.2.4, each x_α may be viewed as a positive element of the K-space $\mathscr{R}{\downarrow}$. The mapping $s : \alpha \mapsto s(\alpha) := x_\alpha$ has the modified ascent $\delta := s{\uparrow}$ which is a set in $\mathscr{R}$, i.e., a numerical net inside $\mathbf{V}^{(B)}$. According to 5.2.3 (3), o-$\lim(x_\alpha) = 0$ if and only if $[\![\lim(\delta) = 0]\!] = \mathbf{1}$, which can be rewritten in an equivalent form as

$$\mathbf{V}^{(B)} \models (\forall\, \varepsilon \in \mathbb{R}^\wedge)(\varepsilon > 0 \to (\exists\, \alpha \in \mathrm{A}^\wedge)(\forall\, \beta \in \mathrm{A}^\wedge)(\beta \geq \alpha \to \delta(\beta) = x_\beta < \varepsilon)).$$

Writing out the Boolean truth values of quantifiers, find another equivalent record:

$$(\forall\, \varepsilon > 0)(\exists\, (b_\alpha))(\forall\, \beta \in \mathrm{A})(\alpha \leq \beta \to b_\alpha \leq [\![\delta(\beta^\wedge) = x_\beta < \varepsilon^\wedge]\!]),$$

where (b_α) is a partition of unity in B.

Finally, applying 5.2.3 (5), infer

$$(\forall\, \varepsilon > 0)(\exists\, (b_\alpha)_{\alpha \in \mathrm{A}})(\forall\, \beta \in \mathrm{A})(\alpha \leq \beta \to \chi(b_\alpha) \leq e_\varepsilon^{x_\beta})$$

or

$$(\forall\, \varepsilon > 0)(\exists\, (b_\alpha)_{\alpha \in \mathrm{A}})(\chi(b_\alpha) \leq \bigwedge \{e_\varepsilon^{x_\beta} : \beta \geq \alpha\}).$$

Since $\bigvee(b_\alpha) = \mathbf{1}$, the equality o-$\lim x_\alpha = 0$ amounts to the following: Granted $\varepsilon > 0$, we have

$$o\text{-}\lim(e_\varepsilon^{x_\alpha}) = \liminf(e_\varepsilon^{x_\alpha}) = \bigvee_{\alpha \in \mathrm{A}} \bigwedge \{e_\varepsilon^{x_\beta} : \beta \geq \alpha\} = \mathbf{1}. \ \triangleright$$

(2) *An order bounded net $(x_\alpha)_{\alpha \in \mathrm{A}}$ in a K-space X with order unit* **1** *converges in order to an element $x \in X$ if and only if to every $\varepsilon > 0$ there is a partition of unity $(\pi_\alpha)_{\alpha \in \mathrm{A}}$ in $\mathfrak{Pr}(X)$ such that*

$$\pi_\alpha |x - x_\beta| \leq \varepsilon \mathbf{1} \quad (\alpha, \beta \in \mathrm{A},\ \beta \geq \alpha).$$

◁ To prove, appeal again to 5.2.4. Take s and δ the same as in (1). Reasoning as above, find out that $x_\alpha \xrightarrow{o} x$ is equivalent to the following: To each $\varepsilon > 0$ there is a partition of unity $(b_\alpha)_{\alpha \in \mathrm{A}}$ in B satisfying

$$b_\alpha \leq [\![|x_\beta - x| \leq \varepsilon^\wedge]\!] \quad (\alpha, \beta \in \mathrm{A},\ \beta \geq \alpha).$$

If $\pi_\alpha := \chi(b_\alpha)$ (see 5.2.2) then the last formula means

$$\pi_\alpha |x_\beta - x| \leq \varepsilon \mathbf{1} \quad (\alpha, \beta \in \mathrm{A},\ \beta \geq \alpha). \ \triangleright$$

(3) *An order bounded net (x_α) in the K-space X with order unit* **1** *converges in order to an element $x \in X$ if and only if to every $\varepsilon > 0$ there is an increasing net of projections (ρ_α) such that o-$\lim(\rho_\alpha) = I_X$ and*

$$\rho_\alpha |x - x_\beta| \leq \varepsilon \mathbf{1} \quad (\alpha, \beta \in \mathrm{A},\ \beta \geq \alpha).$$

◁ Indeed, this is so on putting $\rho_\alpha := \bigvee \{\pi_\beta : \beta \geq \alpha\}$ in (2). ▷

5.2.8. We now turn our attention to results on function representation of vector lattices.

(1) Let B be a complete Boolean algebra. A *resolution of the identity* in B or simply a *resolution of identity* in B is a mapping $e : \mathbb{R} \to B$ having the properties 5.2.6 (1–3) of a spectral function.

Denote by $\mathfrak{R}(B)$ the set of all resolutions of identity in B. Furnish $\mathfrak{R}(B)$ with addition, scalar multiplication, and order by the following rules (cf. 5.2.6 (4–6)):

$$(e_1 + e_2)(\lambda) := \bigvee \{e_1(\mu) \cdot e_2(\nu) : \mu, \nu \in \mathbb{R},\ \mu + \nu = \lambda\};$$
$$(\alpha e)(\lambda) := e(\lambda/\alpha) \quad (\alpha > 0);$$
$$(-e)(\lambda) := \bigvee_{\mu<\lambda} 1 - e(-\mu) = 1 - \bigwedge_{\mu<\lambda} e(-\mu);$$
$$(0 \cdot e)(\lambda) := \mathbf{0}(\lambda) := \begin{cases} \mathbf{1}, & \text{if } \lambda > 0, \\ 0, & \text{if } \lambda \leq 0; \end{cases}$$
$$e_1 \leq e_2 \leftrightarrow (\forall\, \lambda \in \mathbb{R})\, e_1(\lambda) \geq e_2(\lambda).$$

(2) *The set $\mathfrak{R}(B)$ with the above operations and order is a universally complete K-space isomorphic to $\mathscr{R}{\downarrow}$.*

◁ In line with 5.2.2, there is no loss of generality in assuming B to be the base of unit elements of the K-space $\mathscr{R}{\downarrow}$.

Put in correspondence to an element $x \in \mathscr{R}{\downarrow}$ its spectral function $\lambda \mapsto e^x_\lambda$ ($\lambda \in \mathbb{R}$). We have thus obtained an injective lattice homomorphism from $\mathscr{R}{\downarrow}$ to $\mathfrak{R}(B)$, as is seen from Theorem 5.2.6. We are left with justifying that this homomorphism is surjective.

Take an arbitrary resolution of identity $e : \mathbb{R} \to B$. Let Σ be a set of all partitions of the real axis; i.e., $\sigma \in \Sigma$ if $\sigma : \mathbb{Z} \to \mathbb{R}$ is a strictly increasing function, $\lim_{n\to\infty} \sigma(n) = \infty$ and $\lim_{n\to\infty} \sigma(-n) = -\infty$ (as usual, $\mathbb{Z}$ stands for the integers). In the universally complete K-space $\mathscr{R}{\downarrow}$ there is a sum $x_\sigma := \sum_{n\in\mathbb{Z}} \sigma(n+1) b_{n\sigma}$, where $b_{n\sigma} := e(\sigma(n+1)) - e(\sigma(n))$. Put $A := \{x_\sigma : \sigma \in \Sigma\}$ and $x = \inf(A)$. The infimum does exist since $x_\sigma \geq \sum_{n\in\mathbb{Z}} \bar{\sigma}(n) b_{n\sigma}$ for a fixed partition $\bar{\sigma} \in \Sigma$.

Note also that $x_\sigma = \operatorname{mix}(b_{n\sigma}\sigma(n+1)^\wedge)$ and

$$[\![x_\sigma < \lambda^\wedge]\!] = \bigvee \{b_{n\sigma} : \sigma(n+1) < \lambda\} = \bigvee \{e(\sigma(n+1)) : \sigma(n+1) < \lambda\}.$$

Since $[\![x = \inf(A{\uparrow})]\!] = \mathbf{1}$, the following calculations hold:

$$[\![x < \lambda^\wedge]\!] = [\![(\exists\, a \in A{\uparrow}) a < \lambda^\wedge]\!]$$
$$= \bigvee_{a\in A} [\![a < \lambda^\wedge]\!] = \bigvee_{\sigma\in\Sigma} \bigvee_{\sigma(n+1)<\lambda} b_{n\sigma}$$

$$= \bigvee_{\sigma\in\Sigma} \bigvee_{\sigma(n+1)<\lambda} e(\sigma(n)) = \bigvee_{\mu<\lambda} e(\mu) = e(\lambda).$$

Therefore, e is the spectral function of x. ▷

(3) Theorem. *Assume that Q denotes the Stone space of a complete Boolean algebra B, and let $\mathscr{R}$ stand for the reals inside $\mathbf{V}^{(B)}$. The vector lattice $C_\infty(Q)$ is a universally complete K-space linearly and latticially isomorphic to $\mathscr{R}{\downarrow}$. Such an isomorphism may be carried out by sending $\widehat{x} : Q \to \overline{\mathbb{R}}$ to $x \in \mathscr{R}{\downarrow}$ according to the rule*

$$\widehat{x}(q) := \inf\{\lambda \in \mathbb{R} : [\![x < \lambda^\wedge]\!] \in q\}.$$

◁ As was shown in (1), the K-space $\mathscr{R}{\downarrow}$ is isomorphic to the space of all B-valued spectral functions, with the function $\lambda \mapsto [\![x < \lambda^\wedge]\!]$ $(\lambda \in \mathbb{R})$ corresponding to $x \in \mathscr{R}{\downarrow}$. Assume that a clopen subset U_λ of the Stone space Q corresponds to $[\![x < \lambda^\wedge]\!] \in B$. Then, by virtue of 5.1.8 (2), to every element $x \in \mathscr{R}{\downarrow}$ there corresponds a unique continuous function $\widehat{x} : Q \to \overline{\mathbb{R}}$ such that $\{\widehat{x} < \lambda\} \subset U_\lambda \subset \{\widehat{x} \leq \lambda\}$.

In this case, however, $\widehat{x}(q) = \inf\{\lambda \in \mathbb{R} : q \in U_\lambda\} = \inf\{\lambda \in \mathbb{R} : [\![x < \lambda^\wedge]\!] \in q\}$. The formulas $\bigwedge\{[\![x < \lambda^\wedge]\!]\} = 0$ and $\bigvee\{[\![x < \lambda^\wedge]\!]\} = \mathbf{1}$ (cf. 5.2.6 (2)) imply that the interior of the closed set $\bigcap\{U_\lambda : \lambda \in \mathbb{R}\}$ is empty while the open set $\bigcup\{U_\lambda : \lambda \in \mathbb{R}\}$ is dense in Q. Therefore, the function $\widehat{x}$ may assume the values $\pm\infty$ only on a nowhere dense set and so $\widehat{x} \in C_\infty(Q)$.

We omit the elementary demonstration of the fact that $x \mapsto \widehat{x}$ is a linear and lattice isomorphism. ▷

5.2.9. We now list a few corollaries to the above theorem.

(1) *Let X be an arbitrary K-space. Assume further that $\{e_\xi\}_{\xi\in\Xi}$ is a total disjoint positive family in X. Denote by Q the Stone space of the Boolean algebra of bands of $\mathscr{B}(X)$. Then there is a unique linear and lattice isomorphism of X on an order dense ideal of the K-space $C_\infty(Q)$ such that e_ξ transforms into the characteristic function of a clopen subset Q_ξ of Q. This isomorphism sends $x \in X$ to the function $\widehat{x} : Q \to \overline{\mathbb{R}}$ acting by the rule*

$$\widehat{x}(q) := \inf\left\{\lambda \in \mathbb{R} : \{e_\lambda^\xi\}^{\perp\perp} \in q\right\} \quad (q \in Q_\xi),$$

where (e_λ^ξ) is the (value at λ of the) characteristic of the band projection of x to $\{e_\xi\}^{\perp\perp}$ with respect to the order unit e_ξ.

(2) *A space X is a universally complete K-space (K-space of bounded elements) if and only if the image of X under the above isomorphism is all $C_\infty(Q)$ (the subspace $C(Q)$ of all continuous functions on Q).*

(3) *Each Archimedean vector lattice (f-algebra) X is linearly and latticially isomorphic to a vector sublattice (and a subalgebra) of the space $C_\infty(Q)$, where Q is the Stone space of the base $\mathscr{B}(X)$ of X.*

By $C_\infty(Q, S\mathbb{Z})$ we denote the subset of $C_\infty(Q)$ that comprises the functions each of which assumes integer values on a clopen set $S \subset Q$. It is obvious that $C_\infty(Q, S\mathbb{Z})$ is a universally complete f-ring.

(4) *An order complete lattice ordered group G is isomorphic to an order dense ideal of the universally complete lattice ordered group $C_\infty(Q, S\mathbb{Z})$, with Q the Stone space of the base $\mathscr{B}(G)$ of G.*

$\lhd$ If $\mathscr{G}$ is the Boolean valued representation of G then $\mathscr{G}$ is isomorphic to $\mathscr{R}$ or is an infinite cyclic group. Therefore, there is a member b of B such that $b = [\![\mathscr{G} \simeq \mathbb{Z}^\wedge]\!]$ and $b^* = [\![\mathscr{G} \simeq \mathscr{R}]\!]$.

In the same way as in 4.4.13 we establish that G splits into the direct sum of two summands: one is representable as $\mathscr{R}$ in $\mathbf{V}^{([0,b^*])}$ and the other, as $\mathbb{Z}^\wedge$ in $\mathbf{V}^{([0,b])}$.

It suffices to apply (1) to observe that $\mathbb{Z}^\wedge{\downarrow} \simeq B_0(\mathbb{Z}) \simeq C_\infty(S, S\mathbb{Z})$ where S is the clopen set in Q corresponding to $b \in B$. $\rhd$

In an analogous way we may deduce the following proposition.

(5) *Each f-ring is order isomorphic to the product of two f-rings K_1 and K_2 such that K_1 is an order dense ideal and the subring of universally complete f-ring $C_\infty(Q_1, S_1\mathbb{Z})$, while K_2 is an order dense ideal of the universally complete group $C_\infty(Q_2, S_2\mathbb{Z})$ with zero multiplication, where $Q_\imath$ is the Stone space of the algebra $\mathscr{B}(K_\imath)$ and $S_\imath \in \mathscr{B}(Q_\imath)$ $(\imath = 1, 2)$.*

5.2.10. We will construct an integral of Stiltjes type with respect to a spectral measure.

Assume that Ω is a nonempty set and Σ is a σ-algebra of subsets of Ω. Consider the Boolean algebra B of unit elements of some K_σ-space X.

A *spectral measure* is a σ-continuous Boolean homomorphism μ from Σ to B. Here *σ-continuity* means that

$$\mu\Big(\bigvee_{n=0}^{\infty} e_n\Big) = \bigvee_{n=0}^{\infty} \mu(e_n)$$

for every sequence $(e_n)_{n\in\omega}$ of elements of Σ.

Take a measurable function $f : \Omega \to \mathbb{R}$. Given a countable partition of the real axis $\Lambda := (\lambda_k)_{k\in\mathbb{Z}}$, $-\infty \ldots \lambda_{-1} < \lambda_0 < \lambda_1 < \ldots \to$, put $e_n := f^{-1}([\lambda_n, \lambda_{n+1}))$ and arrange the integral sums

$$\underline{\sigma}(f, \Lambda) = \sum_{-\infty}^{\infty} \lambda_n \mu(e_n), \quad \overline{\sigma}(f, \Lambda) = \sum_{-\infty}^{\infty} \lambda_{n+1} \mu(e_n),$$

with all sums calculated in X.

Granted whatever $t_n \in e_n$ $(n \in \mathbb{Z})$, we obviously have

$$\underline{\sigma}(f,\Lambda) \leq \sum_{-\infty}^{\infty} f(t_n)\mu(e_n) \leq \overline{\sigma}(f,\Lambda).$$

Refining a partition Λ, we make $\underline{\sigma}(f,\Lambda)$ increase and $\overline{\sigma}(f,\Lambda)$ decrease. Assume that there is an element x in X satisfying $\sup \underline{\sigma}(f,\Lambda) = x = \inf \overline{\sigma}(f,\Lambda)$, where the supremum and infimum range over all possible partitions $\Lambda := (\lambda_l)_{l\in\mathbb{Z}}$ of the real axis as $\delta(\Lambda) := \sup_{n\in\mathbb{Z}}\{\lambda_n - \lambda_{n-1}\} \to 0$. In this event, call μ a *spectral measure*, say that f is an *integrable function* with respect to μ, and write

$$I(f) := I_\mu(f) := \int\limits_\Omega f d\mu := \int\limits_\Omega f(t)d\mu(t) := x.$$

Since $0 \leq \overline{\sigma}(f,\Lambda) - \underline{\sigma}(f,\Lambda) \leq \sum_{n=-\infty}^{\infty} \delta\mu(e_k) = \delta\mathbf{1}$, where $\delta := \delta(\Lambda)$; for the integral $I_\mu(f)$ to exist it is necessary and sufficient that there exist $\overline{\sigma}(f,\Lambda)$ and $\underline{\sigma}(f,\Lambda)$ for at least one partition of Λ. In particular, a bounded measurable function is integrable.

(1) *Let $X = \mathscr{R}{\downarrow}$ and μ be a spectral measure with values in $B := \mathfrak{C}(X)$. Then $I_\mu(f)$ is a unique element of X satisfying*

$$[\![I_\mu(f) < \lambda^\wedge]\!] = \mu(\{f < \lambda\}) \quad (\lambda \in \mathbb{R})$$

for every measurable function f.

$\lhd$ Take $\lambda \in \mathbb{R}$ and assume that $b \leq [\![\lambda^\wedge \leq I_\mu(f)]\!]$. Given a partition Λ, by Theorem 5.2.2 $b\lambda \leq bI_\mu(f) \leq b\overline{\sigma}(f,\Lambda)$. If $\Lambda := (\lambda_l)_{l\in\mathbb{Z}}$ is such that $\lambda_0 = \lambda$ and $c_n := \{u \in \Omega : \lambda_n \leq f(u) < \lambda_{n+1}\}$, then $\lambda b \wedge \mu(c_n) \leq \lambda_{n+1} b \wedge \mu(c_n)$ for $n < -1$ and so either $b \wedge \mu(c_n) = 0$ or $\lambda_{n+1} < \lambda$. Hence, putting $c := \bigvee_{n=-1}^{-\infty} c_n$, note that $b \wedge \mu(c) = 0$ or $b \leq \mu(c)^* = \mu(\Omega - c) = \mu(\{f \geq \lambda\})$. Therefore, $[\![I_\mu(f) \geq \lambda^\wedge]\!] = \mu(\{f \geq \lambda\})$, which amounts to the desired equality.

Assume that $[\![x < \lambda^\wedge]\!] = \mu(\{f < \lambda\})$ for some $x \in X$. Using the above stated property of $I_\mu(f)$, find that

$$\begin{aligned} &[\![(\forall\, \lambda \in \mathbb{R}^\wedge)(I_\mu(f) < \lambda \leftrightarrow x < \lambda)]\!] \\ &= \bigwedge_{\lambda\in\mathbb{R}} [\![I_\mu(f) < \lambda^\wedge]\!] \Leftrightarrow [\![x < \lambda^\wedge]\!] = \mathbf{1}. \end{aligned}$$

Using the denseness of $\mathbb{R}^\wedge$ in $\mathscr{R}$, conclude that $x = I_\mu(f)$. $\rhd$

(2) *In the hypotheses of (1), the mapping $\lambda \mapsto \mu(\{f < \lambda\})$, with $\lambda \in \mathbb{R}$, is the spectral function of $I_\mu(f)$.*

5.2.11. Theorem. *Assume given a universally complete K_σ-space X and a spectral measure $\mu : \Sigma \to B := \mathfrak{C}(X)$. The spectral integral $I_\mu(\cdot)$ is a sequentially order continuous (linear, multiplicative, and lattice) homomorphism from the f-algebra of measurable functions $\mathscr{M}(\Omega, \Sigma)$ to X.*

$\triangleleft$ Without loss of generality, assume that $X \subset \mathscr{R}\!\downarrow$.

The sums $\bar{\sigma}(f, \Lambda)$ and $\underline{\sigma}(f < \Lambda)$ do exist, since the summands are pairwise disjoint and X is universally complete. As mentioned above, this implies existence of $I_\mu(f)$.

It is evident that I_μ is a positive linear operator. Demonstrate that I_μ is sequentially order continuous.

Take a decreasing sequence $(f_n)_{n\in\omega}$ of measurable functions satisfying the condition $\lim_{n\to\infty} f_n(t) = 0$ for all $t \in \Omega$. Put $x_n := I_\mu(f_n)$ $(n \in \omega)$ and choose $0 < \varepsilon \in \mathbb{R}$. Denote $c_n := \{t \in \Omega : f_n(t) < \varepsilon\}$ to obtain $\Omega = \bigcup_{n=0}^{\infty} c_n$. By 5.2.3 (5) and 5.2.10 (2),

$$o\text{-}\lim_{n\to\infty} e_\varepsilon^{x_n} = \lim_{n\to\infty} \mu(c_n) = \bigvee_{n\in\omega} \mu(c_n) = \mu(\Omega) = \mathbf{1}.$$

Recalling the order convergence test 5.2.7 (1), infer that $o\text{-}\lim_{n\to\infty} x_n = 0$. Moreover, given measurable functions $f, g : \Omega \to \mathbb{R}$ and using 5.2.6 (9) and 5.2.10 (2), note that

$$e_\lambda^{f\vee g} = \mu(\{f \vee g < \lambda\}) = \mu(\{f < \lambda\} \cap \{g < \lambda\})$$
$$= \mu(\{f < \lambda\}) \wedge \mu(\{g < \lambda\}) = e_\lambda^{I(f)} \wedge e_\lambda^{I(g)} = e_\lambda^{I(f)\vee I(g)}.$$

Therefore, $I(f \vee g) = I(f) \vee I(g)$, which means that $I := I_\mu$ is a lattice homomorphism.

By analogy, given $f \geq 0$, $g \geq 0$, and $\lambda \in \mathbb{Q}$, apply 5.2.6 (6) and 5.2.8 (3) to find

$$e_\lambda^{I(fg)} = \mu(\{fg < \lambda\}) = \bigvee \{\mu(\{f < \varkappa\}) \wedge \mu(\{g < \nu\}) : \lambda = \nu\varkappa,$$
$$0 \leq \varkappa, \nu \in \mathbb{Q}\} = \bigvee \{e_\varkappa^{I(f)} \cdot e_\nu^{I(g)} : 0 \leq \varkappa, \nu \in \mathbb{Q}, \nu\varkappa = \lambda\} = e_\lambda^{I(f)\cdot I(g)}.$$

Hence, $I(f) \cdot I(g) = I(fg)$. In the case of arbitrary f and g, the last equality follows from the properties of the spectral integral. Indeed,

$$I_\mu(fg) = I_\mu(f^+g^+) + I_\mu(f^-g^-) - I_\mu(f^+g^-) - I_\mu(f^-g^+)$$
$$= I_\mu(f)^+ I_\mu(g)^+ + I_\mu(f)^- I_\mu(g)^- I_\mu(f)^- I_\mu(g)^+ - I_\mu(f)^+ I_\mu(g)$$
$$= I_\mu(f) \cdot I_\mu(g),$$

which completes the proof. $\triangleright$

5.2.12. *Let $e_0, \dots, e_{n-1} : \mathbb{R} \to B$ be an arbitrary finite set of spectral functions with values in a σ-algebra B. Then there is a unique B-valued spectral measure μ defined on the Borel σ-algebra $\mathscr{B}(\mathbb{R}^n)$ of the space $\mathbb{R}^n$ such that*

$$\mu\left(\prod_{l=0}^{n-1}(-\infty, \lambda_l)\right) = \bigwedge_{l=0}^{n-1} e_l(\lambda_l)$$

for all $\lambda_0, \dots, \lambda_{n-1} \in \mathbb{R}$.

$\lhd$ Without loss of generality, assume that $B = \mathrm{Clop}(Q)$, with Q the Stone space of B. By 5.2.8 (3), there are continuous functions $x_l : Q \to \overline{\mathbb{R}}$ $(l := 0, \dots, n-1)$ satisfying $e_l(\lambda) = \{x_l < \lambda\}$ for all $\lambda \in \mathbb{R}$ and $l = 0, \dots, n-1$.

Assign $f(t) := (x_0(t), \dots, x_{n-1}(t)) \in \mathbb{R}^n$ if all $x_l(t)$ are finite and $f(t) = \infty$ if $x_l(t) = +\infty$ for all least one index l.

We have thus defined a continuous mapping $f : Q \to \mathbb{R}^n \cup \{\infty\}$ (recall that the complements to all balls centered at zero make a base for the neighborhood filter of the point at infinity ∞).

It is obvious that f is measurable with respect to the Borel algebras $\mathrm{Clop}(Q)$ and $\mathscr{B}(\mathbb{R}^n)$. Denote by $\mathrm{Clop}_\sigma(Q)$ the σ-algebra of the subsets of Q which is generated by $\mathrm{Clop}(Q)$ and let Δ stand for the σ-ideal of $\mathrm{Clop}_\sigma(Q)$ consisting of meager sets. In this event there is an isomorphism h of the factor algebra $\mathrm{Clop}_\sigma(Q)/\Delta$ onto the σ-algebra $B := \mathrm{Clop}(Q)$. Denote by $[A]_\Delta$ the coset of A in $\mathrm{Clop}_\sigma(Q)$. We now define a mapping $\mu : \mathscr{B}(\mathbb{R}^n) \to B$ by the formula

$$\mu(A) := h([f^{-1}(A)]_\Delta) \quad (A \in \mathscr{B}(\mathbb{R}^n)).$$

Clearly, μ is a spectral measure.

If $A = \prod_{l=0}^{n-1}(-\infty, \lambda_l)$ then $f^{-1}(A) = \bigcap_{l=0}^{n-1}\{x_l < \lambda_l\} = \bigwedge_{l=0}^{n-1} e_l(\lambda_l)$, and so $\mu(A) = e_0(\lambda_0) \wedge \dots \wedge e_{n-1}(\lambda_{n-1})$.

If μ' is another spectral measure with the same properties as μ then the set $\mathscr{B} := \{A \subset \mathbb{R}^n : \mu(A) = \mu'(A)\}$ is a σ-algebra containing all sets of the type $(-\infty, \lambda_0) \times \dots \times (-\infty, \lambda_{n-1})$. Therefore, $\mathscr{B}(\mathbb{R}^n) \subset \mathscr{B}$ and $\mu = \mu'$. $\rhd$

We now take some elements $x_0, \dots, x_{n-1}$ of a K_σ-space X with unit $\mathbf{1}$. Let $e^{x_l} : \mathbb{R} \to B := \mathfrak{C}(\mathbf{1})$ stand for the spectral function of x_l. By 5.2.12, there is a spectral measure $\mu : \mathscr{B}(\mathbb{R}^n) \to B$ satisfying

$$\mu\left(\prod_{l=0}^{n-1}(-\infty, \lambda_l)\right) = \bigwedge_{l=0}^{n-1} e^{x_l}(\lambda_l).$$

Denote the integral of a measurable function $f : \mathbb{R}^n \to \mathbb{R}$ with respect to μ by $I(f, \mathfrak{x}) := I(f, x_0, \dots, x_{n-1})$, where $\mathfrak{x} := (x_0, \dots, x_{n-1})$.

Recall that $\mathscr{B}(\mathbb{R}^n, \mathbb{R})$, which is the space of all Borel functions from $\mathbb{R}^n$ to $\mathbb{R}$, is a K_σ-space and a universally complete f-algebra.

5.2.13. Theorem. *For every ordered tuple $\mathfrak{r} := (x_0, \dots, x_{n-1})$ of elements of a universally complete K_σ-space X, the mapping $f \mapsto I(f, \mathfrak{r})$ $(f \in \mathscr{B}(\mathbb{R}^n, \mathbb{R}))$ is a homomorphism of the f-algebra $\mathscr{B}(\mathbb{R}^n, \mathbb{R})$ to X meeting the following conditions:*

(1) *$I(d\lambda_l, \mathfrak{r}) = x_l$ for $l < n$ where $d\lambda_l : \mathbb{R}^n \to \mathbb{R}$ is the lth coordinate $(\lambda_0, \dots, \lambda_{n-1}) \mapsto \lambda_l$;*

(2) *If a sequence $(f_k) \subset \mathscr{B}(\mathbb{R}^n, \mathbb{R})$ is such that $\lim_{n\to\infty} f_k(t) = f(t)$ for all $t \in \mathbb{R}^n$ then $o\text{-}\lim_{n\to\infty} I(f_k, \mathfrak{r}) = I(f, \mathfrak{r})$.*

$\lhd$ By Theorem 5.2.11, it suffices to prove (1). For simplicity, we confine exposition to the case of $n = 1$.

So, take $x \in X$, and let μ stand for the spectral measure associated with the spectral function $(e_\lambda^x)_{\lambda\in\mathbb{R}}$ of x. Demonstrate that

$$x = \int_{\mathbb{R}} \lambda d\mu(\lambda) := \int_{\mathbb{R}} \lambda de_\lambda^x.$$

To this end, take an arbitrary $\varepsilon > 0$. Choose a partition $\Lambda := (\lambda_l)$ of the real axis so that $\lambda_{l+1} - \lambda_l < \varepsilon$ for all $l \in \mathbb{Z}$. Put

$$\sigma := \sum_{-\infty}^{\infty} \xi_n \mu([\lambda_{n-1}, \lambda_n)) = \sum_{-\infty}^{\infty} \xi_n (e_{\lambda_n}^x - e_{\lambda_{n-1}}^x),$$

where $\xi_n \in [\lambda_{n-1}, \lambda_n)$.

By 5.2.3 (5),

$$b_n := e_{\lambda_n}^x - e_{\lambda_{n-1}}^x = e_{\lambda_n}^x \wedge (e_{\lambda_{n-1}}^x)^* = [\![\lambda_{n-1}^\wedge \le x < \lambda_n^\wedge]\!].$$

Note that $b_n = [\![\xi_n^\wedge = \sigma]\!]$ (cf. 5.2.2). On the other hand,

$$b_n = [\![\lambda_{n-1}^\wedge \le x < \lambda_n^\wedge]\!] \wedge [\![\lambda_{n-1}^\wedge - \lambda_{n-1}^\wedge \le \varepsilon^\wedge]\!]$$
$$\wedge [\![\lambda_{n-1}^\wedge \le \xi_n < \lambda_n^\wedge]\!] \le [\![\, |x - \xi_n^\wedge| \le \varepsilon^\wedge]\!].$$

Hence, $[\![\, |x - \sigma| \le \varepsilon^\wedge]\!] = \mathbf{1}$, or $|x - \sigma| < \varepsilon\mathbf{1}$. This implies that x is the r-limit of the integral sums in question. $\rhd$

5.2.14. Freudenthal Spectral Theorem. *Suppose that E is a K_σ-space with unity $\mathbf{1}$. Each member x of E may be written down as follows*

$$x = \int_{-\infty}^{\infty} \lambda\, de_\lambda^x,$$

with the integral understood to be the relative uniform limit with regulator $\mathbf{1}$ of the integral sums $x(\beta) := \sum_{n\in\mathbb{Z}} \tau_n (e_{t_{n+1}}^x - e_{t_n}^x)$, where $t_n < \tau_n \le t_{n+1}$, $\beta := (t_n)_{n\in\mathbb{Z}}$, $\mathbb{R} = \bigcup_{n\in\mathbb{Z}} [t_n, t_{n+1}]$, and $\delta(\beta) := \sup_{n\in\mathbb{Z}} (t_{n+1} - t_n) \to 0$.

5.2.15. COMMENTS.

(1) The Gordon Theorem of 5.2.2 was first established in [62] and rediscovered by T. Jech in [85] where a universally complete K-space was defined by another collection of axioms under the alias of a complete Stone algebra. The Gordon Theorem, establishing the Boolean valued status of the concept of K-space, may be paraphrased as follows: *a universally complete K-space is an interpretation of the reals in a suitable Boolean valued universe.* Moreover, each theorem of ZFC about the reals has an analog in every corresponding K-space. This makes precise the Kantorovich motto: "The members of every K-space are generalized reals." Theorem 5.2.5 (1) was proven in [124], cf. [87]. Consult [63, 64, 135, 148] about further Boolean valued analysis of vector lattices.

(2) The results of Section 5.2, with rare exceptions, are well known in vector lattice theory. However, our proofs are far from the tradition: All principal facts are derived by interpreting the simplest properties of the reals inside $\mathbf{V}^{(B)}$ with an appropriate B. It was L. V. Kantorovich who proved in [104] the assertion of 5.2.8 which reads that, for a given complete Boolean algebra B, the set of all resolutions of identity $\mathfrak{R}(B)$ is (the underlying set of) a universally complete K-space with base isomorphic to B. The result of 5.2.9 (1) about representation of an arbitrary K-space as an order dense ideal of $C_\infty(Q)$ was first established by T. Ogasawara and B. Z. Vulikh independently of one another (cf. [104, 253]). Propositions 5.2.9 (3–5) ensue from Theorem 4.4.13 on representation of a K-space. In connection with 5.2.7 and 5.2.5 (3–6) we reverently mention other classical results by L. V. Kantorovich, B. Z. Vulikh, and A. G. Pinsker (cf. [104]) whose enormous discoveries fall beyond the scope of our exposition.

(3) The claim of existence of the isomorphism h in the proof of 5.2.12 is a consequence of the following fact (cf. [220, Theorem 29.1]):

Loomis–Sikorski Theorem. *Let Q be the Stone space of a Boolean σ-algebra B. Denote by* $\mathrm{Clop}_\sigma(Q)$ *the σ-algebra of subsets of Q which is generated by the set* $\mathrm{Clop}(Q)$ *of all clopen subsets of Q. Let Δ stand for the σ-ideal of* $\mathrm{Clop}_\sigma(Q)$ *comprising all meager sets. Then B is isomorphic with the factor algebra* $\mathrm{Clop}_\sigma(Q)/\Delta$. *If $\imath_0$ is an isomorphism of B onto* $\mathrm{Clop}(Q)$ *then the mapping*

$$\imath : b \mapsto [\imath_0(b)]_\Delta \quad (b \in B),$$

with $[A]_\Delta$ the coset containing $A \in \mathrm{Clop}_\sigma(Q)$ in the factor algebra by Δ, is an isomorphism of B onto $\mathrm{Clop}_\sigma(Q)/\Delta$.

In accord with this fact, we may put $h := \imath^{(-1)}$ in the proof of 5.2.12.

(4) Borel functions ranging in an arbitrary K_σ-space with unity seem to be first studied by V. I. Sobolev [223]. The same article claimed that each spectral function with range a σ-algebra defines a spectral measure on the Borel σ-algebra of the real axis. However, this measure is generally impossible to obtain by using the Carathéodory extension. D. A. Vladimirov shown that a complete Boolean algebra B, satisfying the countable chain condition, admits the Carathéodory extension if and only if B is regular. This implies that the extension method of 5.2.12 differs essentially from the Carathéodory extension.

(5) In the case of $n = 1$, J. D. M. Wright established 5.2.12 in [255] as a corollary to the Riesz Theorem he abstracted for the operators with range in a K_σ-space.

(6) The question of whether $\mathbb{R}^\wedge$ and $\mathscr{R}$ coincide inside $\mathbf{V}^{(B)}$ was completely settled by A. E. Gutman in [72]: This property amounts to the σ-distributivity of B (cf. 1.2.7). The same article provides an example of an atomless Boolean algebra B with the desired property.

5.3. Lattice Normed Spaces

A function space X often admits a natural abstraction of a norm. Namely, we may assume that to each vector of X there corresponds some member of another vector lattice called the *norm lattice* of X. The availability of a lattice norm on X is sometimes decisive in studying various structural properties of X. Furthermore, a norm taking values in a vector lattice makes it possible to distinguish an interesting class of the so-called dominated operators. The current section recall preliminaries.

5.3.1. Consider a vector space X and a real vector lattice E. We will assume each vector lattice Archimedean without further stipulations. A mapping $p : X \to E_+$ is called an *(E-valued) vector norm* if p satisfies the following axioms:

(1) $p(x) = 0 \leftrightarrow x = 0$ $(x \in X)$,

(2) $p(\lambda x) = |\lambda| p(x)$ $(\lambda \in \mathbb{R},\ x \in X)$,

(3) $p(x + y) \leq p(x) + p(y)$ $(x, y \in X)$.

A vector norm p is said to be a *decomposable* or *Kantorovich norm* if

(4) for arbitrary $e_1, e_2 \in E_+$ and $x \in X$, the equality $p(x) = e_1 + e_2$ implies the existence of $x_1, x_2 \in X$ such that $x = x_1 + x_2$ and $p(x_k) = e_k$ for $k := 1, 2$.

The 3-tuple (X, p, E) (simpler, X or (X, p) with the implied parameters omitted) is called a *lattice normed space* if p is an E-valued norm on X. If p is a decomposable norm then the space (X, p) itself is called *decomposable*.

5.3.2. Take a net $(x_\alpha)_{\alpha \in \mathrm{A}}$ in X. We say that (x_α) *converges in order* to an element $x \in X$ and write $x = o\text{-}\lim x_\alpha$ provided that there exists a decreasing net $(e_\gamma)_{\gamma \in \Gamma}$ in E such that $\inf_{\gamma \in \Gamma} e_\gamma = 0$ and, to every $\gamma \in \Gamma$, there exists an index $\alpha(\gamma) \in \mathrm{A}$ such that $p(x - x_\alpha) \le e_\gamma$ for all $\alpha \ge \alpha(\gamma)$. Let $e \in E_+$ be an element satisfying the following condition: for an arbitrary $\varepsilon > 0$, there exists an index $\alpha(\varepsilon) \in \mathrm{A}$ such that $p(x - x_\alpha) \le \varepsilon e$ for all $\alpha \ge \alpha(\varepsilon)$. Then we say that (x_α) converges to x relatively uniformly or *r-converges* to x with *regulator* ε and write $x = r\text{-}\lim x_\alpha$. A net (x_α) is *o-fundamental* (*r-fundamental*) if the net $(x_\alpha - x_\beta)_{(\alpha,\beta)\in \mathrm{A}\times\mathrm{A}}$ converges in order (r-converges) to zero. A lattice normed space X is *o-complete* (*r-complete*) if every o-fundamental (r-fundamental) net in it o-converges (r-converges) to some element of X.

Take a net $(x_\xi)_{\xi\in\Xi}$ and relate to it the net $(y_\alpha)_{\alpha\in\mathrm{A}}$, where $\mathrm{A} := \mathscr{P}_{\mathrm{fin}}(\Xi)$ is the collection of all finite subsets of Ξ and $y_\alpha := \sum_{\xi\in\alpha} x_\xi$. If $x := o\text{-}\lim y_\alpha$ exists then we say that (x_ξ) is *o-summable* with *sum* x and write $x = o\text{-}\sum_{\xi\in\Xi} x_\xi$.

5.3.3. Say that elements $x, y \in X$ are *disjoint* and write $x \perp y$ whenever $p(x) \wedge p(y) = 0$. Obviously, the relation $\perp$ satisfies all axioms of disjointness (cf. 4.1.12 (2)). The complete Boolean algebra $\mathscr{B}(X) := \mathfrak{K}_1(X)$ is called the *base* of X. It is easy to see that a band $K \in \mathscr{B}(X)$ is a subspace of X. In fact, $K = h(L) := \{x \in X : p(x) \in L\}$ for some band L in E. The mapping $L \mapsto h(L)$ is a Boolean homomorphism from $\mathscr{B}(E)$ onto $\mathscr{B}(X)$. We call a norm p (or the whole space X) *d-decomposable* provided that, to $x \in X$ and disjoint $e_1, e_2 \in E_+$, there exist $x_1, x_2 \in X$ such that $x = x_1 + x_2$ and $p(x_k) = e_k$ for $k := 1, 2$. Recall that, speaking of a Boolean algebra of projections in a vector space X, we always mean a set of commuting idempotent linear operators with the following Boolean operations:

$$\pi \vee \rho = \pi + \rho - \pi \circ \rho, \quad \pi \wedge \rho = \pi \circ \rho, \quad \pi^* = I_X - \pi.$$

By implication, the zero and identity operators in X serve as the zero and unity of every Boolean algebra of projections.

5.3.4. Theorem. *Let $E_0 := p(X)^{\perp\perp}$ be a lattice with the projection property and let X be a d-decomposable space. Then there exist a complete Boolean algebra $\mathscr{B}$ of projections in X and an isomorphism h from $\mathfrak{P}(E_0)$ onto $\mathscr{B}$ such that*

$$\pi \circ p = p \circ h(\pi) \quad (\pi \in \mathfrak{Pr}(E_0)).$$

$\triangleleft$ The mapping $L \mapsto h(L)$ ($L \in \mathscr{B}(E_0)$) is an isomorphism between the Boolean algebras $\mathscr{B}(E_0)$ and $\mathscr{B}(X)$ since X is d-decomposable and we may project to every band of E_0. Moreover, given $K \in \mathscr{B}(X)$, the band $K^\perp$ is the algebraic complement of K; i.e., $K \cap K^\perp = \{0\}$ and $K + K^\perp = X$. Consequently, there exists a unique projection $\pi_K : X \to X$ onto the band K along $K^\perp$.

Put $\mathscr{B} := \{\pi_K : K \in \mathscr{B}(X)\}$. Then $\mathscr{B}$ is a complete Boolean algebra isomorphic to $\mathscr{B}(X)$. We associate with $\rho \in \mathfrak{Pr}(E_0)$ the projection $\pi_K \in \mathscr{B}$, where $K := h(\rho E_0)$, and the so-obtained mapping $\rho \mapsto \pi_K$ is denoted by the same letter h. Then h is an isomorphism of $\mathfrak{Pr}(E_0)$ onto $\mathscr{B}$.

Take $\pi \in \mathfrak{Pr}(E_0)$ and $x \in X$. Using the definition of h, find that $h(\pi)x \in h(\pi E_0)$ or $p(h(\pi)x) \in \pi E_0$; therefore, $\pi^* p(h(\pi)x) = 0$. Thus, $\pi ph(\pi) = ph(\pi)$. Further, note that $p(x+y) = p(x) + p(y)$ for disjoint $x, y \in X$. Indeed, the inequality $p(x) \le p(x+y) + p(y)$ yields $p(x) \le p(x+y)$, since $p(x) \perp p(y)$. In a similar way, $p(y) \le p(x+y)$. But then $p(x) + p(y) = p(x) \vee p(y) \le p(x+y)$. Given $x \in X$, deduce

$$p(x) = p(h(\pi)x + h(\pi^*)x) = p(h(\pi)x) + p(h(\pi^*)x).$$

Making use of the above proven equality $\pi ph(\pi^*) = 0$, obtain

$$\pi p(x) = \pi p(h(\pi)x) \quad (x \in X);$$

i.e., $\pi p = \pi ph(\pi)$. Finally, $\pi p = \pi ph(\pi) = ph(\pi)$ for all $\pi \in \mathfrak{Pr}(E_0)$. ▷

5.3.5. A *Banach–Kantorovich space* we call a decomposable o-complete lattice normed space. Assume that (Y, q, F) is a Banach–Kantorovich space and $F = q(Y)^{\perp\perp}$. It is easy to show that F is a K-space and $q(Y) = F_+$ (cf. [128]). By 5.3.4, the Boolean algebras $\mathfrak{Pr}(F)$ and $\mathfrak{Pr}(Y)$ may be identified so that $\pi q = q\pi$ for all $\pi \in \mathfrak{Pr}(F)$.

A set $M \subset X$ is called *bounded in norm* or *norm bounded* if there exists $e \in E_+$ such that $p(x) \le e$ for all $x \in M$. A space X is said to be *d-complete* if every bounded set of pairwise disjoint elements in X is o-summable.

To every bounded family $(x_\xi)_{\xi\in\Xi}$ of Y and a partition of unity $(\pi_\xi)_{\xi\in\Xi}$ in $\mathfrak{Pr}(Y)$ there is a unique $x := o\text{-}\sum_{\xi\in\Xi} \pi_\xi x_\xi$ satisfying $\pi_\xi x = \pi_\xi x_\xi$ for all $\xi \in \Xi$.

◁ If $e = \sup p(x_\xi)$ then, given $\alpha, \beta \in \mathscr{P}_{\mathrm{fin}}(\Xi)$, find that

$$q(y_\alpha - y_\beta) = q\Bigg(\sum_{\xi\in\alpha\triangle\beta} \pi_\xi x_\xi\Bigg) \le \Bigg(\sum_{\xi\in\alpha\triangle\beta} \pi_\xi\Bigg) e,$$

where $y_\alpha = \sum_{\xi\in\alpha} \pi_\xi x_\xi$ and $\alpha \triangle \beta$ is the symmetric difference between α and β. Hence, (y_α) is an o-fundamental net. Consequently, it has a limit $x = o\text{-}\lim_\alpha y_\alpha$. ▷

This proposition implies that Y is d-complete. Moreover, it follows from the definitions that Y is r-complete as well.

5.3.6. Let (Y, q, F) be a Banach–Kantorovich space and $F = q(Y)^{\perp\perp}$. Say that Y is *universally complete* if $mF = F$; i.e., if the norm space F of Y is universally complete. This means that Y is a decomposable o-complete space in which every disjoint family is o-summable. A space Y is a *universal completion* of a lattice normed space (X, p, E) provided that

(1) $F = mE$ (in particular, Y is universally complete);

(2) there is a linear isometry $\imath : X \to Y$;

(3) if Z is a decomposable o-complete subspace of Y and $\imath(X) \subset Z$ then $Z = Y$.

We show in the sequel that each lattice normed space possesses a universal completion. Recall again that universal completion is often termed "maximal extension" in the Russian literature.

5.3.7. EXAMPLES.

(1) Put $X := E$ and $p(x) := |x|$ for all $x \in X$. Then p is a decomposable norm.

(2) Assume that Q is a topological space and Y is a normed space. Let $X := C_b(Q, Y)$ be the space of bounded continuous vector valued functions from Q into Y. Put $E := C_b(Q, \mathbb{R})$. Given $f \in X$, define the vector norm $p(f)$ as follows:

$$p(f) : t \mapsto \|f(t)\| \quad (t \in Q).$$

Then p is decomposable and X is r-complete if and only if Y is a Banach space.

(3) Let (Ω, Σ, μ) be a σ-finite measure space. Assume further that Y is a normed space and E is an order dense ideal in $M(\Omega, \Sigma, \mu)$. Denote by $M(\mu, Y)$ the space of cosets of μ-measurable vector valued functions acting from Ω to Y. As usual, vector functions are equivalent if they agree at almost all points of Ω. If $z \in M(\mu, Y)$ is the coset of a measurable function $z_0 : \Omega \to Y$ then denote by $p(z) := |z|$ the equivalence class of the measurable scalar function $t \mapsto \|z_0(t)\|$ with $t \in \Omega$. By definition, assign

$$E(Y) := \{z \in M(\mu, Y) : p(z) \in E\}.$$

Then $(E(Y), p, E)$ is a lattice normed space with decomposable norm. If Y is a Banach space then $E(Y)$ is a Banach–Kantorovich space and $M(\mu, Y)$ is a universal completion of $E(Y)$.

(4) Take the same E and Y as above and consider a *norming* subspace $Z \subset Y'$, i.e., a subspace such that

$$\|y\| = \sup\{\langle y, y' \rangle : y' \in Z, \|y'\| \leq 1\} \quad (y \in Y).$$

Here Y' stands for the dual of Y, and $\langle \cdot, \cdot \rangle$ is the canonical duality bracket $Y \leftrightarrow Y'$. A vector function $z : \Omega \to Y$ is said to be Z-*measurable* if the function $t \mapsto \langle z(t), y' \rangle$, with $t \in \Omega$, is measurable for every $y' \in Z$. Denote by $\langle z, y' \rangle$ the coset of the last function. Let $\mathscr{M}$ be the set of all Z-measurable vector functions z for which the set $\{\langle z, y' \rangle : y' \in Z, \|y'\| \leq 1\}$ is bounded in $M(\Omega, \Sigma, \mu)$. Denote by $\mathscr{N}$ the set of all $z \in \mathscr{M}$ such that the measurable function $t \mapsto \langle z(t), y' \rangle$ equals zero almost everywhere for each $y' \in Z$; i.e., $\langle z', y \rangle = 0$. Given $z \in \mathscr{M}/\mathscr{N}$, put

$$p(z) := |z| := \sup\{\langle u, y' \rangle : y' \in Z, \ \|y'\| \leq 1\},$$

where u_h is an arbitrary representative of the coset z and the supremum is calculated in the K-space $M(\Omega, \Sigma, \mu)$. Now, define the space

$$E_s(Y, Z) := \{z \in \mathscr{M}/\mathscr{N} : p(z) \in E\}$$

with the decomposable E-valued norm p. If Y is a Banach space then $E_s(Y, Z)$ is a Banach–Kantorovich space.

(5) Suppose that E is an order dense ideal in the universally complete K-space $C_\infty(Q)$, where Q is an extremally disconnected compact space.

Recall that a set is *comeager* if its complement is meager. Vector valued functions u and v with comeager domain are *equivalent* if $u(t) = v(t)$ for all $t \in \operatorname{dom}(u) \cap \operatorname{dom}(v)$.

Let $C_\infty(Q, Y)$ comprise the cosets of continuous vector valued functions u from comeager subsets of $\operatorname{dom}(u) \subset Q$ to a normed space Y. To $z \in C_\infty(Q, Y)$, there exists a unique function $z_z \in C_\infty(Q)$ such that $\|u(t)\| = x_z(t)$ for all $t \in \operatorname{dom}(u)$ whatever a representative u of the coset z might be. Put $p(z) := |z| := x_z$ and

$$E(Y) := \{z \in C_\infty(Q) : p(z) \in E\}.$$

(6) Let Z be the same as in (4). Denote by $\mathscr{M}$ the set of all $\sigma(Y, Z)$-continuous vector functions $u : Q_0 := \operatorname{dom}(u) \to Y$ such that $\operatorname{dom}(u)$ is a comeager set in Q and the set $\{\langle u, y' \rangle : y' \in Z, \ \|y'\| \leq 1\}$ is bounded in the K-space $C_\infty(Q)$. Here $\langle u, y' \rangle$ is the unique continuous extension of the function

$$t \mapsto \langle u(t), y' \rangle \quad (t \in Q_0)$$

to the whole Q. Consider the factor set $\mathscr{M}/\sim$, where $u \sim v$ means that $u(t) = v(t)$ for $t \in \operatorname{dom}(u) \cap \operatorname{dom}(v)$. Given $z \in \mathscr{M}/\sim$, put

$$p(z) := \sup\{\langle u, y' \rangle : y' \in Z, \ \|y'\| \leq 1\},$$

$$E_s(Y, Z) := \{z \in \mathscr{M}/\sim : p(z) \in E\}.$$

We can naturally equip the sets $C_\infty(Q, Y)$ and $\mathscr{M}/\sim$ with the structure of a module over the ring $C_\infty(Q)$. Moreover, $E(Y)$ and $E_s(Y, Z)$ are lattice normed spaces with decomposable norm. If Y is a Banach space then $E(Y)$ and $E_s(Y, Z)$ are Banach–Kantorovich spaces. Furthermore, $C_\infty(Q, Y)$ is a universal completion of $E(Y)$.

Take a normed space X and let $\varkappa$ stand for the canonical embedding of X into X''. Put $Y := X'$ and $Z := \varkappa(X)$. In this event we use the notations

$$E_s(X') := E_s(Y, Z), \quad \langle x, u \rangle := \langle u, \varkappa(x) \rangle,$$

where u is an arbitrary member of $E_s(X')$.

5.3.8. Let (X, p, E) and (Y, q, F) be lattice normed spaces. A linear operator $T : X \to Y$ is called *dominated* if there exists a positive operator $S : E \to F$ (called a *dominant* of T) such that

$$q(Tx) \leq S(p(x)) \quad (x \in X).$$

If F is a Kantorovich space and the norm p is decomposable then there exists a least element $|T|$ in the set of all dominants with respect to the order on the space $L^{\sim}(E, F)$ of regular operators. The mapping $T \mapsto |T|$ $(T \in M(X, Y))$ is a vector norm on the space $M(X, Y)$ of all dominated operators from X into Y. This is the so-called *dominant norm.* If Y is a Banach–Kantorovich space and the norm in X is decomposable then $M(X, Y)$ is a Banach–Kantorovich space under the dominant norm (cf. [128, 140]).

5.3.9. Distinguish the following two instances.

(1) Take $E := \mathbb{R}$ and $Y := F$. Then X is a normed space and the fact that $T : X \to F$ is a dominated operator means that the set

$$\{Tx : x \in X,\ \|x\| \leq 1\}$$

is bounded in F. The least upper bound of this set is called the *abstract norm* of T and is denoted by $|T|$ (the notation agrees with what was introduced above provided that the spaces F and $L^{\sim}(\mathbb{R}, F)$ are identified). In this situation we say that T is an *operator with abstract norm.*

(2) Let E and F be order dense ideals in the same K-space. An operator $T \in M(X, Y)$ is *bounded* if $|T| \in \operatorname{Orth}(E, F)$. Denote the space of all bounded operators by $\mathscr{L}_b(X, Y)$. Clearly, T belongs to $\mathscr{L}_b(X, Y)$ if and only if there exists $c \in mE = mF$ such that $c \cdot E \subset F$ and $q(Tx) \leq cp(x)$ for all $x \in X$, where we bear in mind the multiplicative structure on mE that is uniquely determined by the choice of a unity (cf. 5.2.5 (5)).

5.3.10. *Assume that X is a normed space and E is an order dense ideal of the K-space $C_\infty(Q)$. To each operator with abstract norm $T : X \to E$ there is a unique $u_T \in E_s(X')$ satisfying*

$$Tx = \langle x, u_T \rangle \quad (x \in X).$$

The mapping $T \mapsto u_T$ is a linear isometry between the Banach–Kantorovich spaces $L_a(X, E)$ and $E_s(X')$.

$\triangleleft$ If $e := |T|$ then, for every $x \in X$, the function $Tx \in C_\infty(Q)$ takes a finite value at each point of $Q_0 := \{t \in Q : e(t) < +\infty\}$ since $|Tx| \leq e\|x\|$. The last estimate also implies that, for $t \in Q_0$, the functional $v(f) : x \mapsto (Tx)(t)$, with $x \in X$, is bounded and $\|v(f)\| \leq e(t)$. This gives rise to the mapping $v : Q_0 \to X'$ continuous in the weak topology $\sigma(X', X)$. Let u_T denote the coset of v. Then $Tx = \langle x, u_T \rangle$ for all $x \in X$. In particular, the following supremum exists: $\sup\{|\langle Kx, u_T \rangle| : \|x\| \leq 1\} = e$. Hence, $u_T \in E_s(X')$ and $|u_T| = |T|$. We thus see that $T \mapsto u_T$ is an isometry from $L_a(X, E)$ to $E_s(X')$. Clearly, this mapping is linear and surjective. ▷

5.3.11. Take two normed spaces X and Y. Consider $T \in L_a(X \widehat{\otimes} Y, E)$, where $X \widehat{\otimes} Y$ is the projective tensor product of X and Y. It is an easy matter to show that the bilinear operator $b := T\otimes : X \times Y \to E$ has the abstract norm

$$|b| := \sup\{|b(x,y)| : \|x\| \leq 1,\ \|y\| \leq 1\},$$

with $|b| = |T|$. Denote by $\mathscr{B}_a(X \times Y, E)$ the set of all bilinear operators $b : X \times Y \to E$ with abstract norm. We further let $\mathscr{B}(X \times Y)$ denote the set of all bilinear forms on $X \times Y$. Since the isometric isomorphy $(X \widehat{\otimes} Y)' \simeq \mathscr{B}(X \times Y)$ is available, from 5.3.10 we derive the following proposition.

To $b \in \mathscr{B}_a(X \times Y, E)$ there is a unique $u_b \in E_s(\mathscr{B}(X \times Y))$ such that

$$b(x,y) = \langle x \otimes y, u_b \rangle \quad (x \in X,\ y \in Y).$$

The mapping $b \mapsto u_b$ is a linear isometry between $\mathscr{B}_a(X \times Y, E)$ and $E_s(\mathscr{B}(X \times Y))$.

5.3.12. Let G be an order dense ideal of $C_\infty(Q)$. In line with 5.3.7 (5), put $G_s(\mathscr{L}(X, Y')) := G_s(\mathscr{L}(X, Y'), X \otimes Y)$. Consequently, the space $G_s(\mathscr{L}(X, Y'))$ consists of the (cosets of) operator functions $K : \operatorname{dom}(K) \to \mathscr{L}(X, Y')$ such that $\operatorname{dom}(K)$ is a comeager set in Q, the function $t \mapsto \langle y, K(t)x \rangle$, with $t \in \operatorname{dom}(K)$, is continuous for all $x \in X$ and $y \in Y$, and there exists

$$|K| := \sup\{|\langle y, Kx \rangle| \ :\ \|x\| \leq 1,\ \|y\| \leq 1\} \in G.$$

If $K \in G_s(\mathscr{L}(X, Y'))$ and $u \in E(X)$ then the vector function $t \mapsto K(t)\,u(t)$ $(t \in Q_0 := \operatorname{dom}(K) \cap \operatorname{dom}(u))$ is continuous in the weak topology $\sigma(Y', Y)$. Indeed, granted arbitrary $t, t_0 \in Q_0$, observe the estimate

$$\begin{aligned} |\langle y, K(t)\,u(t) - K(t_0)\,u(t_0) \rangle| &\leq |\langle y, (K(t) - K(t_0))\,u(t_0) \rangle| \\ &\quad + |K|(t)\|y\| \cdot \|u(t) - u(t_0)\|. \end{aligned}$$

We may assume that $\operatorname{dom}(K) = \{|K| < +\infty\}$ and so $|K|$ is bounded in a neighborhood about t_0. Considering the strong continuity of u and the weak continuity of K, infer the desired. We denote the coset of a weakly continuous vector function $t \mapsto K(t)\,u(t)$ by Ku and the continuation of $t \mapsto \langle y, K(t)\,u(t) \rangle$ to the whole of Q by $\langle y, Ku \rangle$.

5.3.13. Theorem. *To a bounded operator $T \in L_b(E(X), E_s(Y'))$ there is a unique $K_T \in G_s(\mathscr{L}(X, Y'))$, with $G := \mathrm{Orth}(E)$, satisfying*

$$Tu = K_T u \quad (u \in E(X)).$$

The mapping $T \mapsto K_T$ is a linear isometry between the spaces $L_b(E(X), E_s(Y'))$ and $G_s(\mathscr{L}(X, Y'))$.

$\lhd$ By 5.3.12, it suffices to prove the first claim of the theorem.

Given $x \in X$, $y \in Y$, and $e \in E$, put $S_{x,y}(e) := \langle y, T(x \otimes e)\rangle$. Clearly, $S_{x,y} \in \mathrm{Orth}(E)$. If $b(x,\ y) := S_{x,y}$ then $b : X \times Y \to G$ is a bilinear operator with abstract norm and $|b| = |T|$. By 5.3.11 there is a unique $K_T \in G_s(\mathscr{B}(X, Y))$ such that $|K_T| = |T|$ and

$$\langle y, T(x \otimes e)\rangle = \langle x \otimes y, K_T\rangle e.$$

With the isometric isomorphy $\mathscr{B}(X, Y) \simeq \mathscr{L}(X, Y')$ available, we may assume that $K_T \in G_s(\mathscr{L}(X, Y'))$ and so

$$\langle y, T(x \otimes e)\rangle = \langle y, K_T x\rangle e = \langle y, K_T x \otimes e\rangle.$$

It suffices to note that $X \otimes E$ is order dense in $E(X)$, and T is an order continuous operator (see all details in [125, 128]). $\rhd$

5.3.14. COMMENTS.

(1) It was L. V. Kantorovich who defined a lattice normed vector space as far back as in 1935 (see [96] wherein the bizarre decomposition axiom 5.3.1 (4) appeared for the first time). Curiously, this axiom was treated as inessential and thus omitted in the subsequent publications of other researchers. A. G. Kusraev explained its principal importance in connection with the Boolean valued representation of lattice normed spaces [123] (cf. Section 5.4 to follow).

(2) It was D. Kurepa who had considered the so-called *espaces pseudodistancies*; i.e., the spaces whose metrics take values in an ordered vector space. First applications of vector valued norms and metrics relate to the method of successive approximations in numerical analysis, cf. [99, 104, 110, 215].

(3) The dominated operators of 5.3.8 were also introduced in the article [96] by L. V. Kantorovich, cf. [100]. Their definition had twofold motivation: the theoretical reasons were related to the course of the general development of operator theory on ordered vector spaces, cf. [97, 98, 100, 104]); the applied reasons were tied with what was then called "approximate methods of analysis," cf. [99, 101, 104].

(4) An elaborate theory of dominated operators was propounded in the last decade, cf. [123, 125, 128, 138]. It was A. G. Kusraev who revealed the connection of Theorem 5.3.4 between the decomposition property and existence of a Boolean algebra of projections in a lattice normed space, cf. [122, 123]. The proposition of 5.3.11 belongs to G. N. Shotaev [219]. It implies Theorem 5.3.13 which was demonstrated in [125].

(5) The details of what we sketched in 5.3.7 about measurable and continuous functions with range in a Banach space and particularly the space of bounded linear operators may be found in [36, 37, 42, 128, 152]. Further examples of lattice normed spaces relate to the theory of continuous and measurable Banach fiber bundles (see [71, 128]).

5.4. The Descent of a Banach Space

A Banach–Kantorovich space becomes a Banach space after embedding in an appropriate Boolean valued universe $\mathbf{V}^{(B)}$. The resultant interrelations make the topic of the current section. Recall that $\mathscr{C}$ stands for the field of complex numbers inside $\mathbf{V}^{(B)}$.

5.4.1. *Theorem.* *Let $(\mathscr{X}, \rho)$ be a Banach space inside $\mathbf{V}^{(B)}$. Put $X := \mathscr{X}\!\downarrow$ and $p := \rho\!\downarrow$. Then*

(1) *$(X, p, \mathscr{R}\!\downarrow)$ is a universally complete Banach–Kantorovich space;*

(2) *X admits the structure of a faithful unital module over the ring $\Lambda := \mathscr{C}\!\downarrow$ so that*

(a) $(\lambda\mathbf{1})x = \lambda x \quad (\lambda \in \mathbb{C},\ x \in X)$,

(b) $p(ax) = |a|p(x) \quad (a \in \mathscr{C}\!\downarrow,\ x \in X)$,

(c) $b \leq [\![x = \mathbf{0}]\!] \leftrightarrow \chi(b)x = \mathbf{0} \quad (b \in B,\ x \in X)$,
where χ is some isomorphism from B to $\mathfrak{E}(\mathscr{R}\!\downarrow)$.

$\triangleleft$ Denote the additions of $\mathscr{X}$, $\mathscr{C}$, and $\mathscr{R}$ by the same symbol $\oplus$. Let $\odot$ stand, first, for the scalar multiplication of the complex vector space $\mathscr{X}$ which is an external composition law acting from $\mathscr{C} \times \mathscr{X}$ to $\mathscr{X}$ and, second, for the conventional multiplication in $\mathscr{R}$ and $\mathscr{C}$. Put $+ := \oplus\!\downarrow$ and $\cdot := \odot\!\downarrow$. This means that

$$x + y = z \leftrightarrow [\![x \oplus y = z]\!] = \mathbf{1} \quad (x, y, z \in X);$$
$$a \cdot x = y \leftrightarrow [\![a \odot x = y]\!] = \mathbf{1} \quad (a \in \Lambda;\ x, y \in X).$$

The simplest properties of descent imply that $(X, +)$ is an Abelian group (cf. 4.2.7). For instance, check that $+$ is commutative as follows: Arguing inside $\mathbf{V}^{(B)}$, note $[\![\oplus \circ \jmath = \oplus]\!] = \mathbf{1}$, where $\jmath : \mathscr{X} \times \mathscr{X} \to \mathscr{X} \times \mathscr{X}$ is the transposition of coordinates. But then $\imath := \jmath\!\downarrow$ is the transposition of coordinates in $X \times X$ and

$$+ \circ \imath = (\oplus\!\downarrow) \circ (\jmath\!\downarrow) = (\oplus \circ \jmath)\!\downarrow = \oplus\!\downarrow = +.$$

Given $b \in B$ and $x \in X$, define $\chi(b)x := \operatorname{mix}\{\, bx,\ b^*\mathbf{0}\,\}$, with $\mathbf{0}$ the neutral element of the group $(X, +)$. In other words, $\chi(b)x$ is a unique element of X satisfying $[\![\chi(b)x = x]\!] \geq b$ and $[\![\chi(b)x = \mathbf{0}]\!] \geq b^*$. We have thus defined a mapping $\chi(b) : X \to X$ so that $\chi(b)$ is an additive idempotent. Put $\mathfrak{P} := \{\,\chi(b) : b \in B\,\}$. Then $\mathfrak{P}$ is a complete Boolean algebra and χ is a Boolean isomorphism. Recalling that the axioms of vector space are all satisfied for $\mathscr{X}$ inside $\mathbf{V}^{(B)}$, we may write

$$\begin{aligned} a \cdot (x + y) &= a \odot (x \oplus y) = a \odot x \oplus a \odot y = a \cdot x + a \cdot y, \\ (a + b) \cdot x &= (a \oplus b) \odot x = a \odot x \oplus b \odot x = a \cdot x + b \cdot x, \\ (ab) \cdot x &= (ab) \odot x = a \odot (b \odot x) = a \cdot (b \cdot x), \\ \mathbf{1} \cdot x &= \mathbf{1} \odot x = x \quad (a, b \in \Lambda;\ x, y \in X). \end{aligned}$$

Since $\mathbf{V}^{(B)}$ is a separated universe, the above shows that $+$ and $\cdot$ bring about the structure of a unital Λ-module on X. Letting $\lambda x = (\lambda \mathbf{1}) \cdot x$ $(\lambda \in \mathbb{C}, x \in X)$, arrive at the structure of a complex vector space on X, with (a) holding. Arguing inside $\mathbf{V}^{(B)}$, note that

$$\begin{aligned} \chi(b) = \mathbf{1} &\to \chi(b) \odot x = x, \\ \chi(b) = \mathbf{0} &\to \chi(b) \odot x = \mathbf{0}, \end{aligned}$$

and so, by the definition of χ (cf. 5.2.2),

$$\begin{aligned} b &\leq [\![\chi(b) \odot x = x]\!] = [\![\chi(b) \cdot x = x]\!], \\ b^* &\leq [\![\chi(b) \odot x = \mathbf{0}]\!] = [\![\chi(b) \cdot x = \mathbf{0}]\!]. \end{aligned}$$

Hence, $\chi(b) \cdot x = \operatorname{mix}\{bx,\ b^*\mathbf{0}\} = h(b)x$, which implies (c).

We now examine the Banach properties of the space $(\mathscr{X}, \rho)$. Subadditivity and homogenuity of the norm ρ may be written down as

$$\rho \circ \oplus \leq \oplus \circ (\rho \times \rho), \quad \rho \circ \odot = \odot \circ (|\cdot| \times \rho),$$

where $\rho \times \rho : (x, x) \mapsto (\rho(x), \rho(x))$ and $|\cdot| \times \rho : (a, x) \mapsto (|a|, \rho(x))$. Considering the descent rule for composition 3.2.12, obtain

$$p \circ + \leq + \circ (p \times p), \quad p \circ (\cdot) = (\cdot) \circ (|\cdot| \times p).$$

This means that the operator $p : X \to \mathscr{R}{\downarrow}$ satisfies 5.3.1 (3) and (b). But then 5.3.1 (2) holds in view of (a). If $p(x) = \mathbf{0}$ for some $x \in X$ then the equality $[\![\rho(x) = p(x)]\!] = \mathbf{1}$ implies $[\![\rho(x) = \mathbf{0}]\!] = \mathbf{1}$, and so $[\![x = \mathbf{0}]\!] = \mathbf{1}$; i.e., $x = \mathbf{0}$. Therefore, p is a vector norm. We may derive that p is decomposable on using (b).

Indeed, if $c := p(x) = c_1 + c_2$ $(x \in X;\ c_1, c_2 \in \Lambda_+)$ then there are $a_1, a_2 \in \Lambda_+$ such that $a_k c = c_k$ for $k := 1, 2$ and $a_1 + a_2 = \mathbf{1}$. (Put $a_k = c_k(c + (\mathbf{1} - e_c))^{-1}$, where e_c is the trace of c.) Assigning $x_k := a_k \cdot x$ $(k := 1, 2)$, note $x = x_1 + x_2$ and $p(x_k) = p(a_k \cdot x) = a_k p(x) = c_k$ for $k := 1, 2$.

We are left with demonstrating that X is Dedekind complete. Take an o-fundamental net $s : \mathrm{A} \to X$. If $\bar{s}(\alpha, \beta) = s(\alpha) - s(\beta)$ for all $\alpha, \beta \in \mathrm{A}$ then $o\text{-}\lim_{\alpha,\beta} p \circ \bar{s}(\alpha, \beta) = \mathbf{0}$. Let $\sigma : \mathrm{A}^\wedge \to \mathscr{X}$ stand for the modified ascent of s and put $\bar{\sigma}(\alpha, \beta) := \sigma(\alpha) - \sigma(\beta)$ for all $\alpha, \beta \in \mathrm{A}^\wedge$. Then $\bar{\sigma}$ is the modified ascent of $\bar{s}$ and $\rho \circ \bar{\sigma}$ is the modified ascent of $p \circ \bar{s}$. By 5.2.3, $[\![\lim \rho \circ \bar{\sigma} = \mathbf{0}]\!] = \mathbf{1}$; i.e., $\mathbf{V}^{(B)} \models$ "σ is a fundamental net in $\mathscr{X}$." Since $\mathscr{X}$ is a Banach space inside $\mathbf{V}^{(B)}$; therefore, by the transfer principle there is an element $x \in X$ such that $[\![\lim \rho \circ \sigma_0 = \mathbf{0}]\!] = \mathbf{1}$, where $\sigma_0 : A^\wedge \to \mathscr{X}$ is defined by the formula $\sigma_0(\alpha) := \sigma(\alpha) - x$ for all $\alpha \in \mathrm{A}^\wedge$. The modified descent of σ_0 is the net $s_0 : \alpha \mapsto s(\alpha) - x$ with $\alpha \in \mathrm{A}$. Using 5.2.3, conclude that $o\text{-}\lim p_0 s_0 = \mathbf{0}$ or $o\text{-}\lim_\alpha p(s(\alpha) - x) = \mathbf{0}$. $\triangleright$

The *descent* of $(\mathscr{X}, \rho)$ is defined to be the universally complete Banach–Kantorovich space $\mathscr{X}{\downarrow} := (\mathscr{X}, \rho){\downarrow} := (\mathscr{X}{\downarrow}, \rho{\downarrow}, \mathscr{R}{\downarrow})$.

5.4.2. Theorem. *To each lattice normed space (X, p, E) there is a Banach space $\mathscr{X}$ inside $\mathbf{V}^{(B)}$, with $B \simeq \mathscr{B}(p(X)^{\perp\perp})$, such that the descent $\mathscr{X}{\downarrow}$ of $\mathscr{X}$ is a universal completion of (X, p, E). Moreover, $\mathscr{X}$ is unique up to linear isometry inside $\mathbf{V}^{(B)}$.*

$\triangleleft$ Without loss of generality, assume that $E = p(X)^{\perp\perp} \subset mE = \mathscr{R}{\downarrow}$ and $B = \mathscr{B}(E)$. Put

$$d(x, y) := p(x - y)^{\perp\perp} \quad (x, y \in X).$$

Evidently, d is a B-metric on X. If we equip $\mathbb{C}$ with the discrete B-metric d_0 then the operations of addition $+ : X \times X \to X$ and multiplication $\cdot : \mathbb{C} \times X \to X$ are stabilizers. The vector norm p is a stabilizer too. All these claims are pretty obvious. For instance, regarding multiplication, note that

$$\begin{aligned} d(\alpha x, \beta y) = p(\alpha x - \beta y)^{\perp\perp} &\leq (|\alpha| p(x - y))^{\perp\perp} \vee (|\alpha - \beta| p(y))^{\perp\perp} \\ &\leq d(x, y) \vee d_\circ(\alpha, \beta) \end{aligned}$$

for $\alpha, \beta \in \mathbb{C}$ and $x, y \in X$.

Let $\mathscr{X}_0$ stand for the Boolean valued representation of the B-set (X, d). Put $\rho_0 := \mathscr{F}^\sim(p)$, $\oplus := \mathscr{F}^\sim(+)$, and $\odot := \mathscr{F}^\sim(\cdot)$, with $\mathscr{F}^\sim$ the immersion functor (see Section 3.4). The mappings $\oplus$ and $\odot$ make $\mathscr{X}_0$ into a vector space over $\mathbb{C}^\wedge$ with $\rho_0 : \mathscr{X}_0 \times \mathscr{X}_0 \to \mathscr{R}$ a norm on $\mathscr{X}_0$. By the maximum principle we may find $\mathscr{X}, \rho \in \mathbf{V}^{(B)}$ such that $[\![(\mathscr{X}, \rho)$ is a complex Banach space serving as a completion of the normed space $(\mathscr{X}_0, \rho_0)]\!] = \mathbf{1}$. We may also presume that $[\![\mathscr{X}_0$ is a dense $\mathbb{C}^\wedge$-subspace of $\mathscr{X}]\!] = \mathbf{1}$. Let $\imath : X \to X_0 := \mathscr{X}_0{\downarrow}$ be the canonical injection (cf. 3.5.4).

Since $+$ is a stabilizer acting from $X \times X$ to X; the addition $+ := \oplus\downarrow$ in the space X_0 is uniquely determined from the equality $\imath \circ + = + \circ (\imath \times \imath)$, where $\imath \times \imath : (x, y) \mapsto (\imath x, \imath y)$ is the canonical injection of the B-set $X \times X$ (cf. 3.5.4). In turn, this amounts to the additivity of $\imath$. By analogy, considering $(\cdot) := \odot\downarrow$, obtain $\imath \circ (\cdot) = (\cdot) \circ (\varkappa \times \imath)$, where $\varkappa \times \imath : (\lambda, x) \to (\lambda^{\wedge}, \imath x)$ $(\lambda \in \mathbb{C}, x \in X)$. Therefore, $\imath$ is a linear operator.

Applying the same arguments to $p_0 := \rho_0\downarrow$, infer that $\imath_E \circ p = p_0 \circ \imath$, where $\imath_E$ is the canonical injection of E. Hence, $\imath$ is an isometry, i.e., a mapping preserving the vector norm under study.

Consider some subspace Y such that $\imath X \subset Y \subset \mathscr{X}\downarrow$ and Y is a universally complete Banach–Kantorovich space under the norm $q(y) = \rho\downarrow(y)$ $(y \in Y)$. Since q is a decomposable norm and Y is a disjointly complete space; therefore, $X_0 \subset Y$. Indeed, $X_0 = \operatorname{mix}(\imath(X))$ and by (c) of 5.4.1 (2), given $x \in \mathscr{X}\downarrow$, we have $x = \operatorname{mix}(b_\xi \imath x_\xi)$ if and only if $x = o\text{-}\sum \chi(b_\xi)\imath(x_\xi)$. On the other hand, Y is decomposable and d-complete. Hence, by 5.3.4 and 5.3.5, Y is invariant under every projection $x \mapsto \chi(b)x$, with $x \in \mathscr{X}\downarrow$, containing all sums of the above type. By analogy, $Y = \operatorname{mix}(Y)$. If $\mathscr{Y} := Y\uparrow$ then $[\![\mathscr{X}_0 \subset \mathscr{Y} \subset \mathscr{X}]\!] = \mathbb{1}$, with $\mathscr{Y}\downarrow = Y$. Assume that $\sigma : \omega^{\wedge} \to \mathscr{Y}$ is a fundamental sequence and s is the modified descent of σ. Then s is an o-fundamental sequence in Y and so s has a limit $y := o\text{-}\lim s$. From 5.2.3 (4) it is clear that $[\![y = \lim \sigma]\!] = \mathbb{1}$. This establishes the completeness of $\mathscr{Y}$, and so the equality $\mathscr{X} = \mathscr{Y}$, implying $X = Y$.

Let $\mathscr{Z}$ be a Banach space inside $\mathbf{V}^{(B)}$. Assume that $\mathscr{X}\downarrow$ is the universal completion of the lattice normed space X. If $\imath'$ is the corresponding isometric embedding of X to $\mathscr{Z}\downarrow$ then $\imath' \circ \imath$ extends uniquely to a linear isometry of X_0 to a disjointly complete subspace $Z_0 \subset Z$. The spaces $\mathscr{X}_0$ and $\mathscr{Z}_0 := Z\uparrow$ are isometric. Hence, their completions $\mathscr{X}$ and $\mathscr{Y} \subset \mathscr{Z}$ are isomorphic too. Since $\mathscr{Y}\downarrow$ is a Banach–Kantorovich space and $\imath' X \subset \mathscr{Y}\downarrow \subset \mathscr{Z}\downarrow$; therefore, $\mathscr{Y}\downarrow = \mathscr{Z}\downarrow$. Consequently, $\mathscr{Y} = \mathscr{Z}$ and so $\mathscr{X}$ and $\mathscr{Z}$ are linearly isometric. $\triangleright$

5.4.3. Corollary. *The following hold:*

(1) *Each lattice normed space (X, p, E) possesses a universal completion $(mX, p_m, mE, \imath)$ unique up to linear isometry. Moreover, to all $x \in mX$ and $\varepsilon > 0$ there are a family $(x_\xi)_{\xi \in \Xi}$ in X and a partition of unity $(\pi_\xi x_\xi)_{\xi \in \Xi}$ in $\mathfrak{Pr}(mX)$ such that*

$$p_m\bigg(x - \sum_{\xi \in \Xi} \pi_\xi \imath(x_\xi)\bigg) \le \varepsilon p_m(x).$$

(2) *A lattice normed space X is linearly isometric with an order dense ideal of the universal completion of X if and only if X is decomposable and o-complete; i.e. X is a Banach–Kantorovich space.*

◁ The two claims may conveniently be proven together. With the notations of 5.3.7, assign $mX := \mathscr{X}\downarrow$ and $p_m := \rho\downarrow$. Then $(mX, p_m, mE, \imath)$ is the universal completion of X. Fix an order unit $e \in E^+$ and take $x \in mX$. Clearly, $[\![e \in \mathscr{R}]\!] = [\![e > \mathbf{0}]\!] = [\![x \in \mathscr{X}]\!] = \mathbf{1}$. Since $[\![\mathscr{X}_0$ is dense in $\mathscr{X}]\!] = \mathbf{1}$, to each $\varepsilon > \mathbf{0}$ by the maximum principle there is some element $x_\varepsilon \in \mathbf{V}^{(B)}$ satisfying

$$[\![x_\varepsilon \in \mathscr{X}_\circ]\!] = [\![\rho(x - x_\varepsilon) \leq \varepsilon^\wedge \cdot e]\!] = \mathbf{1}.$$

Hence, $x_\varepsilon \in X_0$ and $p_m(x - x_\varepsilon) \leq \varepsilon e$. Note now that $X_0 = \mathrm{mix}(\imath(X))$ and so x_ε looks like $\sum_{\xi\in\Xi} \pi_\xi \imath(x_\xi)$ where $(x_\xi) \subset X$, and (π_ξ) is a partition of unity in $\mathfrak{Pr}(mX)$.

Evidently, an order dense ideal of a Banach–Kantorovich space is decomposable and o-complete. Conversely, let X be a decomposable and o-complete lattice normed space. It is possible to show that $E_0 := p(X)^{\perp\perp}$ is a K-space. Therefore, we loose no generality in assuming E_0 to be an order dense ideal in $\mathscr{R}\downarrow$. Let $x \in mX$ and $p_m(x) \in E_0$. By (1), there exists a sequence $(x_n) \subset X_0$ such that

$$p_m(x_n - x) \leq \frac{1}{n}e, \quad p_m(x_n) \leq \left(1 + \frac{1}{n}\right) e \quad (n \in \omega).$$

Hence, $x_n \in X$ and $x \in X$, since an o-complete space is d-complete and r-complete. Thereby,

$$X = \{x \in mX : p_m(x) \in E_0\};$$

i.e., X is an order dense ideal in mX.

It remains to establish uniqueness in the claim of (1). Let $(Y, q, mE, \imath_0)$ be a universal completion of X. In view of 5.2.4 and assertion (2) we may assume that $Y = \mathscr{Y}\downarrow$, where $\mathscr{Y}$ is a Banach space inside $\mathbf{V}^{(B)}$. By Theorem 5.3.3, $[\![$ there exists a linear isometry λ of $\mathscr{X}$ onto $\mathscr{Y}]\!] = \mathbf{1}$. But then $\lambda\downarrow$ is a linear isometry of $\mathscr{X}\downarrow$ onto $\mathscr{Y}\downarrow$. ▷

5.4.4. A disjointly complete space (Y, q, dE), where dE stands for a disjoint completion of E, is said to be a *disjoint completion* (*d-completion*) of a lattice normed space (X, p, E) if there exists a linear isometry $\imath : X \to Y$ such that $Y = \mathrm{mix}(\imath X)$.

A Banach–Kantorovich space (Y, q, oE), together with a linear isometry $\imath : X \to Y$, is an *order completion* or *o-completion* of a lattice normed space (X, p, E) provided that every decomposable o-complete subspace $Z \subset Y$, containing $\imath X$, coincides with Y.

If $E = mE$ then an o-completion of X is a universal completion of X (see 5.3.3).

Given a subset $U \subset Y$, introduce the notations

$$rU := \{y := r\text{-}\lim_{n\to\infty} y_n : (y_n)_{n\in\mathbb{N}} \subset U\},$$

$$oU := \{y := o\text{-}\lim y_\alpha : (y_\alpha)_{\alpha\in \mathrm{A}} \subset U\},$$
$$dU := \Big\{y := o\text{-}\sum_{\xi\in\Xi} \pi_\xi y_\xi : (y_\xi)_{\xi\in\Xi} \subset U\Big\},$$

where A is an arbitrary directed set, $(\pi_\xi)_{\xi\in\Xi}$ is an arbitrary partition of unity in $\mathfrak{Pr}(Y)$, and the limits and sum exist in Y.

5.4.5. *Every lattice normed space admits an o-completion and d-completion each unique to within a linear isometry.*

$\lhd$ Recall that $dE \subset oE \subset mE$. Put

$$Y := \{x \in mX : p_m(x) \in oE\}.$$

Then Y is an o-completion and $d\imath X$ is a d-completion of X. $\rhd$

We always assume that a lattice normed space X is contained in an o-completion $\overline{X}$ of X.

5.4.6. *Every o-completion $\overline{X}$ of a space X satisfies the equality $\overline{X} = rdX$. Moreover, if X is decomposable and $E_0 := p(X)^{\perp\perp}$ is a vector lattice with the principal projection property then $\overline{X} = oX$.*

$\lhd$ The first part of the assertion follows from 5.4.3 (1). Take an $x \in \overline{X}$ and find a net $(x_\alpha) \subset X$ converging in order to x. Endow X with the equivalence and preorder by the formulas

$$z \sim y \leftrightarrow p(x - z) = p(y - z),$$
$$z \prec y \leftrightarrow p(x - z) \geq p(y - z).$$

If E_0 is a lattice with the principal projection property then there exists a projection $\pi \in \mathfrak{Pr}(X)$ such that

$$\pi p(x - y) + \pi^* p(x - z) = p(x - y) \wedge p(x - z).$$

Letting $u := \pi y + \pi^* z$, note that

$$p(x - u) = p(x - y) \wedge p(x - z);$$

therefore, $y \prec u$ and $z \prec u$. Thus, the preordered set $(X, \prec)$ is directed upward. Hence, the factor set $\mathrm{A} := X/\sim$ with the factor order is an upward-directed ordered set. Now, consider a net $(x_\alpha)_{\alpha\in\mathrm{A}}$, where $x_\alpha \in \alpha$ $(\alpha \in \mathrm{A})$. The net $(p(x - x_\alpha))_{\alpha\in\mathrm{A}}$ decreases by construction. Put $e := \inf p(x - x_\alpha)$, where the infimum is calculated in oE. By the equality $\overline{X} = rdX$, to $\varepsilon > 0$, there exist a family $(x_\xi) \subset X$ and a partition of unity $(\pi_\xi) \subset \mathfrak{Pr}(X)$ such that

$$p_m\Big(x - o\text{-}\sum \pi_\xi x_\xi\Big) \leq \varepsilon p_m(x).$$

Considering 5.3.4, infer

$$e = \sum \pi_\xi e \leq \sum \pi_\xi p(x - x_\xi) = p\Big(x - o\text{-}\sum \pi_\xi x_\xi\Big) \leq \varepsilon p(x).$$

Hence $e = 0$ and $x = o\text{-}\lim x_\alpha$. $\rhd$

5.4.7. *A decomposable lattice normed space is o-complete if and only if it is d-complete and r-complete.*

$\triangleleft$ Necessity was mentioned in 5.3.5. Sufficiency follows from 5.4.6. $\triangleright$

5.4.8. *Let (X, p, E) be a Banach–Kantorovich space, $E = p(X)^{\perp\perp}$, and $A :=$ Orth(E). Then X admits a unique structure of a faithful unital A-module such that the natural representation of A in X becomes an isomorphism between the Boolean algebras $\mathfrak{Pr}(E) \subset A$ and $\mathfrak{Pr}(X)$. Moreover,*

$$p(ax) = |a|p(x) \quad (x \in X,\ a \in A).$$

$\triangleleft$ We have to apply 5.4.1 (2). In particular, by virtue of 5.4.1 (2) (c), the Boolean algebra $\mathfrak{Pr}(X)$ coincides with the set of the multiplication operators $x \mapsto \chi(b)x$, where $b \in B$. $\triangleright$

A Banach space $\mathscr{X}$ inside $\mathbf{V}^{(B)}$ is said to be a *Boolean valued representation* for a lattice normed space X if $\mathscr{X}{\downarrow}$ is the universal completion of X.

5.4.9. Theorem. *Let $\mathscr{X}$ and $\mathscr{Y}$ be the Boolean valued representations of Banach–Kantorovich spaces X and Y normed by some universally complete K-space E. Let $\mathscr{L}^B(\mathscr{X}, \mathscr{Y})$ be the space of bounded linear operators from $\mathscr{X}$ to $\mathscr{Y}$ inside $\mathbf{V}^{(B)}$, where $B := \mathscr{B}(E)$. The immersion mapping $T \mapsto T^{\sim}$ is a linear isometry between the lattice normed spaces $\mathscr{L}_B(X, Y)$ and $\mathscr{L}^B(\mathscr{X}, \mathscr{Y}){\downarrow}$.*

$\triangleleft$ By Theorem 5.4.3 (2), without loss of generality we may assume that $E = \mathscr{R}{\downarrow}$, $X = \mathscr{X}{\downarrow}$, and $\mathscr{Y}{\downarrow} = Y$. Take a mapping $\mathscr{T} : \mathscr{X} \to \mathscr{Y}$ inside $\mathbf{V}^{(B)}$ and put $T := \mathscr{T}{\downarrow}$. Let ρ and θ be the norms of the Banach spaces $\mathscr{X}$ and $\mathscr{Y}$. Put $p := \rho{\downarrow}$ and $q := \theta{\downarrow}$, and let $+$ stand for addition in each of the spaces $\mathscr{X}$, $\mathscr{Y}$, X, and Y. The linearity and boundedness of $\mathscr{T}$ imply the formulas

$$\mathscr{T} \circ + = + \circ (\mathscr{T} \times \mathscr{T}), \quad \theta \circ \mathscr{T} \le k\rho,$$

where $0 \le k \in \mathscr{R}{\downarrow}$. The rules for ascending and descending composition allow us to paraphrase the above formulas as

$$T \circ + = + \circ (T \times T), \quad q \circ T \le kp.$$

But this means that T is linear and bounded. Let K be the set constituted of $0 \le k \in \mathscr{R}{\downarrow}$ such that $q(Tx) \le kp(x)$ $(x \in X)$. Then $K{\uparrow}$ coincides with $\{k \in \mathscr{R}_+ : \theta \circ \mathscr{T} \le k\rho\}$ inside $\mathbf{V}^{(B)}$.

Appealing to 5.3.2 (2), derive

$$\mathbf{V}^{(B)} \models |T| = \inf(K) = \inf(K{\uparrow}) = \|\mathscr{T}\|.$$

Hence, the mapping $\mathscr{T} \mapsto \mathscr{T}\downarrow$ preserves the vector norm. To justify the linearity of the mapping, it suffices to check its additivity. Given $\mathscr{T}_1, \mathscr{T}_2 \in \mathscr{L}^B(\mathscr{X}, \mathscr{Y})\downarrow$, note that

$$(\mathscr{T}_1 + \mathscr{T}_2)\downarrow(x) = (\mathscr{T}_1 + \mathscr{T}_2)(x) = \mathscr{T}_1 x + \mathscr{T}_2 x$$
$$= \mathscr{T}_1\downarrow x + \mathscr{T}_2\downarrow x = (\mathscr{T}_1\downarrow + \mathscr{T}_2\downarrow)x$$

inside $\mathbf{V}^{(B)}$ for every $x \in X$. Consequently, $(\mathscr{T}_1 + \mathscr{T}_2)\downarrow = \mathscr{T}_1\downarrow + \mathscr{T}_2\downarrow$. So, the descent is a linear isometry of $\mathscr{L}^B(\mathscr{X}, \mathscr{Y})\downarrow$ onto the space of all extensional bounded linear operators from X into Y. It remains to observe that every bounded linear operator from X into Y is a stabilizer or, which is the same, satisfies the inequality $[\![x = 0]\!] \leq [\![Tx = 0]\!]$. Indeed, if $b := [\![x = 0]\!]$ then $\chi(b)x = 0$ by 5.4.1 (2); therefore,

$$\chi(b)q(Tx) \leq \chi(b)p(x) = p(\chi(b)x) = 0.$$

Hence, $q(\chi(b)Tx) = 0$ or $\chi(b)Tx = 0$. Appealing to 5.4.1 (2) again, conclude that $b \leq [\![Tx = 0]\!]$. ▷

5.4.10. *Theorem.* *Assume that X is a normed space and $\widetilde{X}$ is the completion of X. If $\mathscr{X}$ is the completion of the $\mathbb{R}^\wedge$-normed space $X^\wedge$ inside $\mathbf{V}^{(B)}$, then the universally complete Banach–Kantorovich space $\mathscr{X}\downarrow$ is linearly isometric to the space $C_\infty(Q, \widetilde{X})$, with Q the Stone space of $\mathscr{R}\downarrow$.*

◁ Identify the K-spaces $\mathscr{R}\downarrow$ and $C_\infty(Q)$; and apply Theorem 5.4.2 to the lattice normed space $(X, p, \mathscr{R}\downarrow)$, with $p(x) = \|x\| \cdot \mathbf{1}$. Using the notation of the proof of 5.4.2, note that $\mathscr{X}_0 = X^\wedge$. Hence, $\mathscr{X}\downarrow := (\mathscr{X}\downarrow, q, \mathscr{R}\downarrow)$ is the universal completion of $(X, p, \mathscr{R}\downarrow)$.

For simplicity, assume that $X \subset \mathscr{X}\downarrow$. From 5.4.3 we deduce that to $u \in C_\infty(Q, \widetilde{X})$ and $\varepsilon > \mathbf{0}$ there are a family $(x_\xi) \subset X$ and a partition of unity $(Q_\xi) \subset \operatorname{Clop}(Q)$ for which the step function u_ε, equal to x_ξ on the set Q_ξ, obeys the estimate $|u - u_\varepsilon| \leq \varepsilon\mathbf{1}$. Put $\mathscr{T}(u_\varepsilon) := \operatorname{mix}(b_\xi x_\xi)$ where b_ξ stands for the member of B corresponding to the clopen set Q_ξ. Now, $|\mathscr{T}(u_\varepsilon)| = |u_\varepsilon|$. Hence, $\mathscr{T}$ is a linear isometric embedding of the subspace of all vector functions of the shape u_ε. If $\varepsilon \to \mathbf{0}$ then $|u_\varepsilon - u| \xrightarrow{(r)} \mathbf{0}$, and so $\big(\mathscr{T}(u_{1/n})\big)$ is an r-fundamental sequence. Since $\mathscr{X}\downarrow$ is complete, $\mathscr{X}\downarrow$ contains the limit $v := r\text{-}\lim \mathscr{T}(u_{1/n})$. Assigning $\mathscr{T}(U) := v$, obtain a linear isometric embedding $\mathscr{T} : C_\infty(Q, \widetilde{X}) \to \mathscr{X}\downarrow$. If $Z := \operatorname{im}(\mathscr{T})$ then Z is a decomposable o-complete subspace of $\mathscr{X}\downarrow$ and $X \subset Z$. By Theorem 5.4.2 and the definition of 5.3.6, conclude that $Z = \mathscr{X}\downarrow$. ▷

5.4.11. *Assume that X and $\mathscr{X}$ are the same as in 5.4.10. Assume further that $\mathscr{X}'$ is the dual of X inside $\mathbf{V}^{(B)}$. Then the spaces $\mathscr{X}'\downarrow$ and $E_s(X')$, with $E = C_\infty(Q)$, are linearly isometric.*

◁ Apply Theorem 5.4.9 to $Y := E$ and $X := (X, p, E)$, with $p(x) = \|x\|\mathbf{1}$. Deduce that the spaces $\mathscr{X}'{\downarrow} := \mathscr{L}^{(B)}(\mathscr{X}, \mathscr{R}){\downarrow}$ and $L_a(X, E)$ are linearly isometric. To complete the proof, refer to 5.3.10. ▷

5.4.12. COMMENTS.

(1) Theorems 5.4.1, 5.4.2, and 5.4.9, the main results of the current section, belong to A. G. Kusraev, cf. [123, 128].

(2) The completeness criterion of 5.4.7 was formulated by A. G. Kusraev in [122] under the hypothesis that the norm lattice E is Dedekind complete. The article [123] gives a proof in a more general case of a decomposable vector multinorm. The hypothesis about the Dedekind completeness of E was waived in [110]. In the case of an Archimedean vector lattice; i.e., $X = E$, this fact is due to A. I. Veksler and V. A. Geĭler [247].

(3) It was A. G. Pinsker who began studying universal completion for a K-space (cf. [104]). He established in particular that each K-space has a universal completion unique up to isomorphism. Proposition 5.4.3 (1), abstracting the Pinsker Theorem to lattice normed spaces, was in fact established in [123]. About Theorem 5.4.5 on order completion of a lattice normed space, consult [123, 128]. The assertion of 5.4.6 that $X = oX$ belongs to A. E. Gutman. It was A. I. Veksler who proved 5.4.6 for the case of an Archimedean vector lattice (cf. [245]).

(4) Theorem 5.4.10 is a particular case of the general construction of the Boolean completion of a uniform space which was suggested by E. I. Gordon and V. A. Lyubetskiĭ [161]. Theorem 5.4.11 is a simple consequence of 5.3.10 and a relevant result by E. I. Gordon on representation of operators with abstract norm [63].

5.5. Spaces with Mixed Norm

In this section we distinguish an important class of Banach spaces which is connected with the concept of vector norm.

5.5.1. A *normed vector lattice* (*Banach lattice*) we call a vector lattice E that is simultaneously a normed space (Banach space) whose norm is *monotone* in the following sense: For all $x, y \in E$, if $|x| \le |y|$ then $\|x\| \le \|y\|$.

Let (X, p, E) be a lattice normed space, with a normed vector lattice E serving as the norm lattice of X. We may then equip X with the *mixed norm* or *composite norm* by putting

$$|||x||| := \|p(x)\| \quad (x \in X).$$

In this event the normed space $X := (X, |||\cdot|||)$ is also referred to as a *space with mixed norm*, a *composite normed space* or even a *mixed norm space*. Since $|p(x) - p(y)| \le p(x - y)$ and the norm of E is monotone, the vector norm p is a continuous mapping from $(X, |||\cdot|||)$ to E.

5.5.2. *Suppose that E is a Banach lattice. Then $(X, |||\cdot|||)$ is a Banach space if and only if (X, p, E) is complete with respect to relative uniform convergence.*

$\lhd$ Take a fundamental sequence $(x_n) \subset X$. Without loss of generality, assume that $|||x_{n+1} - x_n||| \leq 1/n^3$ $(n \in \mathbb{N})$. Put

$$e_n := p(x_1) + \sum_{k=1}^{n} kp(x_{k+1} - x_k) \quad (n \in \mathbb{N}).$$

Proceed with the estimates

$$\|e_{n+l} - e_n\| = \| \sum_{k=n+1}^{n+l} kp(x_{n+1} - x_k)\|$$

$$\leq \sum_{k=n+1}^{n+l} k|||x_{k+1} - x_k||| \leq \sum_{k=n+1}^{n+l} \frac{1}{k^2} \underset{n,\ l\to 0}{\longrightarrow} 0.$$

Observe that (e_n) is a fundamental sequence and so E contains the limit $e := \lim_{n\to\infty} e_n$. Since $e_{n+k} \geq e_n$ $(n, k \in \mathbb{N})$; therefore, $e = \sup(e_n)$. If $n \geq m$ then

$$mp(x_{n+l} - x_n) \leq \sum_{k=n+1}^{n+l} kp(x_{n+1} - x_k) \leq e_{n+l} - e_n \leq e.$$

Hence, $p(x_{n+l} - x_n) \leq (1/m)e$ implying that (x_n) is an r-fundamental sequence. Using the hypothesis of r-completeness, find $x := r\text{-}\lim_{n\to\infty} x_n$. Clearly, $\lim_{n\to\infty} |||x - x_n||| = 0$.

Suppose that $(x_n) \subset X$ is an r-fundamental sequence; i.e. $p(x_n - x_m) \leq \lambda_k e$ $(m, n, k \in \mathbb{N};\ m, n \geq k)$, where $\mathbf{0} \leq e \in E$ and $\lim_{k\to\infty} \lambda_k = 0$. Then $|||x_n - x_m||| \leq \lambda_k \|e\| \to 0$ as $k \to \infty$. Consequently, we may find $x := \lim_{n\to\infty} x_n$. The vector norm p acts continuously from $(X, |||\cdot|||)$ to $(E, \|\cdot\|)$. Therefore, passage to the limit in norm in the inequality $p(x_m - x_n) \leq \lambda_k e$ as $m \to \infty$ leads to the inequality $p(x - x_n) \leq \lambda_k e$ for all $k \leq n$. Hence, $x = r\text{-}\lim_{n\to\infty} x_n$. $\rhd$

5.5.3. Let F be an ideal of E. Recall that, given $Y := \{x \in X : p(x) \in F\}$ and $q := p \restriction Y$, the 3-tuple (Y, q, F) is the *restriction* of X relative to F or F-*restriction* of X. If X is a Banach–Kantorovich space then so is Y. If X is r-complete and F is a Banach lattice then Y is a Banach space with mixed norm or a *composite Banach space.*

Take a Banach space $(\mathscr{X}, \rho)$ inside $\mathbf{V}^{(B)}$ and an order dense ideal F in $\mathscr{R}{\downarrow}$. The F-restriction of $\mathscr{X}{\downarrow}$ is the F-*descent* of $\mathscr{X}$ or the *descent of* $\mathscr{X}$ *relative to* F.

The F-descent is denoted by $F^{\downarrow}(\mathscr{X})$. More explicitly, the F-descent of $\mathscr{X}$ is the triple $(F^{\downarrow}(\mathscr{X}), p, F)$, where

$$F^{\downarrow}(\mathscr{X}) := \{x \in \mathscr{X}{\downarrow} : \rho{\downarrow}(x) \in F\}, \quad p := (\rho{\downarrow}) \restriction E^{\downarrow}(\mathscr{X}).$$

If a Banach lattice E is an ideal of $\mathscr{R}{\downarrow}$ then $E^{\downarrow}(\mathscr{X})$ is a composite Banach space.

In the case when E is a K-space of bounded elements, i.e., the order ideal of $\mathscr{R}{\downarrow}$ generated by the order unit $\mathbf{1} \in \mathscr{R}{\downarrow}$, we call the E-descent of $\mathscr{X}$ the *bounded descent* of X. Moreover, the *bounded descent functor* we call the resultant functor $E^{\downarrow}$ which sends $\mathscr{X}$ to $E^{\downarrow}(\mathscr{X})$. Also, we use the notation $\mathscr{X}{\downarrow}^{\infty} := E^{\downarrow}(\mathscr{X})$.

5.5.4. The above definitions make it natural to raise the question: What Banach spaces are linearly isometric to E-descents and, in particular, to bounded descents of Banach spaces inside $\mathbf{V}^{(B)}$? Clearly, the answer depends upon geometry. Skipping details, we briefly consider the case of bounded descent which we need in the sequel.

Let X be a normed space. Suppose that $\mathscr{L}(X)$ has a complete Boolean algebra of norm one projections $\mathscr{B}$ which is isomorphic to B. In this event we will identify the Boolean algebras $\mathscr{B}$ and B, writing $B \subset \mathscr{L}(X)$. Say that X is a *normed B-space* if $B \subset \mathscr{L}(X)$ and for every partition of unity $(b_\xi)_{\xi\in\Xi}$ in B the two conditions are met:

(1) If $b_\xi x = 0$ $(\xi \in \Xi)$ for some $x \in X$ then $x = 0$;

(2) If $b_\xi x = b_\xi x_\xi$ $(\xi \in \Xi)$ for $x \in X$ and a family $(x_\xi)_{\xi\in\Xi}$ in X then $\|x\| \le \sup\{\|b_\xi x_\xi\| : \xi \in \Xi\}$.

Conditions (1) and (2) amount to the respective conditions (1′) and (2′):

(1′) To each $x \in X$ there corresponds the greatest projection $b \in B$ such that $bx = 0$;

(2′) If x, (x_ξ), and (b_ξ) are the same as in (2) then $\|x\| = \sup\{\|b_\xi x_\xi\| : \xi \in \Xi\}$.

From (2′) it follows in particular that

$$\left\| \sum_{k=1}^{n} b_k x \right\| = \max_{k=1,\dots,n} \|b_k x\|$$

for $x \in X$ and pairwise disjoint projections $b_1, \dots, b_n$ in B.

Given a partition of unity (b_ξ), we refer to $x \in X$ such that $(\forall\,\xi)\, b_\xi x = b_\xi x_\xi$ as a *mixing* of (x_ξ) by (b_ξ). If (1) holds then there is a unique mixing x of (x_ξ) by (b_ξ). In these circumstances we naturally call x *the* mixing of (x_ξ) by (b_ξ).

Condition (2) may be paraphrased as follows: The unit ball U_X of X is closed under mixing.

5.5.5. Theorem. *For a Banach space X the following are equivalent:*

(1) *X is a decomposable space with mixed norm whose norm lattice is a K-space of bounded elements;*

(2) *X is a Banach B-space.*

$\lhd$ (1) $\to$ (2) Appeal to the appropriate definitions and 5.3.4.

(2) $\to$ (1) Suppose that X is a Banach B-space and $J : B \to \mathscr{B}$ is the corresponding isomorphism of B onto the Boolean algebra of projections $\mathscr{B}$. Denote by E the ideal that is generated by the identity in the universally complete K-space of all B-valued spectral functions (cf. 5.2.8). Take finite valued element $d := \sum_{i=1}^{n} \lambda_i b_i \in E$, where $\lambda_1, \dots, \lambda_n \in \mathbb{R}$, the family $\{b_1, \dots, b_n\}$ is a partition of unity in B, and λb stands for the spectral function $e : \mu \mapsto e(\mu) \in B$ equal to the zero of B for $\mu \leq \lambda$ and equal to the unity of B for $\mu > \lambda$. Put $J(\alpha) := \sum_{i=1}^{n} \lambda_i J(b_i)$ and note that $J(\alpha)$ is a bounded linear operator in X. Calculating the norm of $J(\alpha)$, obtain

$$\|J(\alpha)\| = \sup_{\|x\|\leq 1} \|J(\alpha)x\| = \sup_{\|x\|\leq 1} \sup_{i=1,\dots,n} \{\|\pi_i x\| \cdot |\lambda_l|\}$$
$$= \sup_{i=1,\dots,n} \sup\{\|\pi_i x\| \, |\lambda_l| : \|x\| \leq 1\} = \max\{|\lambda_1|, \dots, |\lambda_n|\}.$$

On the other hand, the norm $\|\alpha\|_\infty$ of a member α of the K-space of bounded elements E coincides with $\max\{|\lambda_1|, \dots, |\lambda_n|\}$ too. Hence, J is a linear isometry of the subspace E_0 of finite valued members of E to the algebra of bounded operators $\mathscr{L}(X)$. It is also clear that $J(\alpha\beta) = J(\alpha) \circ J(\beta)$ for all $\alpha, \beta \in E_0$. Since E_0 is norm dense in E and $\mathscr{L}(X)$ is a Banach algebra; therefore, we may extend J by continuity to an isometric isomorphism of E onto a closed subalgebra of $\mathscr{L}(X)$. Assigning $x\alpha := \alpha x := J(\alpha)x$ for $x \in X$ and $\alpha \in E$, make X into a unital E-module so that

$$\|\alpha x\| \leq \|x\| \, \|\alpha\|_\infty \quad (\alpha \in E, \ x \in X).$$

Moreover, $\alpha U_X + \beta U_X \subset U_X$ for $|\alpha| + |\beta| \leq \mathbf{1}$. Define the mapping $p : X \to E_+$ by the formula

$$p(x) := \inf\{\alpha \in E_+ : x \in \alpha U_X\} \quad (x \in X),$$

with the infimum taken in the K-space E. If $p(x) = \mathbf{0}$ then to $\varepsilon > \mathbf{0}$ there are a partition of unity $(\pi_\xi) \subset B$ and a family $(\alpha_\xi) \subset E_+$ such that $\pi_\xi \alpha_\xi \leq \varepsilon \mathbf{1}$ and $x \in \alpha_\xi U_X$ for all ξ. But then $\pi_\xi x \in \pi_\xi \alpha_\xi U_X \subset \varepsilon U_X$. Since the unit ball U_X is closed under mixing; therefore, $x = \operatorname{mix}(\pi_\xi x) \in \varepsilon U_X$. The arbitrary choice of $\varepsilon > \mathbf{0}$ implies $x = \mathbf{0}$. If $x \in \alpha U_X$ and $y \in \beta U_X$ for some $\alpha, \beta \in E_+$, then, putting $\gamma := \alpha + \beta + \varepsilon \mathbf{1}$, we may write down

$$x + y = \gamma(\gamma^{-1}x + \gamma^{-1}y) \in \gamma(\gamma^{-1}\alpha U_X + \gamma^{-1}\beta U_X) \subset \gamma U_X.$$

Consequently, $p(x+y) \leq \alpha+\beta+\varepsilon\mathbf{1}$; and taking the infimum over α, β, and ε yields $p(x+y) \leq p(x)+p(y)$. Furthermore, granted $\pi \in B$ and $x \in X$, observe the equalities

$$\pi p(x) = \inf\{\pi\alpha : \mathbf{0} \leq \alpha \in E,\ x \in \alpha U_X\}$$
$$= \inf\{\alpha \in E_+ : \pi x \in \alpha U_X\} = p(\pi x).$$

But then, for $\alpha = \sum \lambda_\imath \pi_\imath$, with $\{\pi_1,\dots,\pi_n\}$ a partition of unity in B, infer that

$$p(\alpha x) = \sum \pi_\imath p(\lambda_\imath x) = \sum_{\imath=1}^{n} \pi_\imath |\lambda_\imath| p(x) = |\alpha| p(x).$$

Hence, $p(\alpha x) = |\alpha| p(x)$ for all $\alpha \in E$. Therefore, (X, p, E) is a decomposable lattice normed space.

Show now that the norm of X is a mixed norm; i.e., $\|x\| = \|p(x)\|_\infty$ for all $x \in X$. Take $0 \neq x \in X$ and put $y = x/\|x\|$. Then $y \in U_X$ and $p(y) \leq \mathbf{1}$. Consequently, $p(x) \leq \|x\| \cdot \mathbf{1}$ or $\|p(x)\|_\infty \leq \|x\| \cdot \|\mathbf{1}\|_\infty = \|x\|$. Conversely, given $\varepsilon > 0$, we may find a partition of unity $(\pi_\xi)_{\xi\in\Xi}$ in $\mathfrak{Pr}(E)$ and a family $(\alpha_\xi)_\xi \subset E_+$ such that $\pi_\xi\alpha_\xi \leq p(x)+\varepsilon\mathbf{1} \leq (\|p(x) :_\infty + \varepsilon) \cdot \mathbf{1}$ and $x \in \alpha_\xi U_X \quad (\xi \in \Xi)$. Whence $\pi_\xi x_\xi \in \pi_\xi \alpha_\xi U_X \subset (\|p(x)\|_\infty + \varepsilon) \cdot \pi_\xi \mathbf{1} U_X \subset (\|p(x)\|_\infty + \varepsilon) U_X$. Consequently, $\|\pi_\xi x\| \leq \|p(x)\|_\infty + \varepsilon$. The arbitrary choice of $\varepsilon > 0$, together with 5.5.4 (2), implies that $\|x\| \leq \|p(x)\|_\infty$. $\triangleright$

5.5.6. A normed B-space X is *B-cyclic* if we may find in X a mixing of each norm bounded family by any partition of unity in B. Considering 5.5.4, note that X is a B-cyclic normed space if and only if, to a partition of unity $(b_\xi) \subset B$ and a family $(x_\xi) \subset U_X$, there is a unique element $x \in U_X$ such that $b_\xi x = b_\xi x_\xi$ for all ξ.

(1) *A Banach B-space X is B-cyclic if and only if X is disjointly complete as a lattice normed space.*

$\triangleleft$ This is evident from the definitions. $\triangleright$

An isometry $\imath$ between normed B-spaces is a *B-isometry* if $\imath$ is linear and commutes with every projection in B. Say that Y is a *B-cyclic completion* of a B-space X if Y is B-cyclic and there is a B-isometry $\imath : X \to Y$ such that every B-cyclic subspace of Y containing $\imath(X)$ coincides with Y.

(2) *A normed B-space is a B-cyclic Banach space if and only if the corresponding lattice normed space is o-complete.*

$\triangleleft$ The claim follows from 5.4.7 and (1) on considering that completeness in norm amounts to completeness with respect to relative uniform convergence, cf. 5.5.2. $\triangleright$

(3) *Each Banach B-space possesses a B-cyclic completion unique up to B-isometry.*

$\lhd$ The claim follows from 5.4.5 and (2). $\rhd$

We are now ready to answer the question of 5.5.4.

5.5.7. Theorem. *A Banach space X is linearly isometric to the bounded descent of some Banach space inside $\mathbf{V}^{(B)}$ if and only if X is B-cyclic.*

$\lhd$ Apply 5.4.1, 5.4.2, 5.5.5, and 5.5.6 (2). $\rhd$

Take a normed B-space X. Denote the norm completion of X by $\widetilde{X}$. Note that $\widetilde{X}$ is a Banach B-space, since every projection $b \in B$ admits a unique extension to the whole of $\widetilde{X}$ which preserves the norm of b. By 5.5.6 (3), $\widetilde{X}$ possesses a cyclic B-completion which is denoted by $\overline{X}$. Applying Theorem 5.5.7, take a Banach space $\mathscr{X}$ inside $\mathbf{V}^{(B)}$ whose bounded descent is B-isometric with $\overline{X}$. The element $\mathscr{X} \in \mathbf{V}^{(B)}$ is the *Boolean valued representation* of X.

5.5.8. Let X and Y be normed spaces such that $B \subset \mathscr{L}(X)$ and $B \subset \mathscr{L}(Y)$. An operator $T : X \to Y$ is *B-linear* if T commutes with every projection in B; i.e., $b \circ T = T \circ b$ for all $b \in B$.

Denote by $\mathscr{L}_B(X, Y)$ the set of all bounded B-linear operators from X to Y. In this event $W := \mathscr{L}_B(X, Y)$ is a Banach space and $B \subset W$. If Y is B-cyclic then so is W. A projection $b \in B$ acts in W by the rule $T \mapsto b \circ T$ $(T \in W)$.

We call $X^{\#} := \mathscr{L}_B(X, B(\mathbb{R}))$ the *B-dual* of X. If $X^{\#}$ and Y are B-isometric to each other then we say that Y is a *B-dual space* and X is a *B-predual* of Y. In symbols, $X = Y_{\#}$.

5.5.9. Theorem. *Assume that X is a normed B-space and Y is a B-cyclic Banach space. Let $\mathscr{X}$ and $\mathscr{Y}$ stand for the Boolean valued representations of X and Y. The space $\mathscr{L}_B(X, Y)$ is B-isometric to the bounded descent of the space $\mathscr{L}(\mathscr{X}, \mathscr{Y})$ of all bounded linear operators from $\mathscr{X}$ to $\mathscr{Y}$ inside $\mathbf{V}^{(B)}$. Moreover, to $T \in \mathscr{L}_B(X, Y)$ there corresponds the member $\mathscr{T} := T\uparrow$ of $\mathbf{V}^{(B)}$ determined from the formulas $[\![\mathscr{T} : \mathscr{X} \to \mathscr{Y}]\!] = \mathbf{1}$ and $[\![\mathscr{T} \imath x = \imath T x]\!] = \mathbf{1}$ for all $x \in X$, where $\imath$ stands for the mapping that embeds X into $\mathscr{X}{\downarrow}$ and Y into $\mathscr{Y}{\downarrow}$.*

$\lhd$ Without loss of generality, assume that X and Y are the bounded descents of some Banach spaces $\mathscr{X}$ and $\mathscr{Y}$ (cf. 5.5.6 (3) and 5.5.7). Put $X_0 := \mathscr{X}{\downarrow}$ and $Y_0 := \mathscr{Y}{\downarrow}$. By 5.4.9, the spaces $\mathscr{L}(\mathscr{X}, \mathscr{Y}){\downarrow}$ and $\mathscr{L}_b(X_0, Y_0)$ are linearly isometric. Moreover, the restriction of $\mathscr{L}_b(X_0, Y_0)$ relative to $B(\mathbb{R})$ coincides with the bounded descent of $\mathscr{L}(\mathscr{X}, \mathscr{Y})$. It suffices to note that each member T of $\mathscr{L}_b(X, Y)$ admits a unique extension which preserves the norm of T. $\rhd$

5.5.10. Let $\mathscr{X}^*$ be the dual of $\mathscr{X}$. Denote by $\simeq$ and $\simeq_B$ the relations of isometric isomorphy and isometric B-isomorphy between Banach spaces. Suppose also that X, Y, $\mathscr{X}$, and $\mathscr{Y}$ are the same as in 5.5.9.

(1) *The following equivalence holds:* $X^{\#} \simeq_B Y \leftrightarrow [\![\mathscr{X}^* \simeq \mathscr{Y}]\!] = 1$.

(2) *If* $\overline{X}$ *is the* B*-cyclic completion of* X *then* $X^{\#} = \overline{X}^{\#}$.

5.5.11. COMMENTS.

(1) Spaces with mixed norm in the sense of Section 5.5 were studied in [125]. This article presents various applications of the concept of mixed norm to Banach space geometry and operator theory. The bounded descent of 5.5.3 appeared in the research by G. Takeuti into von Neumann algebras and C^*-algebras within Boolean valued models [236, 237] and in the research by M. Ozawa into Boolean valued interpretation of the theory of Hilbert and Banach spaces [194, 200].

(2) The results of this subsection belong to A. G. Kusraev [125]. Later, analogous properties were established by M. Ozawa [200] in another statement. The difference is in the fact that the article [200] deals with Banach spaces possessing an extra module structure which may be reconstructed in every Banach B-space as seen from 5.4.8 and 5.5.5.

(3) Presenting Theorem 5.5.7, we slightly touch a rich and beautiful direction of research: Banach space geometry; cf. [35, 149, 154]. Banach space with complete Boolean algebra of projections was studied irrespectively of Boolean valued analysis in [44, 214, 246].

Chapter 6

Boolean Valued Analysis of Banach Algebras

The theory of Banach algebras is one of the most attractive traditional sections of functional analysis. This chapter presents the basic results of Boolean valued analysis of involutive Banach algebras.

The possibility of applying Boolean valued analysis to operator algebras rests on the following observation: If the center of an algebra is properly qualified and perfectly located then it becomes a one dimensional subalgebra after immersion in a suitable Boolean valued universe $\mathbf{V}^{(B)}$. This might lead to a simpler algebra. On the other hand, the transfer principle implies that the scope of the formal theory of the initial algebra is the same as that of its Boolean valued representation. Theorems 6.1.5 and 6.1.6 elaborate this claim for a general Banach algebra as well as a C^*-algebra.

Further exposition focuses on analysis of AW^*-algebras and AW^*-modules which transform into AW^*-factors and Hilbert spaces in $\mathbf{V}^{(B)}$; cf. Theorems 6.2.4 and 6.2.8.

The dimension of a Hilbert space becomes a Boolean valued cardinal referred naturally to as the Boolean dimension of the AW^*-module depicting the space. Here a rather delicate effect reveals itself, the so-called *cardinal shift*: standard cardinals may glue together when embedded in $\mathbf{V}^{(B)}$. In other words, the bases of isomorphic AW^*-modules may differ in power. This also implies that every type I AW^*-algebra decomposes into the direct sum of homogeneous subalgebras in many ways. I. Kaplansky conjectured the fact as far back as in 1953. These results are set forth in Sections 6.3 and 6.4.

Leaning on the results about the Boolean valued immersion of AW^*-modules and AW^*-algebras, we further derive the function representations of these objects. To put it more precisely, we prove that every AW^*-module is unitarily equivalent to the direct sum of homogeneous AW^*-modules consisting of continuous vector functions ranging in a Hilbert space. An analogous representation holds for an arbitrary type I AW^*-algebra on replacing continuous vector functions with operator valued functions continuous in the strong operator topology. The relevant facts are presented in Section 6.5.

We call an AW^*-algebra *embeddable* if it is $*$-isomorphic with the double commutant of some type I AW^*-algebra. Each embeddable AW^*-algebra admits a Boolean valued representation that is a von Neumann algebra or factor. We give several characterizations for embeddable AW^*-algebras. In particular, we prove in Section 6.6 that an AW^*-algebra A is embeddable if and only if the center valued normal states of A separate A.

6.1. The Descent of a Banach Algebra

The previous chapter paves a way to Boolean valued representation of Banach spaces. We now proceed to involutive Banach algebras.

6.1.1. We recall the preliminaries on restricting exposition to complex algebras. Note also that by an *algebra* we always mean a unital associative algebra.

An *involutive algebra* or $*$-*algebra* A is an algebra with *involution*; i.e., a mapping $x \mapsto x^*$ $(x \in A)$ satisfying the conditions:

(1) $x^{**} = x$ $(x \in A)$;

(2) $(x + y)^* = x^* + y^*$ $(x, y \in A)$;

(3) $(\lambda x)^* = \lambda^* x^*$ $(\lambda \in \mathbb{C},\ x \in A)$;

(4) $(xy)^* = y^* x^*$ $(x, y \in A)$.

An element x of an involutive algebra is *hermitian* provided that $x^* = x$. A *projection* e in A is a hermitian idempotent; i.e., $e = e^*$ and $e^2 = e$. The symbol $\mathfrak{P}(A)$ stands for the set of all projections of an involutive algebra A. Clearly, the formula

$$c \leq e \leftrightarrow c = ce = ec \quad (c, e \in \mathfrak{P}(X))$$

specifies some order $\leq$ on the set of projections. We call projections e and c *equivalent* and write $e \sim c$ if there is an element x in A satisfying $x^*x = e$ and $xx^* = c$. In this case x is a *partial isometry* with initial projection e and final projection c. The relation $\sim$ is in fact an equivalence over $\mathfrak{P}(A)$.

A projection e is a *central* projection if $ex = xe$ for all $x \in A$. Denote the set of all central projections by $\mathfrak{P}_c(A)$.

6.1.2. Granted a nonempty set $M \subset A$, define

$$M^{\perp} := \{y \in A : (\forall x \in M)\, xy = \mathbf{0}\};$$
$${}^{\perp}M := \{x \in A : (\forall y \in M)\, xy = \mathbf{0}\}.$$

Call $M^\perp$ the *right annihilator* of M and ${}^\perp M$, the *left annihilator* of M.

General properties of annihilators imply that the inclusion ordered sets of all right and all left annihilators are Dedekind complete lattices. The mapping $K \mapsto K^* := \{x^* : x \in K\}$ is an isotonic bijection between these lattices since $(M^\perp)^* = {}^\perp(M^*)$ and $({}^\perp M)^* = (M^*)^\perp$.

A *Baer $*$-algebra* is an involutive algebra A such that, to each nonempty $M \subset A$, there is some e in $\mathfrak{P}(A)$ satisfying $M^\perp = eA$. Clearly, this amounts to the condition that each left annihilator has the form ${}^\perp M = Ac$ for an appropriate projection c. To each left annihilator L in a Baer $*$-algebra there is a unique projection $c_L \in A$ such that $x = xc_L$ for all $x \in L$ and $c_L y = \mathbf{0}$ whenever $y \in L^\perp$. The mapping $L \mapsto c_L$ is an isomorphism between the poset of left annihilators and the poset of all projections. The inverse isomorphism has the form $c \mapsto {}^\perp(\mathbf{1} - c)$, with $c \in \mathfrak{P}(A)$. An analogous claim is true for right annihilators. This implies in particular that the poset $\mathfrak{P}(A)$ is a Dedekind complete lattice. The mapping $e \mapsto e^\perp := \mathbf{1} - e$, with $e \in \mathfrak{P}(A)$, satisfies the conditions:

$$e^{\perp\perp} = e,\ e \wedge e^\perp = \mathbf{0},\ e \vee e^\perp = \mathbf{1},$$
$$(e \wedge c)^\perp = e^\perp \vee c^\perp,\ (e \vee c)^\perp = e^\perp \wedge c^\perp,$$
$$e \leq c \rightarrow e \vee (e^\perp \wedge c) = c.$$

In other words, $(\mathfrak{P}(A), \wedge, \vee, \perp)$ is an *orthomodular lattice* (cf. [14]).

6.1.3. A norm $\|\cdot\|$ on an algebra A is *submultiplicative* if

$$\|xy\| \leq \|x\|\,\|y\| \quad (x, y \in A).$$

A *Banach algebra* A is an algebra furnished with a submultiplicative norm making A into a Banach space. If A is an involutive Banach algebra satisfying

$$\|xx^*\| = \|x\|^2 \quad (x \in A)$$

then A is called a *C^*-algebra.*

An element x of a C^*-algebra A is *positive* whenever $x = y^*y$ for some $y \in A$. The set A_+ of all positive elements is an ordering cone and so (A, A_+) is an ordered vector space. Treating a C^*-algebra as an ordered vector space, we always imply the order that is conventionally induced by A_+.

6.1.4. A Banach algebra A is *B-cyclic* with respect to a complete Boolean algebra B of projections of A provided that A is a B-cyclic Banach space in the sense of 5.5.6 and every member of B is a *multiplicative projection*, i.e.

$$\pi(xy) = \pi(x)\pi(y) = x\pi y = \pi(x)y \quad (x, y \in A,\ \pi \in B).$$

The definition of *B-cyclic involutive algebra* appears on requiring additionally that every member of B is *$*$-preserving*, i.e.

$$\pi(x^*) = (\pi x)^* \quad (x \in A,\ \pi \in B).$$

Finally, the definition of *B-cyclic C^*-algebra* is evident.

Recall that we consider only unital algebras. With this in mind, let $\mathbf{1}$ be the unity of A and identify each projection $b \in B$ with the element $b\mathbf{1}$. If A is involutive then $b\mathbf{1}$ is a central projection of A in the sense of 6.1.1. In this event we write $B \subset \mathfrak{P}_c(A)$. The record $B \sqsubset A$ means that A is a B-cyclic Banach algebra. Note that a C^*-algebra A is *B-cyclic* whenever to every partition of unity $(b_\xi)_{\xi\in\Xi}$ and to every bounded family $(x_\xi)_{\xi\in\Xi} \subset A$ there is a unique $x \in A$ satisfying $b_\xi x = b_\xi x_\xi$ for all $\xi \in \Xi$.

Each complex K-space of bounded elements with base B and fixed unity exhibits an example of a B-cyclic C^*-algebra (cf. 5.1.3 and 5.2.5 (5)). This algebra is clearly unique up to $*$-isomorphism. We denote this algebra by $B(\mathbb{C})$. We will often identify $B(\mathbb{C})$ with the bounded part of the descent $\mathscr{C}{\downarrow}$ of the field $\mathscr{C}$ of complex numbers inside $\mathbf{V}^{(B)}$. The algebra $B(\mathbb{C})$ is also referred to as the *Stone algebra with base* B denoted sometimes by $\mathscr{S}(B)$.

Take B-cyclic algebras A_1 and A_2. A bounded operator $\Phi : A_1 \to A_2$ is a B-homomorphism whenever Φ is B-linear in the sense of 5.5.8 and multiplicative: $\Phi(xy) = \Phi(x) \cdot \Phi(y)$ for all $x, y \in A$. If A_1 and A_2 are involutive algebras and some B-homomorphism Φ is $*$-preserving, i.e., $\Phi(x^*) = \Phi(x)^*$ for all $x \in A_1$; then Φ is a *$*$-B-homomorphism*. Hence, A_1 and A_2 are B-isomorphic whenever there is an isomorphism acting from A_1 to A_2 and commuting with projections in B. If a B-isomorphism is $*$-preserving then we call it a $*$-B-isomorphism.

6.1.5. Theorem. *The bounded descent of a Banach algebra inside $\mathbf{V}^{(B)}$ is a B-cyclic Banach algebra. Conversely, to each B-cyclic Banach algebra A there is a Banach algebra $\mathscr{A}$ inside $\mathbf{V}^{(B)}$ such that A is isometrically B-isomorphic to the bounded descent of $\mathscr{A}$. Moreover, this algebra $\mathscr{A}$ is unique up to isomorphism.*

$\triangleleft$ Take a B-cyclic Banach algebra A. By Theorem 5.5.7 there is a Banach space $\mathscr{A}$ in $\mathbf{V}^{(B)}$ whose bounded descent A_0 is a B-cyclic Banach space admitting an isometric B-isomorphism with A. Without loss of generality, assume that $A_0 = A$. Multiplication on A is extensional. Indeed, if $b \le [\![x = u]\!] \wedge [\![y = v]\!]$ with $x, y, u, v \in A$ then, by item (b) of 5.4.1 (2),

$$\begin{aligned} &\mathbf{0} = x\chi(b)(y - v) + \chi(b)(x - u)v \\ &\to \chi(b)(xy - uv) = \mathbf{0} \to \chi(b)(xy) = \chi(b)uv \to b \le [\![xy = uv]\!]. \end{aligned}$$

Let $\odot$ stand for the ascent of the multiplication $\cdot$ on A. It is easy that $\odot$ is a binary operation on $\mathscr{A}$ and the space $\mathscr{A}$ with the operation $\odot$ is an algebra. If p is the vector norm of the space A then $\|a\| = \|p(a)\|_\infty$ and $[\![p(a) = \rho(a)]\!] = \mathbf{1}$ for all $a \in \mathscr{A}$ where ρ is the norm of $\mathscr{A}$ (cf. 5.5.5). Show that p is a submultiplicative norm, i.e. $p(xy) \leq p(x)p(y)$. To this end, recall (cf. 5.4.1 (2) and 5.5.5) that A is a Banach module over the ring $B(\mathbb{R})$, with $B(\mathbb{R})$ the bounded part of $\mathscr{R}{\downarrow}$. Furthermore, p maintains the equality

$$p(x) = \inf\{\alpha \in E^+ : x \in \alpha U_A\} \quad (x \in A).$$

Hence, submultiplicativity of p follows from the fact that the unit ball U_A is closed under multiplication; i.e., the containments $x,\ y \in U_A$ imply $xy \in U_A$. Therefore, $p \circ (\cdot) \leq (\cdot) \circ (p \times p)$. Using the rules for ascending mappings (cf. 3.3.11), we see that $[\![\rho \circ \odot \leq \odot \circ (\rho \times \rho)]\!] = \mathbf{1}$, i.e., $[\![\rho$ is a submultiplicative norm $]\!] = \mathbf{1}$. We finally infer that $\mathscr{A}$ is a Banach algebra inside $\mathbf{V}^{(B)}$. To show uniqueness of A argue as follows: Assume that $\mathscr{A}_1$ and $\mathscr{A}_2$ are Banach algebras inside $\mathbf{V}^{(B)}$. Let g be an isometric isomorphism between the bounded descents of $\mathscr{A}_1$ and $\mathscr{A}_2$. Then g is an extensional mapping and $\psi := g{\uparrow}$, the ascent of g, is a linear isometry between the Banach spaces $\mathscr{A}_1$ and $\mathscr{A}_2$. Multiplicativity of ψ follows from the formulas

$$\psi \circ \odot = g{\uparrow} \circ (\cdot){\uparrow} = (g \circ (\cdot)){\uparrow} = ((\cdot) \circ (g \times g)){\uparrow} = (\cdot){\uparrow} \circ (g{\uparrow} \times g{\uparrow}) = \odot \circ (\psi \times \psi)$$

with $\odot$ standing for the operations of multiplication on $\mathscr{A}_1$ and $\mathscr{A}_2$ and $(\cdot)$ symbolizing the operations of multiplication in the bounded descents of $\mathscr{A}_1$ and $\mathscr{A}_2$.

Assume now that $\mathscr{A}$ is a Banach algebra inside $\mathbf{V}^{(B)}$ and A is the bounded descent of $\mathscr{A}$. We know that A is a B-cyclic Banach space (cf. 5.5.11). If χ is the canonical isomorphism of B onto the base $\mathfrak{E}(E)$ then $b \leq [\![x = \mathbf{0}]\!] \leftrightarrow \chi(b)x = \mathbf{0}$ for all $x \in A$ (cf. 5.4.1 (2)). Considering the definition of χ and the obvious implication

$$\chi(b) = \mathbf{0} \vee \chi(b) = \mathbf{1} \rightarrow (\forall\, x \in A)(\forall\, y \in A)(\chi(b)xy = (\chi(b)x)y = x\,(\chi(b)y)),$$

take arbitrary $x, y \in A$ and deduce that

$$[\![\chi(b)xy = x\chi(b)y = (\chi(b)x)y]\!] \geq [\![\chi(b) = \mathbf{1}]\!] \vee [\![\chi(b) = \mathbf{0}]\!] = b \vee b^* = \mathbf{1}.$$

This shows that the projection $\pi_b : x \mapsto \chi(b)x$, with $x \in A$, enjoys the needed property: $\pi_b xy = (\pi_b x)y = x\,(\pi_b y)$ for all $x, y \in A$. Thus, A is a B-cyclic algebra. $\triangleright$

6.1.6. Theorem. *The bounded descent of a C^*-algebra inside $\mathbf{V}^{(B)}$ is a B-cyclic C^*-algebra. Conversely, to each B-cyclic C^*-algebra A there is a C^*-algebra $\mathscr{A}$ inside $\mathbf{V}^{(B)}$ such that the bounded descent of $\mathscr{A}$ is $*$-B-isomorphic with A. Moreover, this C^*-algebra $\mathscr{A}$ is unique up to $*$-isomorphism.*

$\lhd$ If A is a B-cyclic C^*-algebra then the structure of a Banach $\mathscr{S}(B)$-module on A possesses the additional property that $(\alpha x)^* = \alpha x^*$ for all $\alpha \in B(\mathbb{R})$ and $x \in A$. As before, $B(\mathbb{R})$ stands for the real part of the complex Banach algebra $\mathscr{S}(B)$. Indeed, if $\alpha := \sum_{k=1}^{n} \lambda_k \pi_k$ with $\lambda_1, \dots, \lambda_n \in \mathbb{R}$ and $\pi_1, \dots, \pi_n \in \mathfrak{E}(\mathscr{S}(B))$, then

$$(\alpha x)^* = \sum_{k=1}^{n} \lambda_k (\pi_k x)^* = \sum_{k=1}^{n} \lambda_k \pi_k x^* = \alpha x^*.$$

Involution is an isometry on every C^*-algebra and so $U_A^* = U_A$. We may conclude that

$$x \in \alpha U_A \leftrightarrow x x^* \in \alpha^2 U_A \quad (x \in A,\ \alpha \in \mathscr{S}(B)).$$

Hence, $p(xx^*) = p(x)^2$. In particular, the involution on A is an isometry with respect to the vector norm p, i.e. $p(x^*) = p(x)$ for all $x \in A$. Note also that if $(\mathscr{A}, \rho)$ is a Banach algebra inside $\mathbf{V}^{(B)}$, A is the bounded descent of $\mathscr{A}$, and p is the restriction of $\rho{\downarrow}$ to A; then the descent of the involution on A obeys the conditions $[\![(\forall x \in \mathscr{A}) \rho(xx^*) = \rho(x)^2]\!] = \mathbf{1}$ if and only if $p(xx^*) = p(x)^2$ for all $x \in A$. It suffices to appeal to Theorem 6.1.5 and offer some elementary arguments. $\rhd$

6.1.7. Theorem. *Let A be a B-cyclic Banach algebra such that every element $x \in A$, satisfying the condition $(\forall b \in B)(bx = \mathbf{0} \to b = \mathbf{0})$, is invertible. Then A is isometrically B-isomorphic to the Stone algebra with base B.*

$\lhd$ By Theorem 6.1.5 we may treat A as the bounded descent of some Banach algebra $\mathscr{A} \in \mathbf{V}^{(B)}$. By hypothesis, every nonzero element of $\mathscr{A}$ is invertible. Indeed, we may calculate the following Boolean truth value:

$$\begin{aligned} c &:= [\![(\forall x)(x \in \mathscr{A} \wedge x \neq \mathbf{0} \to (\exists z)(z = x^{-1}))]\!] \\ &= \bigwedge \{ [\![(\exists z)(z = x^{-1})]\!] : x \in A,\ [\![x \neq \mathbf{0}]\!] = \mathbf{1} \}. \end{aligned}$$

By 5.4.1 (2) (c), the equality $[\![x \neq \mathbf{0}]\!] = \mathbf{1}$ amounts to the condition $\chi(b)x = \mathbf{0} \leftrightarrow b = \mathbf{0}$. Hence, if $[\![x \neq \mathbf{0}]\!] = \mathbf{1}$ then we may find the inverse x^{-1} of x in A and $[\![(\exists z)(z = x^{-1})]\!] = \mathbf{1}$. Consequently, $c = \mathbf{1}$. Applying the Gelfand–Mazur Theorem inside $\mathbf{V}^{(B)}$, note that the algebra $\mathscr{A}$ is isometrically isomorphic to the field of complex numbers $\mathscr{C}$ inside $\mathbf{V}^{(B)}$. But then A is isometrically B-isomorphic with the bounded descent of $\mathscr{C}$ which is the Stone algebra with base B (cf. 6.1.4). $\rhd$

6.1.8. Theorem. *Assume that A is a B-cyclic Banach algebra, $\mathscr{S}(B)$ is the Stone algebra with base B, and $\Phi : A \to \mathscr{S}(B)$ is a B-linear operator. Assume further that $\Phi(1) = \mathbf{1}$ and $e_{\Phi(x)} = \mathbf{1}$ for every invertible element $x \in A$. Then Φ is multiplicative, i.e., $\Phi(xy) = \Phi(x)\Phi(y)$ for all $x, y \in A$.*

$\lhd$ Arguing like in 6.1.7, let $\varphi := \Phi\uparrow$. Then $[\![\varphi : \mathscr{A} \to \mathscr{C}$ is a linear functional $]\!] = \mathbf{1}$. Moreover, $[\![\varphi(x) \neq \mathbf{0}$ for every invertible $x \in A]\!] = \mathbf{1}$. By the Gleason–Želazko–Kahane Theorem $[\![\varphi$ is a multiplicative functional $]\!] = \mathbf{1}$. This implies multiplicativity for Φ in much the same way as we use in 6.1.5 while proving submultiplicativity for the norm p. $\rhd$

6.1.9. *Theorem.* *Assume that A is a B-cyclic commutative Banach algebra with involution, $\mathscr{S}(B)$ is the Stone algebra with base B, and $\Phi : A \to \mathscr{S}(B)$ is a B-linear operator. Assume further that K is the set of all positive B-linear operators $\Psi : A \to \mathscr{S}(B)$ satisfying $\Psi(1) \leq \mathbf{1}$. If $\Phi \in K$ then the following are equivalent:*

(1) $\Phi(xy) = \Phi(x)\Phi(y)$ $(x, y \in A)$;

(2) $\Phi(xx^*) = \Phi(x)\Phi(x^*)$ $(x \in A)$;

(3) $\Phi \in \operatorname{ext}(K)$, *where* $\operatorname{ext}(K)$ *denotes the set of extreme points of* K.

$\lhd$ With the notation of the proof of Theorem 6.1.8, we may assert that $[\![\mathscr{A}$ is a commutative Banach algebra with involution, and $\varphi : \mathscr{A} \to \mathscr{C}$ is a positive functional satisfying $\varphi(1) \leq \mathbf{1}]\!] = \mathbf{1}$. Let $\mathscr{K}$ consist of all positive linear functionals ψ on $\mathscr{A}$ satisfying $\psi(1) \leq \mathbf{1}$. It is evident that $\psi \mapsto (\psi\downarrow) \restriction A$ is an affine bijection λ between the convex sets $\mathscr{K}\downarrow$ and $\overline{K} := \{\Psi\uparrow : \Psi \in K\}$. Moreover, $[\![\psi \in \operatorname{ext}(\mathscr{K})]\!] = \mathbf{1} \leftrightarrow \lambda\psi \in \operatorname{ext}(K)$. We are left with applying the scalar version (in which case $\mathscr{S}(B) = \mathscr{C}$) of the claim inside $\mathbf{V}^{(B)}$. This ends the proof. $\rhd$

6.1.10. We agree to let B-$\operatorname{Hom}(A_1, A_2)$ stand for the set comprising B-homomorphisms from A_1 to A_2. We also agree that $\operatorname{Hom}^B(\mathscr{A}_1, \mathscr{A}_2)$ symbolizes the element of $\mathbf{V}^{(B)}$ which depicts the set of all homomorphisms from $\mathscr{A}_1$ to $\mathscr{A}_2$.

(1) *Assume that $\mathscr{A}_1$ and $\mathscr{A}_2$ are Banach algebras inside $\mathbf{V}^{(B)}$ and A_1 and A_2 are their respective bounded descents. If $\Phi \in B\text{-}\operatorname{Hom}^B(A_1, A_2)$ and $\varphi := \Phi\uparrow$ then $[\![\varphi \in \operatorname{Hom}^B(\mathscr{A}_1, \mathscr{A}_2)]\!] = \mathbf{1}$ and $[\![\|\varphi\| \leq C]\!] = \mathbf{1}$ for some $C \in \mathbb{R}$. The mapping $\Phi \mapsto \varphi$ is an isometric bijection between $B\text{-}\operatorname{Hom}(A_1, A_2)$ and $\operatorname{Hom}^B(\mathscr{A}_1, \mathscr{A}_2)\downarrow^\infty$.*

$\lhd$ All claims with the exception of multiplicativity ensue from 5.4.9. The fact that φ and Φ are multiplicative may be justified in much the same way as in the proof of uniqueness in 6.1.5. $\rhd$

(2) *Assume that $\mathscr{A}_1$ and $\mathscr{A}_2$ are involutive Banach algebras inside $\mathbf{V}^{(B)}$, while $\Phi \in B\text{-}\operatorname{Hom}(A_1, A_2)$ and $\varphi \in \operatorname{Hom}^B(\mathscr{A}_1, \mathscr{A}_2)$ correspond to one another under the bijection of* (1). *Then the equality $[\![\varphi$ is $*$-preserving$]\!] = \mathbf{1}$ holds if and only if Φ is $*$-preserving.*

$\lhd$ Appealing to 5.5.4 and 6.1.6 completes the proof. $\rhd$

6.1.11. *Assume that $\mathscr{A}$ is an involutive Banach algebra inside $\mathbf{V}^{(B)}$ and A is the bounded descent of $\mathscr{A}$. Then $x \in A$ is a hermitian element or a positive (central) projection if and only if $[\![x$ is a hermitian element or a positive (central) projection$)]\!] = \mathbf{1}$.*

◁ This is obvious. ▷

6.1.12. Comments.

(1) J. von Neumann started the study of involutive operator algebras, inspired by the mathematical problems of quantum mechanics, cf. [177, 178]. This traditional connection with theoretical physics is still alive (cf., for instance, [19]). However, the present-day theory of involutive topological algebras contains several rather abstract and esoteric fields of research, raising many subtle mathematical problems. To enter this field, the reader may consult [6, 38, 40, 95, 151, 174, 207, 211, 224, 229, 230, 257].

The study of C^*-algebras was originated by I. M. Gelfand and M. A. Naĭmark in 1943. The principal structural properties of C^*-algebra are connected with positivity. As regards the preliminaries of involutive algebras, consult [12]. See [6, 38, 39, 146, 173] for the details of C^*-algebras and [40, 211, 230], for the details of von Neumann algebras.

(2) G. Takeuti started studying C^*-algebras and von Neumann algebras by using Boolean valued models with [236, 237]. Theorem 6.1.6 belongs to him. Theorems 6.1.7 and 6.1.8 are Boolean valued interpretations of the Gelfand–Mazur Theorem and the Gleason–Želazko–Kahane Theorem, ranking as classical facts of functional analysis (see, for instance, [146, 210]).

Note also the monograph [33] which deals with applications of Boolean valued models to independence problem in the related section of analysis.

6.2. AW^*-Algebras and AW^*-Modules

In this section we present results on Boolean valued representation of the objects indicated in the title.

6.2.1. An *AW^*-algebra* is a C^*-algebra presenting a Baer $*$-algebra. More explicitly, an AW^*-algebra is a C^*-algebra whose every right annihilator has the form eA with e a projection. By the way, we note that a better term for an AW^*-algebra is a Baer C^*-algebra.

A C^*-algebra A is an AW^*-algebra if and only if the following conditions are satisfied:

(1) *Every orthogonal family in $\mathfrak{P}(A)$ has a supremum;*

(2) *Every maximal commutative $*$-subalgebra A_0 of A is a complex K-space of bounded elements.*

The space $\mathscr{L}(H)$ of all bounded linear endomorphisms of a complex Hilbert space H exhibits an example of an AW^*-algebra. Recall that the structure of a Banach algebra in $\mathscr{L}(H)$ results from the conventional addition and composition of operators and the routine operator norm. The involution in $\mathscr{L}(H)$ is the taking of the adjoint of an operator. Observe that every commutative AW^*-algebra, referred also to as a *Stone algebra*, is a complex K-space of bounded elements with the unity of multiplication serving as a strong order unit.

6.2.2. Spectral Theorem. *To each hermitian element A of an AW^*-algebra A there is a unique resolution of identity $\lambda \mapsto e_\lambda$, with $\lambda \in \mathbb{R}$, in $\mathfrak{P}(A)$ such that*

$$a = \int\limits_{-\|a\|}^{\|a\|} \lambda \, de_\lambda.$$

Moreover, an element $x \in A$ commutes with a if and only if $xe_\lambda = e_\lambda x$ for all $\lambda \in \mathbb{R}$.

$\lhd$ Recall that the term "resolution of identity" in $\mathfrak{P}(A)$ means the same as in the case of a Boolean algebra; i.e., we call so every function $\lambda \mapsto e_\lambda$ that satisfies 5.2.6 (1–3) (cf. 5.2.8). Every maximal commutative $*$-subalgebra of A which contains a is a complex K-space by 6.2.1 (2). Hence, the sought representation ensues from the Freudenthal Theorem 5.2.14. The claim about commuting follows from the fact that the element a and the set $\{e_\lambda : \lambda \in \mathbb{R}\}$ generate the same maximal $*$-subalgebra. $\rhd$

6.2.3. Theorem. *An AW^*-algebra A is a B-cyclic C^*-algebra for every order closed subalgebra B of the complete Boolean algebra $\mathfrak{P}_c(A)$.*

$\lhd$ Let U denote the unit ball of A. It suffices to demonstrate that to every partition of unity $(b_\xi)_{\xi\in\Xi} \subset B$ and every family $(a_\xi)_{\xi\in\Xi} \subset U$ there is a unique element a in U satisfying $b_\xi a_\xi = b_\xi a$ for all $\xi \in \Xi$. Assume first that a_ξ is hermitian for all $\xi \in \Xi$. Then the family $(b_\xi a_\xi)$ consists of pairwise commuting hermitian elements since $(b_\xi a_\xi) \cdot (b_\eta a_\eta) = (b_\xi b_\eta) \cdot (a_\xi a_\eta)$ for $\xi \neq \eta$.

Denote by A_0 the maximal commutative $*$-subalgebra of A that includes $(b_\xi a_\xi)$. By 6.2.1 (2), A_0 is a complex K-space of bounded elements. Hence, A_0 contains the element $a = o\text{-}\sum_{\xi\in\Xi} b_\xi a_\xi$, where o-summation is done in A_0. Clearly, $b_\xi a_\xi = b_\xi a$ for all $\xi \in \Xi$. On the other hand, $-\mathbf{1} \le a_\xi \le \mathbf{1}$ implies that $-\mathbf{1} \le a \le \mathbf{1}$, and so $\|a\| \le 1$.

Uniqueness is now in order. Assume that for some hermitian element $d \in A$ we have $b_\xi d = 0$ for all $\xi \in \Xi$. By 5.2.6 (10),

$$e_\lambda^{b_\xi d} = b_\xi^\perp \vee e_\lambda^d = \mathbf{1} = e_\lambda^{\mathbf{1}} \quad (\lambda \in \mathbb{R}, \lambda > 0),$$

$$e_\lambda^{b_\xi d} = b_\xi \wedge e_\lambda^d = \mathbf{0} = e_\lambda^{\mathbf{0}} \quad (\lambda \in \mathbb{R}, \lambda \le 0).$$

The equalities $b_\xi^\perp \vee e_\lambda^d = \mathbf{1}$ and $b_\xi \wedge e_\lambda^d = \mathbf{0}$ are equivalent with the respective inequalities $e_\lambda^d \ge b_\xi$ and $e_\lambda^d \le b_\xi^\perp$. Therefore, $e_\lambda^d = \mathbf{1}$ for $\lambda > 0$ and $e_\lambda^d = \mathbf{0}$ for $\lambda \le 0$; i.e. the spectral function of d is that of the zero element. Consequently, $d = 0$.

In the general case of arbitrary $a_\xi \in U$, use the presentation $a_\xi = u_\xi + iv_\xi$, where i stands for the imaginary unity, and u_ξ and v_ξ are uniquely determined hermitian elements of U. The above shows that there are hermitian elements $u, v \in U$ satisfying $b_\xi u = b_\xi u_\xi$ and $b_\xi v = b_\xi v_\xi$ for all $\xi \in \Xi$. Observe that $a = u + iv$ is a sought element. Indeed, $b_\xi a = b_\xi a_\xi$ for all $\xi \in \Xi$. Moreover, the hermitian elements $a_\xi^* a_\xi$ belong to U, and $b_\xi a^* a = b_\xi a_\xi^* a_\xi$ for all $\xi \in \Xi$. Only one element meets these conditions. Since a^*a fits in, we see that $a^*a \in U$. Whence $a \in U$ for $\|a\|^2 = \|a^*a\| \leq 1$. $\triangleright$

6.2.4. Theorem. *Assume that $\mathscr{A}$ is an AW^*-algebra inside $\mathbf{V}^{(B)}$ and A is the bounded descent of $\mathscr{A}$. Then A is also an AW^*-algebra and, moreover, $\mathfrak{P}_c(A)$ has an order closed subalgebra isomorphic with B. Conversely, let A be an AW^*-algebra such that B is an order closed subalgebra of the Boolean algebra $\mathfrak{P}_c(A)$. Then there is an AW^*-algebra $\mathscr{A}$ in $\mathbf{V}^{(B)}$ whose bounded descent is $*$-B-isomorphic with A. This algebra $\mathscr{A}$ is unique up to isomorphism inside $\mathbf{V}^{(B)}$.*

$\triangleleft$ By Theorems 6.1.6 and 6.2.3 we only need to prove that the C^*-algebras A and $\mathscr{A}$ are Baer. The last claim is immediate on using 6.1.11 and the rules for ascending and descending polars which are annihilators in our case (cf. 3.2.13 (2) and 3.3.12 (6)). $\triangleright$

6.2.5. The *center* of an AW^*-algebra A is the set $\mathscr{Z}(A)$ comprising z in A, that commute with every member of A; i.e. $\mathscr{Z}(A) := \{z \in A : (\forall x \in A)\, xz = zx\}$. Clearly, $\mathscr{Z}(A)$ is a commutative AW^*-subalgebra of A, with $\lambda\mathbf{1} \in \mathscr{Z}(A)$ for all $\lambda \in \mathbb{C}$. If $\mathscr{Z}(A) = \{\lambda\mathbf{1} : \lambda \in \mathbb{C}\}$ then the AW^*-algebra A is an *AW^*-factor.*

Theorem. *If $\mathscr{A}$ is an AW^*-factor inside $\mathbf{V}^{(B)}$ then the bounded descent A of $\mathscr{A}$ is an AW^*-algebra whose Boolean algebra of central projections is isomorphic with B. Conversely, if A is an AW^*-algebra and $B := \mathfrak{P}_c(A)$ then there is an AW^*-factor $\mathscr{A}$ in $\mathbf{V}^{(B)}$ whose bounded descent is isomorphic with A. This factor is unique up to $*$-isomorphism inside $\mathbf{V}^{(B)}$.*

$\triangleleft$ Apply 6.2.4 and recall that the descent of the two-element Boolean algebra $\{\mathbf{0}, \mathbf{1}\}$ is isomorphic with B (cf. 4.2.2). $\triangleright$

6.2.6. Suppose that Λ is a commutative AW^*-algebra and B is a complete Boolean algebra of projections of Λ. Consider a unital Λ-module X. The mapping $\langle\cdot\,|\,\cdot\rangle : X \times X \to \Lambda$ is a *Λ-valued inner product* if for all $x, y, z \in X$ and $a \in \Lambda$ the following conditions are satisfied

(1) $\langle x\,|\,x\rangle \geq \mathbf{0}$; $\langle x\,|\,x\rangle = \mathbf{0} \leftrightarrow x = \mathbf{0}$;

(2) $\langle x\,|\,y\rangle = \langle y,\ x\rangle^*$;

(3) $\langle ax\,|\,y\rangle = a\langle x\,|\,y\rangle$;

(4) $\langle x + y\,|\,z\rangle = \langle x\,|\,z\rangle + \langle y\,|\,z\rangle$.

Using a Λ-valued inner product, we may introduce the norm in X by the formula

$$\textbf{(5)}\ |||x||| := \sqrt{\|\langle x \mid x\rangle\|}\ (x \in X),$$

and the vector norm

$$\textbf{(6)}\ |x| := \sqrt{\langle x \mid x\rangle}\ (x \in X).$$

In this event $|||x||| = \||x|\|$ for all $x \in X$, since $\|a\| = \|(\sqrt{a})^2\| = \|\sqrt{a}\|^2$ for every positive $a \in \Lambda$. Therefore, the formula (5) defines a mixed norm on X (cf. 5.5.1).

6.2.7. Theorem. *The pair $(X, |||\cdot|||)$ is a B-cyclic Banach space if and only if $(X, |\cdot|)$ is a Banach–Kantorovich space.*

◁ Note that 6.2.6 (6) gives a decomposable norm since $|bx| = b|x|$ for all $x \in X$ and $b \in B$ according to 6.2.6 (3). By Theorem 5.5.2, the normed space $(X, |||\cdot|||)$ is complete if and only if $(X, |\cdot|)$ is r-complete. Furthermore, it is clear that the B-cyclicity of $(X, |||\cdot|||)$ amounts to the disjoint completeness of $(X, |\cdot|)$. The above remarks justify 5.4.7, so completing the proof. ▷

An *AW^*-module* over Λ is a unital Λ-module equipped with a Λ-valued inner product and possessing each of the properties whose equivalence is stated in

6.2.8. Theorem. *The bounded descent of an arbitrary Hilbert space in $\mathbf{V}^{(B)}$ is an AW^*-module over the Stone algebra $\mathscr{S}(B)$. Conversely, if X is an AW^*-module over $\mathscr{S}(B)$, then there is a Hilbert space $\mathscr{X}$ in $\mathbf{V}^{(B)}$ whose bounded descent is unitarily equivalent with X. This space is unique to within unitary equivalence inside $\mathbf{V}^{(B)}$.*

◁ Without loss of generality, we may assume that $\mathscr{S}(B) \subset \mathscr{C}{\downarrow}$. Suppose that $\mathscr{X}$ is a Hilbert space inside $\mathbf{V}^{(B)}$ and X is the bounded descent of $\mathscr{X}$. Then the pair $(X, |\cdot|)$, with $|\cdot|$ the descent of the norm of $\mathscr{X}$, is a Banach–Kantorovich space and the pair $(X, |||\cdot|||)$, with $|||x||| = \||x|\|$ for all $x \in X$, is a B-cyclic Banach space (cf. 5.5.7). In particular, X is a unital module over $\mathscr{S}(B)$. Suppose that $(\cdot\,|\,\cdot) \in \mathbf{V}^{(B)}$ is the inner product in $\mathscr{X}$ and $\langle\cdot \mid \cdot\rangle$ is the descent of $(\cdot\,|\,\cdot)$. It is easy to check that $\langle\cdot \mid \cdot\rangle$ satisfies 6.2.6 (1–4) for all $x,\ y,\ z \in \mathscr{X}{\downarrow}$ and $a \in \mathscr{C}{\downarrow}$. If $x,\ y \in X$ then $[\![\, |(x\,|\,y)| \le \|x\| \cdot \|y\| \,]\!] = \mathbf{1}$. Hence, $|\langle x \mid y\rangle| \le |x| \cdot |y|$. Since $|x|, |y| \in \mathscr{S}(B)$; therefore, $\langle x \mid y\rangle \in \mathscr{S}(B)$. Thus, the restriction of $\langle\cdot\,|\,\cdot\rangle$ to $X \times X$, denoted by the same symbol, is a $\mathscr{S}(B)$-valued inner product on X. It suffices to note that $|x| = \sqrt{\langle x \mid x\rangle}$, since $[\![\, \|x\| = \sqrt{(x\,|\,x)} \,]\!] = \mathbf{1}$ and the descent of the function $\surd : \mathscr{R}^+ \to \mathscr{R}^+$ depicts the square root in $\mathscr{S}(B)$.

Now, consider an AW^*-module X over $\mathscr{S}(B)$. By Theorem 5.4.2, the Boolean valued representation $\mathscr{X} \in \mathbf{V}^{(B)}$ of the Banach–Kantorovich space $(X, |\cdot|, \mathscr{S}(B))$ is a Banach space inside $\mathbf{V}^{(B)}$. We may thus assume that $X \subset \mathscr{X}{\downarrow}$. Let $(\cdot\,|\,\cdot)$ stand for the ascent of the $\mathscr{S}(B)$-valued inner product $\langle\cdot\,|\,\cdot\rangle$ in X. Then $(\cdot\,|\,\cdot)$ is an inner product on $\mathscr{X}$ inside $\mathbf{V}^{(B)}$. Arguing as above, we see that $[\![\, \|x\| = \sqrt{(x \mid x)} \text{ for all } x \in \mathscr{X} \,]\!] = \mathbf{1}$, since $|x| = \sqrt{\langle x \mid x \rangle}$ for all $x \in X$.

Suppose that $\mathscr{Y}$ is another Hilbert space inside $\mathbf{V}^{(B)}$ and the bounded descent Y of $\mathscr{Y}$ is unitarily equivalent with X. If $U : X \to Y$ is a unitary isomorphism then $u := U\!\uparrow$ is a linear bijection from $\mathscr{X}$ to $\mathscr{Y}$. Since U enjoys the property $\langle \cdot \mid \cdot \rangle \circ (U \times U) = \langle \cdot \mid \cdot \rangle$, note inside $\mathbf{V}^{(B)}$ that

$$(\cdot \mid \cdot) \circ (u \times u) = \langle \cdot \mid \cdot \rangle \uparrow \circ (U\!\uparrow \times U\!\uparrow) = (\langle \cdot \mid \cdot \rangle \circ (U \times U))\!\uparrow = \langle \cdot \mid \cdot \rangle\!\uparrow = (\cdot \mid \cdot).$$

Hence, u is a unitary equivalence between $\mathscr{X}$ and $\mathscr{Y}$. This ends the proof. ▷

As usual, we call $\mathscr{X}$ the *Boolean valued representation* of an AW^*-module X.

Suppose that $\mathscr{L}^B(\mathscr{X}, \mathscr{Y})$ is the space of bounded linear operators from $\mathscr{X}$ to $\mathscr{Y}$ inside $\mathbf{V}^{(B)}$ (cf. 5.4.9). Let $\mathrm{Hom}(X, Y)$ stand for the space of all bounded Λ-linear operators from X to Y where X and Y are AW^*-modules over the commutative AW^*-algebra $\mathscr{S}(B)$. As before, we let $\mathscr{S}(B)$ stand for the bounded descent of the field $\mathbb{C}$. It is easy that $\mathrm{Hom}(X, Y) = \mathscr{L}_B(X, Y)$ (cf. 5.5.9).

6.2.9. Theorem. *Suppose that $\mathscr{X}$ and $\mathscr{Y}$ are Hilbert spaces inside $\mathbf{V}^{(B)}$. Let X and Y stand for the bounded descents of $\mathscr{X}$ and $\mathscr{Y}$. For every bounded Λ-linear operator $\Phi : X \to Y$ the element $\varphi := \Phi\!\uparrow$ is a bounded linear operator from $\mathscr{X}$ to $\mathscr{Y}$ inside $\mathbf{V}^{(B)}$. Moreover, $[\![\, \|\varphi\| \le c^{\wedge} \,]\!] = \mathbf{1}$ for some $c \in \mathbb{R}$. The mapping $\Phi \mapsto \varphi$ is a B-linear isometry between the B-cyclic Banach spaces* $\mathrm{Hom}(X, Y)$ *and* $\mathscr{L}^B(\mathscr{X}, \mathscr{Y})\!\downarrow^{\infty}$.

◁ Appealing to 5.4.9 and 5.5.9 completes the proof. ▷

6.2.10. We now state some corollaries.

(1) Denote by AW^*-mod-$\mathscr{S}(B)$ the category of AW^*-modules over the Stone algebra $\mathscr{S}(B)$ and bounded $\mathscr{S}(B)$-linear operators. Also, consider the category $\mathrm{Hilbert}_{\infty}^{(B)}$ whose objects are Hilbert spaces inside $\mathbf{V}^{(B)}$ and whose morphisms are bounded linear operators $f : \mathscr{X} \to \mathscr{Y}$ inside $\mathbf{V}^{(B)}$ satisfying $[\![\, \|f\| \le c^{\wedge} \,]\!]$ for some $c \in \mathbb{R}$. Theorems 6.2.8 and 6.2.9 may be paraphrased as follows:

Theorem. *The bounded descent and immersion functors establish equivalence of the categories* $\mathrm{Hilbert}_{\infty}^{(B)}$ *and* AW^**-mod-*$\mathscr{S}(B)$.

(2) Put $\mathrm{End}(X) := \mathrm{Hom}(X, X)$ and $\mathscr{L}(\mathscr{X}) := \mathscr{L}^B(\mathscr{X}, \mathscr{X})$. From 6.2.9 it follows that $\mathrm{End}(X)$ and $\mathscr{L}(\mathscr{X})\!\downarrow^{\infty}$ are isometrically B-isomorphic. Since the space of all bounded operators $\mathscr{L}(\mathscr{X})$ is an AW^*-factor inside $\mathbf{V}^{(B)}$ then $\mathscr{L}(X)\!\downarrow^{\infty}$ is an AW^*-algebra (cf. 6.2.5). An isometric B-isomorphism between $\mathrm{End}(X)$ and $\mathscr{L}(\mathscr{X})\!\downarrow^{\infty}$ is an isomorphism between algebras provided that the multiplication of $\mathrm{End}(X)$ is the composition of operators and the adjoint of an operator T in $\mathrm{End}(X)$ is defined by the rule $\langle Tx \mid y \rangle = \langle x \mid T^* y \rangle$ for all $x, y \in X$.

Theorem. *The space* End(X), *furnished with above operations, is an* AW^*-*algebra.*

6.2.11. We will now demonstrate that immersion in a Boolean valued universe preserves the type of an AW^*-algebra. The type of such an algebra A is determined from the structure of the lattice of projections in A. Consequently, we have to trace changes with the qualification of a projection which happens in the process of Boolean valued representation.

Recall the relevant definitions. Take an AW^*-algebra A. A projection $\pi \in A$ is called: (a) *abelian* if the algebra $\pi A \pi$ is commutative; (b) *finite*, if for every projection $\rho \in A$ from $\pi \sim \rho \le \pi$ it follows that $\rho = \pi$; (c) *infinite*, if π is not finite; (d) *purely infinite*, if π does not contain nonzero finite projections. As usual, the phrase "a projection π contains a projection ρ" stands for the inequality $\rho \le \pi$.

An algebra A has *type* I if each nonzero projection in A contains nonzero abelian projection. An algebra A has *type* II if A does not contain nonzero abelian projections and each nonzero projection in A contains a nonzero finite projection. An algebra A has *type* III if the unity of A is a purely infinite projection. An algebra A is *finite* if the unity of A is a finite projection.

6.2.12. Theorem. *Suppose that* $\mathscr{A}$ *is an* AW^*-*algebra inside* $\mathbf{V}^{(B)}$ *and* A *is the bounded descent of* $\mathscr{A}$. *For every projection* $\pi \in \mathfrak{P}(A)$ *the following hold:*

(1) π *is abelian* $\leftrightarrow$ $[\![\pi$ *is abelian*$]\!] = \mathbf{1}$;

(2) π *is finite* $\leftrightarrow$ $[\![\pi$ *is finite*$]\!] = \mathbf{1}$;

(3) π *is purely infinite* $\leftrightarrow$ $[\![\pi$ *is purely infinite*$]\!] = \mathbf{1}$.

$\lhd$ The claim of (1) is obvious. Furthermore, note that for $\pi, \rho \in \mathfrak{P}(A)$ the formulas $\pi \sim \rho$, $\pi \le \rho$, and $\pi \lesssim \rho$ may be rewritten as algebraic identities (cf. 6.1.1):

$$\pi \sim \rho \leftrightarrow xx^* = \pi \wedge x^*x = \rho,$$
$$\pi \le \rho \leftrightarrow \pi\rho = \rho\pi = \pi,$$
$$\pi \lesssim \rho \leftrightarrow \pi \sim \pi_0 \wedge \pi_0 \le \rho.$$

Multiplication, involution, and equality in A appear as the descents of the corresponding objects in $\mathscr{A}$. Therefore,

$$\pi \sim \rho \leftrightarrow [\![\pi \sim \rho]\!] = \mathbf{1},$$
$$\pi \le \rho \leftrightarrow [\![\pi \le \rho]\!] = \mathbf{1},$$
$$\pi \lesssim \rho \leftrightarrow [\![\pi \lesssim \rho]\!] = \mathbf{1}.$$

To prove (2), recall the formula

$$[\![(\forall x \in \mathscr{A})\varphi(x) \to \psi(x)]\!] = \bigwedge\{[\![\psi(x)]\!] : x \in \mathscr{A}{\downarrow},\ [\![\varphi(x)]\!] = \mathbf{1}\},$$

and the equality $\mathfrak{P}(\mathscr{A}){\downarrow} = \mathfrak{P}(A)$. Then, write down the chain of equivalences:

$$
\begin{aligned}
&[\![\, \pi \text{ is finite} \,]\!] = \mathbf{1}\\
&\leftrightarrow [\![(\forall \rho \in \mathfrak{P}(\mathscr{A}))\pi \sim \rho \leq \pi \to \pi = \rho]\!] = \mathbf{1}\\
&\leftrightarrow (\forall \rho \in \mathfrak{P}(A))[\![\, \pi \sim \rho \leq \pi \,]\!] = \mathbf{1} \to [\![\, \pi = \rho \,]\!] = \mathbf{1}\\
&\leftrightarrow (\forall \rho \in \mathfrak{P}(A))\pi \sim \rho \leq \pi \to \pi = \rho.
\end{aligned}
$$

We arrive at (3) similarly, thus completing the proof. ▷

6.2.13. *Theorem.* *Suppose that algebras A and $\mathscr{A}$ are the same as in* 6.2.12. *Then the following hold:*

(1) *A is finite $\leftrightarrow$ $[\![\, \mathscr{A}$ is finite$]\!] = \mathbf{1}$;*

(2) *A has type I $\leftrightarrow$ $[\![\, \mathscr{A}$ has type I$]\!] = \mathbf{1}$;*

(3) *A has type II $\leftrightarrow$ $[\![\, \mathscr{A}$ has type II$]\!] = \mathbf{1}$;*

(4) *A has type III $\leftrightarrow$ $[\![\, \mathscr{A}$ has type III$]\!] = \mathbf{1}$.*

◁ All claims are immediate from 6.2.12 and definitions. ▷

6.2.14. Comments.

(1) The modern structural theory of AW^*-algebras and AW^*-modules originates with the articles [105–107] by I. Kaplansky. These objects appear naturally by way of algebraization of the theory of von Neumann operator algebras.

(2) The main results of the current section, Theorems 6.2.4, 6.2.8, and 6.2.9, belong to M. Ozawa [194–200]. Our exposition is somewhat different as basing on the representation theorems of Chapter 5. Theorems 6.2.12 and 6.2.13 must be attributed to G. Takeuti [236].

(3) JB-algebras serve as real nonassociative analogs of C^*-algebras and von Neumann operator algebras. The theory of JB-algebras stems from the article [92] by P. Jordan, J. von Neumann and E. Wigner. This theory is an established section of functional analysis from the mid sixties. The articles [2] and [243] reflect the stages of progress in JB-algebras. This theory flourishes and expands its applications. Among the main directions of research we may list a few: structural properties and general classification of JB-algebras, nonassociative integration and quantum probability, geometry of states of JB-algebras, etc. (cf. [7, 8, 75, 213] and the bibliography therein).

(4) We now state a typical result on Boolean valued representation of JB-algebras by analogy with Theorem 6.2.4. Assume that B is a subalgebra of the Boolean algebra of central idempotents of a JB-algebra A. We call A a *B-JB-algebra* provided that to every partition of unity $(e_\xi)_{\xi\in\Xi}$ in B and every family $(x_\xi)_{\xi\in\Xi}$ there is a unique B-mixing $x := \operatorname{mix}_{\xi\in\Xi}(e_\xi x_\xi)$ in A. The following result is available on Boolean valued representation of a JB-algebra (cf. [127]).

Theorem. *The bounded descent of a JB-algebra inside $\mathbf{V}^{(B)}$ is a B-JB-algebra. Conversely, for each B-JB-algebra A there is a JB-algebra $\mathscr{A}$ whose bounded descent is B-isomorphic with A. This algebra $\mathscr{A}$ is unique up to isomorphism inside $\mathbf{V}^{(B)}$. Moreover, $[\![\mathscr{A}$ is a JB-factor$]\!] = \mathbf{1}$ if and only if the bounded part of the descent $\mathscr{R}{\downarrow}$ coincides with $\mathscr{Z}(A)$.*

6.3. The Boolean Dimension of an AW^*-Module

To each AW^*-module A we may uniquely assign some nonstandard cardinal, the *Hilbert dimension* of the Boolean valued representation of A. The external deciphering of this leads to the concept of Boolean dimension.

6.3.1. Suppose that X is a unital AW^*-module over a commutative AW^*-algebra Λ. A subset $\mathscr{E}$ of X is a *basis* for X provided that

(1) $\langle x \mid y \rangle = \mathbf{0}$ for all distinct $x, y \in \mathscr{E}$;

(2) $\langle x \mid x \rangle = \mathbf{1}$ for every $x \in \mathscr{E}$;

(3) the condition $(\forall e \in \mathscr{E}) \langle x \mid e \rangle = \mathbf{0}$ implies $x = \mathbf{0}$.

We say that an AW^*-module X is *λ-homogeneous* if λ is a cardinal and X has a basis of cardinality λ.

Granted $\mathbf{0} \neq b \in B$, denote by $\varkappa(b)$ the least cardinal γ such that an AW^*-module bX is γ-homogeneous. If X is homogeneous then $\varkappa(b)$ is defined for all $\mathbf{0} \neq b \in B$. Hence, $\varkappa$ is a mapping of $B^+ := \{b \in B : b \neq \mathbf{0}\}$ to some set of cardinals. We can demonstrate that $\varkappa$ is a *multiplicity function*; i.e., $\varkappa(\sup(b_\xi)) = \sup(\varkappa(b_\xi))$ for every family $(b_\xi) \subset B$. We shall say that an AW^*-module X is *strictly γ-homogeneous* if X is homogeneous and $\gamma = \varkappa(b)$ for all nonzero $b \in B$. If γ is a finite cardinal then the properties of γ-homogenuity and strict γ-homogenuity of an AW^*-module are equivalent. It is convenient to assume that $\varkappa(\mathbf{0}) = \mathbf{0}$.

Denote by $|M|$ the *cardinality* of M; i.e., a cardinal bijective with M. The record $[\![\dim(\mathscr{X}) = \lambda]\!] = \mathbf{1}$ signifies that $\mathbf{V}^{(B)} \models$ "the cardinality of every orthonormal basis for a space $\mathscr{X}$ equals λ." We now present the Boolean valued interpretation of homogenuity and strict homogenuity.

6.3.2. Theorem. *For an AW^*-module X to be λ-homogeneous it is necessary and sufficient that $[\![\dim(\mathscr{X}) = |\lambda^\wedge|]\!] = \mathbf{1}$.*

$\lhd$ By Theorem 5.4.2, assume that $X \subset \mathscr{X}{\downarrow}$. The mapping $\langle \cdot \mid \cdot \rangle$ and the descent of the form $(\cdot \mid \cdot)$ agree on $X \times X$. Therefore, for all $x, y \in X$ and $a \in \Lambda$, the following are equivalent: $\langle x \mid y \rangle = a$ and $[\![(x \mid y) = a]\!] = \mathbf{1}$. We thus see that the orthogonality relation on X is the restriction to X of the descent of the orthogonality relation on $\mathscr{X}$. From these observations it follows that a subset $\mathscr{E}$ of X is orthonormal if and only if $[\![\mathscr{E}{\uparrow}$ is an orthonormal set in $\mathscr{X}]\!] = \mathbf{1}$. Applying the descent rules for polars to orthogonal complements in X and $\mathscr{X}$, infer $(\mathscr{E}{\uparrow})^{\perp}{\downarrow} = (\mathscr{E}{\uparrow\downarrow})^{\perp}$. Observe also that $\mathscr{E}^\perp = (\mathscr{E}{\uparrow\downarrow})^\perp$. Hence, $\mathscr{E}^\perp{\uparrow} = (\mathscr{E}{\uparrow})^\perp$. In particular, $\mathscr{E}^\perp = \mathbf{0}$ if and only if $[\![(\mathscr{E}{\uparrow})^\perp = \{\mathbf{0}\}]\!] = \mathbf{1}$. Hence, $\mathscr{E}$ is a basis for X only on condition that $[\![\mathscr{E}$ is a basis for $\mathscr{X}]\!] = \mathbf{1}$. If $|\mathscr{E}| = \lambda$ and $\varphi : \lambda \to \mathscr{E}$ are bijections then the modified ascent $\varphi{\uparrow}$ is a bijection of $\lambda^\wedge$ to $\mathscr{E}{\uparrow}$. Conversely, suppose that $\mathscr{D}$ is a basis for $\mathscr{X}$ and $[\![\psi : \lambda^\wedge \to \mathscr{D}$ is a bijection$]\!] = \mathbf{1}$ for some cardinal λ. In this case the modified descent $\varphi := \psi{\downarrow} : \lambda \to \mathscr{D}{\downarrow}$ is injective. Consequently, the set $\mathscr{E} := \mathrm{im}(\varphi)$ has cardinality λ. Moreover, as shown above, $\mathscr{E}$ is orthonormal. We are left with observing that $\mathscr{D}{\downarrow} = \mathrm{mix}(\mathscr{E}) = \mathscr{E}{\uparrow\downarrow}$, i.e., $[\![\mathscr{E}{\uparrow} = \mathscr{D}]\!] = \mathbf{1}$. Finally, $\mathscr{E}$ is a basis for X, which completes the proof. ▷

6.3.3. ***Theorem.*** *For an AW^*-module X to be strictly λ-homogeneous it is necessary and sufficient that* $[\![\dim(\mathscr{X}) = \lambda^\wedge]\!] = \mathbf{1}$.

◁ Suppose that X is a strictly λ-homogeneous module. By Theorem 6.3.2 $[\![\dim(\mathscr{X}) = |\lambda^\wedge|]\!] = \mathbf{1}$. On the other hand, there is a partition of unity $(b_\alpha)_{\alpha<\beta}$ in the Boolean algebra B such that $|\lambda^\wedge| = \mathrm{mix}_{\alpha<\beta}(b_\alpha \alpha^\wedge)$. Since $b_\alpha \leq [\![\mathscr{X} = b_\alpha \mathscr{X}]\!]$; therefore, $b_\alpha \leq [\![\dim(b_\alpha \mathscr{X}) = \alpha^\wedge]\!]$. Consider the set $B_\alpha := [\mathbf{0}, b_\alpha] := \{b' \in B : b' \leq b_\alpha\}$. If $b_\alpha \neq \mathbf{0}$ then B_α is a complete Boolean algebra. In the respective universe $\mathbf{V}^{(B_\alpha)}$ we observe that $\mathbf{V}^{(B_\alpha)} \models$ "$b_\alpha \mathscr{X}$ is a Hilbert space and $\alpha^\wedge = \dim(b_\alpha \mathscr{X})$." The space $b_\alpha X$ is the bounded descent of the Hilbert space $b_\alpha \mathscr{X}$ inside $\mathbf{V}^{(B_\alpha)}$. Consequently, $b_\alpha X$ is an α-homogeneous AW^*-module. Furthermore, $\mathbf{V}^{(B_\alpha)} \models$ "$\alpha^\wedge$ is a cardinal" and so α is a cardinal too. By the definition of strict homogeneity, $\lambda \leq \alpha$. Hence, $b_\alpha = \mathbf{0}$ for $\alpha < \lambda$. Therefore, $[\![\lambda^\wedge \leq |\lambda^\wedge|]\!] = \mathbf{1}$. Thus, $[\![\lambda^\wedge = |\lambda^\wedge|]\!] = \mathbf{1}$; since the formula $[\![|\lambda^\wedge| \leq \lambda^\wedge]\!] = \mathbf{1}$ holds by the definition of cardinality. We may now conclude that $[\![\dim(\mathscr{X}) = \lambda^\wedge]\!] = \mathbf{1}$.

Assume the last equality holding. Note that λ is a cardinal, since $\lambda^\wedge$ is a cardinal inside $\mathbf{V}^{(B)}$. By 6.3.2 X is λ-homogeneous. If X is γ-homogeneous for some γ then, appealing to 6.3.2 again, we obtain $[\![\dim(\mathscr{X}) = |\gamma^\wedge|]\!] = \mathbf{1}$. Hence, $[\![\lambda^\wedge = |\gamma^\wedge| \leq \gamma^\wedge]\!] = \mathbf{1}$ and so $\lambda \leq \gamma$. The same arguments will apply to every AW^*-algebra bX with $\mathbf{0} \neq b \in B$ provided that we substitute $\mathbf{V}^{([\mathbf{0},b])}$ for $\mathbf{V}^{(B)}$. Thus, X is a strictly λ-homogeneous AW^*-module. ▷

6.3.4. We now introduce the main concept of the current section. We call a partition of unity $(b_\gamma)_{\gamma\in\Gamma}$ in B the *B-dimension* of an AW^*-module X provided that Γ is a nonempty set of cardinals, $b_\gamma \neq 0$ for all $\gamma \in \Gamma$, and $b_\gamma X$ is a strictly γ-homogeneous AW^*-module for every $\gamma \in \Gamma$. In this event we write $B\text{-}\dim(X) = (b_\gamma)_{\gamma\in\Gamma}$. Note that the members of the B-dimension of an AW^*-module are pairwise distinct by the definition of strict homogenuity. We say that the B-dimension of X equals γ (in symbols, $B\text{-}\dim(X) = \gamma$) if $\Gamma = \{\gamma\}$ and $b_\gamma = \mathbf{1}$. The equality $B\text{-}\dim(X) = \gamma$ means evidently that X is strictly γ-homogeneous. We may define the multiplicity function $\varkappa$ of 6.3.1 in the case of an arbitrary AW^*-module X by the formula $\varkappa(b) = \sup\{\varkappa(b') : b' \leq b,\ b' \in hb\}$, where the set hb comprises

$b' \le b$ such that $b'X$ is homogeneous. Clearly, if $B\text{-dim}(X) = (b_\gamma)_{\gamma\in\Gamma}$ then $\varkappa(b) = \sup\{\gamma \in \Gamma : b \wedge b_\gamma \ne \mathbf{0}\}$.

6.3.5. Theorem. *Suppose that $(b_\gamma)_{\gamma\in\Gamma}$ is a partition of unity in B, with $b_\gamma \ne \mathbf{0}$ $(\gamma \in \Gamma)$ and Γ a set of cardinals. Then $B\text{-dim}\, X = (b_\gamma)_{\gamma\in\Gamma}$ if and only if $[\![\, \dim(\mathscr{X}) = \text{mix}_{\gamma\in\Gamma}(b_\gamma\gamma^\wedge)\,]\!] = \mathbf{1}$.*

$\lhd$ As was noted above, we may identify $b_\gamma X$ with the bounded descent of the Hilbert space $b_\gamma \mathscr{X}$ inside $\mathbf{V}^{(B_\gamma)}$ where $B_\gamma := [\mathbf{0}, b_\gamma]$. By virtue of 6.3.4 γ-homogenuity for $b_\gamma X$ amounts to the formula $b_\gamma = [\![\, \dim(b_\gamma\mathscr{X}) = \gamma^\wedge \,]\!]^{B_\gamma} \le [\![\, \dim(\mathscr{X}) = \gamma^\wedge \,]\!]^B$. But then the equality $B\text{-dim}(X) = (b_\gamma)_{\gamma\in\Gamma}$ holds if and only if $b_\gamma \le [\![\, \dim(\mathscr{X}) = \gamma^\wedge \,]\!]$ $(\gamma \in \Gamma)$, since $b_\gamma \le [\![\, \mathscr{X} = b_\gamma\mathscr{X} \,]\!] = [\![\, \dim(\mathscr{X}) = \dim(b_\gamma\mathscr{X}) \,]\!]$. In turn, the last formulas imply that $[\![\, \dim(X) = \text{mix}_{\gamma\in\Gamma}(b_\gamma\gamma^\wedge) \,]\!] = \mathbf{1}$. This ends the proof. $\rhd$

6.3.6. We will now find which partition of unity may serve as the B-dimension of an AW^*-module. Take some cardinal λ. Granted $b \in B$ and $\beta \in \text{On}$, denote by $b(\beta)$ the set of all partitions of b having the form $(b_\alpha)_{\alpha\in\beta}$. Define the $[\mathbf{0}, b]$-valued metric d on $b(\beta)$ by the formula

$$d(u,v) := \Big(\bigvee_{\alpha\in\beta} u_\alpha \wedge v_\alpha\Big)^* \quad \big(u = (u_\alpha),\ \ v = (v_\alpha) \in b(\beta)\big).$$

Observe that $\big(b(\beta), d\big)$ is a Boolean set. Granted $\gamma \in \text{On}$, write $b(\beta) \simeq b(\gamma)$ if there is a bijection between $b(\beta)$ and $b(\gamma)$ which preserves the Boolean metric; i.e., there is a B-isometry between these B-sets. We call the Boolean algebra B and its Stone space *λ-stable* provided that $\lambda \le \alpha$ for all nonzero $b \in B$ and each ordinal α in $b(\lambda) \simeq b(\alpha)$. A nonzero element $b \in B$ is λ-stable by definition whenever $[\mathbf{0}, b]$ is a λ-stable Boolean algebra.

6.3.7. Theorem. *A partition of unity $(b_\gamma)_{\gamma\in\Gamma}$ in a complete Boolean algebra B, which consists of pairwise distinct elements serves as the B-dimension of some AW^*-module if and only if Γ consists of cardinals and b_γ is a γ-stable element for every $\gamma \in \Gamma$.*

$\lhd$ Put $\lambda := \text{mix}_{\gamma\in\Gamma}(b_\gamma\gamma^\wedge)$. Inside $\mathbf{V}^{(B)}$ we may find a Hilbert space $\mathscr{X}$, satisfying $[\![\, \dim(\mathscr{X}) = |\lambda| \,]\!] = \mathbf{1}$. By 6.3.5, $B\text{-dim}(X) = (b_\gamma)_{\gamma\in\Gamma}$ if and only if $[\![\, |\lambda| = \lambda \,]\!] = \mathbf{1}$. The last relation amounts to the estimates

$$b_\gamma \le [\![\, |\gamma^\wedge| = \gamma^\wedge \,]\!] \quad (\gamma \in \Gamma).$$

The inequality $b_\gamma \le [\![\, |\gamma^\wedge| = \gamma^\wedge \,]\!]$ for a nonzero b_γ means that $\mathbf{V}^{([\mathbf{0},b_\gamma])} \models \gamma^\wedge = |\gamma^\wedge|$. Consequently, it remains to demonstrate that the γ-stability of the Boolean algebra

$B_0 = [\mathbf{0}, b]$ and the formula $\mathbf{V}^{(B_0)} \models \gamma^\wedge = |\gamma^\wedge|$ hold or fail simultaneously. Note that

$$\begin{aligned}[\![\gamma^\wedge = |\gamma^\wedge|]\!] &= [\![(\forall \alpha \in \mathrm{On})(\gamma^\wedge \sim \alpha \to \gamma^\wedge \leq \alpha)]\!] \\ &= \bigwedge \{[\![\gamma^\wedge \sim \alpha^\wedge]\!] \Rightarrow [\![\gamma^\wedge \leq \alpha]\!] : \alpha \in \mathrm{On}\}.\end{aligned}$$

Clearly, $[\![\gamma^\wedge = |\gamma^\wedge|]\!] = \mathbf{1}$ if and only if $c := [\![\gamma^\wedge \sim \alpha^\wedge]\!] \leq [\![\gamma^\wedge \leq \alpha^\wedge]\!]$ for every ordinal α. If $c \neq \mathbf{0}$ then $\gamma \leq \alpha$. Furthermore, the inequality $c \leq [\![\gamma^\wedge \sim \alpha^\wedge]\!]$ means that $c(\gamma) \simeq c(\alpha)$. Thus, the equality $[\![\gamma^\wedge = |\gamma^\wedge|]\!] = \mathbf{1}$ amounts to the γ-stability of the Boolean algebra B_0. $\rhd$

6.3.8. COMMENTS.

A. G. Kusraev studied the Boolean dimension of an AW^*-module in [126], using the same definition as in 6.3.4. Prior to this research, M. Ozawa had defined the Boolean dimension of an AW^*-module as the dimension of any Hilbert space serving as a Boolean valued representation of the module in question, i.e., as an internal object of a Boolean valued universe [195]. So, the definition of B-dimension in 6.3.4 is an external decoding of the definition by M. Ozawa. Theorems 6.3.2 and 6.3.3 are demonstrated in [126] and [195]. Theorem 6.3.7 may be found in [126, 195].

6.4. Representation of an AW^*-Module

In this section we prove that every AW^*-module may be represented as the direct sum of a family of modules of continuous vector functions. Moreover, this representation is unique in a definite sense. Denote by $C_\#(Q, H)$ the part of $C_\infty(Q, H)$ that consists of vector functions z satisfying $|z| \in C(Q)$ (cf. 5.3.7 (5)).

6.4.1. *Suppose that Q is an extremally disconnected compact space, and H is a Hilbert space of dimension λ. The space $C_\#(Q, H)$ is a λ-homogeneous AW^*-module over the algebra $\Lambda := C(Q, \mathbb{C})$.*

$\lhd$ Let $(\cdot\,|\,\cdot)$ stand for the inner product of H. Introduce some Λ-valued inner product in $C_\#(Q, H)$ as follows. Take continuous vector functions $u : \mathrm{dom}(u) \to H$ and $v : \mathrm{dom}(v) \to H$. The function $q \mapsto \langle u(q)|v(q)\rangle$, with $q \in \mathrm{dom}(u) \cap \mathrm{dom}(v)$, is continuous and admits a unique continuation $z \in C(Q)$ to the whole of Q. If x and y are the cosets containing vector functions u and v then assign $(x\,|\,y) := z$. Clearly, $(\cdot\,|\,\cdot)$ is a Λ-valued inner product and $|x| = \sqrt{(x\,|\,x)}$ for all $x \in C_\#(Q, H)$. Since $C_\#(Q, H)$ is a Banach–Kantorovich space; therefore, $C_\#(Q, H)$ is disjointly complete. Moreover, $C_\#(Q, H)$ is a Banach space whose norm satisfies the equalities

$$\|x\| = \||x|\|_\infty = \sqrt{\|(x\,|\,x)\|_\infty} \quad (x \in C_\#(Q, H)).$$

Suppose that $\mathscr{E}$ is a basis for H. Given $e \in \mathscr{E}$, introduce the vector function $\bar{e} : q \mapsto e$, with $q \in Q$, and put $\overline{\mathscr{E}} := \{\bar{e} : e \in \mathscr{E}\}$. It is easy to note that $\overline{\mathscr{E}}$ is a basis for $C_\#(Q, H)$. Summarizing, conclude that $C_\#(Q, H)$ is a λ-homogeneous AW^*-module, with $\lambda = \dim(H)$. $\rhd$

6.4.2. We need another auxiliary fact. Denote by $\mathbb{P}$-lin(A) the set of all linear combinations of the members of A with coefficients in $\mathbb{P}$.

Suppose that X is a vector space over $\mathbb{F}$ and $\mathbb{P}$ is a subfield of $\mathbb{F}$. Then $X^\wedge$ is a vector space over the field $\mathbb{F}^\wedge$ and $(\mathbb{P}\text{-lin}(A))^\wedge = \mathbb{P}^\wedge\text{-lin}(A^\wedge)$ for every $A \subset X$.

$\lhd$ The first claim is evident, since the proposition "X is a vector space over $\mathbb{F}$" presents a bounded formula. By the same reason, $(\mathbb{P}\text{-lin}(A))^\wedge$ is a $\mathbb{P}^\wedge$-linear subspace in $X^\wedge$ which contains $A^\wedge$. Therefore, $\mathbb{P}^\wedge\text{-lin}(A^\wedge) \subset (\mathbb{P}\text{-lin}(A))^\wedge$. Conversely, suppose that an element x in X has the form $\sum_{k\in n} \alpha(k)\, u(k)$, where $n \in \mathbb{N}$, $\alpha : n \to \mathbb{P}$, and $u : n \to A$. Then $\alpha^\wedge : n^\wedge \to \mathbb{P}^\wedge$, $u^\wedge : n^\wedge \to A^\wedge$, and $x^\wedge = \sum_{k\in n^\wedge} \alpha^\wedge(k) u^\wedge(k)$. Consequently, $x^\wedge \in \mathbb{P}^\wedge\text{-lin}(A^\wedge)$, which proves the inclusion $(\mathbb{P}\text{-lin}(A))^\wedge \subset \mathbb{P}^\wedge\text{-lin}(A^\wedge)$. $\rhd$

6.4.3. Theorem. *Suppose that H is a Hilbert space and $\lambda = \dim(H)$. Suppose further that $\mathscr{H}$ is the completion of the metric space $H^\wedge$ inside $\mathbf{V}^{(B)}$. Then $[\![\mathscr{H}$ is a Hilbert space and $\dim(\mathscr{H}) = |\lambda^\wedge|]\!] = \mathbf{1}$.*

$\lhd$ By definition, $\mathscr{H}$ is a Banach space. If $b(\cdot\,, \cdot)$ is the inner product on H then $b^\wedge : H^\wedge \times H^\wedge \to \mathbb{C}^\wedge$ is a uniformly continuous function admitting a unique continuation on the whole of $\mathscr{H} \times \mathscr{H}$. We let $(\cdot \,|\, \cdot)$ stand for this continuation. Clearly, $(\cdot \,|\, \cdot)$ is an inner product on $\mathscr{H}$ and

$$\mathbf{V}^{(B)} \models \|x\| = \sqrt{(x \,|\, x)} \quad (x \in \mathscr{H}).$$

Hence, $[\![\mathscr{H}$ is a Hilbert space$]\!] = \mathbf{1}$. Suppose that $\mathscr{E}$ is a Hilbert basis for H. Show that $[\![\mathscr{E}^\wedge$ is a basis for $\mathscr{H}]\!] = \mathbf{1}$. Orthonormality for $\mathscr{E}^\wedge$ ensues from the definition of inner product on $\mathscr{H}$. Indeed, this is seen from the following calculations:

$$[\![(\forall x \in \mathscr{E}^\wedge)\, (x \,|\, x) = 1]\!] = \bigwedge_{x\in\mathscr{E}} [\![(x^\wedge | x^\wedge) = 1]\!] = \bigwedge_{x\in\mathscr{E}} [\![b(x,x)^\wedge = \mathbf{1}^\wedge]\!] = \mathbf{1};$$

$$[\![(\forall x, y \in \mathscr{E}^\wedge)\, (x \neq y \to (x \,|\, y) = 0)]\!] = \bigwedge_{x,y\in\mathscr{E}} [\![x^\wedge \neq y^\wedge]\!]$$

$$\Rightarrow [\![(x^\wedge | y^\wedge) = 0]\!] = \bigwedge_{\substack{x,y\in\mathscr{E}\\ x\neq y}} [\![b^\wedge(x^\wedge, y^\wedge) = 0]\!] = \bigwedge_{\substack{x,y\in\mathscr{E}\\ x\neq y}} [\![b(x,y)^\wedge = 0^\wedge]\!] = \mathbf{1}.$$

Since $H^\wedge$ is dense in $\mathscr{H}$ and $\mathbb{C}^\wedge\text{-lin}(\mathscr{E}^\wedge) \subset \mathscr{C}\text{-lin}(\mathscr{E}^\wedge)$; therefore, we are left with showing only that $\mathbb{C}^\wedge\text{-lin}(\mathscr{E}^\wedge)$ is dense in $H^\wedge$. Take $x \in H$ and $\varepsilon > 0$. Since $\mathscr{E}$ is a basis for H, there is $x_\varepsilon \in \mathbb{C}\text{-lin}(\mathscr{E})$ satisfying $\|x - x_\varepsilon\| < \varepsilon$. Hence, $[\![\, \|x^\wedge - x_\varepsilon{}^\wedge\| < \varepsilon^\wedge]\!] = \mathbf{1}$ and $[\![x_\varepsilon^\wedge \in (\mathscr{C}\text{-lin}(\mathscr{E}))^\wedge]\!] = \mathbf{1}$. Recalling 6.4.2, conclude that the formula

$$(\forall x \in H)\, (\forall 0 < \varepsilon \in \mathbb{R}^\wedge)\, (\exists x_\varepsilon \in \mathbb{C}^\wedge\text{-}\operatorname{lin}(\mathscr{E}^\wedge)\, (\|x - x_\varepsilon\| < \varepsilon)$$

is satisfied inside $\mathbf{V}^{(B)}$; i.e., $[\![\mathbb{C}^\wedge\text{-lin}(\mathscr{E}^\wedge)$ is dense in $H^\wedge]\!] = \mathbf{1}$. It remains to note that if φ is a bijection between the set $\mathscr{E}$ and the cardinal λ then $\varphi^\wedge$ is a bijection between $\mathscr{E}^\wedge$ and $\lambda^\wedge$ inside $\mathbf{V}^{(B)}$. This ends the proof. $\triangleright$

We list a few corollaries.

6.4.4. *In the hypotheses of Theorem* 6.4.3 *the bounded descent of a Hilbert space* $\mathscr{H}$ *inside* $\mathbf{V}^{(B)}$ *is unitarily equivalent to the* AW^**-module* $C_\#(\mathrm{St}(B), H)$, *where* $\mathrm{St}(B)$ *is the Stone space of* B.

$\triangleleft$ This ensues for 5.4.10 and 6.4.1. $\triangleright$

6.4.5. *Let* M *be a nonempty set. The bounded descent of the Hilbert space* $l_2(M^\wedge)$ *inside* $\mathbf{V}^{(B)}$ *is unitarily equivalent to the* AW^**-module* $C_\#(\mathrm{St}(B), l_2(M))$, *where* $\mathrm{St}(B)$ *is the Stone space of* B.

$\triangleleft$ Assign $H = l_2(M)$ in Theorem 6.4.3 and recall the formula $[\![\dim(\mathscr{H}) = |M^\wedge|]\!] = \mathbf{1}$. We now see that $[\![\mathscr{H}$ and $l_2(M^\wedge)$ are unitarily equivalent $]\!] = \mathbf{1}$. This completes the proof. $\triangleright$

6.4.6. *Suppose that* $\lambda = \dim(H)$ *is an infinite cardinal. The* AW^**-module* $C_\#(Q, H)$ *is strictly* λ*-homogeneous if and only if* Q *is a* λ*-stable compact space.*

$\triangleleft$ Apply 6.3.3, 6.3.7, and 6.4.3 to complete the proof. $\triangleright$

6.4.7. *To an arbitrary infinitely dimensional Hilbert spaces* H_1 *and* H_2, *there is an extremally disconnected compact space* Q *so that the* AW^**-modules* $C_\#(Q,\ H_1)$ *and* $C_\#(Q, H_2)$ *are unitarily equivalent.*

$\triangleleft$ Put $\lambda_k := \dim(H_k)$ $(k = 1, 2)$. There exists a complete Boolean algebra B such that the ordinals $\lambda_1^\wedge$ and $\lambda_2^\wedge$ have the same cardinality inside $\mathbf{V}^{(B)}$ (cf. [11, 83]). The claim follows from 6.4.3 and 6.4.4. $\triangleright$

6.4.8. *Let* $k = 1, 2$. *Suppose that* H_k *is a Hilbert space and* $\lambda_k := \dim(H_k) \geq \omega$. *Suppose further that the* AW^**-module* $C_\#(Q,\ H_k)$ *is strictly* λ_k*-homogeneous. If the modules* $C_\#(Q, H_1)$ *and* $C_\#(Q, H_2)$ *are unitarily equivalent then the Hilbert spaces* H_1 *and* H_2 *are unitarily equivalent too.*

$\triangleleft$ From 6.3.3, 6.4.3, and 6.4.4 we see that $[\![\lambda_1^\wedge = |\lambda_1^\wedge| = |\lambda_2^\wedge| = \lambda_2^\wedge]\!] = \mathbf{1}$. Therefore, $\lambda_1 = \lambda_2$. $\triangleright$

6.4.9. An AW^*-module X is *B-separable* if there is a sequence $(x_n) \subset X$ such that the AW^*-submodule of X, generated by the set $\{bx_n : n \in \mathbb{N},\ b \in B\}$, coincides with X. Obviously, if H is a B-separable Hilbert space then the AW^*-module $C_\#(Q, H)$ is B-separable.

6.4.10. *To every infinitely dimensional Hilbert space* H, *there exists an extremally disconnected compact space* Q *such that the* AW^**-module* $C_\#(Q, H)$ *is* B*-separable, with* B *standing for the Boolean algebra of the characteristic functions of clopen subsets of* Q.

◁ Put $H_1 := l_2(\omega)$ and $H_2 := H$ in 6.4.7 and use the separability of $l_2(\omega)$ to complete the proof. ▷

6.4.11. Theorem. *To each AW^*-module X there is a family of nonempty extremally disconnected compact spaces $(Q_\gamma)_{\gamma\in\Gamma}$, with Γ a set of cardinals, such that Q_γ is γ-stable for all $\gamma \in \Gamma$ and the following unitary equivalence holds:*

$$X \simeq \sum_{\gamma\in\Gamma}^{\oplus} C_{\#}(Q_\gamma, l_2(\gamma)).$$

If some family $(P_\delta)_{\delta\in\Delta}$ of extremally disconnected compact spaces satisfies the above conditions then $\Gamma = \Delta$, and P_γ is homeomorphic with Q_γ for all $\gamma \in \Gamma$.

◁ By Theorem 6.2.8 we may assume that X is the bounded descent of a Hilbert space $\mathscr{X}$ inside $\mathbf{V}^{(B)}$. Suppose further that B-dim$(X) = (b_\gamma)_{\gamma\in\Gamma}$ and Q_γ is the clopen subset of the Stone space of B which corresponds to $b_\gamma \in B$; i.e., the support of b. We make use of the fact that X is the direct sum of the spaces of the form $b_\gamma X$, with $b_\gamma X$ unitarily equivalent to the bounded descent of the space $b_\gamma \mathscr{X}$ inside $\mathbf{V}^{(B_\gamma)}$, where $B_\gamma = [\mathbf{0}, b_\gamma]$. By 6.3.5, note that $b_\gamma \leq [\![\dim(b_\gamma \mathscr{X}) = \gamma^\wedge]\!]$. Consequently, given a nonzero b_γ, conclude that $\mathbf{V}^{(B_\gamma)} \models$ "$b_\gamma \mathscr{X}$ is a Hilbert space of dimension $\gamma^\wedge$." Appealing to the transfer principle, infer that $\mathbf{V}^{(B_\gamma)} \models$ "$b_\gamma \mathscr{X}$ is unitarily equivalent to $l_2(\gamma^\wedge)$." By virtue of 6.4.5, the bounded descent of $l_2(\gamma^\wedge)$ in $\mathbf{V}^{(B_\gamma)}$ is unitarily equivalent to the AW^*-module $C_{\#}(Q_\gamma, l_2(\gamma))$. Suppose that $u_\gamma \in \mathbf{V}^{(B_\gamma)}$ is a unitary isomorphism from $b_\gamma \mathscr{X}$ onto $l_2(\gamma^\wedge)$ inside $\mathbf{V}^{(B_\gamma)}$, and U_γ is the bounded descent of u_γ. Then U_γ establishes unitary equivalence between the AW^*-modules $b_\gamma X$ and $C_{\#}(Q_\gamma, l_2(\gamma))$. By definition, the element $b_\gamma \in B$, together with the compact space Q_γ, is γ-stable.

Assume now that some family of extremally disconnected compact spaces $(P_\delta)_{\delta\in\Delta}$ obeys the same conditions as $(Q_\gamma)_{\gamma\in\Gamma}$. Then P_δ is homeomorphic with some clopen subset P'_δ of the Stone space of B. Moreover, P'_δ is δ-stable. If $P_{\delta\gamma} := P'_\delta \cap Q_\gamma$ and $b_{\gamma\delta}$ is the corresponding element of B then the AW^*-modules $C_{\#}(P_{\delta\gamma}, l_2(\delta))$ and $C_{\#}(P_{\delta\gamma}, l_2(\gamma))$ are unitarily equivalent to the same member $b_{\delta\gamma} X$. Furthermore, the compact space $P_{\delta\gamma}$ must be δ- and γ-stable simultaneously. According to 6.4.6 and 6.4.8, $P_{\delta\gamma} = \varnothing$ or $l_2(\delta) \sim l_2(\gamma)$, implying $\delta = \gamma$. Therefore, $P'_\gamma = Q_\gamma$ $(\gamma \in \Gamma)$. ▷

6.4.12. COMMENTS.

All results of the current section are taken from [126]. Propositions 6.4.7 and 6.4.11 show that for infinite cardinals $\alpha < \beta$ there is a AW^*-module that is γ-homogeneous for all $\alpha \leq \gamma \leq \beta$. The last fact was established by M. Ozawa [195, 197].

6.5. Representation of a Type I AW^*-Algebra

Using the results of the preceding section, we now obtain a function representation of a type I AW^*-algebra. Throughout this section we assume that A stands for an arbitrary type I AW^*-algebra, Λ denotes the center of A, and B is a complete Boolean algebra of central idempotents of A so that $B \subset \Lambda \subset A$.

6.5.1. Suppose that B_h is the set comprising $b \in B$ such that bA is a homogeneous algebra. Given $b \in B_h$, denote by $\varkappa(b)$ the least cardinal λ for which bA is a λ-homogeneous AW^*-algebra. Granted an arbitrary $b \in B$, put $\varkappa(b) := \sup\{\varkappa(b') : b' \leq b,\ b' \in B_h\}$. We thus define some function $\varkappa$ on B that takes values in a set of cardinals. Call $\varkappa$ the *multiplicity function* of A. An element $b \in B$, as well as the algebra bA, are called *strictly λ-homogeneous* provided that $\varkappa(b') = \lambda$ for $0 \neq b' \leq b$. We also say that b and bA are of *strict multiplicity* λ. There exists a unique mapping $\bar{\varkappa} : \Gamma \to B$ such that Γ is some set of cardinals each of which is at most $\varkappa(\mathbf{1})$, the family $(\bar{\varkappa}(\gamma))_{\gamma\in\Gamma}$ is a partition of unity in B, and the element $\bar{\varkappa}(\gamma)$ has strict multiplicity γ for all $\gamma \in \Gamma$. This partition of unity $(\bar{\varkappa}(\gamma))_{\gamma\in\Gamma}$ is a *strict decomposition series* of an AW^*-algebra A. It is easy to note that if $A = \operatorname{End}(X)$ for an AW^*-module X, then the strict decomposition series of A coincides with B-$\dim(X)$, and $\varkappa$ coincides with the multiplicity function of 6.4.1. The multiplicity functions $\varkappa$ and $\varkappa'$ on the Boolean algebras B and B', together with the corresponding partitions of unity $\bar{\varkappa}$ and $\overline{\varkappa'}$, are referred to as *congruent* if there is an isomorphism π of B onto B' such that $\varkappa' \circ \pi = \varkappa$. As we see, the congruency between $\bar{\varkappa}$ and $\overline{\varkappa'}$ implies that these functions have the same domain. Moreover, $\pi \circ \bar{\varkappa} = \overline{\varkappa'}$.

6.5.2. Suppose that Q is some extremally disconnected compact space, H is a Hilbert space, and $\mathscr{L}(H)$ is the space of bounded linear endomorphisms of H.

Denote by $\mathfrak{C}(Q, \mathscr{L}(H))$ the set of all operator functions $u : \operatorname{dom}(u) \to \mathscr{L}(H)$ defined on the comeager sets $\operatorname{dom}(u) \subset Q$ and continuous in the strong operator topology.

If $u \in \mathfrak{C}(Q, \mathscr{L}(H))$ and $h \in H$, then the vector function $uh : q \mapsto u(q)h$, with $q \in \operatorname{dom}(u)$, is continuous thus determining a unique element $\widetilde{uh} \in C_\infty(Q, H)$ from the condition $uh \in \widetilde{uh}$ (cf. 5.3.7 (5)). Introduce an equivalence on $\mathfrak{C}(Q, \mathscr{L}(H))$ by putting $u \sim v$ if and only if u and v agree on $\operatorname{dom}(u) \cap \operatorname{dom}(v)$. If $\tilde{u}$ is a coset of the operator function $u : \operatorname{dom}(u) \to \mathscr{L}(H)$ then $\tilde{u}h := \widetilde{uh}$ $(h \in H)$ by definition.

Denote by $SC_\infty(Q, \mathscr{L}(H))$ the set of all cosets $\tilde{u}$ such that $u \in \mathfrak{C}(Q, \mathscr{L}(H))$ and the set $\{|\tilde{u}h| : \|h\| \leq 1\}$ is bounded in $C_\infty(Q)$.

Since $|\tilde{u}h|$ agrees with the function $q \mapsto \|u(q)h\|$ on some comeager set; the containment $\tilde{u} \in SC_\infty(Q, \mathscr{L}(H))$ means that the function $q \mapsto \|u(q)\|$, with $q \in \operatorname{dom}(u)$, is continuous on a comeager set. Hence, there are an element $|\tilde{u}| \in C_\infty(Q)$ and a comeager set $Q_0 \subset Q$ satisfying $|\tilde{u}|(q) = \|u(q)\|$ for all $q \in Q_0$. Moreover,

$|\tilde{u}| = \sup\{|\tilde{u}h| : \|h\| \leq 1\}$, where the supremum is taken over $C_\infty(Q)$. We naturally equip $SC_\infty(Q, \mathscr{L}(H))$ with the structure of a $*$-algebra and a unital $C_\infty(Q)$-module by means of the operations

$$\begin{aligned} (u+v)(q) &:= u(q) + v(q) \quad (q \in \mathrm{dom}(u) \cap \mathrm{dom}(v)), \\ (uv)(q) &:= u(q) \circ v(q) \quad (q \in \mathrm{dom}(u) \cap \mathrm{dom}(v)), \\ (av)(q) &:= a(q)v(q) \quad (q \in \mathrm{dom}(a) \cap \mathrm{dom}(v)), \\ u^*(q) &:= u(q)^* \quad (q \in \mathrm{dom}(u)), \end{aligned}$$

with $u, v \in \mathfrak{C}(Q, \mathscr{L}(H))$ and $a \in C_\infty(Q)$. Furthermore, we note the following

$$\begin{gathered} |\tilde{u} + \tilde{v}| \leq |\tilde{u}| + |\tilde{v}|, \\ |\tilde{u}\tilde{v}| \leq |\tilde{u}| \cdot |\tilde{v}|, \\ |a\tilde{v}| = |a||\tilde{v}|, \ |\tilde{u} \cdot \tilde{u}^*| = |\tilde{u}|^2. \end{gathered}$$

If $\tilde{u} \in SC_\infty(Q, \mathscr{L}(H))$ and the element $\tilde{x} \in C_\infty(Q, H)$ is determined by a continuous vector function $x : \mathrm{dom}(x) \to H$ then we can define $\tilde{u}\tilde{x} := \widetilde{ux} \in C_\infty(Q, H)$, with $ux : q \mapsto u(q)x(q)$ where $q \in \mathrm{dom}(u) \cap \mathrm{dom}(x)$, since the last function is continuous. We also have

$$|\tilde{u}x| \leq |\tilde{u}| \cdot |x| \quad (x \in C_\infty(Q, H)).$$

It follows in particular that

$$|\tilde{u}| = \sup\left\{|\tilde{u}x| : x \in C_\infty(Q, H), |x| \leq 1\right\}.$$

Denote the operator $x \mapsto \tilde{u}x$ by $S_{\tilde{u}}$.

We now introduce the following normed $*$-algebra

$$\begin{gathered} SC_\#(Q, \mathscr{L}(H)) := \{v \in SC_\infty(Q, \mathscr{L}(H)) : |v| \in C(Q)\}, \\ \|v\| = \||v|\|_\infty \quad (v \in SC_\#(Q, \mathscr{L}(H))). \end{gathered}$$

6.5.3. Theorem. *To each operator $U \in \mathrm{End}(C_\#(Q, H))$ there is a unique element $u \in SC_\#(Q, \mathscr{L}(H))$ satisfying $U = S_u$. The mapping $U \mapsto u$ is a $*$-B-isomorphism of* $\mathrm{End}(C_\#(Q, H))$ *onto $A := SC_\#(Q, \mathscr{L}(H))$. In particular, A is a λ-homogeneous algebra. Moreover, if Q is a λ-stable compact space then A is a strictly λ-homogeneous AW^*-algebra, with $\lambda = \dim(H)$.*

$\lhd$ First of all note that the operator S_u obeys the inequality $|S_u x| \leq |u| \cdot |x|$ for all $x \in C_\#(Q, H)$. Consequently, given $u \in SC_\#(Q, \mathscr{L}(H))$, we see that S_u acts in $C_\#(Q, H)$ as a bounded linear operator. Moreover,

$$\|S_u\| = \sup_{\|x\| \leq 1} \||S_u x|\|_\infty = \sup_{|x| \leq 1} \sup_{q \in Q} |ux|(q) = \sup_{q \in Q} |u|(q) = \|u\|.$$

Clearly, $S_{u^*} = S_u{}^*$ for all $u \in SC_\#(Q, \mathscr{L}(H))$. Therefore, the mapping $u \mapsto S_u$ is a $*$-B-isomorphic embedding of $SC_\#(Q, \mathscr{L}(H))$ into $\mathrm{End}(C_\#(Q, H))$. Prove that this embedding is surjective. The mapping $U \in \mathrm{End}(C_\#(Q, H))$ is a *dominated operator*; i.e., U obeys the inequality $|Ux| \le f \cdot |x|$ for all $x \in C_\#(Q, H)$, where $f := \sup\{|Ux| : |x| \le 1\} \in C(Q)$. By Theorem 5.3.13 there is an operator function $u : \mathrm{dom}(u) \to \mathscr{L}(H)$ satisfying the conditions: (1) the function $q \mapsto \langle u(q)h|g\rangle$ is continuous for all $g, h \in H$; (2) there is a function $\varphi \in C_\infty(Q)$ such that $\|u(q)\| \le \varphi(q)$ for all $q \in \mathrm{dom}(u)$; (3) $Ux = \tilde{u}x$ for all $x \in C_\#(Q, H)$ and $|u| = f$. Thus, $U = S_{\tilde{u}}$ and we are left with justifying only that u is continuous in the strong operator topology. Recalling the definition of the least upper bound of a set in the K-space $C_\infty(Q)$, we may observe that $\|u(q)\| = |u|(q)$ for all $q \in Q_0$ where Q_0 is some comeager subset of Q. Therefore, substituting $Q_0 \cap \mathrm{dom}(u)$ for $\mathrm{dom}(u)$ if need be, we may assume that $q \mapsto \|u(q)\|$ is a continuous function. Together with the above condition (1), this implies the continuity of u in the strong operator topology; i.e., $u \in SC_\#(Q, \mathscr{L}(H))$. The rest of the theorem ensues from 5.3.4 (3). $\triangleright$

We say that the families of nonempty compact sets $(Q_\gamma)_{\gamma\in\Gamma}$ and $(P_\delta)_{\delta\in\Delta}$ are *congruent* provided that $\Gamma = \Delta$, and Q_γ and P_γ are homeomorphic to one another for all $\gamma \in \Gamma$.

6.5.4. Theorem. *To each type I AW^*-algebra A there is a family of nonempty extremally disconnected compact spaces $(Q_\gamma)_{\gamma\in\Gamma}$ such that the following conditions are met:*

(1) *Γ is a set of cardinals and Q_γ is γ-stable for each $\gamma \in \Gamma$;*

(2) *There is a $*$-isomorphism:*

$$A \simeq \sum_{\gamma\in\Gamma}{}^{\oplus} SC_\#(Q_\gamma, \mathscr{L}(l_2(\gamma))).$$

This family is unique up to congruence.

$\triangleleft$ By Theorem 6.2.5 we can assume that A is the bounded descent of an AW^*-factor $\mathcal{A}$ in $\mathbf{V}^{(B)}$. In this event $\mathcal{A}$ has type I, and so $\mathcal{A} \simeq B(\mathscr{X})$ where $\mathscr{X}$ is a Hilbert space inside $\mathbf{V}^{(B)}$. Hence, we see that A is $*$-isomorphic with $\mathrm{End}(X)$, where X stands for the bounded descent of $\mathscr{X}$. Suppose that $B\text{-}\mathrm{dim}(X) = (b_\gamma)_{\gamma\in\Gamma}$, and Q_γ is the clopen subset of the Stone space of the Boolean algebra B which corresponds to $b_\gamma \in B$. By virtue of 6.3.7, Q_γ is a γ-stable compact space. So (1) holds. Theorem 6.4.11 yields the following unitary equivalence $X \simeq \sum^{\oplus}_{\gamma\in\Gamma} C_\#(Q_\gamma, l_2(\gamma))$. Hence, note the next $*$-isomorphism of AW^*-algebras:

$$\mathrm{End}(X) \simeq \sum_{\gamma\in\Gamma}{}^{\oplus} \mathrm{End}(C_\#(Q_\gamma, l_2(\gamma))).$$

Appealing to Theorem 6.5.3, we arrive at the sought condition (2).

Uniqueness ensues from 6.4.11. $\triangleright$

6.5.5. Corollary. *The following hold:*

(1) *Every type I AW^*-algebra splits into the direct sum of strictly homogeneous components. This decomposition is unique up to $*$-isomorphism;*

(2) *Two type I AW^*-algebras are $*$-isomorphic to one another if and only if they have congruent multiplicity functions or, which is the same, congruent strict decomposition series.*

$\lhd$ This assertion ensues from (1) on observing that in the representation of 6.5.4 the dimension of A is congruent to the partition of unity $(\chi_\gamma)_{\gamma\in\Gamma}$, with χ_γ the characteristic function of the set Q_γ in the disjoint sum of the family (Q_γ). $\rhd$

(3) *Suppose that Γ is a set of cardinals and (b_γ) is a partition of unity in B which consists of nonzero pairwise distinct elements. Then $(b_\gamma)_{\gamma\in\Gamma}$ is a strict decomposition series of some AW^*-algebra if and only if b_γ is γ-stable for all $\gamma \in \Gamma$.*

$\lhd$ This ensues from 6.3.7 and 6.5.3. $\rhd$

6.5.6. COMMENTS.

(1) The main results on function representation, Theorems 6.4.11 and 6.5.4, were established by A. G. Kusraev in [62]. M. Ozawa had classified the type I AW^*-algebras in somewhat different form in [195] (cf. 6.5.5 (2)). The true distinction lies in the fact that the invariant, characterizing a type I AW^*-algebra to within $*$-isomorphism in the M. Ozawa research, is a Boolean valued cardinal; i.e., an internal object of the Boolean valued universe in question. The definition of 6.5.1 does not appeal to the construction of the Boolean valued universe.

(2) Observe that 6.4.8 and 6.5.5 (2) imply a negative solution to the I. Kaplansky problem of unique decomposition of a type I AW^*-algebra into the direct sum of homogeneous components. M. Ozawa gave this solution in [196, 197]. As we can see from 6.4.8, the failure of uniqueness is tied with the effect of the cardinal shift that may happens during immersion into $\mathbf{V}^{(B)}$ (cf. 3.1.13 (1)). The cardinal shift is impossible in the case when the Boolean algebra of central idempotents B under study satisfies the countable chain condition (cf. 3.1.13 (2)) and so the decomposition in question is unique. I. Kaplansky established uniqueness of the decomposition on assuming that B satisfies the countable chain condition and conjectured that uniqueness fails in general [107].

6.6. Embeddable C^*-Algebras

Type I algebras have the simplest structure in the class of all AW^*-algebras. Most attractive is an algebra that may be presented as the double commutant of a type I AW^*-algebra. Such an algebra is called *embeddable*. Moreover, as we may deduce from the results of Section 6.2, an embeddable algebra transforms into a von Neumann algebra if embedded in a suitable Boolean valued universe. That is how a possibility opens up of translating the facts of von Neumann algebras into the corresponding results about embeddable algebras. The current section demonstrates this technique by a few examples.

6.6.1. We give the necessary definitions and facts.

(1) Assume that H is again a Hilbert space, and $\mathscr{L}(H)$ is the space of bounded endomorphisms of H. Granted $M \subset \mathscr{L}(H)$, recall that the *commutant* M' of M is the set of all members of $\mathscr{L}(H)$ commuting with every element of M (cf. 6.2.5). The *double commutant* or *bicommutant* of M is the set $M'' := (M')'$. Clearly, M' is a Banach operator algebra with unity the identity operator $\mathbf{1} := I_H$. A *von Neumann algebra* over H is a $*$-subalgebra A' of $\mathscr{L}(H)$ coinciding with the double commutant of itself, i.e. $A = A''$. The *center* of a von Neumann algebra A is the set $\mathscr{Z}(A) := A \cap A'$. A von Neumann algebra A is a *factor* provided that the center of A is trivial; i.e., in the case when $\mathscr{Z}(A) = \mathbb{C} \cdot \mathbf{1} := \{x \cdot I_H : \lambda \in \mathbb{C}\}$.

(2) Double Commutant Theorem. *Let A be an involutive operator algebra over a Hilbert space H and $I_H \in A$. Then A coincides with the double commutant A'' of A if and only if A is closed with respect to the strong operator topology of $\mathscr{L}(H)$ or, which is the same, closed with respect to the weak operator topology of $\mathscr{L}(H)$.*

(3) Sakai Theorem. *A C^*-algebra A is $*$-isomorphic with a von Neumann algebra if and only if A is the dual of some Banach space.*

(4) A C^*-algebra A is *B-embeddable* if there are a type I AW^*-algebra N and a $*$-monomorphism $\imath : A \to N$ such that $B = \mathfrak{P}_c(N)$ and $\imath(A) = \imath(A)''$, with $\imath(A)''$ standing for the double commutant of $\imath(A)$ in N. Note that in this event A is an AW^*-algebra and B is a regular subalgebra of $\mathfrak{P}_c(A)$. In particular, A is a B-cyclic algebra (cf. 6.2.3). If $B = \mathfrak{P}_c(A)$ and A is B-embeddable, then A is called *centrally embeddable.* Granted a C^*-algebra A, say that another C^*-algebra $\bar{A}$ is *embeddable* if $\bar{A}$ is B-embeddable, with B some regular subalgebra of $\mathfrak{P}_c(A)$.

Recall that we always assume that each of the C^*-algebras under study is unital. Also, the record $B \sqsubset A$ means that A is a B-cyclic algebra.

6.6.2. Theorem. *Suppose that $\mathscr{A}$ is a C^*-algebra inside $\mathbf{V}^{(B)}$ and A is the bounded descent of $\mathscr{A}$. Then A is a B-embeddable AW^*-algebra if and only if $\mathscr{A}$ is a von Neumann algebra inside $\mathbf{V}^{(B)}$. Moreover, A is centrally embeddable if and only if $\mathscr{A}$ is a factor inside $\mathbf{V}^{(B)}$.*

$\lhd$ Suppose that A coincides with the double commutant of A in a type I AW^*-algebra N and, moreover, $\mathfrak{P}_c(N) = B$. By 6.2.5 and 6.2.13, we may assume that N is the bounded descent of some type I AW^*-factor $\mathscr{N}$ inside $\mathbf{V}^{(B)}$. Since $A'' \subset N$ and $A'' = A$; therefore, we clearly see that $[\![\mathscr{A} = A{\uparrow} \subset \mathscr{N}]\!] = \mathbf{1}$ and

$[\![\mathscr{A}'' = (A\uparrow)'' = A''\uparrow = \mathscr{A}]\!] = \mathbf{1}$. Hence, $\mathscr{A}$ is the double commutant of $\mathscr{A}$ in $\mathscr{N}$. It remains to note that every type I AW^*-factor of $\mathscr{N}$ is isomorphic with the algebra $B(\mathscr{H})$ for some Hilbert space $\mathscr{H}$.

Conversely, suppose that $[\![\mathscr{A}$ is a von Neumann algebra$]\!] = \mathbf{1}$. This means that $[\![\mathscr{A}$ is the double commutant of $\mathscr{A}$ in the endomorphism space $\mathscr{L}(\mathscr{H})$ of $\mathscr{H}]\!] = \mathbf{1}$ for some Hilbert space $\mathscr{H}$ inside $\mathbf{V}^{(B)}$. Let N stand for the bounded descent of $\mathscr{L}(\mathscr{H})$. Then N is a type I AW^*-algebra, cf. 6.2.13 (2). Moreover, A is the double commutant of A in N and $\mathfrak{P}_c(N) = B$, cf. 6.2.5. The second claim follows from Theorem 6.2.5 which reads that $\mathscr{A}$ is a factor inside $\mathbf{V}^{(B)}$ if and only if $\mathfrak{P}_c(A) = B$. $\rhd$

6.6.3. We now characterize an embeddable C^*-algebra. Recall that, given a normed B-space, we denote by $X^{\#}$ the B-dual of X (cf. 5.5.8). Say that a C^*-algebra A is B-dual if A includes a Boolean algebra B of central projections and A is B-isometric with the B-dual $X^{\#}$ of some normed B-space X. In this event, say that X is ***B-predual*** to A and write $A_{\#} = X$.

6.6.4. Theorem. *A C^*algebra is B-embeddable if and only if A is B-dual. Every B-predual space is unique to within B-isometry in the class of B-cyclic Banach spaces.*

$\lhd$ Suppose that A is a C^*-algebra and $B \sqsubset \mathfrak{P}_c(A)$. By Theorem 6.1.6, we may assume that A coincides with the bounded descent of some C^*-algebra $\mathscr{A}$ in $\mathbf{V}^{(B)}$.

Using the Sakai Theorem inside $\mathbf{V}^{(B)}$ and applying the transfer principle, observe that $[\![\mathscr{A}$ is a von Neumann algebra$]\!] = [\![\mathscr{A}$ is linearly isometric with the dual $\mathscr{X}'$ of some Banach space $\mathscr{X}]\!]$. If X is the bounded descent of the Banach space $\mathscr{X}$ then $X^{\#}$ is B-linearly isometric with the bounded descent of $\mathscr{X}'$ (cf. 5.5.10). By Theorem 6.6.2 note now that if A is B-embeddable then A is also a B-dual algebra. Moreover, $A_{\#} = X$ is a B-cyclic space.

Conversely, assume that A is B-dual and $A_{\#} = X_0$ for some normed B-space X_0. If X is the B-cyclic completion of X_0 then $X_0^{\#} = X^{\#}$, which implies $A_{\#} = X$. Denote by $\mathscr{X}$ the Boolean valued representation of X. Then $\mathscr{A} \simeq \mathscr{X}^{\#}$. By Theorem 6.6.2, A is a B-embeddable algebra.

Suppose now that each of the B-cyclic spaces X and Y serves as B-predual of A. Denote by $\mathscr{X}$ and $\mathscr{Y}$ the representations of X and Y in $\mathbf{V}^{(B)}$. Observe that $[\![\mathscr{X}$ and $\mathscr{Y}$ are predual to $\mathscr{A}]\!] = \mathbf{1}$. A predual of a von Neumann algebra is unique up to linear isometry. Consequently, $[\![\mathscr{X}$ and $\mathscr{Y}$ are linearly isometric$]\!] = \mathbf{1}$. Since X and Y coincide with the bounded descents of $\mathscr{X}$ and $\mathscr{Y}$ respectively, conclude that X and Y are B-isometric. $\rhd$

6.6.5. Theorem. *Assume that N is some type I AW^*-algebra and A is an AW^*-subalgebra of N including the center $\mathscr{Z}(N)$ of N. Then the algebra A and the commutant A' of A in N are of the same type I, II, or III.*

◁ According to 6.2.5 and 6.2.13 we may suppose that N and A are the bounded descents of some algebras $\mathscr{N}$ and $\mathscr{A}$ inside $\mathbf{V}^{(B)}$, where $B = \mathfrak{P}_c(N)$, $[\![\mathscr{N} = \mathscr{L}(\mathscr{H})$ for some Hilbert space $\mathscr{H}]\!] = \mathbb{1}$ and $[\![\mathscr{A}$ is an AW^*-subalgebra of $\mathscr{N}]\!] = \mathbb{1}$. Thus, $\mathscr{A}$ is a von Neumann algebra inside $\mathbf{V}^{(B)}$. The claims in question hold for every von Neumann algebra (cf. [211]); i.e., $\mathscr{A}$ and $\mathscr{A}'$ have the same type I, II, or III. Furthermore, A' coincides with the bounded descent of $\mathscr{A}'$ since $\mathscr{A}'{\downarrow} = (\mathscr{A}{\downarrow})^\circ$ where $(\cdot)^\circ$ is the taking of the commutant of a subset of the algebra $\mathscr{N}{\downarrow}$. We complete the proof by appealing to Theorem 6.2.13 again. ▷

6.6.6. Theorem. *Suppose that a C^*-algebra A is B_0-embeddable for some regular subalgebra B_0 of $\mathfrak{P}_c(A)$. Then A is B-embeddable for every regular subalgebra B_0 of $\mathfrak{P}_c(A)$.*

◁ Suppose that A is the double commutant in a type I AW^*-algebra N and $\mathfrak{P}_c(N) = B_0$. Suppose that B is a regular subalgebra of the Boolean algebra $\mathfrak{P}_c(A)$ and, moreover, $B_0 \subset B$. Denote by $\mathscr{C}(B)$ the C^*-algebra that is generated by B. Since B is a regular subalgebra, $\mathscr{C}(B)$ is an AW^*-subalgebra in N (cf. 6.2.1 (1, 2)). Furthermore, $\mathscr{C}(B)$ includes the center of N since $B_0 = \mathfrak{P}_c(N)$. By Theorem 6.6.4, the commutant $\mathscr{C}(B)' = B'$ of the algebra $\mathscr{C}(B)$ in N has the same type as $\mathscr{C}(B)$. But $\mathscr{C}(B)$ is a commutative AW^*-algebra. Hence, $\mathscr{C}(B)'$ is a type I algebra. Since $\mathscr{C}(B)$ is commutative, we see that the center of $\mathscr{C}(B)'$ coincides with $\mathscr{C}(B)$. Since $\mathscr{C}(B)$ lies in the center of A; therefore, the commutant A' of A, calculated in N, is included in $\mathscr{C}(B)'$. Consequently, the double commutant of A in $\mathscr{C}(B)'$ coincides with the double commutant of A in N; i.e., A is a double commutant in $\mathscr{C}(B)$. Whence A is a B-embeddable algebra. ▷

6.6.7. Corollary. *The following hold:*

(1) *A C^*-algebra A is embeddable if and only if A is centrally embeddable;*

(2) *A von Neumann algebra A is B-embeddable for every regular subalgebra B of $\mathfrak{P}_c(A)$.*

6.6.8. Suppose that A is a C^*-algebra and $B \sqsubset A$. A linear operator $T : A \to B(\mathbb{C})$ is *positive* if $T(x^*x) \geq 0$ for all $x \in A$. A positive B-linear operator T is a *state* if $\|T\| = 1$. We call a state T *normal* if $T(\sup(x_\alpha)) = \sup(T(x_\alpha))$ for every increasing net (x_α) of hermitian operators which has a supremum. The set of all $B(\mathbb{C})$-valued states of A *separates* A if the positivity of $x \in A$ amounts to the fact that $Tx \geq 0$ for every $B(\mathbb{C})$-valued normal state T. We call a $B(\mathbb{C})$-valued state a *center valued* state when this leads to no confusion.

Monotone completeness of a C^*-algebra A means that every upper bounded increasing net of hermitian elements of A has a least upper bound. It is an easy matter to check that A is monotone complete whenever so is the Boolean valued representation of A.

6.6.9. Theorem. *Assume that $\mathscr{A}$ is a C^*-algebra inside $\mathbf{V}^{(B)}$ and A is the bounded descent of $\mathscr{A}$. Given a $B(\mathbb{C})$-valued state Φ on A, note that $[\![\varphi := \Phi\!\uparrow$ is a state on $\mathscr{A}]\!] = \mathbf{1}$. Every state on $\mathscr{A}$ has the form $\Phi\!\uparrow$, where Φ is some $B(\mathbb{C})$-valued state on A. Moreover, a state Φ is normal if and only if $[\![\varphi := \Phi\!\uparrow$* is a normal state$]\!] = \mathbf{1}$.

$\lhd$ The first part of the theorem follows from 5.5.9. Suffice it to say that the mapping $\Phi \mapsto \varphi := \Phi\!\uparrow$ preserves positivity since $\Phi(A_+)\!\uparrow = \varphi(A_+\!\uparrow) = \varphi(A_+)$. The claim about normal states is easy on recalling the rules for ascending and descending polars (cf. 5.2.13 and 5.3.12). $\rhd$

6.6.10. Theorem. *For a B-cyclic C^*-algebra A the following are equivalent:*

(1) *A is a B-embeddable algebra;*

(2) *A is monotone complete and the set of all $B(\mathbb{C})$-valued states on A separates A.*

$\lhd$ By Theorem 6.1.6, we may assume that A is the bounded descent of a C^*-algebra $\mathscr{A}$ inside $\mathbf{V}^{(B)}$. By Theorem 6.6.2 A is B-embeddable if and only if $[\![\mathscr{A}$ is a von Neumann algebra$]\!] = \mathbf{1}$. Now we make use of the following fact: a C^*-algebra A is a von Neumann algebra if and only if A is monotone complete and the normal states of A separate A. Omitting a few details, we expatiate upon existence of normal states. Suppose that $\mathscr{S}_n(\mathscr{A})$ is the set comprising the normal states of the algebra $\mathscr{A}$ inside $\mathbf{V}^{(B)}$ and $\mathscr{S}_n(\mathscr{A}, B)$ is the set comprising all normal $B(\mathbb{C})$-valued states on A. The mapping $\Phi \mapsto \varphi := \Phi\!\uparrow$ is a bijection between $\mathscr{S}_n(\mathscr{A})\!\downarrow$ and $\mathscr{S}_n(A, B)$ (cf. 6.6.9).

Assume that $\mathscr{S}_n(A, B)$ separates A. Granted a nonzero $x \in A$, find $\Phi_0 \in \mathscr{S}_n(A, B)$ satisfying $\Phi_0 x \neq 0$. Since Φ is B-linear; therefore, $[\![0 \neq x]\!] \leq [\![\Phi_0(x) \neq 0]\!]$. Recalling the rules for calculating Boolean truth values, write

$$
\begin{aligned}
&[\![\mathscr{S}_n(\mathscr{A}) \text{ separates } \mathscr{A}]\!] \\
&= [\![(\forall x \in \mathscr{A})\,(x \neq 0 \to (\exists \varphi \in \mathscr{S}_n(\mathscr{A}))\ \varphi(x) \neq 0)]\!] \\
&= \bigwedge_{x \in A} [\![x \neq 0]\!] \Rightarrow \bigvee_{\Phi \in \mathscr{S}_n(A,B)} [\![\Phi\!\uparrow\!(x) \neq 0]\!] \\
&\geq \bigwedge_{x \in A} [\![x \neq 0]\!] \Rightarrow [\![\Phi_0\!\uparrow\!(x) \neq 0]\!] = \mathbf{1}.
\end{aligned}
$$

Consequently, $\mathscr{S}_n(\mathscr{A})$ separates $\mathscr{A}$ inside $\mathbf{V}^{(B)}$.

Conversely, assume the last assertion true. Given $x \in A$, note that $b := [\![x \neq 0]\!] > \mathbf{0}$. By the maximum principle, there is some φ in $\mathscr{S}_n(\mathscr{A})\!\downarrow$ such that $b \leq [\![\varphi(x) \neq 0]\!]$. Suppose that Φ is the restriction to $A \subset \mathscr{A}\!\downarrow$ of the operator $\varphi\!\downarrow$. Then $\Phi \in \mathscr{S}_n(A, B)$ and $b \leq [\![\Phi(x) \neq 0]\!]$. Consequently, the trace $e_{\Phi(x)}$ of $\Phi(x)$ is greater than or equal to b (cf. 5.2.3 (5)). Hence, $\Phi(x) \neq 0$. $\rhd$

6.6.11. Theorem. *For an AW^*-algebra A the following are equivalent:*

(1) *A is embeddable;*

(2) *A is centrally embeddable;*

(3) *The center valued normal states of A separate A;*

(4) *A is a $\mathfrak{P}_c(A)$-dual space.*

$\lhd$ Appeal to 6.6.4, 6.6.7 (1), and 6.6.10. $\rhd$

6.6.12. COMMENTS.

(1) All results of this section belong to M. Ozawa [196, 199, 200].

(2) There exist other various classes of ordered and involutive algebras to which we may apply the technique of Sections 6.2–6.6 (see [28, 213]). Among the most important of them we mention the class of JB-algebras.

(3) By way of illustration, we state a simple Jordan analog of Theorem 6.6.10 which was established in [127].

Theorem. *Let A be a B-JB-algebra. Then A is B-dual if and only if A is monotone complete and the $B(\mathbb{C})$-valued normal states on A separate A.*

Moreover, if A is B-dual then the B-predual of A is the part of the B-dual $A^{\#}$ of A which consists of order continuous operators.

(4) Regarding other applications of Boolean valued analysis which are close to the topic of the current chapter, consult [93, 112, 134–136, 179, 180, 183–190, 194–200, 236, 237].

Other applications of Boolean valued analysis are reflected also in [67, 68, 117, 121, 123–126, 128–137, 143–145].

Appendix

This Appendix contains some preliminaries to set theory and category theory.

A.1. The Language of Set Theory

Axiomatic set theories prescribe the bylaws of sound set formation. In evocative words, every axiomatics of set theory describes a world or universe that consists of all sets we need for adequate expression of our intuitive conception of the treasure-trove of the "Cantorian paradise," the all-embracing universe of naive set theory. Present-day mathematics customarily expounds and studies any attractive axiomatics as a formal theory. We readily acknowledge that a formal approach has proven itself to be exceptionally productive and successful in spite of its obvious limitations stemming from the fact that mathematics reduces only in part to the syntax of mathematical texts. This success is in many respects due to the paucity of formal means since the semiotic aspects, if properly distinguished, invoke the insurmountable problem of meaning. The list of achievements of the formal approach contains the celebrated Gödel completeness and incompleteness theorems, independence of the continuum hypothesis and of the axiom of choice, Boolean valued analysis, etc.

The cornerstone of a formal theory is its language. Intending to give the latter an exact description and to study the properties of the theory, we are impelled to use another language that differs in general from the original language. It is in common parlance to call this extra language the *metalanguage* of our theory. The mostly presents a collection of fragments of natural languages trimmed and formalized slightly but heavily enriched with numerous technical terms. The tools of the metalanguage of a theory are of utmost importance for metamathematics. Since we are interested in applicable rather than metamathematical aspects of an axiomatic set theory, we never impose extremely stringent constraints on the metalanguage of the theory. In particular, we use the expressive means and level of rigor that are common to every-day mathematics.

A.1.1. Each axiomatic set theory is a *formal system*. The ingredients of the latter are its alphabet, formulas, axioms, and rules of inference. The alphabet of a formal theory is a fixed set A of symbols of an arbitrary nature, i.e., a Cantorian set of letters. Finite sequences of letters of A, possibly with blanks, are called *expressions*, or *records*, or *texts*. If we somehow choose the set of $\Phi(A)$ of the so-called "well formed" expressions by giving detailed prescriptions, algorithms, etc.; then we declare given a language with alphabet A and call the chosen expressions well formed formulas. The next step consists in selection of some finite (or infinite) families of formulas called axioms in company with explicit description of the admissible rules of inference which might be viewed as abstract relations on $\Phi(A)$. A *theorem* is a formula that results from axioms by successively using finitely many rules of inference. Using common parlance, we express this in a freer and cozier fashion as follows: the theorems of a formal theory comprise the least set of formulas which contains all axioms and is closed under the rules of inference of the theory.

A.1.2. Of primary interest for us is a special formal language called a *first-order language* of predicate calculus.

Recall that the *signature* σ of a language is a 3-tuple (F, P, a), where F and P are some sets called the set of *function* or *operation symbols* and the set of *predicate symbols*, respectively, while a is a mapping of $F \cup P$ into the set of natural numbers. Say that $u \in F \cup P$ is an *n-ary symbol* or *n-place symbol* whenever $a(u) = n$. Regarding the *alphabet* of a first-order language of *signature* σ, we usually distinguish

(1) the set of *symbols* of signature σ, i.e., the set $F \cup P$;

(2) the set of *variables* composed of lower case or upper case Latin letters possibly with indices;

(3) the set of *propositional connectives*: $\wedge$, conjunction; $\vee$, disjunction; $\rightarrow$, implication; and $\neg$, negation;

(4) the set of the symbols of *quantifiers*: $\forall$, the symbol of a universal quantifier, and $\exists$, the symbol of an existential quantifier;

(5) the sign of *equality* $=$;

(6) the set of *auxiliary symbols*: (, which is the opening parenthesis;), which is the closing parenthesis; and , which is a comma.

A.1.3. In the language of set theory we distinguish terms and formulas.

(1) A *term* of signature σ is an element of the least set of expressions of the language (of the same signature σ) obeying the following conditions:

(a) Each variable is a term;

(b) Each nullary function symbol is a term;

(c) If $f \in F$, $a(f) = n$, and $t_1, \dots, t_n$ are terms then $f(t_1, \dots, t_n)$ is a term.

(2) An *atomic formula* of signature σ is an expression of the kind

$$t_1 = t_2, \quad p(y_1, \ldots, y_n), \quad q,$$

where $t_1, t_2, y_1, \ldots, y_n$ are terms of signature σ, the letter p stands for some n-ary predicate symbol, and q is a nullary predicate symbol.

(3) *Formulas of signature* σ constitute the least set of records obeying the following conditions:

(a) Each atomic formula of signature σ is a formula of signature σ;

(b) If φ and ψ are formulas of signature σ then $(\varphi \wedge \psi)$, $(\varphi \vee \psi)$, $(\varphi \to \psi)$, and $\neg\varphi$ are formulas of signature σ, too;

(c) If φ is a formula of signature σ and x is a variable then $(\forall x)\varphi$ and $(\exists x)\varphi$ are formulas of signature σ too.

A variable x is *bound* in some formula φ or belongs to the domain of a quantifier provided that x appears in a subformula of φ of the kind $(\forall x)\psi$ or $(\exists x)\varphi$. In the opposite case, x is *free* in φ. We also speak about *free* or *bound occurrence* of a variable in a formula. Intending to stress that only the variables $x_1, \ldots, x_n$ are free in the formula φ, we write $\varphi = \varphi(x_1, \ldots, x_n)$, or simply $\varphi(x_1, \ldots, x_n)$. The words "proposition" and "statement" are informally treated as synonyms of "formula." A formula with no free variables is a *sentence*. Speaking about *verity* or *falsity* of φ, we imply the *universal closure* of φ which results from generalization of φ by every free variable of φ. It is also worth observing that quantification is admissible only by variables. In fact, the words "first-order" distinguish this syntactic feature of the formal languages we discuss.

A.1.4. The *language of set theory* is a first-order language whose signature contains only one binary predicate symbol $\in$ and so it has neither predicates other than $\in$ nor any function symbols. So, set theory is a simple instance of the abstract *first-order theories*. We agree to write $x \in y$ instead of $\in(x, y)$ and say that x is an *element* of y or a *member* of y. It is also in common parlance to speak of *membership* or *containment*. As usual, a formula of set theory is a formal text resulting from the atomic formulas like $x \in y$ and $x = y$ by appropriate usage of propositional connectives and quantifiers.

Set theory (or strictly speaking, *the* set theory we profess in this book) bases upon the laws of classical logic. In other words, set theory uses the common logical axioms and rules of inference of predicate calculus which are listed in nearly every manual on mathematical logic (see, for instance, [48, 108, 217]). Note also that the instance of predicate calculus we use in this book appears often with some of the epithets *classical*, or *lower*, or *narrow*, or *first-order* and is formally addressed as the *first-order classical predicate calculus with equality*. In addition, a particular set theory contains some special nonlogical axioms that legitimize the conceptions of sets and classes we want to explicate. By reasonably varying the special axioms, we may come to axiomatic set theories that differ in the power of expression. This appendix describes one of the most popular axiomatic set theories, *Zermelo–Fraenkel theory* symbolized as ZF or ZFC if the axiom of choice is available.

A.1.5. Among the best conveniences of any metalanguage we must mention abbreviations. The point is that formalization of the simplest fragments of workable mathematics leads to bulky texts whose recording and playing back is problematic for both physical and psychological reasons. This is why we must introduce many abbreviations, building a more convenient abridged dialect of the initial symbolic language. Naturally, this is reasonable only if we ensure a principal possibility of unambiguous translation from the dialect to the original and vise versa. In accord with our intentions, we will not expatiate on the exact technique of abbreviation and translation and adhere to every-day practice of doing Math. For instance, we use the *assignment operator* or *definor* := throughout the book, with no fuss about accompanying formal subtleties.

A.1.6. We now give some examples of abbreviated texts in the language of set theory. These examples rely to intuition of naive set theory. We start with the most customary instances. Here they are

$$(\exists!\, x)\ \varphi(x) := (\exists\, x)\varphi(x) \wedge (\forall\, x)(\forall\, y)(\varphi(x) \wedge \varphi(y) \to x = y);$$
$$(\exists\, x \in y)\varphi := (\exists\, x)\ (x \in y \wedge \varphi);$$
$$(\forall\, x \in y)\varphi := (\forall\, x)\ (x \in y \to \varphi),$$

with φ a formula. As usual, we put $x \neq y := \urcorner(x = y)$ and $x \notin y := \urcorner(x \in y)$. Also, we use the routine conventions about the traditional operations on sets:

$$x \subset y := (\forall\, z)(z \in x \to z \in y);$$
$$u = \cup x = \cup(x) := (\forall\, z)(z \in u \leftrightarrow (\exists\, y \in x) z \in y);$$
$$u = \cap x = \cap(x) := (\forall\, z)(z \in u \leftrightarrow (\forall\, y \in x) z \in y);$$
$$u = y - x = y \setminus x := (\forall\, z)(z \in u \leftrightarrow (z \in y \wedge z \notin x)).$$

Given a formula φ, we introduce the collection $\mathscr{P}_\varphi(x)$ of all subsets of x which satisfies φ as follows

$$u = \mathscr{P}_\varphi(x) := (\forall\, z)(z \in u \leftrightarrow (z \subset x) \wedge \varphi(z)).$$

We call a set u *empty* and denote it by $\varnothing$ if u contains no elements. In other words,

$$u = \varnothing := (\forall\, x)(x \in u \leftrightarrow x \neq x).$$

An empty set is unique practically in every set theory and so we refer to $\varnothing$ as *the* empty set.

These examples use one of the commonest methods of abbreviation, namely, omission of some parentheses.

A.1.7. The statement that x is the *unordered pair* of elements y and z is formalized as follows:

$$(\forall u)(u \in x \leftrightarrow u = y \vee u = z).$$

In this event we put $\{y, z\} := x$ and speak about the *pair* $\{y, z\}$. Note that braces do not belong to the original alphabet and so they are *metasymbols*, i.e. symbols of metalanguage.

An *ordered pair* and an *ordered n-tuple* result from the Kuratowski trick:

$$(x, y) := \langle x, y\rangle := \{\{x\}, \{x, y\}\};$$
$$(x_1, \dots, x_n) := \langle x_1, \dots, x_n\rangle := \langle\langle x_1, \dots, x_{n-1}\rangle, x_n\rangle,$$

where $\{x\} := \{x, x\}$. Observe the overuse of parentheses. This is inevitable and must never be regarded as pretext for introducing new symbols.

The agreements we made enable us to ascribe a formal meaning to the expression "X *is the Cartesian product* $Y \times Z$ of Y and Z." Namely, we assign $X := \{(y, z) : y \in Y, z \in Z\}$. Note also that the nickname "product" is in common parlance for "Cartesian product."

A.1.8. Consider the following propositions:

(1) $\mathrm{Rel}(X)$;

(2) $Y = \mathrm{dom}(X)$;

(3) $Z = \mathrm{im}(X)$.

Putting these formally, find

(1′) $(\forall u)\ (u \in X \to (\exists v)(\exists w)\ u = (v, w))$;

(2′) $(\forall u)\ (u \in Y \leftrightarrow (\exists v)(\exists w)\ w = (u, v) \wedge w \in X)$;

(3′) $(\forall u)\ (u \in Z \leftrightarrow (\exists v)(\exists w)\ w = (v, u) \wedge w \in X)$.

In other words, we state in (1)–(3) that the members of X are ordered pairs, Y is the collection of the first coordinates of the members of X, and Z comprises the second coordinates of the members of X. It is in common parlance to say that Y is the *domain* of X, and Z is the *range* or *image* of X. In this event we refer to X as an *abstract relation.*

We express the fact that X is *single-valued* or $\mathrm{Un}(X)$ by the formula

$$\mathrm{Un}(X) := (\forall u)(\forall v_1)(\forall v_2)((u, v_1) \in X \wedge (u, v_2) \in X \to v_1 = v_2).$$

We put $\mathrm{Fnc}(X) := \mathrm{Func}(X) := \mathrm{Un}(x) \wedge \mathrm{Rel}(X)$. In case $\mathrm{Fnc}(X)$ is valid, we have many obvious reasons to call X a *function* or even a *class-function.* Paraphrasing the membership $(u, v) \in X$, we write $v = X(u)$, $X : u \mapsto v$, etc. We say that F is a *mapping* or *function* from X to Y, implying that every member of F belongs to $X \times Y$, while F is single-valued, and the domain of F coincides with X; that is,

$$F : X \to Y := F \subset X \times Y \wedge \mathrm{Fnc}(F) \wedge \mathrm{dom}(F) = X.$$

The term *class-function* is also applied to F if we want to stress that F is a class.

The *restriction* of X to U is by definition $X \cap (U \times Z)$. We denote it by $X \restriction U$. If there is a unique z satisfying $(y, z) \in X$ then we put $X`y := z$. We finally let $X``Y := \mathrm{im}(X \restriction Y)$. Instead of $X``\{y\}$ we write $X(y)$ or even Xy when this does not lead to misunderstanding. It is worth emphasizing that we always exercise a liberal view on placing and removing parentheses. In other words, we insert or eliminate parentheses, influenced as a rule by what is convenient or needed for a formal presentation of a record we discuss.

Abstract relations deserve special attention. Relevant details follow.

A *correspondence* Φ from X to Y is an ordered 3-tuple $\Phi := (F, X, Y)$, where F is some subset of the product $X \times Y$. Clearly, $\mathrm{Rel}(F)$ holds. It is in common parlance to say that F is the *graph* of Φ, in symbols, $\mathrm{Gr}(\Phi) = F$; while X is the *domain of departure* and Y is the *domain of arrival* or *target* of Φ. Recall that a *relation* or a *binary relation* on X is a correspondence whose domain of departure and target are the same set X.

The *image* of $A \subset X$ under Φ is the projection of $(A \times Y) \cap F$ to Y. The image of A under F is denoted by $\Phi(A)$ or simply $F(A)$. Thus,

$$\Phi(A) := F(A) := \{y \in Y : (\exists x \in A)((x, y) \in F)\}.$$

To define a correspondence Φ amounts to describing the mapping

$$\widetilde{\Phi} : x \mapsto \Phi(\{x\}) \in \mathscr{P}(Y) \quad (x \in X),$$

where $\mathscr{P}(Y)$ stands for the *powerset* of Y which is the collection of all subsets of Y. This enables us to identify a correspondence Φ with the mapping $\widetilde{\Phi}$. Abusing the language, we often identify the mapping $\widetilde{\Phi}$, the correspondence Φ, and the graph of Φ, denoting these three objects by the same letter.

The *domain of definition* or simply *domain* of Φ is the domain of the graph of Φ. In other words,

$$\mathrm{dom}(\Phi) := \{x \in X : \Phi(x) \neq \varnothing\}.$$

By analogy, the image of a correspondence is the image of its graph.

A.1.9. Assume that X and Y are abstract relations; i.e., $\mathrm{Rel}(X)$ and $\mathrm{Rel}(Y)$. We may arrange the *composition* of X and Y, denoted by the symbol $Y \circ X$, by collecting all ordered pairs (x, z) such that $(x, y) \in X$ and $(y, z) \in Y$ for some y:

$$(\forall u)(u \in Y \circ X \leftrightarrow (\exists x)(\exists y)(\exists z)(x, y) \in X \wedge (y, z) \in Y \wedge u = (x, z)).$$

The *inverse* of X, in symbols X^{-1}, is defined as

$$(\forall u)(u \in X^{-1} \leftrightarrow (\exists x)(\exists y)(x, y) \in X \wedge u = (y, x)).$$

The symbol I_X denotes the *identity relation* or the *identity mapping* on X, i.e.,

$$(\forall u)(u \in I_X \leftrightarrow (\exists x)(x \in X \wedge u = (x, x))).$$

We elaborate the above for correspondences.

So, assume that $\Phi := (F, X, Y)$ is a correspondence from X to Y. Assign $F^{-1} := \{(y, x) \in Y \times X : (x, y) \in F\}$. The correspondence $\Phi^{-1} := (F^{-1}, Y, X)$ is the *inverse* of Φ. Consider another correspondence $\Psi := (G, Y, Z)$. Denote by H the image of $(F \times Z) \cap (X \times G)$ under the mapping $(x, y, z) \mapsto (x, z)$. Clearly,

$$H = \{(x, z) \in X \times Z : (\exists y \in Y)((x, y) \in F \wedge (y, z) \in G)\}.$$

Hence, H coincides with the composition $G \circ F$ of the graphs G and F. The correspondence $\Psi \circ \Phi := (G \circ F, X, Z)$ is the *composition*, or *composite*, or *superposition* of Φ and Ψ. We have the following obvious equalities:

$$(\Psi \circ \Phi)^{-1} = \Phi^{-1} \circ \Psi^{-1}, \quad \Theta \circ (\Psi \circ \Phi) = (\Theta \circ \Psi) \circ \Phi.$$

A few words about another abbreviation related to correspondences. Consider $\Phi := (F, X, Y)$. The *polar* $\pi_\Phi(A)$ of $A \subset X$ under Φ is the collection of all $y \in Y$ satisfying $A \times \{y\} \subset \mathscr{F}$. In other words,

$$\pi_\Phi(A) := \pi_F(A) := \{y \in Y : (\forall x \in A)((x, y) \in F)\}.$$

If Φ is fixed then we abbreviate $\pi_\Phi(A)$ to $\pi(A)$ and $\pi_{\Phi^{-1}}(A)$ to $\pi^{-1}(A)$.

The simplest properties of polars are as follows:

(1) If $A \subset B \subset X$ then $\pi(A) \supset \pi(B)$;

(2) For every $A \subset X$ the inclusions hold:

$$A \subset \pi^{-1}(\pi(A)); \quad A \times \pi(A) \subset F;$$

(3) If $A \times B \subset F$ then $B \subset \pi(A)$ and $A \subset \pi^{-1}(B)$;

(4) If $(A_\xi)_{\xi \in \Xi}$ is a nonempty family of subsets of X then $\pi(\bigcup_{\xi \in \Xi} A_\xi) = \bigcap_{\xi \in \Xi} \pi(A_\xi)$;

(5) If $A \subset X$ and $B \subset Y$ then $\pi(A) = \pi(\pi^{-1}(\pi(A)))$ and $\pi^{-1}(B) = \pi^{-1}(\pi(\pi^{-1}(B)))$.

A.1.10. Provided that $\mathrm{Rel}(X) \wedge ((X \cap Y^2) \circ (X \cap Y^2) \subset X)$, we call X a *transitive* relation on Y. A relation X is *reflexive* (over Y) if $\mathrm{Rel}(X) \wedge (I_Y \subset X)$. A relation X is *symmetric* if $X = X^{-1}$. Finally, we say that "X is an *antisymmetric* relation on Y" if $\mathrm{Rel}(X) \wedge ((X \cap X^{-1}) \cap Y^2 \subset I_Y)$. As usual, we use the conventional abbreviation $Y^2 := Y \times Y$.

A reflexive and transitive relation on Y is a *preorder* on Y. An antisymmetric preorder on Y is an *order* or *ordering* on Y. A symmetric preorder is an *equivalence*.

Other terms are also applied that are now in common parlance. Recall in particular that an order X on Y is *total* or *linear*, while Y itself is called a *chain* (relative to X) whenever $Y^2 \subset X \cup X^{-1}$. If each nonempty subset of the set Y has a least element (relative to the order of X) then we say that X *well orders* Y or that Y is *well ordered* with respect to the order of X.

A.1.11. Quantifiers are *bounded* if they appear in the text as $(\forall x \in y)$ or $(\exists x \in y)$. The formulas of set theory (and, generally speaking, of every first-order theory) are classified according to how they use bounded and unbounded quantifiers.

Of especial importance to our exposition are the class of bounded formulas or Σ_0-*formulas* and the class of the so-called Σ_1-*formulas*. Recall that a formula φ is *bounded* provided that each quantifier in φ is bounded. Say that φ is of *class* Σ_1 or a Σ_1-*formula* if φ results from atomic formulas and their negations by using only the logical operations $\wedge$, $\vee$, $(\forall x \in y)$, and $(\exists x)$.

Clearly, every bounded formula is of class σ_1. However, it is false that every σ_1-formula is bounded. Moreover, there are formulas not belonging to the class σ_1. The corresponding examples follow. We start with bounded formulas.

A.1.12. The proposition $z = \{x, y\}$ amounts to the bounded formula

$$x \in z \wedge y \in z \wedge (\forall u \in z)(u = x \vee u = y).$$

So, the definition of ordered pair is a bounded formula. The same holds for the definition of product since we may rewrite $Z = X \times Y$ as

$$(\forall z \in Z)(\exists x \in X)(\exists y \in Y)(z = (x, y))$$
$$\wedge(\forall x \in X)(\forall y \in Y))\ (\exists z \in Z)\ (z = (x, y)).$$

Another bounded formula reads "a mapping F from X to Y" (see A.1.8). Indeed, the above shows that $F \subset X \times Y$ is a bounded formula. Moreover, bounded are the expressions $\mathrm{dom}(F) = X$ and $\mathrm{Un}(F)$, equivalent to the respective formulas

$$(\forall x \in X)(\exists y \in Y)(\exists z \in F)(z = (x, y));$$
$$(\forall z_1 \in F)(\forall z_2 \in F)(\forall x \in X)(\forall y_1 \in Y)(\forall y_2 \in Y)$$

$$(z_1 = (x, y_1) \wedge z_2 = (x, y_2) \to y_1 = y_2).$$

A.1.13. The statements, that x and y are ***equipollent***, or ***equipotent***, or x and y have the same ***cardinality*** (symbolically, $x \simeq y$), each implying that there is a bijection between x and y, are all equivalent to the following Σ_1-formula:

$$(\exists f)(f : x \to y \wedge \mathrm{im}(f) = y \wedge \mathrm{Un}(f^{-1})).$$

However, this fact is not expressible by a bounded formula. The notion of abstract relation gives another Σ_1-formula:

$$\mathrm{Rel}(X) := (\forall u \in X)(\exists v)(\exists w)(u = (v, w)).$$

Out of the class σ_1 lies the following formula stating that a set y is equipollent to none of its members:

$$(\forall x \in y)\ \neg(x \simeq y).$$

A.1.14. COMMENTS.

(1) It goes without saying that we may vary not only the special axioms of a first-order theory (see A.1.4) but also its logical part, i.e., the logical axioms and rules of inference. The collections of the so-resulting theorems may essentially differ from each other. For instance, eliminating the law of the excluded middle from the axioms of propositional calculus, we arrive at intuitionistic propositional calculus. Intuitionistic predicate calculus (see [60, 90]) appears in a similar way.

(2) The modern formal logic was grown in the course of the evolution of philosophical and mathematical thought with immense difficulties. The classical predicate calculus originates with the Aristotle syllogistic whereas the origin of intuitionistic logic belongs elsewhere. Other logical systems, different essentially from the two systems, were invented in various times for various purposes. For instance, an ancient Indian logic had three types of negation, expressing the ideas: something has never exist and cannot happen now, something was but is absent now, and something happens now but will disappear soon.

(3) As is seen from A.1.6 and A.1.7, abbreviations may appear in formulas, in other abbreviations, in abbreviations of abbreviations, etc. Invention of abbreviating symbols is an art in its own right, and as such it can never be formalized completely. Nevertheless, systematization and codification of the rules for abbreviation is at the request of both theory and practice. Some of these systems of rules (exact descriptions, introduction of function letters, etc.) can be found in the literature [29, 77, 108].

A.2. Zermelo–Fraenkel Set Theory

As has been noted in A.1.4, the axioms of set theory include the general logical axioms of predicate calculus which fix the classical rules for logical inference. Below we list the special axioms of set theory, ZF_1–ZF_6 and AC. The theory, proclaiming ZF_1–ZF_6 as special axioms, is called *Zermelo–Fraenkel set theory* and denoted by ZF. Enriching ZF with the axiom of choice AC, we come to a wider theory denoted by ZFC and still called Zermelo–Fraenkel set theory. Note that we supply the formal axioms below with their verbal statements in the wake of Cantor's ideas of sets.

A.2.1. We often encounter the terms "*property*" and "*class*" dealing with ZFC. We now elucidate their formal status. Consider a formula $\varphi = \varphi(x)$ of ZFC (in symbols, $\varphi \in (\mathrm{ZFC})$). Instead of the text $\varphi(y)$ we write $y \in \{x : \varphi(x)\}$. In other words, we use the so-called *Church schema* for classification:

$$y \in \{x : \varphi(x)\} := \varphi(y).$$

The expression $y \in \{x : \varphi(x)\}$ means in the language of ZFC that y has the property φ or, in other words, y belongs to the class $\{x : \varphi(x)\}$. Bearing this in mind, we say that a property, a formula, and a class mean the same in ZFC. We has already applied the Church schema in A.1.6 and A.1.7. Working with ZFC, we conveniently use many current abbreviations:

$$\mathbf{U} := \{x : x = x\} \text{ is the } \textit{universe of discourse} \text{ or the } \textit{class of all sets};$$
$$\{x : \varphi(x)\} \in \mathbf{U} := (\exists z)(\forall y)\varphi(y) \leftrightarrow y \in z;$$
$$\{x : \varphi(x),\ \psi(x)\} := \{x : \varphi(x)\} \cap \{x : \psi(x)\};$$
$$x \cup y := \cup\{x, y\}, \quad x \cap y \cap z := \cap\{x, y, z\} \dots .$$

We are now ready to formulate the special axioms of ZFC.

A.2.2. *Axiom of Extensionality* $\mathbf{ZF_1}$. *Two sets are equal if and only if they consist of the same elements*:

$$(\forall x)(\forall y)(\forall z)((z \in x \leftrightarrow z \in y) \leftrightarrow x = y).$$

Note that we may replace the last equivalence by $\rightarrow$ without loss of scope, since the reverse implication is a theorem of predicate calculus.

A.2.3. *Axiom of Union* $\mathbf{ZF_2}$. *The union of a set of sets is also a set*:

$$(\forall x)(\exists y)(\forall z)(\exists u)((u \in z \wedge z \in x) \leftrightarrow z \in y).$$

With the abbreviations of A.1.6 and A.2.1, ZF_2 takes the form

$$(\forall x) \ \cup x \in \mathbf{U}.$$

A.2.4. Axiom of Powerset ZF_3. *All subsets of every set comprise a new set:*

$$(\forall x)(\exists y)(\forall z)(z \in y \leftrightarrow (\forall u)(u \in z \to u \in x)).$$

In short,

$$(\forall x)\mathscr{P}(x) \in \mathbf{U}.$$

This axiom is also referred to as the *axiom of powers*.

A.2.5. Axiom of Replacement $ZF_{4\varphi}$. *The image of a set under every bijective mapping is a set again:*

$$(\forall x)(\forall y)(\forall z)(\varphi(x,y)) \wedge \varphi(x,z) \to y = z)$$
$$\to (\forall a)(\exists b)((\exists s \in x)(\exists t)\varphi(s,t) \leftrightarrow t \in y).$$

In short,

$$(\forall x)(\forall y)(\forall z)(\varphi(x,y) \wedge \varphi(x,z) \to y = z)$$
$$\to (\forall a)(\{v : (\exists u \in a)\varphi(u,v)\} \in \mathbf{U}).$$

Here φ is a formula of ZFC containing no free occurrences of a. Note that $ZF_{4\varphi}$ is a schema for infinitely many axioms since a separate axiom appears with an arbitrary choice of $\varphi \in$ (ZFC). Bearing in mind this peculiarity, we often abstain from using a more precise term "axiom-schema" and continue speaking about the axiom of replacement for the sake of brevity and uniformity.

Note a few useful corollaries of $ZF_{4\varphi}$.

A.2.6. Let $\psi = \psi(z)$ be a formula of ZFC. Given a set x, we may arrange a subset of x by selecting the members of x with the property ψ, namely,

$$(\forall x)\{z \in x : \psi(x)\} \in \mathbf{U}.$$

Our claim is $ZF_{4\varphi}$, with $\psi(u) \wedge (u = v)$ playing the role of φ. This particular form of the axiom of replacement is often called the *axiom of separation* or *comprehension.*

A.2.7. Applying $ZF_{4\varphi}$ with the formula

$$\varphi(u,v) := (u = \varnothing \to v = x) \wedge (u \neq \varnothing \to v = y)$$

to the set $z := \mathscr{P}(\mathscr{P}(\varnothing))$, we deduce that the unordered pair $\{x,y\}$ of two sets (cf. A.1.7) is also a set. This assertion is often referred to as the *axiom of pairing*.

A.2.8. *Axiom of Infinity* ZF$_5$. *There is at least one infinite set:*

$$(\exists x)(\varnothing \in x \wedge (\forall y)(y \in x \to y \cup \{y\} \in x)).$$

In other words, there is a set x such that $\varnothing \in x$, $\{\varnothing\} \in x$, $\{\varnothing, \{\varnothing\}\} \in x$, $\{\varnothing, \{\varnothing\}, \{\varnothing, \{\varnothing\}\}\} \in x$, etc. The cute reader will observe a tiny gap between formal and informal statements of the axiom of infinity. The vigilant reader might suspect the abuse of the term "infinity." In fact, the axiom of infinity belongs to the basic Cantorian doctrines and so some mystery is inevitable and welcome in this respect.

A.2.9. *Axiom of Regularity* ZF$_6$. *Each nonempty set x has a member having no common elements with x:*

$$(\forall x)(x \neq \varnothing \to (\exists y)\ (y \in x \wedge y \cap x = \varnothing)).$$

Another name for the axiom of regularity is the *axiom of foundation*.

Applying ZF$_6$ to a *singleton*, i.e., a one-point set $x := \{y\}$, we see that $y \notin y$. Speaking a bit prematurely, we may note, on taking $x := \{x_1, \dots, x_n\}$, that there are no infinitely decreasing $\in$-sequences $x_1 \ni x_2 \ni \dots \ni x_n \ni \dots$.

A.2.10. *Axiom of Choice* AC. *To each set x there is a choice function on x; i.e., a single-valued correspondence assigning an element of X to each nonempty member of X; i.e.,*

$$(\forall x)(\exists f)(\mathrm{Fnc}\,(f) \wedge x \subset \mathrm{dom}(f)) \wedge (\forall y \in x) y \neq \varnothing \to f(y) \in y.$$

Set theory has many propositions equivalent to AC (cf. [84]). We recall the two most popular among them.

Zermelo Theorem (the well-ordering principle). *Every set may be well ordered.*

Kuratowski–Zorn Lemma (the maximality principle). *Let M be a (partially) ordered set whose every chain has an upper bound. Then to every $x \in M$ there is a maximal element $m \in M$ such that $m \geq x$.*

A.2.11. The axiomatics of ZFC enables us to find a concrete presentation for the class of all sets in the form of the "von Neumann universe." The starting point of our construction is the empty set. An elementary step consists in forming the union of sets of the subsets of available sets, thus making new sets from those available. Transfinite repetition of these steps exhausts the class of all sets. Classes (in a "Platonic" sense) may be viewed as external objects lying beyond the von Neumann universe. Pursuing this approach, we consider a class as a family of sets obeying a set-theoretic property given by a formula of Zermelo–Fraenkel theory. Therefore, the class consisting of some members of a certain set is a set itself (by the axiom of replacement). A formally sound definition of the von Neumann universe requires preliminary acquaintance with the notions of ordinal and cumulative hierarchy. Below we give a minimum of information on these objects sufficient for a "naive" definition. A more explicit presentation is given in Section 1.5.

A.2.12. A set x is *transitive* if every member of x is a subset of x. A set x is an *ordinal* if x is transitive and totally ordered by the membership relation $\in$. These definitions look in symbolic form as follows:

$$\begin{gathered}\mathrm{Tr}\,(x) := (\forall\, y \in x)(y \subset x) := \text{ "x is a transitive set"};\\ \mathrm{Ord}\,(x) := \mathrm{Tr}\,(x) \wedge (\forall\, y \in x)(\forall\, z \in x)\\ (y \in z \vee z \in y \vee z = y) := \text{ "x is an ordinal."}\end{gathered}$$

We commonly denote ordinals by lower case Greek letters. Every ordinal is endowed with the natural order by membership: given $\beta,\ \gamma \in \alpha$, we put

$$\gamma \leq \beta \leftrightarrow \gamma \in \beta \vee \gamma = \beta.$$

The class of all ordinals is denoted by On. So, $\mathrm{On} := \{\alpha : \mathrm{Ord}\,(\alpha)\}$.

An ordinal is a *well ordered set*; i.e., it is totally ordered and its every subset has the least element (which is ensured by the axiom of regularity). We can easily see that

$$\begin{gathered}\alpha \in \mathrm{On} \wedge \beta \in \mathrm{On} \to \alpha \in \beta \vee \alpha = \beta \vee \beta \in \alpha;\\ \alpha \in \mathrm{On} \wedge \beta \in \alpha \to \beta \in \mathrm{On};\\ \alpha \in \mathrm{On} \to \alpha \cup \{\alpha\} \in \mathrm{On};\\ \mathrm{Ord}\,(\varnothing).\end{gathered}$$

The ordinal $\alpha + 1 := \alpha \cup \{\alpha\}$ is called the *successor* of α or the *son* of α. A nonzero ordinal other than a successor is a *limit* ordinal. The following notation is common:

$$\begin{gathered}K_{\mathrm{I}} := \{\alpha \in \mathrm{On} : (\exists\, \beta)\ \mathrm{Ord}\,(\beta) \wedge \alpha = \beta + 1 \vee \alpha = \varnothing\};\\ K_{\mathrm{II}} := \{\alpha \in \mathrm{On} : \alpha \text{ is a limit ordinal}\};\\ 0 := \varnothing, \quad 1 := 0 + 1, \quad 2 := 1 + 1, \dots,\\ \omega := \{0, 1, 2, \dots\}.\end{gathered}$$

This is a right place to recall that the *continuum* we talk about so much in this book is simply the powerset of ω.

A.2.13. It is worth observing that ZFC enables us to prove the properties of ordinals well known at a naive level. In particular, ZFC legitimizes transfinite induction and recursion. We now define the von Neumann universe, purposefully omitting formalities.

Given an ordinal α, put

$$V_\alpha := \bigcup_{\beta<\alpha} \mathscr{P}(V_\beta),$$

i.e., $V_\alpha = \{x : (\exists\, \beta)\ (\beta \in \alpha \wedge x \subset V_\beta)\}$. More explicitly,

$$V_0 := \varnothing;$$
$$V_{\alpha+1} := \mathscr{P}(V_\alpha);$$
$$V_\beta := \bigcup_{\alpha<\beta} V_\alpha, \quad \text{if} \quad \beta \in K_{\mathrm{II}}.$$

Assign

$$\mathbf{V} := \bigcup_{\alpha\in\mathrm{On}} V_\alpha.$$

Of principal importance is the following theorem, ensuing from the axiom of regularity:

$$(\forall\, x)\ (\exists\, \alpha)\ (\mathrm{Ord}\,(\alpha) \wedge x \in V_\alpha).$$

In shorter symbols,

$$\mathbf{U} = \mathbf{V}.$$

Alternatively, we express this fact as follows: "*The class of all sets is the von Neumann universe*," or "*every set is well founded*."

The von Neumann universe $\mathbf{V}$, also called the *sets*, is customarily viewed as a pyramid "upside down," that is, a pyramid standing on its vertex which is the empty set. It is helpful to look at a few "lower floors" of the von Neumann universe:

$$V_0 = \varnothing,\ V_1 = \{\varnothing\},\ V_2 = \{\varnothing, \{\varnothing\}\}, \dots,$$
$$V_\omega = \{\varnothing, \{\varnothing\}, \{\varnothing, \{\varnothing\}\}, \dots\}, \dots .$$

A.2.14. The representation of the von Neumann universe $\mathbf{V}$ as the "*cumulative hierarchy*" of $(V_\alpha)_{\alpha\in\mathrm{On}}$ makes it possible to introduce the concept of the *ordinal rank* or simply the *rank* of a set. Namely, given a set x, put

$$\mathrm{rank}(x) := \text{ a least ordinal } \alpha \text{ such that } x \in V_{\alpha+1}.$$

It is easy to prove that

$$a \in b \to \operatorname{rank}(a) < \operatorname{rank}(b);$$
$$\operatorname{Ord}(\alpha) \to \operatorname{rank}(\alpha) = \alpha;$$
$$(\forall x)(\forall y) \operatorname{rank}(y) < \operatorname{rank}(x) \to (\varphi(y) \to \varphi(x)) \to (\forall x)\varphi(x),$$

where φ is a formula of ZFC. The preceding theorem (or, more precisely, the schema of theorems) is called the *principle of induction on rank.*

A.2.15. COMMENTS.

(1) E. Zermelo suggested in 1908 an axiomatics that coincides practically with ZF_1–ZF_3, ZF_5, A.2.5, and A.2.6. This system, together with the B. Russel theory of types, is among the first formal axiomatics for set theory.

The axioms of extensionality ZF_1 and union ZF_2 were proposed earlier by G. Frege (1883) and G. Cantor (1899). The idea of the axiom of infinity ZF_5 belongs to J. W. R. Dedekind.

(2) The axiom of choice AC seems to be in use implicitly for a long time before it was distinguished by G. Peano in 1890 and B. Levy in 1902. This axiom was formally introduced by E. Zermelo in 1904 and remained most disputable for many years. The axiom of choice is part and parcel of the most vital fragments of modern mathematics. So, it is no wonder that AC is accepted by the working majority of present-day mathematicians. Discussions of the place and role of the axiom of choice may be found elsewhere [30, 55, 59, 84, 153].

(3) The axiomatics of ZFC was completely elaborated at the beginning of the 1920s. By that time the formalization of the set-theoretic language had been completed, which made it possible to clarify the vague description of the type of properties admissible in the axiom of comprehension. On the other hand, the Zermelo axioms do not yield the Cantor claim that each bijective image of a set is a set. This drawback was eliminated by A. Fraenkel in 1922 and T. Scolem in 1923 who suggested variations of the axiom of replacement. This moment seems to pinpoint the birth of ZFC.

(4) The axiom of foundation ZF_6 was in fact suggested by von Neumann in 1925. This axiom is independent of the other axioms of ZFC.

(5) The system of the axioms of ZFC is infinite as noted in A.2.4. Absence of finite axiomatizability for ZFC was proven by R. Montague in 1960 (see [55, 73, 153, 254]).

A.3. Categories and Functors

Category theory, alongside set theory, provides a universal language for contemporary mathematics. The present book uses categories and functors as convenient tools for treating various mathematical constructions and arguments in a uniform manner when we formulate general properties of mathematical structures. We just sketch the basic concepts of category theory. Details may be found elsewhere; e.g., [21, 163, 244].

A.3.1. A category $\mathscr{K}$ consists of the classes $\operatorname{Ob}\mathscr{K}$, $\operatorname{Mor}\mathscr{K}$, and Com, called the *class of objects*, *class of morphisms*, and *law of composition* of $\mathscr{K}$ and satisfying the conditions:

(1) There are mappings $\mathscr{D}$ and $\mathscr{R}$ from $\operatorname{Mor}\mathscr{K}$ to $\operatorname{Ob}\mathscr{K}$ such that the class

$$H_{\mathscr{K}}(a,b) := \{\alpha \in \operatorname{Mor}\mathscr{K} : \mathscr{D}(\alpha) = a, \mathscr{R}(\alpha) = b\},$$

called the *class of morphisms* from a to b, is a set for all $a, b \in \operatorname{Ob}\mathscr{K}$;

(2) Com is an associative partial binary operation on $\operatorname{Mor}\mathscr{K}$ satisfying

$$\operatorname{dom}(\mathrm{Com}) = \big\{(\alpha,\beta) \in (\operatorname{Mor}\mathscr{K}) \times (\operatorname{Mor}\mathscr{K}) : \mathscr{D}(\beta) = \mathscr{R}(\alpha)\big\};$$

(3) To each object $a \in \operatorname{Ob}\mathscr{K}$, there is a morphism 1_a called the *identity morphism* of a such that $\mathscr{D}(1_a) = a = \mathscr{R}(1_a)$ and, moreover, $\mathrm{Com}(1_a, \alpha) = \alpha$ for $\mathscr{R}(\alpha) = a$ and $\mathrm{Com}(\beta, 1_a) = \beta$ for $\mathscr{D}(\beta) = a$.

Clearly, the class $\operatorname{Mor}\mathscr{K}$ is the union of the sets $H_{\mathscr{K}}(a,b)$, where a and b range over $\operatorname{Ob}\mathscr{K}$, and the sets $H_{\mathscr{K}}(a,b)$ and $H_{\mathscr{K}}(c,d)$ are disjoint for $(a,b) \neq (c,d)$. Given $\alpha, \beta \in \operatorname{Mor}\mathscr{K}$, we usually write $\beta \circ \alpha$ or $\beta\alpha$ instead of $\mathrm{Com}(\alpha,\beta)$. The containment $\alpha \in H_{\mathscr{K}}(a,b)$ is often written down as $\alpha : a \to b$; in words,“α is a morphism from a to b.”

A category $\mathscr{H}$ is a *subcategory* of a category $\mathscr{K}$ if the following are satisfied:

(1) $\operatorname{Ob}\mathscr{H} \subset \operatorname{Ob}\mathscr{K}$ and $H_{\mathscr{H}}(a,b) \subset H_{\mathscr{K}}(a,b)$ for every pair $a, b \in \operatorname{Ob}\mathscr{H}$;

(2) the composition of $\mathscr{H}$ is the restriction of the composition of $\mathscr{K}$ to the class $(\operatorname{Mor}\mathscr{H}) \times (\operatorname{Mor}\mathscr{H})$. In this event the identity morphism of each object $a \in \operatorname{Ob}\mathscr{H}$ coincides with the identity morphism of this object in the category $\mathscr{K}$.

A subcategory $\mathscr{H}$ of a category $\mathscr{K}$ is *full* provided that $H_{\mathscr{K}}(a,b) = H_{\mathscr{H}}(a,b)$ for all $a, b \in \operatorname{Ob}\mathscr{H}$.

The *product* $\mathscr{H} \times \mathscr{K}$ of categories $\mathscr{H}$ and $\mathscr{K}$ is defined by the formulas:

$$\operatorname{Ob}\mathscr{H} \times \mathscr{K} := (\operatorname{Ob}\mathscr{H}) \times (\operatorname{Ob}\mathscr{K});$$
$$H_{\mathscr{H}\times\mathscr{K}}((a,b),(a',b')) := H_{\mathscr{H}}(a,a') \times H_{\mathscr{K}}(b,b'),$$
$$(\alpha',\beta') \circ (\alpha,\beta) := (\alpha'\alpha, \beta'\beta),$$

where $a, a' \in \operatorname{Ob} \mathscr{H}$; $b, b' \in \operatorname{Ob} \mathscr{K}$; $\alpha, \alpha' \in \operatorname{Mor} \mathscr{H}$, and $\beta, \beta' \in \operatorname{Mor} \mathscr{K}$.

The *dual category* $\mathscr{K}^{\circ}$ of an arbitrary category $\mathscr{K}$ has the same objects and morphisms as $\mathscr{K}$. The law of composition Com° of the category $\mathscr{K}^{\circ}$ is defined by the rule

$$(\alpha, \beta, \gamma) \in \mathrm{Com}^{\circ} \leftrightarrow (\beta, \alpha, \gamma) \in \mathrm{Com}.$$

In applications the classes of objects and morphisms of a category may and usually do intersect. However, we lose no generality in assuming that these classes are disjoint in every category. Indeed, we may mark every object with some extra label, thus distinguishing objects from morphisms. We presume this agreement effective throughout.

A.3.3. Consider two categories $\mathscr{H}$ and $\mathscr{K}$. A *covariant functor* $\mathscr{F} : \mathscr{H} \to \mathscr{K}$ from $\mathscr{H}$ to $\mathscr{K}$ is a mapping whose domain comprises all objects and morphisms of $\mathscr{K}$ and which satisfies the conditions:

(1) If $\alpha : a \to b$ is a morphism of $\mathscr{H}$ then $\mathscr{F}(\alpha) : \mathscr{F}(a) \to \mathscr{F}(b)$;

(2) If $\alpha : a \to b$ and $\beta : b \to c$ are morphisms of $\mathscr{H}$ then $\mathscr{F}(\beta\alpha) = \mathscr{F}(\beta)\mathscr{F}(\alpha)$;

(3) If $a \in \operatorname{Ob} \mathscr{H}$ then $\mathscr{F}(1_a) = 1_{\mathscr{F}(a)}$.

Hence, given a pair of objects $a, b \in \operatorname{Ob} \mathscr{H}$, a functor $\mathscr{F}$ defines the mapping $\mathscr{F}_{a,b} : H_{\mathscr{H}}(a, b) \to H_{\mathscr{K}}(a, b)$. If $\mathscr{F}_{a,b}$ is injective (surjective) for all a and b then $\mathscr{F}$ is a *faithful* (*full*) functor. A covariant functor from $\mathscr{H}^{\circ}$ to $\mathscr{K}$ (or from $\mathscr{H}$ to $\mathscr{K}^{\circ}$) is a *contravariant functor* from $\mathscr{H}$ to $\mathscr{K}$.

A.3.4. Let $\mathscr{H}$ and $\mathscr{K}$ be categories. Assume given covariant functors $\mathscr{F} : \mathscr{H} \to \mathscr{K}$ and $\mathscr{G} : \mathscr{H} \to \mathscr{K}$. A *natural transformation* of $\mathscr{F}$ to $\mathscr{G}$, in symbols $\varphi : \mathscr{F} \to \mathscr{G}$, is a mapping $\varphi : \operatorname{Ob} \mathscr{H} \to \operatorname{Mor} \mathscr{K}$ such that

(1) $\varphi_a := \varphi(a) \in H_{\mathscr{K}}(\mathscr{F}(a), \mathscr{F}(b))$ for all $a \in \operatorname{Ob} \mathscr{H}$;

(2) for each morphism $\alpha : a \to b$ of $\mathscr{H}$ the following diagram commutes

$$\begin{array}{ccc} \mathscr{F}(a) & \xrightarrow{\varphi_a} & \mathscr{G}(a) \\ {\scriptstyle \mathscr{F}(\alpha)}\downarrow & & \downarrow{\scriptstyle \mathscr{G}(\alpha)} \\ \mathscr{F}(b) & \xrightarrow[\varphi_b]{} & \mathscr{G}(b) \end{array}$$

In other words, $\mathscr{G}(\alpha)\varphi_a = \varphi_b\mathscr{F}(\alpha)$. In this event, φ is also called a *functor morphism*.

A natural transformation $\varphi : \mathscr{F} \to \mathscr{G}$ is a *natural equivalence* of $\mathscr{F}$ and $\mathscr{G}$, or a *functor equivalence*, or a *functor isomorphism* between $\mathscr{F}$ and $\mathscr{G}$ provided that φ_a is an isomorphism in $\mathscr{K}$ for every $a \in \operatorname{Ob} \mathscr{H}$. The mappings φ_a^{-1} give rise to the natural transformation of $\mathscr{G}$ to $\mathscr{F}$ which is denoted by φ^{-1}. Recall that a morphism $\alpha : a \to b$ is an *isomorphism* if there is a morphism $\beta : b \to a$ satisfying $\alpha\beta = 1_b$ and $\beta\alpha = 1_a$.

A.3.5. Categories $\mathscr{H}$ and $\mathscr{K}$ are *equivalent* if there are functors $\mathscr{F} : \mathscr{H} \to \mathscr{K}$ and $\mathscr{G} : \mathscr{K} \to \mathscr{H}$ such that the functor $\mathscr{F}\mathscr{G}$ is naturally equivalent to the identity functor $I_{\mathscr{H}}$, while the functor $\mathscr{G}\mathscr{F}$ is naturally equivalent to the identity functor $I_{\mathscr{K}}$. Say that each of the functors $\mathscr{F}$ and $\mathscr{G}$ *implements equivalence* or is an *equivalence* between $\mathscr{H}$ and $\mathscr{K}$; the latter usage slightly abuses the language, of course.

The equivalence relation between categories is reflexive, symmetric, and transitive.

A.3.6. *Categories $\mathscr{H}$ and $\mathscr{K}$ are equivalent if and only if there is a full and faithful functor $\mathscr{F}$ from $\mathscr{H}$ to $\mathscr{K}$ such that, to each object $b \in \operatorname{Ob}\mathscr{K}$, there corresponds an isomorphic object of the type $\mathscr{F}(a)$, where $a \in \operatorname{Ob}\mathscr{H}$.*

A.3.7. Consider two functors $\mathscr{F} : \mathscr{H} \to \mathscr{K}$ and $\mathscr{G} : \mathscr{K} \to \mathscr{H}$. Assign to these functors another two functors $H^{\mathscr{F}}$ and $H_{\mathscr{G}}$ from the category $\mathscr{H}^{\circ} \times \mathscr{K}$ to the category of sets and mappings. Namely, given $a \in \operatorname{Ob}\mathscr{H}$, $b \in \operatorname{Ob}\mathscr{K}$, $\alpha \in H_{\mathscr{H}}(a, a')$, and $\beta \in H_{\mathscr{K}}(b, b')$, put

$$
\begin{gathered}
H^{\mathscr{F}}(a, b) := H_{\mathscr{K}}(\mathscr{F}(a), b), H_{\mathscr{G}}(a, b) := H_{\mathscr{H}}(a, \mathscr{G}(b)), \\
H^{\mathscr{F}}(\alpha, \beta) : f \to \beta f \mathscr{F}(\alpha), H_{\mathscr{G}}(\alpha, \beta) : g \to \mathscr{G}(\beta) g \alpha,
\end{gathered}
$$

where $f \in H_{\mathscr{K}}(\mathscr{F}(\alpha), b)$ and $g \in H_{\mathscr{H}}(a, \mathscr{G}(b))$.

Say that the functors $\mathscr{F}$ and $\mathscr{G}$ are an *adjoint pair* if the functors $H^{\mathscr{F}}$ and $H_{\mathscr{G}}$ are isomorphic. In this event, $\mathscr{F}$ is *left adjoint* to $\mathscr{G}$, and $\mathscr{G}$ is *right adjoint* to $\mathscr{F}$.

Two left adjoints of $\mathscr{F}$ are naturally equivalent. This enables us to speak about *the* left adjoint of $\mathscr{F}$. The same relates to right adjoints.

The isomorphism $\varphi : H^{\mathscr{F}} \to H_{\mathscr{G}}$ is referred to as *adjunction*; while the inverse isomorphism φ^{-1}, as *coadjunction*.

A.3.8. Let $\mathscr{K}$ be a subcategory of a category $\mathscr{H}$. An object $b \in \operatorname{Ob}\mathscr{K}$ is a *$\mathscr{K}$-reflector* of an object $a \in \operatorname{Ob}\mathscr{H}$ provided that there is a morphism $\varphi : a \to b$, such that each morphism $\alpha : a \to c$, where $c \in \operatorname{Ob}\mathscr{K}$, has the form $\alpha = \varphi\beta$ with a uniquely determined morphism $\beta : b \to c$. Say that $\mathscr{K}$ is *reflective* if each object of $\mathscr{H}$ possesses a $\mathscr{K}$-reflector.

A.3.9. *A subcategory $\mathscr{K}$ of a category $\mathscr{H}$ is reflective if and only if the inclusion functor $\mathscr{K} \to \mathscr{H}$* has a right adjoint $\mathscr{R} : \mathscr{H} \to \mathscr{K}$.

The functor $\mathscr{R}$ is called the *$\mathscr{K}$-reflector* of $\mathscr{H}$.

A.3.10. By way of example, we consider Sets, the *category of sets*. The objects of Sets are all sets, while the morphisms of Sets are arbitrary mappings. Composition of morphisms in Sets is the routine composition of mappings. Clearly, given $f \in \operatorname{Mor}\mathrm{Sets}$ we see that $\mathscr{D}(f)$ and $\mathscr{R}(f)$ are the domain and target of f. The morphism 1_a is the identity mapping of a.

Various examples of categories appear as *subcategories of structured sets*. The objects of such a subcategory are sets furnished with some structure σ (which might include algebraic operations, relations, norms, topologies, etc.). The morphisms in this event are mappings that preserve the structure σ at least partly. Evidently, Sets is a category of structured sets with the empty structure.

A.3.11. It also stands to reason to consider a wider *category of sets and correspondences* Sets_*. The classes of objects of Sets and Sets_* are the same, whereas the morphisms of Sets_* are all available correspondences. The identity morphism on a set in Sets_* is the identity relation on A. Clearly, Sets_* is a category while Sets is a subcategory of Sets.

A.3.11. Comments.

Categories and functors were suggested by S. MacLane and S. Eilenberg in 1945 in connection with their research into homological algebra. In the subsequent decades, category theory expanded far beyond the limits of algebraic topology and began to play a visible role in various branches of mathematics. Our exposition deals with the minimum minimorum of categories and functors we need in Boolean valued analysis. More details about categories and functors may be found, for instance, in [21, 60, 244].

References

1. Akilov G. P. and Kutateladze S. S., Ordered Vector Spaces [in Russian], Nauka, Novosibirsk (1978).
2. Alfsen E. M., Shultz F. W., and Störmer E., "A Gelfand–Neumark theorem for Jordan algebras," Adv. in Math., **28**, No. 1, 11–56 (1978).
3. Aliprantis C. D. and Burkinshaw O., Locally Solid Riesz Spaces, Academic Press, New York etc. (1978).
4. Aliprantis C. D. and Burkinshaw O., Positive Operators, Academic Press, New York (1985).
5. Arens R. F. and Kaplansky I., "Topological representation of algebras," Trans. Amer. Math. Soc., **63**, No. 3, 457–481 (1948).
6. Arveson W., An Invitation to C^*-Algebras, Springer-Verlag, Berlin (1976).
7. Ayupov S. A., "Jordan operator algebras," in: Mathematical Analysis [in Russian], VINITI, Moscow, 1985, pp. 67–97. (Itogi Nauki i Tekhniki, **27**.)
8. Ayupov S. A., Classification and Representation of Ordered Jordan Algebras [in Russian], Fan, Tashkent (1986).
9. Beĭdar K. I. and Mikhalëv A. V., "Orthogonal completeness and algebraic systems," Uspekhi Mat. Nauk, **40**, No. 6, 79–115 (1985).
10. Bell J. L. and Slomson A. B., Models and Ultraproducts: an Introduction, North-Holland, Amsterdam etc. (1969).
11. Bell J. L., Boolean-Valued Models and Independence Proofs in Set Theory, Clarendon Press, New York etc. (1985).
12. Berberian S. K., Baer $*$-Rings, Springer-Verlag, Berlin (1972).
13. Bigard A., Keimel K., and Wolfenstein S., Groupes et Anneaux Réticulés [in French], Springer-Verlag, Berlin etc. (1977).
14. Birkhoff G., Lattice Theory, Amer. Math. Soc., Providence (1967).
15. Blumenthal L. M., Theory and Applications of Distance Geometry, Clarendon Press, Oxford (1953).
16. Boole G., An Investigation of the Laws of Thought on Which Are Founded the Mathematical Theories of Logic and Probabilities, Dover, New York (1957).

17. Boole G., Selected Manuscripts on Logic and Its Philosophy, Birkhäuser-Verlag, Basel (1997).

18. Bourbaki N., Set Theory [in French], Hermann, Paris (1958).

19. Bratteli O. and Robertson D., Operator Algebras and Quantum Statistical Mechanics. Vol. 1, Springer-Verlag, New York etc. (1982).

20. Browder F. (ed.), Mathematical Developments Arising from Hilbert Problems, Amer. Math. Soc., Providence (1976).

21. Bucur I. and Deleanu A., Introduction to the Theory of Categories and Functors, Wiley Interscience, London etc. (1968).

22. Bukhvalov A. V., "Order bounded operators in vector lattices and spaces of measurable functions," in: Mathematical Analysis [in Russian], VINITI, Moscow, 1988, pp. 3–63. (Itogi Nauki i Tekhniki, **26**.)

23. Bukhvalov A. V., Veksler A. I., and Geĭler V. A., "Normed lattices," in: Mathematical Analysis [in Russian], VINITI, Moscow, 1980, pp. 125–184. (Itogi Nauki i Tekhniki, **18**.)

24. Bukhvalov A. V., Veksler A. I., and Lozanovskiĭ G. Ya., "Banach lattices: some Banach aspects of the theory," Uspekhi Mat. Nauk, **34**, No. 2, 137–183 (1979).

25. Burden C. W. and Mulvey C. J., "Banach spaces in categories of sheaves," in: Applications of Sheaves (Proc. Res. Sympos. Appl. Sheaf Theory to Logic, Algebra and Anal., Univ. Durham, Durham, 1977), Springer-Verlag, Berlin, 1979, pp. 169–196.

26. Cantor G., Works on Set Theory [in Russian], Nauka, Moscow (1985).

27. Chang S. S. and Keisler H. J., Model Theory, North-Holland, Amsterdam etc. (1990).

28. Chilin V. I., "Partially ordered involutive Baer algebras," in: Contemporary Problems of Mathematics. Newest Advances [in Russian], VINITI, Moscow, **27**, 1985, pp. 99–128.

29. Church A., Introduction to Mathematical Logic, Princeton University Press, Princeton (1956).

30. Cohen P. J., Set Theory and the Continuum Hypothesis, Benjamin, New York etc. (1966).

31. Cohen P. J., "On foundations of set theory," Uspekhi Mat. Nauk, **29**, No. 5, 169–176 (1974).

32. Ciesielski K., Set Theory for the Working Mathematician, Cambridge University Press, Cambridge etc. (1997).

33. Dales H. and Woodin W., An Introduction to Independence for Analysts, Cambridge University Press, Cambridge (1987).

34. Day M., Normed Linear Spaces, Springer-Verlag, New York and Heidelberg (1973).

35. Diestel J., Geometry of Banach Spaces: Selected Topics, Springer-Verlag, Berlin etc. (1975).

36. Diestel J. and Uhl J. J., Vector Measures, Amer. Math. Soc., Providence (1977).

37. Dinculeanu N., Vector Measures, VEB Deutscher Verlag der Wissenschaften, Berlin (1966).

38. Dixmier J., C^*-Algebras and Their Representations [in French], Gauthier-Villars, Paris (1964).

39. Dixmier J., C^*-Algebras, North-Holland, Amsterdam etc. (1977).

40. Dixmier J., Algebras of Operators in Hilbert Space (Algebras of von Neumann) [in French], Gauthier-Villars, Paris (1996).

41. Dragalin A. G., "An explicit Boolean-valued model for nonstandard arithmetic," Publ. Math. Debrecen, **42**, No. 3–4, 369–389 (1993).

42. Dunford N. and Schwartz J. T., Linear Operators. Vol. 1: General Theory, John Wiley & Sons, New York (1988).

43. Dunford N. and Schwartz J. T., Linear Operators. Vol. 2: Spectral Theory. Selfadjoint Operators in Hilbert Space, John Wiley & Sons, New York (1988).

44. Dunford N. and Schwartz J. T., Linear Operators. Vol. 3: Spectral Operators, John Wiley & Sons, New York (1988).

45. Eda K., "Boolean power and a direct product of abelian groups," Tsukuba J. Math., **6**, No. 2, 187–194 (1982).

46. Eda K., "On a Boolean power of a torsion free abelian group," J. Algebra, **82**, No. 1, 84–93 (1983).

47. Ellis D., "Geometry in abstract distance spaces," Publ. Math. Debrecen, **2**, 1–25 (1951).

48. Ershov Yu. L. and Palyutin E. A., Mathematical Logic [in Russian], Nauka, Moscow (1987).

49. Espanol L., "Dimension of Boolean valued lattices and rings," J. Pure Appl. Algebra, No. 42, 223–236 (1986).

50. Faith C., Algebra: Rings, Modules and Categories. Vol. 1, Springer-Verlag, Berlin etc. (1981).

51. Foster A. L., "Generalized 'Boolean' theory of universal algebras. I. Subdirect sums and normal representation theorems," Math. Z., **58**, No. 3, 306–336 (1953).

52. Foster A. L., "Generalized 'Boolean' theory of universal algebras. II. Identities and subdirect sums of functionally complete algebras," Math. Z., **59**, No. 2, 191–199 (1953).

53. Fourman M. P., "The logic of toposes," in: Handbook of Mathematical Logic, North-Holland, Amsterdam (1977).

54. Fourman M. P. and Scott D. S., "Sheaves and logic," in: Applications of Sheaves (Proc. Res. Sympos. Appl. Sheaf Theory to Logic, Algebra and Anal., Univ. Durham, Durham, 1977), Springer-Verlag, Berlin, 1979, pp. 302–401.

55. Fraenkel A. and Bar-Hillel I., Foundations of Set Theory, North-Holland, Amsterdam (1958).

56. Fuchs L., Partially Ordered Algebraic Systems, Pergamon Press, Oxford (1963).

57. Georgescu G. and Voiculescu I., "Eastern model theory for Boolean-valued theories," Z. Math. Logik Grundlag. Math., No. 31, 79–88 (1985).

58. Gödel K.,"What is Cantor's continuum problem," Amer. Math. Monthly, **54**, No. 9, 515–525 (1947).

59. Gödel K., "Compatibility of the axiom of choice and the generalized continuum hypothesis with the axioms of set theory," Uspekhi Mat. Nauk, **8**, No. 1, 96–149 (1948).

60. Goldblatt R., Toposes. Categorical Analysis of Logic, North-Holland, Amsterdam etc. (1979).

61. Goodearl K. R., Von Neumann Regular Rings, Pitman, London (1979).

62. Gordon E. I., "Real numbers in Boolean-valued models of set theory and K-spaces," Dokl. Akad. Nauk SSSR, **237**, No. 4, 773–775 (1977).

63. Gordon E. I., "K-spaces in Boolean-valued models of set theory," Dokl. Akad. Nauk SSSR, **258**, No. 4, 777–780 (1981).

64. Gordon E. I., "To the theorems of identity preservation in K-spaces," Sibirsk. Mat. Zh., **23**, No. 5, 55–65 (1982).

65. Gordon E. I., Rationally Complete Semiprime Commutative Rings in Boolean Valued Models of Set Theory, Gor′kiĭ, VINITI, No. 3286-83 Dep (1983).

66. Gordon E. I., Elements of Boolean Valued Analysis [in Russian], Gor′kiĭ State University, Gor′kiĭ (1991).

67. Gordon E. I. and Lyubetskiĭ V. A., "Some applications of nonstandard analysis in the theory of Boolean valued measures," Dokl. Akad. Nauk SSSR, **256**, No. 5, 1037–1041 (1981).

68. Gordon E. I. and Morozov S. F., Boolean Valued Models of Set Theory [in Russian], Gor′kiĭ State University, Gor′kiĭ (1982).

69. Grätzer G., General Lattice Theory, Birkhäuser-Verlag, Basel (1978).

70. Grayson R. J., "Heyting-valued models for intuitionistic set theory," in: Applications of Sheaves (Proc. Res. Sympos. Appl. Sheaf Theory to Logic, Algebra and Anal., Univ. Durham, Durham, 1977), Springer-Verlag, Berlin, 1977.

71. Gutman A. E., "Banach fibering in lattice normed space theory," in: Linear Operators Compatible with Order [in Russian], Sobolev Institute Press,

Novosibirsk, 1995, pp. 63–211.

72. Gutman A. E., "Locally one-dimensional K-spaces and σ-distributive Boolean algebras," Siberian Adv. Math., **5**, No. 2, 99–121 (1995).

73. Hallet M., Cantorian Set Theory and Limitation of Size, Clarendon Press, Oxford (1984).

74. Halmos P. R., Lectures on Boolean Algebras, Van Nostrand, Toronto etc. (1963).

75. Hanshe-Olsen H. and Störmer E., Jordan Operator Algebras, Pitman Publ., Boston etc. (1984).

76. Hernandez E. G., "Boolean-valued models of set theory with automorphisms," Z. Math. Logik Grundlag. Math., **32**, No. 2, 117–130 (1986).

77. Hilbert D. and Bernays P., Foundations of Mathematics. Logical Calculi and the Formalization of Arithmetic [in Russian], Nauka, Moscow (1979).

78. Hoehle U., "Almost everywhere convergence and Boolean-valued topologies," in: Topology, Proc. 5th Int. Meet., Lecce/Italy 1990, Suppl. Rend. Circ. Mat. Palermo, II. Ser. **29**, 1992, pp. 215–227.

79. Hofstedter D. R., Gödel, Escher, Bach: an Eternal Golden Braid, Vintage Books, New York (1980).

80. Horiguchi H., "A definition of the category of Boolean-valued models," Comment. Math. Univ. St. Paul., **30**, No. 2, 135–147 (1981).

81. Horiguchi H., "The category of Boolean-valued models and its applications," Comment. Math. Univ. St. Paul., **34**, No. 1, 71–89 (1985).

82. Ionescu Tulcea A. and Ionescu Tulcea C., Topics in the Theory of Lifting, Springer-Verlag, Berlin etc. (1969).

83. Jech T. J., Lectures in Set Theory with Particular Emphasis on the Method of Forcing, Springer-Verlag, Berlin (1971).

84. Jech T. J., The Axiom of Choice, North-Holland, Amsterdam etc. (1973).

85. Jech T. J., "Abstract theory of abelian operator algebras: an application of forcing," Trans. Amer. Math. Soc., **289**, No. 1, 133–162 (1985).

86. Jech T. J., "First order theory of complete Stonean algebras (Boolean-valued real and complex numbers)," Canad. Math. Bull., **30**, No. 4, 385–392 (1987).

87. Jech T. J., "Boolean-linear spaces," Adv. in Math., **81**, No. 2, 117–197 (1990).

88. Jech T. J., Set Theory, Springer-Verlag, Berlin etc. (1997).

89. Johnstone P. T., Topos Theory, Academic Press, London etc. (1977).

90. Johnstone P. T., Stone Spaces, Cambridge University Press, Cambridge (1982).

91. de Jonge E. and van Rooij A. C. M., Introduction to Riesz Spaces, Mathematisch Centrum, Amsterdam (1977).

92. Jordan P., von Neumann J., and Wigner E., "On an algebraic generalization of the quantum mechanic formalism," Ann. Math., **35**, 29–64 (1944).

93. Judah H. (ed.), Set Theory of the Reals, Bar-Ilan University, Ramat-Gan (1993).

94. Just W. and Weese M., Discovering Modern Set Theory. The Basics, Amer. Math. Soc., Providence (1996).

95. Kadison R. V. and Ringrose J. R. Fundamentals of the Theory of Operator Algebras.—Vol. 1, 2.—Providence, RI: Amer. Math. Soc., 1997. Vol. 3, 4.—Boston: Birkhäuser Boston, 1991–1992.

96. Kantorovich L. V., "On semi-ordered linear spaces and their applications to the theory of linear operations," Dokl. Akad. Nauk SSSR, **4**, No. 1–2, 11–14 (1935).

97. Kantorovich L. V., "On some classes of linear operations," Dokl. Akad. Nauk SSSR, **3**, No. 1, 9–13 (1936).

98. Kantorovich L. V., "General forms of some classes of linear operations," Dokl. Akad. Nauk SSSR, **3**, No. 9, 101–106 (1936).

99. Kantorovich L. V., "On one class of functional equations," Dokl. Akad. Nauk SSSR, **4**, No. 5, 211–216 (1936).

100. Kantorovich L. V., "To the general theory of operations in semi-ordered spaces," Dokl. Akad. Nauk SSSR, **1**, No. 7, 271–274 (1936).

101. Kantorovich L. V., "On functional equations," Transactions of Leningrad State University, **3**, No. 7, 17–33 (1937).

102. Kantorovich L. V., Selected Works. Part 1, Gordon and Breach, London etc. (1996).

103. Kantorovich L. V. and Akilov G. P., Functional Analysis, Pergamon Press, Oxford and New York (1982).

104. Kantorovich L. V., Vulikh B. Z., and Pinsker A. G., Functional Analysis in Semiordered Spaces [in Russian], Gostekhizdat, Moscow and Leningrad (1950).

105. Kaplansky I., "Projections in Banach algebras," Ann. of Math. (2), **53**, 235–249 (1951).

106. Kaplansky I., "Algebras of type I," Ann. of Math. (2), **56**, 460–472 (1952).

107. Kaplansky I., "Modules over operator algebras," Amer. J. Math., **75**, No. 4, 839–858 (1953).

108. Kleene S., Mathematical Logic, John Wiley & Sons, New York etc. (1967).

109. Kolesnikov E. V., Kusraev A. G., and Malyugin S. A., On Dominated Operators [Preprint], Institute of Mathematics, Novosibirsk (1988).

110. Kollatz L., Functional Analysis and Numerical Mathematics [in German], Springer-Verlag, Berlin etc. (1964).

111. Kopytov V. M., Lattice Ordered Groups [in Russian], Nauka, Moscow (1984).

112. Korol′ A. I. and Chilin V. I., "Measurable operators in a Boolean valued model of set theory," Dokl. Akad. Nauk UzSSR, No. 3, 7–9 (1989).

113. Kramosil I., "Comparing alternative definitions of Boolean-valued fuzzy sets," Kybernetika, **28**, No. 6, 425–443 (1992).

114. Krasnosel′skiĭ M. A., Positive Solutions to Operator Equations. Chapters of Nonlinear Analysis [in Russian], Fizmatgiz, Moscow (1962).

115. Kuratowski K. and Mostowski A., Set Theory, North-Holland, Amsterdam etc. (1967).

116. Kusraev A. G., Some Applications of the Theory of Boolean Valued Models in Functional Analysis [Preprint], Institute of Mathematics, Novosibirsk (1982).

117. Kusraev A. G., "General desintegration formulas," Dokl. Akad. Nauk SSSR, **265**, No. 6, 1312–1316 (1982).

118. Kusraev A. G., "Boolean valued analysis of duality between universally complete modules," Dokl. Akad. Nauk SSSR, **267**, No. 5, 1049–1052 (1982).

119. Kusraev A. G., "On some categories and functors of Boolean valued analysis," Dokl. Akad. Nauk SSSR, **271**, No. 2, 283–286 (1983).

120. Kusraev A. G., "On Boolean valued convex analysis," in: Mathematische Optimiering. Theorie und Anwendungen, Wartburg/Eisenach, 1983, pp. 106–109.

121. Kusraev A. G., "Order continuous functionals in Boolean valued models of set theory," Sibirsk. Mat. Zh., **25**, No. 1, 69–79 (1984).

122. Kusraev A. G., "On Banach–Kantorovich spaces," Sibirsk. Mat. Zh., **26**, No. 2, 119–126 (1985).

123. Kusraev A. G., Vector Duality and Its Applications [in Russian], Nauka, Novosibirsk (1985).

124. Kusraev A. G., "Numerical systems in Boolean-valued models of set theory," in: Proceedings of the VIII All-Union Conference in Mathematical Logic (Moscow), Moscow, 1986, p. 99.

125. Kusraev A. G., "Linear operators in lattice-normed spaces," in: Studies on Geometry in the Large and Mathematical Analysis. Vol. 9 [in Russian], Trudy Inst. Mat. (Novosibirsk), Novosibirsk, 1987, pp. 84–123.

126. Kusraev A. G., "On function representation of type I AW^*-algebras," Sibirsk. Mat. Zh., **32**, No. 3, 78–88 (1991).

127. Kusraev A. G., "Boolean valued analysis and JB-algebras," Sibirsk. Mat. Zh., **35**, No. 1, 124–134 (1994).

128. Kusraev A. G., "Dominated operators," in: Linear Operators Compatible with Order [in Russian], Sobolev Institute Press, Novosibirsk, 1995, pp. 212–292.

129. Kusraev A. G., Boolean Valued Analysis of Duality Between Involutive Banach Algebras [in Russian], North Ossetian University Press, Vladikavkaz (1996).

130. Kusraev A. G. and Kutateladze S. S.,"Analysis of subdifferentials by means of Boolean valued models," Dokl. Akad. Nauk SSSR, **265**, No. 5, 1061–1064 (1982).

131. Kusraev A. G. and Kutateladze S. S., Notes on Boolean Valued Analysis [in Russian], Novosibirsk University Press, Novosibirsk (1984).

132. Kusraev A. G. and Kutateladze S. S., "Nonstandard methods for Kantorovich spaces," Siberian Adv. Math., **2**, No. 2, 114–152 (1992).

133. Kusraev A. G. and Kutateladze S. S., "Nonstandard methods in geometric functional analysis," Amer. Math. Soc. Transl. Ser. 2, **151**, 91–105 (1992).

134. Kusraev A. G. and Kutateladze S. S., "Boolean-valued introduction to the theory of vector lattices," Amer. Math. Soc. Transl. Ser. 2, **163**, 103–126 (1992).

135. Kusraev A. G. and Kutateladze S. S., Nonstandard Methods of Analysis [in Russian], Nauka, Novosibirsk (1990); [Translation into English] Kluwer Academic Publishers, Dordrecht (1994).

136. Kusraev A. G. and Kutateladze S. S., "Nonstandard methods in functional analysis," in: Interaction Between Functional Analysis, Harmonic Analysis, and Probability Theory, Marcel Dekker, New York, 1995, pp. 301–306.

137. Kusraev A. G. and Kutateladze S. S., Subdifferentials: Theory and Applications [in Russian], Nauka, Novosibirsk (1992); [Translation into English] Kluwer Academic Publishers, Dordrecht (1995).

138. Kusraev A. G. and Malyugin S. A., Some Questions of the Theory of Vector Measures [in Russian], Institute of Mathematics, Novosibirsk (1978).

139. Kusraev A. G. and Malyugin S. A., "On atomic decomposition for vector measures," Sibirsk. Mat. Zh., **30**, No. 5, 101–110 (1989).

140. Kusraev A. G. and Strizhevskiĭ V. Z., "Lattice normed spaces and dominated operators," in: Studies on Geometry and Functional Analysis. Vol. 7 [in Russian], Trudy Inst. Mat. (Novosibirsk), Novosibirsk, 1987, pp. 132–158.

141. Kutateladze S. S., "Descents and ascents," Dokl. Akad. Nauk SSSR, **272**, No. 3, 521–524 (1983).

142. Kutateladze S. S., "On the technique of descending and ascending," Optimization, **33**, 17–43 (1983).

143. Kutateladze S. S., "Cyclic monads and their applications," Sibirsk. Mat. Zh., **27**, No. 1, 100–110 (1986).

144. Kutateladze S. S., "Monads of ultrafilters and extensional filters," Sibirsk. Mat. Zh., **30**, No. 1, 129–133 (1989).

145. Kutateladze S. S., "On fragments of positive operators," Sibirsk. Mat. Zh., **30**, No. 5, 111–119 (1989).

146. Kutateladze S. S., Fundamentals of Functional Analysis, Kluwer Academic Publishers, Dordrecht (1996).

147. Kutateladze S. S., "Nonstandard tools for convex analysis," Math. Japon., **43**, No. 2, 391–410 (1996).

148. Kutateladze S. S. (ed.), Vector Lattices and Integral Operators, Kluwer Academic Publishers, Dordrecht (1996).

149. Lacey H. E., The Isometric Theory of Classical Banach Spaces, Springer-Verlag, Berlin etc. (1974).

150. Lambek J., Lectures on Rings and Modules, Blaisdell, Waltham (1966).

151. Larsen R., Banach Algebras: an Introduction, Dekker, New York (1973).

152. Levin V. L., Convex Analysis in Spaces of Measurable Functions and Its Application in Mathematics and Economics [in Russian], Nauka, Moscow (1985).

153. Levy A., Basic Set Theory, Springer-Verlag, Berlin etc. (1979).

154. Lindenstrauss J. and Tzafriri L., Classical Banach Spaces. Vol. 2: Function Spaces, Springer-Verlag, Berlin etc. (1979).

155. Li N., "The Boolean-valued model of the axiom system of GB," Chinese Sci. Bull., **36**, No. 2, 99–102 (1991).

156. Locher J. L. (ed.), The World of M. C. Escher, Abradale Press, New York (1988).

157. Lowen R., "Mathematics and fuzziness," in: Fuzzy Sets Theory and Applications (Louvain-la-Neuve, 1985), NATO Adv. Sci. Inst. Ser. C: Math. Phys. Sci., 177, Reidel, Dordrecht, and Boston, 1986, pp. 3–38.

158. Luxemburg W. A. J. and Zaanen A. C., Riesz Spaces. Vol. 1, North-Holland, Amsterdam and London (1971).

159. Luzin N. N., "Real function theory: state of the art," in: Proceedings of the All-Russian Congress of Mathematicians (Moscow, April 27–May 4, 1927), Glavnauka, Moscow and Leningrad, 1928, pp. 11–32.

160. Lyubetskiĭ V. A., "On some algebraic problems of nonstandard analysis," Dokl. Akad. Nauk SSSR, **280**, No. 1, 38–41 (1985).

161. Lyubetskiĭ V. A. and Gordon E. I., "Boolean completion of uniformities," in: Studies on Nonclassical Logics and Formal Systems [in Russian], Nauka, Moscow, 1983, pp. 82–153.

162. Lyubetskiĭ V. A. and Gordon E. I., Immersing bundles into a Heyting valued universe," Dokl. Akad. Nauk SSSR, **268**, No. 4, 794–798 (1983).

163. MacLane S., Categories for the Working Mathematician, Springer-Verlag, New York (1971).

164. Maltsev A. I. Algebraic Systems. Posthumous Edition, Springer-Verlag, New York and Heidelberg (1973).

165. Malykhin V. I., "New aspects in general topology pertaining to forcing," Uspekhi Mat. Nauk, **43**, No. 4, 83–94 (1988).

166. Manin Yu. I., A Course in Mathematical Logic, Springer-Verlag, Berlin etc. (1977).

167. Melter R., "Boolean valued rings and Boolean metric spaces," Arch. Math., No. 15, 354–363 (1964).

168. Mendelson E., Introduction to Mathematical Logic, Van Nostrand, Princeton etc. (1963).

169. Milvay C. J., "Banach sheaves," J. Pure Appl. Algebra, **17**, No. 1, 69–84 (1980).

170. Molchanov I. S., "Set-valued estimators for mean bodies related to Boolean models," Statistics, **28**, No. 1, 43–56 (1996).

171. Monk J. D. and Bonnet R. (eds.), Handbook of Boolean Algebras. Vol. 1–3, North-Holland, Amsterdam etc. (1989).

172. Mostowski A., Constructible Sets with Applications, North-Holland, Amsterdam (1969).

173. Murphy G., C^*-Algebras and Operator Theory, Academic Press, Boston, (1990).

174. Naĭmark M. A., Normed Rings, Noordhoff, Groningen (1959).

175. Namba K., "Formal systems and Boolean valued combinatorics," in: Southeast Asian Conference on Logic (Singapore, 1981), pp. 115–132. Stud. Logic Found. Math., 111, North-Holland, Amsterdam and New York, 1983.

176. von Neumann J., Collected Works. Vol. 3: Rings of Operators, Pergamon Press, New York etc. (1961).

177. von Neumann J., Collected Works. Vol. 4: Continuous Geometry and Other Topics, Pergamon Press, Oxford etc. (1962).

178. von Neumann J., Selected Works on Functional Analysis [in Russian], Nauka, Moscow (1987).

179. Nishimura H., "An approach to the dimension theory of continuous geometry from the standpoint of Boolean valued analysis," Publ. Res. Inst. Math. Sci., **20**, No. 5, 1091–1101 (1984).

180. Nishimura H., "Boolean valued decomposition theory of states," Publ. Res. Inst. Math. Sci., **21**, No. 5, 1051–1058 (1985).

181. Nishimura H., "Heyting valued set theory and fibre bundles," Publ. Res. Inst. Math. Sci., **24**, No. 2, 225–247 (1988).

182. Nishimura H., "On the absoluteness of types in Boolean valued lattices," Z. Math. Logik Grundlag. Math., **36**, No. 3, 241–246 (1990).

183. Nishimura H., "Some connections between Boolean valued analysis and topological reduction theory for C^*-algebras," Z. Math. Logik Grundlag. Math., **36**, No. 5, 471–479 (1990).

184. Nishimura H., "Boolean valued Dedekind domains," Z. Math. Logik Grundlag. Math., **37**, No. 1, 65–76 (1991).

185. Nishimura H., "Boolean valued Lie algebras," J. Symbolic Logic, **56**, No. 2, 731–741 (1991).

186. Nishimura H., "Some Boolean valued commutative algebra," Z. Math. Logik Grundlag. Math., **37**, No. 4, 367–384 (1991).

187. Nishimura H., "Foundations of Boolean-valued algebraic geometry," Z. Math. Logik Grundlag. Math., **37**, No. 5, 421–438 (1991).

188. Nishimura H., "On a duality between Boolean valued analysis and topological reduction theory," Math. Logic Quart., **39**, No. 1, 23–32 (1993).

189. Nishimura H., "On the duality between Boolean-valued analysis and reduction theory under the assumption of separability," Internat. J. Theoret. Phys., **32**, No. 3, 443–488 (1993).

190. Nishimura H., "A Boolean-valued approach to Gleason's theorem," Rep. Math. Phys., **34**, No. 2, 125–132 (1994).

191. Nishimura H., "Boolean valued and Stone algebra valued measure theories," Math. Logic Quart., **40**, No. 1, 69–75 (1994).

192. Novikov P. S., Constructive Mathematical Logic from the Point of View of Classical Logic [in Russian], Nauka, Moscow (1977).

193. Ozawa M., "Boolean valued analysis and type I AW^*-algebras," Proc. Japan Acad. Ser. A Math. Sci., **59A**, No. 8, 368–371 (1983).

194. Ozawa M., "Boolean valued interpretation of Hilbert space theory," J. Math. Soc. Japan, **35**, No. 4, 609–627 (1983).

195. Ozawa M., "A classification of type I AW^*-algebras and Boolean valued analysis," J. Math. Soc. Japan, **36**, No. 4, 589–608 (1984).

196. Ozawa M., "A transfer principle from von Neumann algebras to AW^*-algebras," J. London Math. Soc. (2), **32**, No. 1, 141–148 (1985).

197. Ozawa M., "Nonuniqueness of the cardinality attached to homogeneous AW^*-algebras," Proc. Amer. Math. Soc., **93**, 681–684 (1985).

198. Ozawa M., "Boolean valued analysis approach to the trace problem of AW^*-algebras," J. London Math. Soc. (2), **33**, No. 2, 347–354 (1986).

199. Ozawa M., "Embeddable AW^*-algebras and regular completions," J. London Math. Soc., **34**, No. 3, 511–523 (1986).

200. Ozawa M., "Boolean-valued interpretation of Banach space theory and module structures of von Neumann algebras," Nagoya Math. J., **117**, 1–36 (1990).

201. Pedersen G. K., Analysis Now, Springer-Verlag, Berlin etc. (1995).

202. Pinus A. G., Boolean Constructions in Universal Algebras, Kluwer Academic Publishers, Dordrecht etc. (1993).

203. Pirce R. S., Modules over Commutative Regular Rings, Amer. Math. Soc., Providence (1967).

204. Rasiowa E. and Sikorski R., The Mathematics of Metamathematics, PWN, Warsaw (1962).

205. Rema P. S., "Boolean metrization and topological spaces," Math. Japon., **9**, No. 9, 19–30 (1964).

206. Repicky M., "Cardinal characteristics of the real line and Boolean-valued models," Comment. Math. Univ. Carolin., **33**, No. 1, 184 (1992).

207. Rickart Ch., General Theory of Banach Algebras, Van Nostrand, Princeton (1960).

208. Rosser J. B., Logic for Mathematicians, New York etc., McGraw-Hill Book Company, (1953).

209. Rosser J. B., Simplified Independence Proofs. Boolean Valued Models of Set Theory, Academic Press, New York and London (1969).

210. Rudin W., Functional Analysis, McGraw-Hill, New York (1991).

211. Sakai S. C^*-algebras and W^*-algebras, Springer-Verlag, Berlin etc. (1971).

212. Saracino D. and Weispfenning V., "On algebraic curves over commutative regular rings," in: Model Theory and Algebra (a Memorial Tribute to Abraham Robinson), Springer-Verlag, New York etc., 1969.

213. Sarymsakov T. A. et al., Ordered Algebras [in Russian], Fan, Tashkent (1983).

214. Schaefer H. H., Banach Lattices and Positive Operators, Springer-Verlag, Berlin etc. (1974).

215. Schröder J., "Das Iterationsverfahren bei allgemeinierem Abshtandsbegriff," Math. Z., **66**, 111–116 (1956).

216. Schwarz H.-V., Banach Lattices and Operators, Teubner, Leipzig (1984).

217. Shoenfield G. R., Mathematical Logic, Addison-Wesley, Reading (1967).

218. Shoenfield G. R., "Axioms of set theory," in: Handbook of Mathematical Logic, North-Holland, Amsterdam, 1977.

219. Shotaev G. N., "On bilinear operators in lattice-normed spaces," Optimization, **37**, 38–50 (1986).

220. Sikorski R., Boolean Algebras, Springer-Verlag, Berlin etc. (1964).

221. Sikorskiĭ M. R., "Some applications of Boolean-valued models to study operators on polynormed spaces," Sov. Math., **33**, No. 2, 106–110 (1989).

222. Smith K., "Commutative regular rings and Boolean-valued fields," J. Symbolic Logic, **49**, No. 1, 281–297 (1984).

223. Sobolev V. I., "On a poset valued measure of a set, measurable functions, and some abstract integrals," Dokl. Akad. Nauk SSSR, **91**, No. 1, 23–26 (1953).

224. Solov'ëv Yu. P. and Troitskiĭ E. V., C^*-Algebras and Elliptic Operators in Differential Topology [in Russian], Factorial, Moscow (1996).

225. Solovay R. M., "A model of set theory in which every set of reals is Lebesgue measurable," Ann. of Math. (2), **92**, No. 2, 1–56 (1970).

226. Solovay R. M., "Real-valued measurable cardinals," in: Axiomatic Set Theory (Proc. Sympos. Pure Math., Vol. 13, Part 1, Univ. California, Los Angeles, Calif., 1967), Amer. Math. Soc., Providence, 1971, pp. 397–428.

227. Solovay R. and Tennenbaum S., "Iterated Cohen extensions and Souslin's problem," Ann. Math., **94**, No. 2, 201–245 (1972).

228. Spivak M. D., The Joy of TEX, Amer. Math. Soc., Providence (1990).

229. Sunder V. S., An Invitation to von Neumann Algebras, Springer-Verlag, New York etc. (1987).

230. Takesaki M., Theory of Operator Algebras. Vol. 1, Springer-Verlag, New York (1979).

231. Takeuti G., Two Applications of Logic to Mathematics, Iwanami and Princeton University Press, Tokyo and Princeton (1978).

232. Takeuti G., "A transfer principle in harmonic analysis," J. Symbolic Logic, **44**, No. 3, 417–440 (1979).

233. Takeuti G., "Boolean valued analysis," in: Applications of Sheaves (Proc. Res. Sympos. Appl. Sheaf Theory to Logic, Algebra and Anal., Univ. Durham, Durham, 1977), Springer-Verlag, Berlin etc., 1979.

234. Takeuti G., "Quantum set theory," in: Current Issues in Quantum Logic (Erice, 1979), Plenum, New York and London, 1981, pp. 303–322.

235. Takeuti G., "Boolean completion and m-convergence," in: Categorical Aspects of Topology and Analysis (Ottawa, Ont., 1980), Springer-Verlag, Berlin etc., 1982, pp. 333–350.

236. Takeuti G., "Von Neumann algebras and Boolean valued analysis," J. Math. Soc. Japan, **35**, No. 1, 1–21 (1983).

237. Takeuti G., "C^*-algebras and Boolean valued analysis," Japan. J. Math. (N.S.), **9**, No. 2, 207–246 (1983).

238. Takeuti G. and Titani S., "Heyting-valued universes of intuitionistic set theory," in: Logic Symposia (Hakone, 1979/1980), Springer-Verlag, Berlin and New York, 1981, pp. 189–306.

239. Takeuti G. and Titani S., "Globalization of intuitionistic set theory," Ann. Pure Appl. Logic, **33**, No. 2, 195–211 (1987).

240. Takeuti G. and Zaring W. M., Introduction to Axiomatic Set Theory, Springer-Verlag, New York etc. (1971).

241. Takeuti G. and Zaring W. M., Axiomatic Set Theory, Springer-Verlag, New York (1973).

242. Tkadlec J., "Boolean orthoposets and two-valued Jauch–Piron states," Tatra Mt. Math. Publ., No. 3, 155–160 (1993).

243. Topping D. M., Jordan Algebras of Selfadjoint Operators, Amer. Math. Soc., Providence (1965).

244. Tsalenko M. Sh. and Shul′geĭfer E. F., Fundamentals of Category Theory [in Russian], Nauka, Moscow (1974).

245. Veksler A. I., "On a new construction of the Dedekind completions of vector lattices and divisible l-groups," Sibirsk. Mat. Zh., **10**, No. 6, 70–73 (1969).

246. Veksler A. I., "Cyclic Banach spaces and Banach lattices," Dokl. Akad. Nauk SSSR, **213**, No. 4, 770–773 (1973).

247. Veksler A. I. and Geĭler V. A.,"On Dedekind and disjoint completeness of semiordered linear spaces," Sibirsk. Mat. Zh., **13**, No. 1, 43–51 (1972).

248. Venkataraman K., "Boolean valued almost periodic functions: existence of the mean," J. Indian Math. Soc. (N.S.), **43**, No. 1–4, 275–283 (1979).

249. Venkataraman K., "Boolean valued almost periodic functions on topological groups," J. Indian Math. Soc. (N.S.), **48**, No. 1–4, 153–164 (1984).

250. Vladimirov D. A., Boolean Algebras [in Russian], Nauka, Moscow (1969).

251. Vopěnka P., "The limits of sheaves over extremally disconnected compact Hausdorff spaces," Bull. Acad. Polon. Sci. Ser. Math. Astronom. Phys., **15**, No. 1, 1–4 (1967).

252. Vopěnka P., "General theory of ∇-models," Comment. Math. Univ. Carolin., **7**, No. 1, 147–170 (1967).

253. Vulikh B. Z., Introduction to the Theory of Partially Ordered Spaces, Noordhoff, Groningen (1967).

254. Wang Hao and McNaughton R., Axiomatic Systems of Set Theory [in French], Gauthier-Villars, Paris; E. Nauwelaerts, Louvain (1953).

255. Wright J. D. M., "Vector lattice measures on locally compact spaces," Math. Z., **120**, No. 3, 193–203 (1971).

256. Yamaguchi J., "Boolean $[0,1]$-valued continuous operators," Internat. J. Comput. Math., **68**, No. 1–2, 71–79 (1998).

257. Yood B., Banach Algebras: an Introduction, Carleton University Press, Ottawa (1988).

258. Zaanen A. C., Riesz Spaces. Vol. 2, North-Holland, Amsterdam etc. (1983).

259. Zaanen A. C., Introduction to Operator Theory in Riesz Spaces, Springer-Verlag, Berlin etc. (1997).

260. Zadeh L., The Concept of a Linguistic Variable and Its Application to Approximate Reasoning, Elsevier, New York etc. (1973).

261. Zhang Jin-wen, "A unified treatment of fuzzy set theory and Boolean valued set theory, fuzzy set structures and normal fuzzy set structures," J. Math. Anal. Appl., **76**, No. 1, 297–301 (1980).

262. Zhang Jin-wen, "Between fuzzy set theory and Boolean valued set theory," in: Fuzzy Information and Decision Processes, North-Holland, Amsterdam etc., 1982, pp. 143–147.

Subject Index

Other *Mathematics and Its Applications* titles of interest:

J. Chaillou: *Hyperbolic Differential Polynomials and their Singular Perturbations.* 1979, 184 pp. ISBN 90-277-1032-5

S. Fučik: *Solvability of Nonlinear Equations and Boundary Value Problems.* 1981, 404 pp. ISBN 90-277-1077-5

V.I. Istrăţescu: *Fixed Point Theory. An Introduction.* 1981, 488 pp. *out of print,* ISBN 90-277-1224-7

F. Langouche, D. Roekaerts and E. Tirapegui: *Functional Integration and Semiclassical Expansions.* 1982, 328 pp. ISBN 90-277-1472-X

N.E. Hurt: *Geometric Quantization in Action. Applications of Harmonic Analysis in Quantum Statistical Mechanics and Quantum Field Theory.* 1982, 352 pp. ISBN 90-277-1426-6

F.H. Vasilescu: *Analytic Functional Calculus and Spectral Decompositions.* 1983, 392 pp. ISBN 90-277-1376-6

W. Kecs: *The Convolution Product and Some Applications.* 1983, 352 pp. ISBN 90-277-1409-6

C.P. Bruter, A. Aragnol and A. Lichnerowicz (eds.): *Bifurcation Theory, Mechanics and Physics. Mathematical Developments and Applications.* 1983, 400 pp. *out of print,* ISBN 90-277-1631-5

J. Aczél (ed.) : *Functional Equations: History, Applications and Theory.* 1984, 256 pp. *out of print,* ISBN 90-277-1706-0

P.E.T. Jørgensen and R.T. Moore: *Operator Commutation Relations.* 1984, 512 pp. ISBN 90-277-1710-9

D.S. Mitrinović and J.D. Keckić: *The Cauchy Method of Residues. Theory and Applications.* 1984, 376 pp. ISBN 90-277-1623-4

R.A. Askey, T.H. Koornwinder and W. Schempp (eds.): *Special Functions: Group Theoretical Aspects and Applications. 1984, 352 pp.* ISBN 90-277-1822-9

R. Bellman and G. Adomian: *Partial Differential Equations. New Methods for their Treatment and Solution.* 1984, 308 pp. ISBN 90-277-1681-1

S. Rolewicz: *Metric Linear Spaces.* 1985, 472 pp. ISBN 90-277-1480-0

Y. Cherruault: *Mathematical Modelling in Biomedicine.* 1986, 276 pp. ISBN 90-277-2149-1

R.E. Bellman and R.S. Roth: *Methods in Approximation. Techniques for Mathematical Modelling.* 1986, 240 pp. ISBN 90-277-2188-2

R. Bellman and R. Vasudevan: *Wave Propagation. An Invariant Imbedding Approach.* 1986, 384 pp. ISBN 90-277-1766-4

Other *Mathematics and Its Applications* titles of interest:

A.G. Ramm: *Scattering by Obstacles.* 1986, 440 pp. ISBN 90-277-2103-3

C.W. Kilmister (ed.): *Disequilibrium and Self-Organisation.* 1986, 320 pp. ISBN 90-277-2300-1

A.M. Krall: *Applied Analysis.* 1986, 576 pp. ISBN 90-277-2328-1 (hb), ISBN 90-277-2342-7 (pb)

J.A. Dubinskij: *Sobolev Spaces of Infinite Order and Differential Equations.* 1986, 164 pp. ISBN 90-277-2147-5

H. Triebel: *Analysis and Mathematical Physics.* 1987, 484 pp. ISBN 90-277-2077-0

B.A. Kupershmidt: *Elements of Superintegrable Systems. Basic Techniques and Results.* 1987, 206 pp. ISBN 90-277-2434-2

M. Greguš: *Third Order Linear Differential Equations.* 1987, 288 pp. ISBN 90-277-2193-9

M.S. Birman and M.Z. Solomjak: *Spectral Theory of Self-Adjoint Operators in Hilbert Space.* 1987, 320 pp. ISBN 90-277-2179-3

V.I. Istrăţescu: *Inner Product Structures. Theory and Applications.* 1987, 912 pp. ISBN 90-277-2182-3

R. Vich: *Z Transform Theory and Applications.* 1987, 260 pp. ISBN 90-277-1917-9

N.V. Krylov: *Nonlinear Elliptic and Parabolic Equations of the Second -Order.* 1987, 480 pp. ISBN 90-277-2289-7

W.I. Fushchich and A.G. Nikitin: *Symmetries of Maxwell's Equations.* 1987, 228 pp. ISBN 90-277-2320-6

P.S. Bullen, D.S. Mitrinović and P.M. Vasić (eds.): *Means and their Inequalities.* 1987, 480 pp. ISBN 90-277-2629-9

V.A. Marchenko: *Nonlinear Equations and Operator Algebras.* 1987, 176 pp. ISBN 90-277-2654-X

Yu.L. Rodin: *The Riemann Boundary Problem on Riemann Surfaces.* 1988, 216 pp. ISBN 90-277-2653-1

A. Cuyt (ed.): *Nonlinear Numerical Methods and Rational Approximation.* 1988, 480 pp. ISBN 90-277-2669-8

D. Przeworska-Rolewicz: *Algebraic Analysis.* 1988, 640 pp. ISBN 90-277-2443-1

V.S. Vladimirov, YU.N. Drozzinov and B.I. Zavialov: *Tauberian Theorems for Generalized Functions.* 1988, 312 pp. ISBN 90-277-2383-4

G. Moroşanu: *Nonlinear Evolution Equations and Applications.* 1988, 352 pp. ISBN 90-277-2486-5

A.F. Filippov: *Differential Equations with Discontinuous Righthand Sides.* 1988, 320 pp. ISBN 90-277-2699-X

Other *Mathematics and Its Applications* titles of interest:

A.T. Fomenko: *Integrability and Nonintegrability in Geometry and Mechanics.* 1988, 360 pp. ISBN 90-277-2818-6

G. Adomian: *Nonlinear Stochastic Systems Theory and Applications to Physics.* 1988, 244 pp. ISBN 90-277-2525-X

A. Tesár and Ludovt Fillo: *Transfer Matrix Method.* 1988, 260 pp. ISBN 90-277-2590-X

A. Kaneko: *Introduction to the Theory of Hyperfunctions.* 1989, 472 pp. ISBN 90-277-2837-2

D.S. Mitrinović, J.E. Pećarič and V. Volenec: *Recent Advances in Geometric Inequalities.* 1989, 734 pp. ISBN 90-277-2565-9

A.W. Leung: *Systems of Nonlinear PDEs: Applications to Biology and Engineering.* 1989, 424 pp. ISBN 0-7923-0138-2

N.E. Hurt: *Phase Retrieval and Zero Crossings: Mathematical Methods in Image Reconstruction.* 1989, 320 pp. ISBN 0-7923-0210-9

V.I. Fabrikant: *Applications of Potential Theory in Mechanics. A Selection of New Results.* 1989, 484 pp. ISBN 0-7923-0173-0

R. Feistel and W. Ebeling: *Evolution of Complex Systems. Selforganization, Entropy and Development.* 1989, 248 pp. ISBN 90-277-2666-3

S.M. Ermakov, V.V. Nekrutkin and A.S. Sipin: *Random Processes for Classical Equations of Mathematical Physics.* 1989, 304 pp. ISBN 0-7923-0036-X

B.A. Plamenevskii: *Algebras of Pseudodifferential Operators.* 1989, 304 pp. ISBN 0-7923-0231-1

N. Bakhvalov and G. Panasenko: *Homogenisation: Averaging Processes in Periodic Media. Mathematical Problems in the Mechanics of Composite Materials.* 1989, 404 pp. ISBN 0-7923-0049-1

A.Ya. Helemskii: *The Homology of Banach and Topological Algebras.* 1989, 356 pp. ISBN 0-7923-0217-6

M. Toda: *Nonlinear Waves and Solitons.* 1989, 386 pp. ISBN 0-7923-0442-X

M.I. Rabinovich and D.I. Trubetskov: *Oscillations and Waves in Linear and Nonlinear Systems.* 1989, 600 pp. ISBN 0-7923-0445-4

A. Crumeyrolle: *Orthogonal and Symplectic Clifford Algebras. Spinor Structures.* 1990, 364 pp. ISBN 0-7923-0541-8

V. Goldshtein and Yu. Reshetnyak: *Quasiconformal Mappings and Sobolev Spaces.* 1990, 392 pp. ISBN 0-7923-0543-4

I.H. Dimovski: *Convolutional Calculus.* 1990, 208 pp. ISBN 0-7923-0623-6

Other *Mathematics and Its Applications* titles of interest:

Y.M. Svirezhev and V.P. Pasekov: *Fundamentals of Mathematical Evolutionary Genetics.* 1990, 384 pp. ISBN 90-277-2772-4

S. Levendorskii: *Asymptotic Distribution of Eigenvalues of Differential Operators.* 1991, 297 pp. ISBN 0-7923-0539-6

V.G. Makhankov: *Soliton Phenomenology.* 1990, 461 pp. ISBN 90-277-2830-5

I. Cioranescu: *Geometry of Banach Spaces, Duality Mappings and Nonlinear Problems.* 1990, 274 pp. ISBN 0-7923-0910-3

B.I. Sendov: *Hausdorff Approximation.* 1990, 384 pp. ISBN 0-7923-0901-4

A.B. Venkov: *Spectral Theory of Automorphic Functions and Its Applications.* 1991, 280 pp. ISBN 0-7923-0487-X

V.I. Arnold: *Singularities of Caustics and Wave Fronts.* 1990, 274 pp. ISBN 0-7923-1038-1

A.A. Pankov: *Bounded and Almost Periodic Solutions of Nonlinear Operator Differential Equations.* 1990, 232 pp. ISBN 0-7923-0585-X

A.S. Davydov: *Solitons in Molecular Systems. Second Edition.* 1991, 428 pp. ISBN 0-7923-1029-2

B.M. Levitan and I.S. Sargsjan: *Sturm-Liouville and Dirac Operators.* 1991, 362 pp. ISBN 0-7923-0992-8

V.I. Gorbachuk and M.L. Gorbachuk: *Boundary Value Problems for Operator Differential Equations.* 1991, 376 pp. ISBN 0-7923-0381-4

Y.S. Samoilenko: *Spectral Theory of Families of Self-Adjoint Operators.* 1991, 309 pp. ISBN 0-7923-0703-8

B.I. Golubov A.V. Efimov and V.A. Scvortsov: *Walsh Series and Transforms.* 1991, 382 pp. ISBN 0-7923-1100-0

V. Laksmikantham, V.M. Matrosov and S. Sivasundaram: *Vector Lyapunov Functions and Stability Analysis of Nonlinear Systems.* 1991, 250 pp. ISBN 0-7923-1152-3

F.A. Berezin and M.A. Shubin: *The Schrdinger Equation.* 1991, 556 pp. ISBN 0-7923-1218-X

D.S. Mitrinović, J.E. Pečarić and A.M. Fink: *Inequalities Involving Functions and their Integrals and Derivatives.* 1991, 588 pp. ISBN 0-7923-1330-5

Julii A. Dubinskii: *Analytic Pseudo-Differential Operators and their Applications.* 1991, 252 pp. ISBN 0-7923-1296-1

V.I. Fabrikant: *Mixed Boundary Value Problems in Potential Theory and their Applications.* 1991, 452 pp. ISBN 0-7923-1157-4

A.M. Samoilenko: *Elements of the Mathematical Theory of Multi-Frequency Oscillations.* 1991, 314 pp. ISBN 0-7923-1438-7

Other *Mathematics and Its Applications* titles of interest:

Yu.L. Dalecky and S.V. Fomin: *Measures and Differential Equations in Infinite-Dimensional Space.* 1991, 338 pp. ISBN 0-7923-1517-0

W. Mlak: *Hilbert Space and Operator Theory.* 1991, 296 pp. ISBN 0-7923-1042-X

N.Ja. Vilenkin and A.U. Klimyk: *Representation of Lie Groups and Special Functions. Volume 1: Simplest Lie Groups, Special Functions, and Integral Transforms.* 1991, 608 pp. ISBN 0-7923-1466-2

N.Ja. Vilenkin and A.U. Klimyk: *Representation of Lie Groups and Special Functions. Volume 2: Class I Representations, Special Functions, and Integral Transforms.* 1992, 630 pp. ISBN 0-7923-1492-1

N.Ja. Vilenkin and A.U. Klimyk: *Representation of Lie Groups and Special Functions. Volume 3: Classical and Quantum Groups and Special Functions.* 1992, 650 pp ISBN 0-7923-1493-X
(Set ISBN for Vols. 1, 2 and 3: 0-7923-1494-8)

K. Gopalsamy: *Stability and Oscillations in Delay Differential Equations of Population Dynamics.* 1992, 502 pp. ISBN 0-7923-1594-4

N.M. Korobov: *Exponential Sums and their Applications.* 1992, 210 pp. ISBN 0-7923-1647-9

Chuang-Gan Hu and Chung-Chun Yang: *Vector-Valued Functions and their Applications.* 1991, 172 pp. ISBN 0-7923-1605-3

Z. Szmydt and B. Ziemian: *The Mellin Transformation and Fuchsian Type Partial Differential Equations.* 1992, 224 pp. ISBN 0-7923-1683-5

L.I. Ronkin: *Functions of Completely Regular Growth.* 1992, 394 pp. ISBN 0-7923-1677-0

R. Delanghe, F. Sommen and V. Soucek: *Clifford Algebra and Spinor-valued Functions. A Function Theory of the Dirac Operator.* 1992, 486 pp. ISBN 0-7923-0229-X

A. Tempelman: *Ergodic Theorems for Group Actions.* 1992, 400 pp. ISBN 0-7923-1717-3

D. Bainov and P. Simenov: *Integral Inequalities and Applications.* 1992, 426 pp. ISBN 0-7923-1714-9

I. Imai: *Applied Hyperfunction Theory.* 1992, 460 pp. ISBN 0-7923-1507-3

Yu.I. Neimark and P.S. Landa: *Stochastic and Chaotic Oscillations.* 1992, 502 pp. ISBN 0-7923-1530-8

H.M. Srivastava and R.G. Buschman: *Theory and Applications of Convolution Integral Equations.* 1992, 240 pp. ISBN 0-7923-1891-9

A. van der Burgh and J. Simonis (eds.): *Topics in Engineering Mathematics.* 1992, 266 pp. ISBN 0-7923-2005-3

Other *Mathematics and Its Applications* titles of interest:

F. Neuman: *Global Properties of Linear Ordinary Differential Equations.* 1992, 320 pp. ISBN 0-7923-1269-4

A. Dvurečenskij: *Gleason's Theorem and its Applications.* 1992, 334 pp. ISBN 0-7923-1990-7

D.S. Mitrinović, J.E. Pečarić and A.M. Fink: *Classical and New Inequalities in Analysis.* 1992, 740 pp. ISBN 0-7923-2064-6

H.M. Hapaev: *Averaging in Stability Theory.* 1992, 280 pp. ISBN 0-7923-1581-2

S. Gindinkin and L.R. Volevich: *The Method of Newton's Polyhedron in the Theory of PDE's.* 1992, 276 pp. ISBN 0-7923-2037-9

Yu.A. Mitropolsky, A.M. Samoilenko and D.I. Martinyuk: *Systems of Evolution Equations with Periodic and Quasiperiodic Coefficients.* 1992, 280 pp. ISBN 0-7923-2054-9

I.T. Kiguradze and T.A. Chanturia: *Asymptotic Properties of Solutions of Nonautonomous Ordinary Differential Equations.* 1992, 332 pp. ISBN 0-7923-2059-X

V.L. Kocic and G. Ladas: *Global Behavior of Nonlinear Difference Equations of Higher Order with Applications.* 1993, 228 pp. ISBN 0-7923-2286-X

S. Levendorskii: *Degenerate Elliptic Equations.* 1993, 445 pp. ISBN 0-7923-2305-X

D. Mitrinović and J.D. Kečkić: *The Cauchy Method of Residues, Volume 2.* Theory and Applications. 1993, 202 pp. ISBN 0-7923-2311-8

R.P. Agarwal and P.J.Y Wong: *Error Inequalities in Polynomial Interpolation and Their Applications.* 1993, 376 pp. ISBN 0-7923-2337-8

A.G. Butkovskiy and L.M. Pustyl'nikov (eds.): *Characteristics of Distributed-Parameter Systems.* 1993, 386 pp. ISBN 0-7923-2499-4

B. Sternin and V. Shatalov: *Differential Equations on Complex Manifolds.* 1994, 504 pp. ISBN 0-7923-2710-1

S.B. Yakubovich and Y.F. Luchko: *The Hypergeometric Approach to Integral Transforms and Convolutions.* 1994, 324 pp. ISBN 0-7923-2856-6

C. Gu, X. Ding and C.-C. Yang: *Partial Differential Equations in China.* 1994, 181 pp. ISBN 0-7923-2857-4

V.G. Kravchenko and G.S. Litvinchuk: *Introduction to the Theory of Singular Integral Operators with Shift.* 1994, 288 pp. ISBN 0-7923-2864-7

A. Cuyt (ed.): *Nonlinear Numerical Methods and Rational Approximation II.* 1994, 446 pp. ISBN 0-7923-2967-8

G. Gaeta: *Nonlinear Symmetries and Nonlinear Equations.* 1994, 258 pp. ISBN 0-7923-3048-X

Other *Mathematics and Its Applications* titles of interest:

V.A. Vassiliev: *Ramified Integrals, Singularities and Lacunas*. 1995, 289 pp. ISBN 0-7923-3193-1

N.Ja. Vilenkin and A.U. Klimyk: *Representation of Lie Groups and Special Functions. Recent Advances*. 1995, 497 pp. ISBN 0-7923-3210-5

Yu. A. Mitropolsky and A.K. Lopatin: *Nonlinear Mechanics, Groups and Symmetry*. 1995, 388 pp. ISBN 0-7923-3339-X

R.P. Agarwal and P.Y.H. Pang: *Opial Inequalities with Applications in Differential and Difference Equations*. 1995, 393 pp. ISBN 0-7923-3365-9

A.G. Kusraev and S.S. Kutateladze: *Subdifferentials: Theory and Applications*. 1995, 408 pp. ISBN 0-7923-3389-6

M. Cheng, D.-G. Deng, S. Gong and C.-C. Yang (eds.): *Harmonic Analysis in China*. 1995, 318 pp. ISBN 0-7923-3566-X

M.S. Livšic, N. Kravitsky, A.S. Markus and V. Vinnikov: *Theory of Commuting Nonselfadjoint Operators*. 1995, 314 pp. ISBN 0-7923-3588-0

A.I. Stepanets: *Classification and Approximation of Periodic Functions*. 1995, 360 pp. ISBN 0-7923-3603-8

C.-G. Ambrozie and F.-H. Vasilescu: *Banach Space Complexes*. 1995, 205 pp. ISBN 0-7923-3630-5

E. Pap: *Null-Additive Set Functions*. 1995, 312 pp. ISBN 0-7923-3658-5

C.J. Colbourn and E.S. Mahmoodian (eds.): *Combinatorics Advances*. 1995, 338 pp. ISBN 0-7923-3574-0

V.G. Danilov, V.P. Maslov and K.A. Volosov: *Mathematical Modelling of Heat and Mass Transfer Processes*. 1995, 330 pp. ISBN 0-7923-3789-1

A. Laurinčikas: *Limit Theorems for the Riemann Zeta-Function*. 1996, 312 pp. ISBN 0-7923-3824-3

A. Kuzhel: *Characteristic Functions and Models of Nonself-Adjoint Operators*. 1996, 283 pp. ISBN 0-7923-3879-0

G.A. Leonov, I.M. Burkin and A.I. Shepeljavyi: *Frequency Methods in Oscillation Theory*. 1996, 415 pp. ISBN 0-7923-3896-0

B. Li, S. Wang, S. Yan and C.-C. Yang (eds.): *Functional Analysis in China*. 1996, 390 pp. ISBN 0-7923-3880-4

P.S. Landa: *Nonlinear Oscillations and Waves in Dynamical Systems*. 1996, 554 pp. ISBN 0-7923-3931-2

A.J. Jerri: *Linear Difference Equations with Discrete Transform Methods*. 1996, 462 pp. ISBN 0-7923-3940-1

Other *Mathematics and Its Applications* titles of interest:

I. Novikov and E. Semenov: *Haar Series and Linear Operators*. 1997, 234 pp. ISBN 0-7923-4006-X

L. Zhizhiashvili: *Trigonometric Fourier Series and Their Conjugates*. 1996, 312 pp. ISBN 0-7923-4088-4

R.G. Buschman: *Integral Transformation, Operational Calculus, and Generalized Functions*. 1996, 246 pp. ISBN 0-7923-4183-X

V. Lakshmikantham, S. Sivasundaram and B. Kaymakcalan: *Dynamic Systems on Measure Chains*. 1996, 296 pp. ISBN 0-7923-4116-3

D. Guo, V. Lakshmikantham and X. Liu: *Nonlinear Integral Equations in Abstract Spaces*. 1996, 350 pp. ISBN 0-7923-4144-9

Y. Roitberg: *Elliptic Boundary Value Problems in the Spaces of Distributions*. 1996, 427 pp. ISBN 0-7923-4303-4

Y. Komatu: *Distortion Theorems in Relation to Linear Integral Operators*. 1996, 313 pp. ISBN 0-7923-4304-2

A.G. Chentsov: *Asymptotic Attainability*. 1997, 336 pp. ISBN 0-7923-4302-6

S.T. Zavalishchin and A.N. Sesekin: *Dynamic Impulse Systems*. Theory and Applications. 1997, 268 pp. ISBN 0-7923-4394-8

U. Elias: *Oscillation Theory of Two-Term Differential Equations*. 1997, 226 pp. ISBN 0-7923-4447-2

D. O'Regan: *Existence Theory for Nonlinear Ordinary Differential Equations*. 1997, 204 pp. ISBN 0-7923-4511-8

Yu. Mitropolskii, G. Khoma and M. Gromyak: *Asymptotic Methods for Investigating Quasiwave Equations of Hyperbolic Type*. 1997, 418 pp. ISBN 0-7923-4529-0

R.P. Agarwal and P.J.Y. Wong: *Advanced Topics in Difference Equations*. 1997, 518 pp. ISBN 0-7923-4521-5

N.N. Tarkhanov: *The Analysis of Solutions of Elliptic Equations*. 1997, 406 pp. ISBN 0-7923-4531-2

B. Riečan and T. Neubrunn: *Integral, Measure, and Ordering*. 1997, 376 pp. ISBN 0-7923-4566-5

N.L. Gol'dman: *Inverse Stefan Problems*. 1997, 258 pp. ISBN 0-7923-4588-6

S. Singh, B. Watson and P. Srivastava: *Fixed Point Theory and Best Approximation: The KKM-map Principle*. 1997, 230 pp. ISBN 0-7923-4758-7

A. Pankov: G-*Convergence and Homogenization of Nonlinear Partial Differential Operators*. 1997, 263 pp. ISBN 0-7923-4720-X

Other *Mathematics and Its Applications* titles of interest:

S. Hu and N.S. Papageorgiou: *Handbook of Multivalued Analysis.* Volume I: Theory. 1997, 980 pp. ISBN 0-7923-4682-3 (Set of 2 volumes: 0-7923-4683-1)

L.A. Sakhnovich: *Interpolation Theory and Its Applications.* 1997, 216 pp. ISBN 0-7923-4830-0

G.V. Milovanović: *Recent Progress in Inequalities.* 1998, 531 pp. ISBN 0-7923-4845-1

V.V. Filippov: *Basic Topological Stru a ures of Ordinary Differential Equations.* 1998, 530 pp. ISBN 0-7293-4951-2

S. Gong: *Convex and Starlike Mappings in Several Complex Variables.* 1998, 208 pp. ISBN 0-7923-4964-4

A.B. Kharazishvili: *Applications of Point Set Theory in Real Analysis.* 1998, 244 pp. ISBN 0-7923-4979-2

R.P. Agarwal: *Focal Boundary Value Problems for Differential and Difference Equations.* 1998, 300 pp. ISBN 0-7923-4978-4

D. Przeworska-Rolewicz: *Logarithms and Antilogarithms.* An Algebraic Analysis Approach. 1998, 358 pp. ISBN 0-7923-4974-1

Yu. M. Berezansky and A.A. Kalyuzhnyi: *Harmonic Analysis in Hypercomplex Systems.* 1998, 493 pp. ISBN 0-7923-5029-4

V. Lakshmikantham and A.S. Vatsala: *Generalized Quasilinearization for Nonlinear Problems.* 1998, 286 pp. ISBN 0-7923-5038-3

V. Barbu: *Partial Differential Equations and Boundary Value Problems.* 1998, 292 pp. ISBN 0-7923-5056-1

J. P. Boyd: *Weakly Nonlocal Solitary Waves and Beyond-All-Orders Asymptotics.* Generalized Solitons and Hyperasymptotic Perturbation Theory. 1998, 610 pp. ISBN 0-7923-5072-3

D. O'Regan and M. Meehan: *Existence Theory for Nonlinear Integral and Integrodifferential Equations.* 1998, 228 pp. ISBN 0-7923-5089-8

A.J. Jerri: *The Gibbs Phenomenon in Fourier Analysis, Splines and Wavelet Approximations.* 1998, 364 pp. ISBN 0-7923-5109-6

C. Constantinescu, W. Filter and K. Weber, in collaboration with A. Sontag: *Advanced Integration Theory.* 1998, 872 pp. ISBN 0-7923-5234-3

V. Bykov, A. Kytmanov and M. Lazman, with M. Passare (ed.): *Elimination Methods in Polynomial Computer Algebra.* 1998, 252 pp. ISBN 0-7923-5240-8

W.-H. Steeb: *Hilbert Spaces, Wavelets, Generalised Functions and Modern Quantum Mechanics.* 1998, 234 pp. ISBN 0-7923-5231-9

Other *Mathematics and Its Applications* titles of interest:

X. Xu: *Introduction to Vertex Operator Superalgebras and Their Modules*. 1998, 356 pp.
ISBN 0-7923-5242-4

E.E. Rosinger: *Parametric Lie Group Actions on Global Generalised Solutions of Nonlinear PDFs* including a Solution to Hilbert's Fifth Problem. 1998, 234 pp. ISBN 0-7923-5232-7

A. Khrennikov: *Superanalysis*. 1999, 370 pp. ISBN 0-7923-5607-1

V. Koshmanenko: *Singular Quaaratic Forms in Perturbation Theory*. 1999, 316 pp.
ISBN 0-7923-5625-X

Z. Kamont: *Hyperbolic Functional Differential Inequalities and Applications*. 1999, 318 pp.
ISBN 0-7923-5791-4

A.G. Kusraev and S.S. Kutateladze: *Boolean Valued Analysis*. 1999, 334 pp.
ISBN 0-7923-5921-6

Zeitfracht Medien GmbH
Ferdinand-Jühlke-Straße 7
99095 Erfurt, Deutschland
produktsicherheit@kolibri360.de